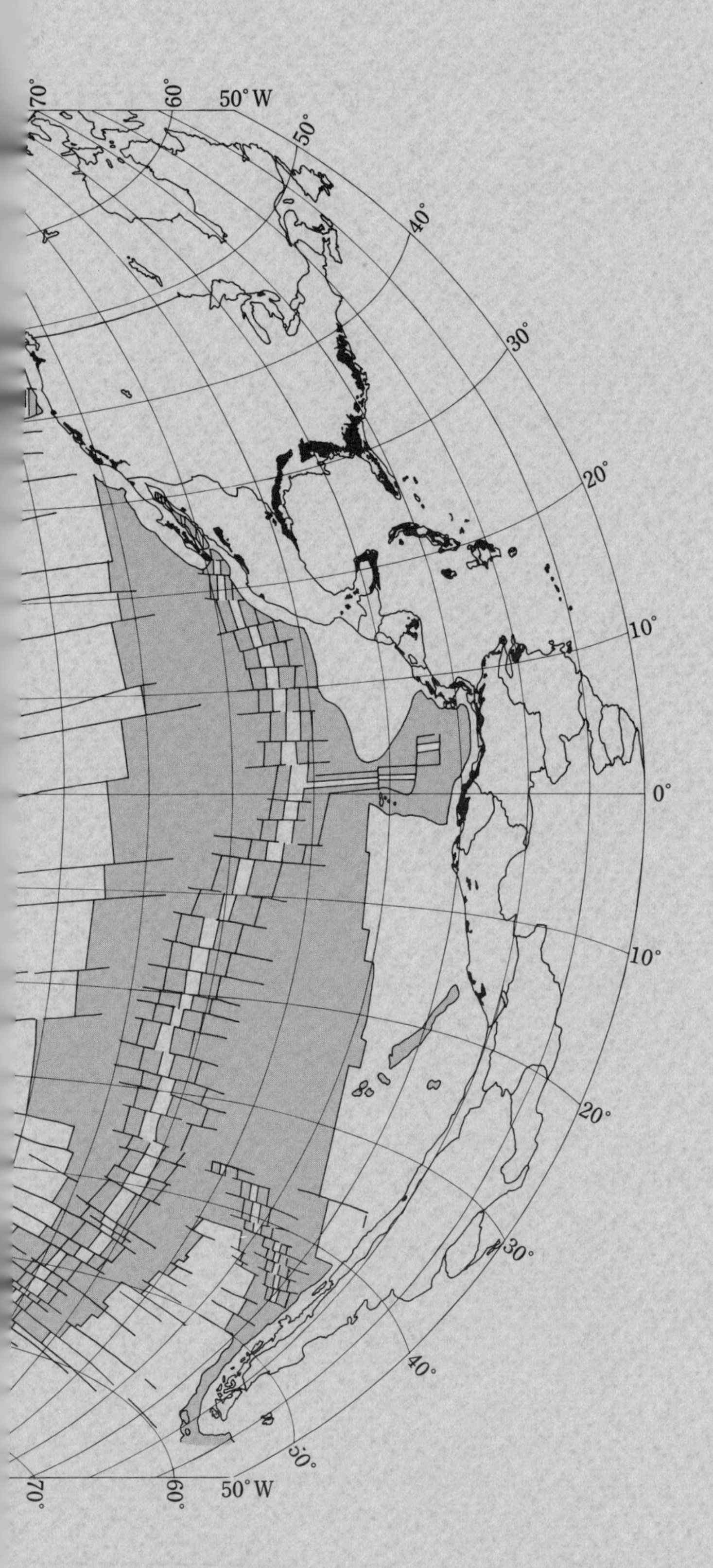

Distribution of Neogene marine strata in the Pacific region

Neogene marine strata on land may include some Paleogene or early Pleistocene sequences in places. The map is compiled from published and unpublished data mainly supplied through the courtesy of the staff of the Circum-Pacific Map project: W. O. Addicott, general chairman; K. J. Drummond (1983), Geologic Map of the Circum-Pacific Region, NE Quadrant, American Association of Petroleum Geologists, scale 1:10,000,000; J. Corvalán (1984), Geologic Map of the SE Quadrant; C. Nishiwaki, Geologic Map of the NW Quadrant (in press). The published source of the ocean Neogene basement data is B. C. Heezen and D. J. Fornari (1975).

PACIFIC NEOGENE DATUM PLANES

Pacific Neogene Datum Planes

— Contributions to Biostratigraphy and Chronology —

Editors

Nobuo Ikebe	(Tezukayama University) 6–4, Hama-cho, Ashiya 659, Japan
Ryuichi Tsuchi	Geoscience Institute, Faculty of Science, Shizuoka University, Shizuoka 422, Japan

PACIFIC NEOGENE DATUM PLANES

Contributions to Biostratigraphy and Chronology

Edited by
NOBUO IKEBE and RYUICHI TSUCHI

UNIVERSITY OF TOKYO PRESS

Publication of this book was supported by a grant-in-aid from the Nippon Life Insurance Foundation.

ISBN 4–13–066085–3 (UTP 66856)
ISBN 0–86008–354–3

Printed in Japan

Contents

Preface

A sequence of well-defined biostratigraphic datum planes (or datum levels) provides a grid for establishing precise correlation, with its geographic effectiveness varying in accordance with the reliability of datum plane species. Critical examination of each of the important datum planes in terms of the taxonomy, phylogeny, paleoenvironment and geochronological position of datum plane species is necessary to evaluate its reliability in stratigraphic correlation.

IGCP Project 114, "Evaluation of the biostratigraphic datum planes of the Pacific Neogene for the purpose of global-scale correlation," attempted to obtain some convincing evidence useful for cross-correlation on the Neogene of the Circum-Pacific areas. The first meeting of the working group of IGCP-114 was held in Tokyo in May 1976, followed by meetings in Bandung (Indonesia), Stanford (California), Khabarovsk (USSR), Paris during the 26th IGC, and finally Osaka-Kobe in 1981.

With the kind cooperation of all the working group members of IGCP-114, the present volume, the final report of IGCP-114, has been completed. This final report is based mainly on data presented and discussed during 1976-1981, which were summarized by the working group members at the Osaka-Kobe meeting at the end of 1981. Some additional data obtained in 1982 are also included and discussed.

The first part of this report is devoted to the evaluation of planktonic microfossil datum planes of the Pacific Neogene. Datum planes defined by various microfossil groups are cross-correlated, and references are made to biostratigraphic data based on larger foraminifers and molluscs. Chronological bases (radiometric and magnetostratigraphic) for dating these biostratigraphic events are also discussed.

In the second part, the relations among the Neogene biostratigraphies of the Pacific, Atlantic and the Mediterranean are discussed by W. A. Berggren in his two papers. His first paper, which appeared in the Proceedings of the 1981 Osaka International Workshop on Pacific Neogene Biostratigraphy, is reprinted here because of its importance to the discussion in this volume. His second paper presents the basic data used for the formulation of the first paper.

The third part consists of summaries of regional biostratigraphies of some important areas in the Pacific. These fundamental data formed the basis for the discussion in the first part.

The activities and achievements of IGCP-114 are summarized in the fourth part, with a list of the members of the international working group and a list of publications related to IGCP-114.

Many important contributions summarizing up-to-date data are included in this final report. The editors express their cordial thanks to all the members of IGCP-114, especially to those members who have contributed their valuable work to this volume. They are grateful to the Editorial Board members—Dr. W. O. Addicott, Dr. L. H. Burckle, Dr. M. Chiji, Prof. J. C. Ingle, Jr., Prof. T. Saito and Prof. Y. Takayanagi—for their generous cooperation, and to Miss Masako Ibaraki, Secretary, and Miss Tokiko Kawata, Assistant, for their assistance in the editorial work.

Acknowledgements are also due to the Board, Scientific Committee and Secretariat of the International Geological Correlation Programme (IGCP), UNESCO, and the Secretariat of the International Union of Geological Sciences (IUGS) for their support of this project. Finally, the editors express their gratitude to the Nippon Life Insurance Foundation for its financial aid toward this publication.

January 23, 1984

Nobuo Ikebe
Leader of IGCP–114

Ryuichi Tsuchi
Leader of National Working Group of Japan

I

Evaluation of Planktonic Datum Planes of the Pacific Neogene

I - 1

Planktonic Foraminiferal Datum Planes for Biostratigraphic Correlation of Pacific Neogene Sequences–1982 Status Report

Tsunemasa SAITO

Introduction

The datum level or datum plane concept has received growing acceptance among biostratigraphers during the past decade. In practice, it has been used to provide an instantaneous and synchronous time plane for the establishment of intercontinental as well as interoceanic stratigraphic correlation. On the other hand, the more traditional concept of biostratigraphic zones has recently been used more restrictively to subdivide local or regional stratigraphic sequences. One exception is the numerical zonal scheme first devised by Blow in 1969 for zonations based on planktonic foraminifera. The popularity of Blow's zonal scheme has since led to the erection of numerical zones based on other microfossil groups including calcareous nannoplankton. The problems associated with the use of traditional zonal schemes have been discussed by several authors (e.g. Saito, 1977; Burckle and Opdyke, 1977) and will not be repeated here.

The concept of the datum level was originally defined by Hornibrook (1966) to be "a correlation plane joining levels in rock sequences which on paleontological or other grounds appear to be isochronous," and the most-favored "paleontological grounds" have been evolutionary appearances and extinctions of planktonic microfossil taxa. In order for any given datum level to be effective in inter-regional stratigraphic correlation, close scrutiny of the following three subjects should be made; 1) identification of the main biostratigraphic events in broad latitudinal belts of the Pacific region and evaluation of the degree of synchroneity of these events, 2) determination of the extent to which these events maintain an invariant sequential order in various areas of the Pacific, and 3) determination of which of these events are evolutionary.

The three-day workshop meeting on Pacific Neogene Biostratigraphy held at the Inter-University Seminar House of Kansai in Kobe, Japan, on 27-29 of November 1981, and associated meeting addressed the above-mentioned three subjects. Initial discussions covered procedures to be used in evaluating provincial ranges of key taxa, recognition of the Neogene zones of Blow (1969), and synthesis of regional datum levels. The meeting concluded that charts of datum levels based on stratigraphic ranges of age-diagnostic taxa would be prepared for the following six regions of the Pacific: 1) the eastern North Pacific, 2) equatorial Pacific, 3) eastern South Pacific, 4) western and central Pacific—DSDP Sites 208 and 206, 5) New Zealand onshore sections, and 6) Indonesian region. This summary report is intended to present the datum level charts covering the major regions of the Pacific which were compiled by the meeting participants.

The author wishes to take this opportunity to thank Professor James C. Ingle, Jr., who served as co-chairman of the planktonic foraminifera discussion group during the workshop.

Background

With the increasing acceptance of the datum level concept in biostratigraphic correlation, a number of papers have appeared which either review the datum level concept or discuss a sequence of datum levels recognized in the Pacific region for various Neogene time intervals. This status report would not be complete without introducing these earlier works, which give more extensive coverage on the subject and, in fact, from which many of the basic range data were drawn during preparation of the regional datum level charts.

In 1971 Hornibrook and Edwards presented a status report on Cenozoic datum levels useful in the temperate New Zealand region. Saito (1977) discussed equatorial Pacific datum levels applicable to Late Cenozoic stratigraphic correlation. Detailed and comprehensive discussions on Late Cenozoic planktonic foraminiferal datum levels for the Pacific were attempted by Keller in her two companion papers published in 1980 with one paper covering the Early to Middle Miocene interval and the other Middle to Late Miocene time.

The reliability of datum levels for establishing time-

Department of Earth Sciences, Faculty of Science, Yamagata University, Yamagata 990, Japan

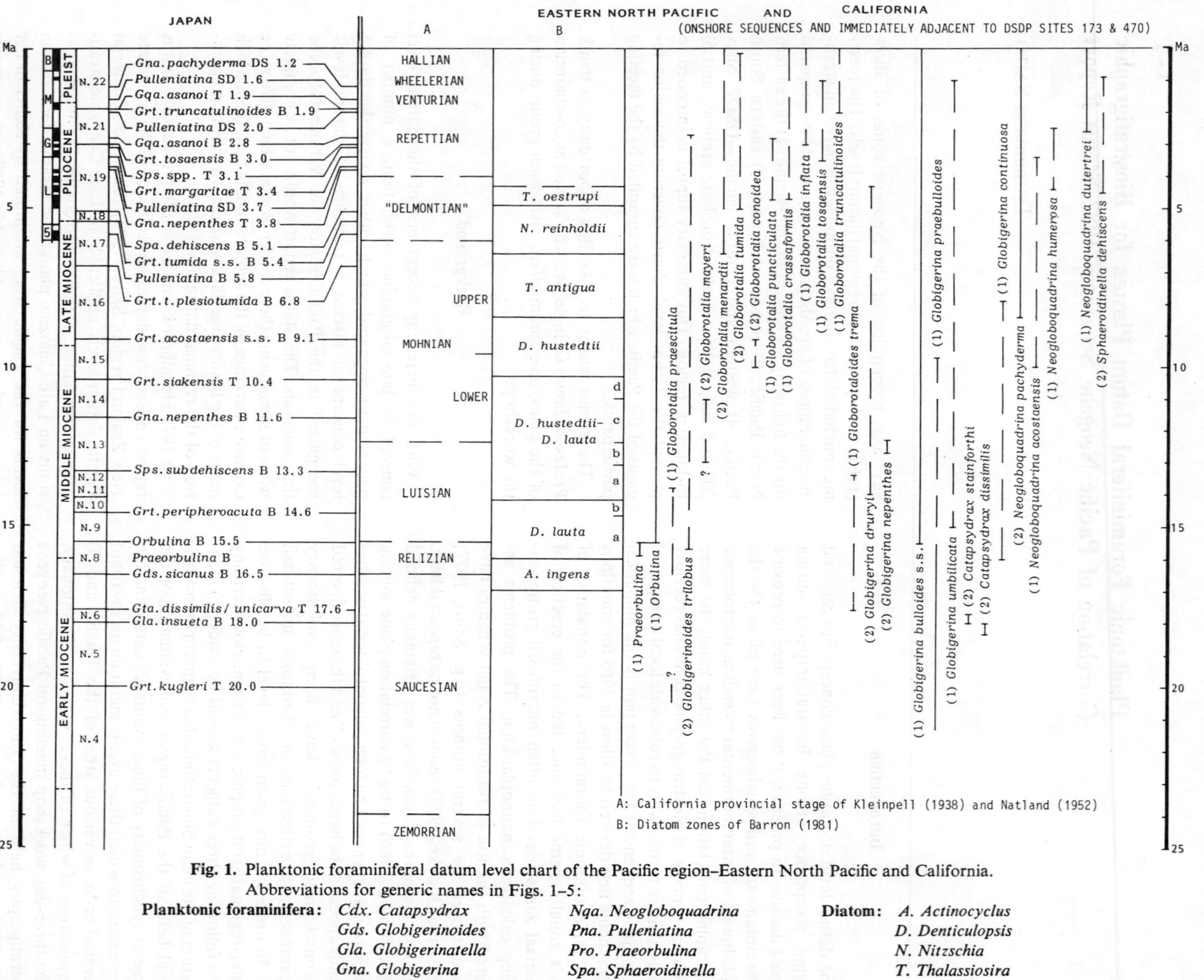

Fig. 1. Planktonic foraminiferal datum level chart of the Pacific region–Eastern North Pacific and California. Abbreviations for generic names in Figs. 1–5:

Planktonic foraminifera: *Cdx. Catapsydrax*, *Gds. Globigerinoides*, *Gla. Globigerinatella*, *Gna. Globigerina*, *Grt. Globorotalia*, *Gqa. Globoquadrina*, *Gta. Globigerinita*, *Nqa. Neogloboquadrina*, *Pna. Pulleniatina*, *Pro. Praeorbulina*, *Spa. Sphaeroidinella*, *Sps. Sphaeroidinellopsis*

Diatom: *A. Actinocyclus*, *D. Denticulopsis*, *N. Nitzschia*, *T. Thalassiosira*

correlative levels in rock sequences can be proven unequivocally only when faunal events that characterize each datum level are checked against other independent means of geochronology. In recent years, significant advances in establishing true synchroneity of faunal events have come from the combined use of datum levels with paleomagnetic stratigraphy and isotope records. Along this line of research, Burckle and Opdyke (1977) and Burckle (1977, 1979) examined diatom datum levels, and Theyer and Hammond (1974) and Theyer *et al.* (1978) calibrated radiolarian datum levels against paleomagnetic stratigraphy. Thierstein *et al.* (1977) established global synchroneity of late Quaternary coccolith datum levels by comparing them with oxygen isotope records. Keller, Barron and Burckle (1982) attempted the intercalibration of four major microfossil groups (planktonic foraminifera, radiolarians, calcareous nannofossils, and diatoms) by the combined use of stable isotopes, variation in percent of calcium carbonate, and paleomagnetic stratigraphy.

The Deep Sea Drilling Project has sampled sedimentary layers throughout the oceans during the 13 years of this international effort providing deep sea carbonate sequences suitable for detailed analyses of planktonic foraminiferal biostratigraphy and evolution. Initial reports of the Deep Sea Drilling Project nearly always include discussions on Neogene planktonic foraminifera as this microfossil represents one of the three major fossil groups continuously employed for dating and correlating the cored sequences. Comprehensive reviews of the current status of Neogene planktonic foraminiferal biostratigraphy as revealed by various DSDP projects, including zonations, datum levels, and paleooceanography, have been presented by Srinivasan and Kennett (1981a, b).

Discussion and Presentation of Datum Level Charts

The datum level charts included in this report were prepared under the direction of the following authors: James C. Ingle, Jr. and Richard S. Poore (Eastern North Pacific) Ruben E. Martinez-Pardo (Eastern South Pacific), James P. Kennett (equatorial Pacific-DSDP Site 289), James P. Kennett (Western and Central Pacific-DSDP Sites 208 and 206), N. de B. Hornibrook (New Zealand onshore sections) and Darwin Kadar and Soemoenar Soeka (Indonesian region).

In these charts, each datum level is followed by one or two kinds of suffix. Suffix T and B (or EB) after species names denote those datum levels defined by the first evolutionary appearance and the extinction of a given species, respectively. However, charts for temperate areas are coded differently to differentiate the first and last evolutionary appearances versus those controlled by paleooceanographic-paleoclimatic events. Numerals 1, 2, and 3 indicate three orders of reliability in stratigraphic correlation as not all microfossil species are equally useful in practice for correlation purposes. The distinction between these three orders was originally proposed by Hornibrook and Edwards (1977) and their criteria are adopted herein.

First order: Easily identifiable, common and persistent, either rapid first evolutionary appearance or highly consistent range.

Second order: Easily identifiable, fairly consistently present but not necessarily common; or gradual first evolutionary appearance; fairly consistent range.

Third order: Consistency of identification difficult, or rare, or well-defined ecological or local geographical restrictions, or known in very few sections, or limits of range uncertain due to sampling gaps or disconformities.

Examination of these charts clearly indicates that datum levels common to all areas of the Pacific are indeed very few. A broad distribution of subtropical climates over much of the Pacific during the Early and Middle Miocene permits the use of a single set of datum levels for inter-regional correlation. However, the declining temperature trend beginning in the Middle Miocene initiated a process of biogeographic provincialization by effecting a distinct latitudinal control over the distribution of marine microfaunas. For the post-Middle Miocene interval, an independent set of datum levels will be required for these separate biogeographic provinces, and to establish correlation between them would require calibration with other kinds of geochronology such as oxygen isotope and paleomagnetic stratigraphy.

References

Blow, W. H., 1969: Late Middle Eocene to Recent planktonic foraminiferal biostratigraphy. *In* Bronnimann, P. and Renz, H. H., eds., *1st Internat. Conf. Planktonic Microfossils, Geneva 1967, Proc.*, v. **1**, p. 199-422, pls. 1–54.

Burckle, L. H., 1977: Pliocene and Pleistocene diatom datum levels for the equatorial Pacific. *Quaternary Res.*, v. **7**, no. 3, p. 330-340.

Burckle, L. H., 1979: Early Miocene to Pliocene diatom datum levels for the equatorial Pacific. *In* Wiryosujono, S. and Marks, E., eds., *2nd Working Group Meet., Biost. Datum Pacific Neogene, IGCP Project 114, Proc.*, p. 25–44.

Burckle, L. H. and Opdyke, N. D., 1977: Late Neogene diatom correlations in the Circum-Pacific. *In* Saito, T. and Ujiie, H., eds., *1st CPNS, Tokyo, 1976, Proc.*, p. 255–284.

Hornibrook, N. de B., 1966: New Zealand Tertiary microfossil zonation, correlation and climate. *In* Hatai, K.,

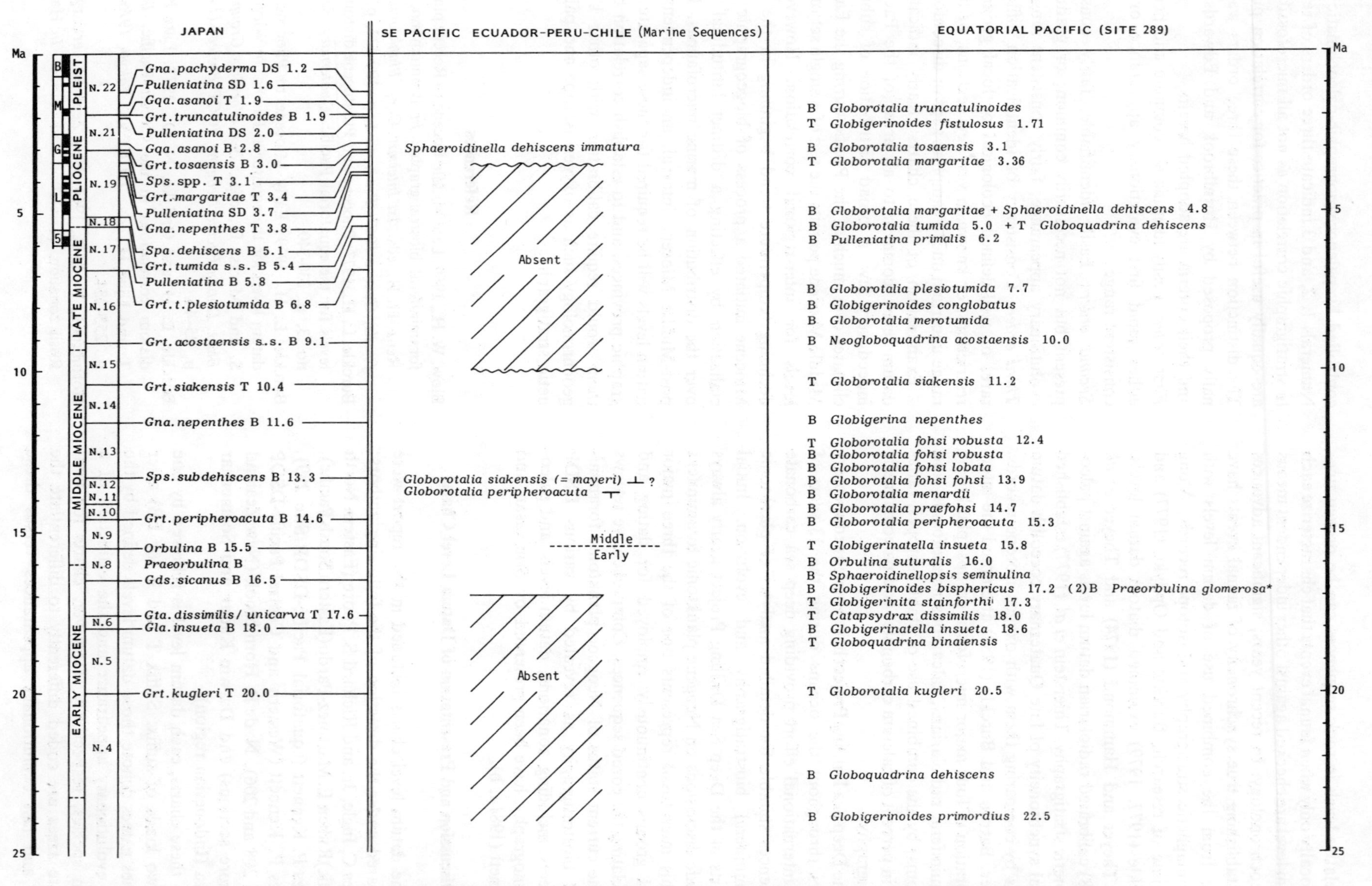

Fig. 2. Planktonic foraminiferal datum level chart of the Pacific region—Southeastern Pacific and equatorial Pacific. (above)
Fig. 3. Planktonic foraminiferal datum level chart of the Pacific region—Western and central Pacific. (below)

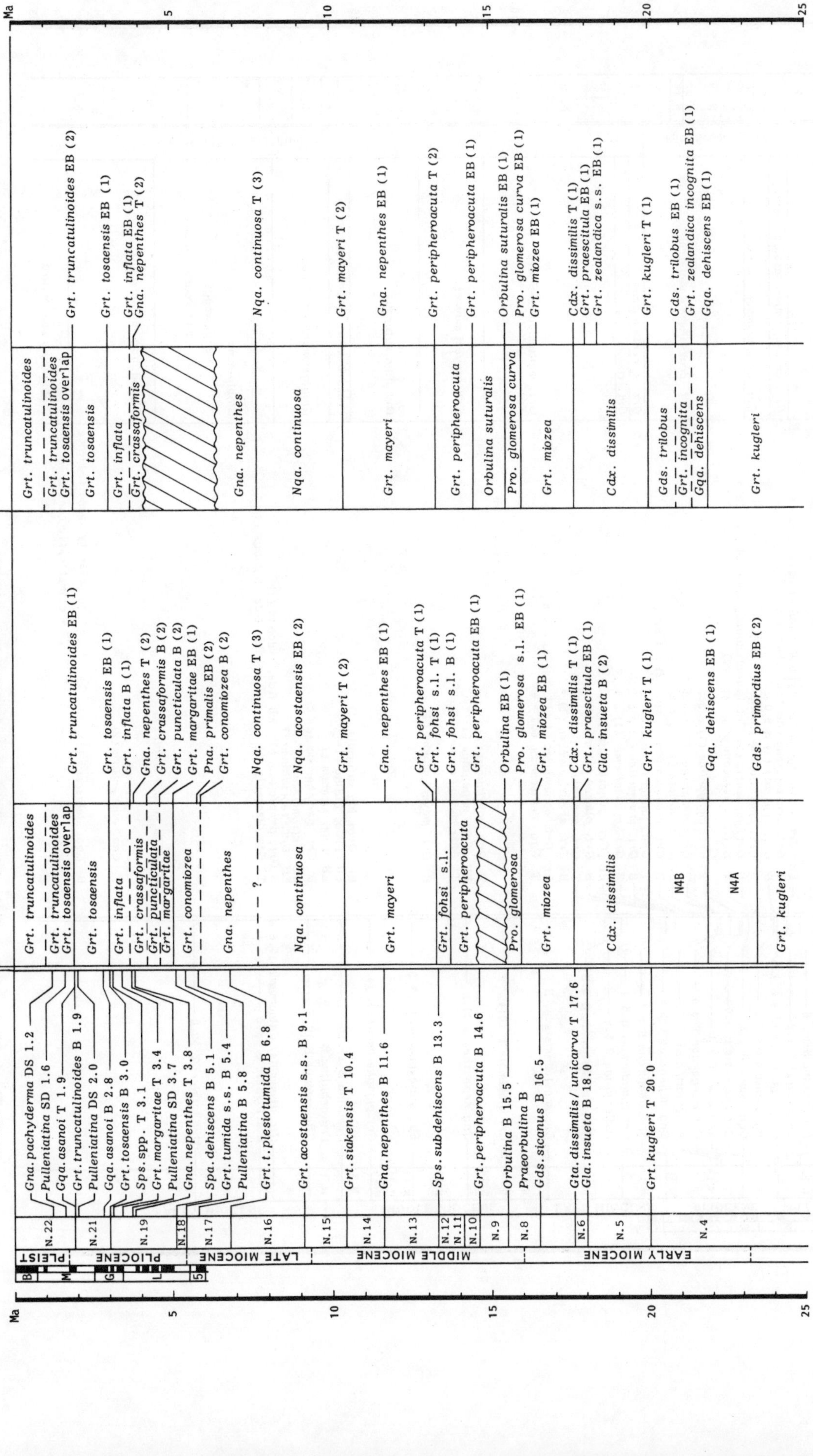

WESTERN AND CENTRAL PACIFIC

JAPAN

Ma: 5, 10, 15, 20, 25

B, M, G, L, 5

PLEIST, PLIOCENE, LATE MIOCENE, MIDDLE MIOCENE, EARLY MIOCENE

N.22, N.21, N.19, N.18, N.17, N.16, N.15, N.14, N.13, N.12, N.11, N.10, N.9, N.8, N.6, N.5, N.4

- *Gna. pachyderma* DS 1.2
- *Pulleniatina* SD 1.6
- *Gqa. asanoi* T 1.9
- *Grt. truncatulinoides* B 1.9
- *Pulleniatina* DS 2.0
- *Gqa. asanoi* B 2.8
- *Grt. tosaensis* B 3.0
- *Sps.* spp. T 3.1
- *Grt. margaritae* T 3.4
- *Pulleniatina* SD 3.7
- *Gna. nepenthes* T 3.8
- *Spa. dehiscens* B 5.1
- *Grt. tumida* s.s. B 5.4
- *Pulleniatina* B 5.8
- *Grt. t. plesiotumida* B 6.8
- *Grt. acostaensis* s.s. B 9.1
- *Grt. siakensis* T 10.4
- *Gna. nepenthes* B 11.6
- *Sps. subdehiscens* B 13.3
- *Grt. peripheroacuta* B 14.6
- *Orbulina* B 15.5
- *Praeorbulina* B
- *Gds. sicanus* B 16.5
- *Gta. dissimilis / unicarva* T 17.6
- *Gla. insueta* B 18.0
- *Grt. kugleri* T 20.0

WARM SUBTROPICS (SITE 208)

Zones:
- *Grt. truncatulinoides*
- *Grt. truncatulinoides* / *Grt. tosaensis* overlap
- *Grt. tosaensis*
- *Grt. inflata*
- *Grt. crassaformis*
- *Grt. puncticulata*
- *Grt. margaritae*
- *Grt. conomiozea*
- *Gna. nepenthes*
- ?
- *Nqa. continuosa*
- *Grt. mayeri*
- *Grt. fohsi* s.l.
- *Grt. peripheroacuta*
- *Pro. glomerosa*
- *Grt. miozea*
- *Cdx. dissimilis*
- N4B
- N4A
- *Grt. kugleri*

Datums:
- *Grt. truncatulinoides* EB (1)
- *Grt. tosaensis* EB (1)
- *Grt. inflata* B (1)
- *Gna. nepenthes* T (2)
- *Grt. crassaformis* B (2)
- *Grt. puncticulata* B (2)
- *Grt. margaritae* EB (1)
- *Pna. primalis* EB (2)
- *Grt. conomiozea* B (2)
- *Nqa. continuosa* T (3)
- *Nqa. acostaensis* EB (2)
- *Grt. mayeri* T (2)
- *Gna. nepenthes* EB (1)
- *Grt. peripheroacuta* T (1)
- *Grt. fohsi* s.l. T (1)
- *Grt. fohsi* s.l. B (1)
- *Grt. peripheroacuta* EB (1)
- *Orbulina* EB (1)
- *Pro. glomerosa* s.l. EB (1)
- *Grt. miozea* EB (1)
- *Cdx. dissimilis* T (1)
- *Grt. praescitula* EB (1)
- *Gla. insueta* B (2)
- *Grt. kugleri* T (1)
- *Gqa. dehiscens* EB (1)
- *Gds. primordius* EB (2)

TRANSITIONAL (SITE 206)

Zones:
- *Grt. truncatulinoides*
- *Grt. truncatulinoides* / *Grt. tosaensis* overlap
- *Grt. tosaensis*
- *Grt. inflata*
- *Grt. crassaformis*
- *Gna. nepenthes*
- *Nqa. continuosa*
- *Grt. mayeri*
- *Grt. peripheroacuta*
- *Orbulina suturalis*
- *Pro. glomerosa curva*
- *Grt. miozea*
- *Cdx. dissimilis*
- *Gds. trilobus*
- *Grt. incognita*
- *Gqa. dehiscens*
- *Grt. kugleri*

Datums:
- *Grt. truncatulinoides* EB (2)
- *Grt. tosaensis* EB (1)
- *Grt. inflata* EB (1)
- *Gna. nepenthes* T (2)
- *Nqa. continuosa* T (3)
- *Grt. mayeri* T (2)
- *Gna. nepenthes* EB (1)
- *Grt. peripheroacuta* T (2)
- *Grt. peripheroacuta* EB (1)
- *Orbulina suturalis* EB (1)
- *Pro. glomerosa curva* EB (1)
- *Grt. miozea* EB (1)
- *Cdx. dissimilis* T (1)
- *Grt. praescitula* EB (1)
- *Grt. zealandica* s.s. EB (1)
- *Grt. kugleri* T (1)
- *Gds. trilobus* EB (1)
- *Grt. zealandica incognita* EB (1)
- *Gqa. dehiscens* EB (1)

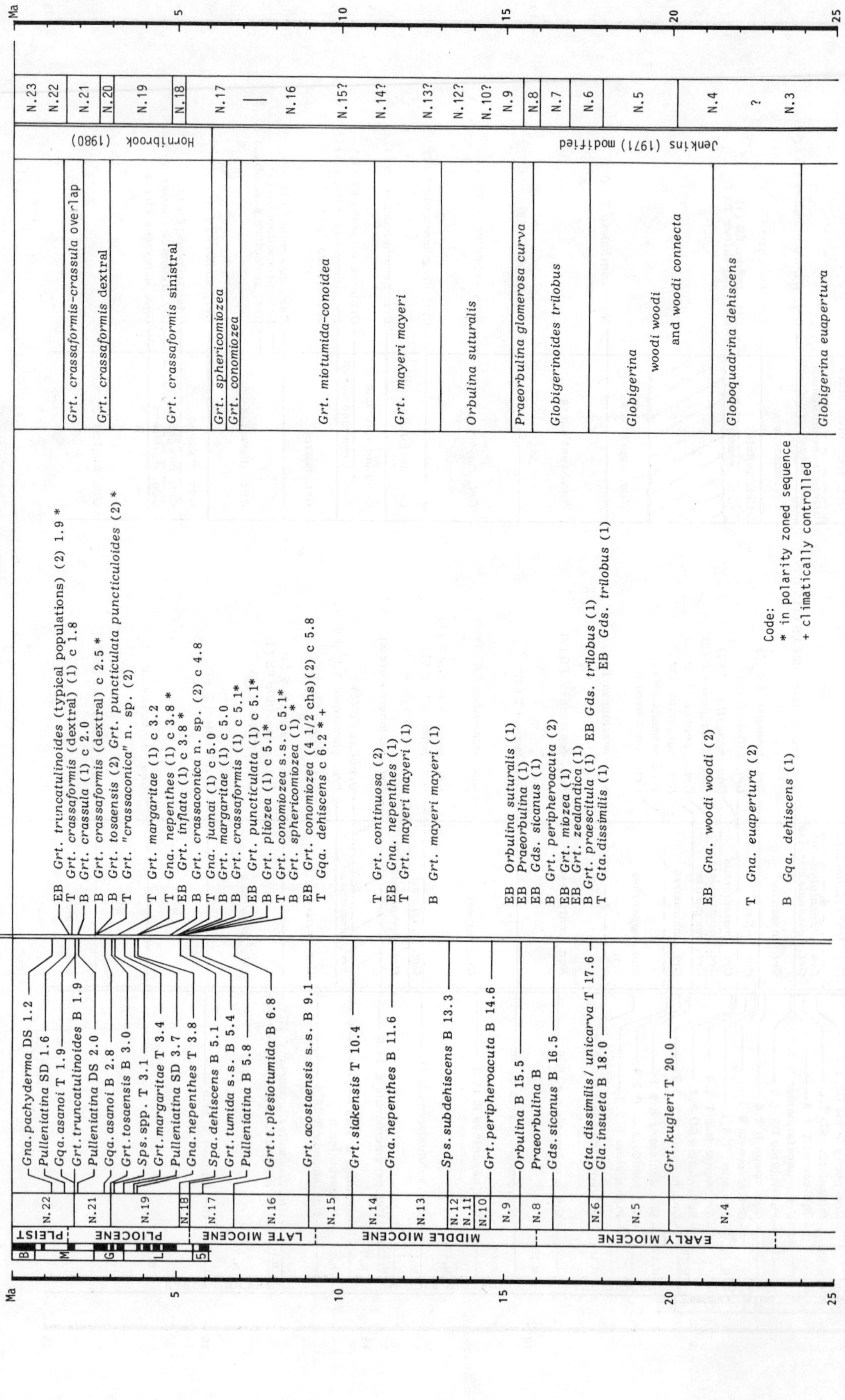

JAPAN

Ma

PLEIST
PLIOCENE
LATE MIOCENE
MIDDLE MIOCENE
EARLY MIOCENE

N.22
Gna. pachyderma DS 1.2
Pulleniatina SD 1.6
Gqa. asanoi T 1.9
Grt. truncatulinoides B 1.9
N.21
Pulleniatina DS 2.0
Gqa. asanoi B 2.8
Grt. tosaensis B 3.0
N.19
Sps. spp. T 3.1
Grt. margaritae T 3.4
Pulleniatina SD 3.7
N.18
Gna. nepenthes T 3.8
N.17
Spa. dehiscens B 5.1
Grt. tumida s.s. B 5.4
Pulleniatina B 5.8
N.16
Grt. t. plesiotumida B 6.8
Grt. acostaensis s.s. B 9.1
N.15
Grt. siakensis T 10.4
N.14
Gna. nepenthes B 11.6
N.13
Sps. subdehiscens B 13.3
N.12
N.11
N.10
Grt. peripheroacuta B 14.6
N.9
Orbulina B 15.5
N.8
Praeorbulina B
Gds. sicanus B 16.5
N.6
Gta. dissimilis / unicarva T 17.6
Gla. insueta B 18.0
N.5
Grt. kugleri T 20.0
N.4

NEW ZEALAND ON-LAND

EB Grt. truncatulinoides (typical populations) (2) 1.9 *
T Grt. crassaformis (dextral) (1) c 1.8
B Grt. crassula (1) c 2.0
B Grt. crassaformis (dextral) c 2.5 *
B Grt. tosaensis (2) Grt. puncticulata puncticuloides (2) *
T Grt. "crassaconica" n. sp. (2)
T Grt. margaritae (1) c 3.2
T Gna. nepenthes (1) c 3.8 *
EB Grt. inflata (1) c 3.8 *
B Grt. crassaconica n. sp. (2) c 4.8
T Gna. juanai (1) c 5.0
B Grt. margaritae (1) c 5.0
B Grt. crassaformis (1) c 5.1*
EB Grt. puncticulata (1) c 5.1*
B Grt. pliozea (1) c 5.1*
T Grt. conomiozea s.s. c 5.1*
B Grt. sphericomiozea (1) *
EB Grt. conomiozea (4 1/2 chs)(2) c 5.8
T Gqa. dehiscens c 6.2 * +
T Grt. continuosa (2)
EB Gna. nepenthes (1)
T Grt. mayeri mayeri (1)
B Grt. mayeri mayeri (1)
EB Orbulina suturalis (1)
EB Praeorbulina (1)
EB Gds. sicanus (1)
B Grt. peripheroacuta (2)
EB Grt. miozea (1)
EB Grt. zealandica (1)
B Grt. praescitula (1) EB Gds. trilobus (1)
T Gta. dissimilis (1) EB Gds. trilobus (1)
EB Gna. woodi woodi (2)
T Gna. euapertura (2)
B Gqa. dehiscens (1)

Code:
* in polarity zoned sequence
+ climatically controlled

Grt. crassaformis-crassula overlap
Grt. crassaformis dextral
Grt. crassaformis sinistral
Grt. sphericomiozea
Grt. conomiozea
Grt. miotumida-conoidea
Grt. mayeri mayeri
Orbulina suturalis
Praeorbulina glomerosa curva
Globigerinoides trilobus
Globigerina woodi woodi and woodi connecta
Globoquadrina dehiscens
Globigerina euapertura

Hornibrook (1980)
Jenkins (1971) modified

N.23
N.22
N.21
N.20
N.19
N.18
N.17
N.16
N.15?
N.14?
N.13?
N.12?
N.10?
N.9
N.8
N.7
N.6
N.5
N.4
?
N.3

Ma
5
10
15
20
25

Fig. 4. Planktonic foraminiferal datum level chart of the Pacific region—New Zealand on-shore sections. (above)
Fig. 5. Planktonic foraminiferal datum level chart of the Pacific region—Indonesian region. (below)

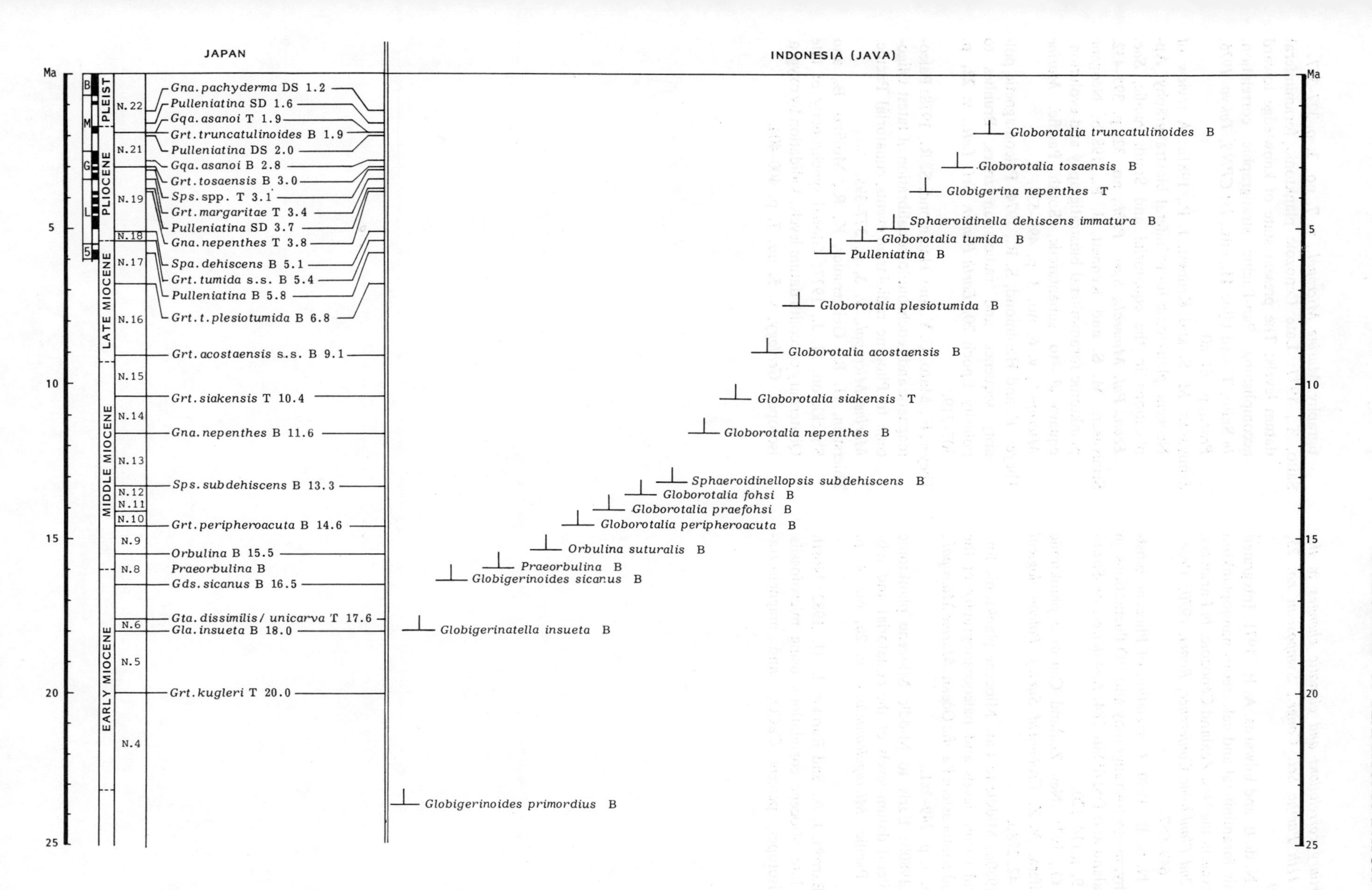
JAPAN
INDONESIA (JAVA)
Ma
5
10
15
20
25
B
M
G
L
5
PLEIST
PLIOCENE
LATE MIOCENE
MIDDLE MIOCENE
EARLY MIOCENE
N.22
N.21
N.19
N.18
N.17
N.16
N.15
N.14
N.13
N.12
N.11
N.10
N.9
N.8
N.6
N.5
N.4
Gna. pachyderma DS 1.2
Pulleniatina SD 1.6
Gqa. asanoi T 1.9
Grt. truncatulinoides B 1.9
Pulleniatina DS 2.0
Gqa. asanoi B 2.8
Grt. tosaensis B 3.0
Sps. spp. T 3.1
Grt. margaritae T 3.4
Pulleniatina SD 3.7
Gna. nepenthes T 3.8
Spa. dehiscens B 5.1
Grt. tumida s.s. B 5.4
Pulleniatina B 5.8
Grt. t. plesiotumida B 6.8
Grt. acostaensis s.s. B 9.1
Grt. siakensis T 10.4
Gna. nepenthes B 11.6
Sps. subdehiscens B 13.3
Grt. peripheroacuta B 14.6
Orbulina B 15.5
Praeorbulina B
Gds. sicanus B 16.5
Gta. dissimilis / unicarva T 17.6
Gla. insueta B 18.0
Grt. kugleri T 20.0
Globorotalia truncatulinoides B
Globorotalia tosaensis B
Globigerina nepenthes T
Sphaeroidinella dehiscens immatura B
Globorotalia tumida B
Pulleniatina B
Globorotalia plesiotumida B
Globorotalia acostaensis B
Globorotalia siakensis T
Globorotalia nepenthes B
Sphaeroidinellopsis subdehiscens B
Globorotalia fohsi B
Globorotalia praefohsi B
Globorotalia peripheroacuta B
Orbulina suturalis B
Praeorbulina B
Globigerinoides sicanus B
Globigerinatella insueta B
Globigerinoides primordius B

ed., *Tertiary correlations and climatic changes in the Pacific, 11th Pacific Sci. Congr., Symposium no. 25*, p. 29–39.

Hornibrook, N. de B. and Edwards, A. R., 1971: Integrated planktonic foraminiferal and calcareous nannoplankton datum levels in the New Zealand Cenozoic. *In* Farinacci, A., ed., *2nd Planktonic Conference, Roma, 1970, Proc.*, v. **1**, p. 649–657.

Hornibrook, N. de B., 1980: Correlation of Pliocene biostratigraphy, magnetostratigraphy and ^{18}O fluctuations in New Zealand and DSDP Site 284. *Newsletters on Stratigraphy*, 9, p.114–120.

Jenkins, D. G., 1971: New Zealand Cenozoic planktonic foraminifera. *N. Z. Geological Survey Paleontological Bulletin*, 42, 278p.

Keller, G., 1980a: Middle to Late Miocene planktonic foraminiferal datum levels and paleoceanography of the north and southeastern Pacific Ocean. *Marine Micropal.*, v. **5**, no. 3, p. 249–281.

Keller, G., 1980b: Early to Middle Miocene planktonic foraminiferal datum levels of the equatorial and subtropical Pacific. *Micropaleontology*, v. **26**, no. 4, p. 372–391.

Keller, G., Barron, J. A., and Burckle, L. H., 1982: North Pacific late Miocene correlations using microfossils, stable isotopes, percent $CaCO_3$, and magnetostratigraphy. *Marine Micropal.*, v. **7**, no. 3, p. 327–357.

Saito, T., 1977: Late Cenozoic planktonic foraminiferal datum levels: The present state of knowledge toward accomplishing Pan-Pacific stratigraphic correlation. *In* Saito, T. and Ujiie, H., eds., *1st CPNS, Tokyo, 1976, Proc.*, p. 61–80.

Srinivasan, M. S. and Kennett, J. P., 1981a: A review of Neogene planktonic foraminiferal biostratigraphy: Applications in the equatorial and South Pacific, *Soc. Econ. Pal. Mineral., Spec. Publ.*, no. 32, p. 395–432.

Srinivasan, M. S. and Kennett, J. P., 1981b: Neogene planktonic foraminiferal biostratigraphy and evolution: equatorial to subantarctic, South Pacific. *Marine Micropal.*, v. **6**, no. 3, p. 499–533.

Theyer, F. and Hammond, S. R., 1974: Paleomagnetic polarity sequence and radiolarian zones, Brunhes to polarity Epoch 20. *Earth Planet. Sci. Lett.*, v. **22**, p. 307–319.

Theyer, F., Mato, C. Y. and Hammond, S. R., 1978: Paleomagnetic and geochronologic calibration of latest Oligocene to Pliocene radiolarian events, equatorial Pacific. *Marine Micropal.*, v. **3**, p. 377–395.

Thierstein, H. R., Geitzenauer, K. R., Molfino, B., and Shackleton, N. J., 1977: Global synchroneity of late Quaternary coccolith datum levels: Validation by oxygen isotopes. *Geology*, v. **5**, no. 7, p. 400–404.

I - 2

Neogene Planktonic Foraminiferal Datum Planes of the South Pacific: Mid to Equatorial Latitudes

James P. KENNETT* and M. S. SRINIVASAN**

Abstract

In southwest Pacific DSDP sites, two broad biostratigraphic schemes are recognized—tropical at Site 289 and temperate at Sites 284 and 207A. Tropical and temperate zonations are linked by a warm subtropical biostratigraphic scheme at Site 208 which includes a mixture of tropical and temperate elements. Although the two broad schemes are recognized, none of the biostratigraphic sequences are identical between any of the sites. For instance, Site 206 is transitional between the warm subtropical (Site 208) and the temperate (Site 207A) schemes. This reflects differences in biogeography and evolution and diachronous extinction at various latitudes during the Neogene.

A series of 34 datum levels based on evolutionary appearances or extinctions are described for the Neogene in three sites considered here (Sites 289, 208, 206). The most widely applicable datums include the following: latest Oligocene—*Globigerinoides* F.A.; Early Miocene—*Globoquadrina dehiscens,* F.A., *Globorotalia kugleri* L.A., *Catapsydrax dissimilis* L.A. and *Praeorbulina glomerosa* F.A., Middle Miocene—*Orbulina suturalis* F.A., *Globorotalia peripheroacuta* F.A., *Fohsella* lineage L.A., *Globorotalia mayeri* L.A., Late Miocene—*"Neogloboquadrina" continuosa* L A.; Early Pliocene—*Globorotalia puncticulata* F.A., *Globorotalia margaritae* F.A.; Early Pleistocene—*Globorotalia truncatulinoides* F.A. The other datums assist with correlation over more restricted latitudinal ranges. Few datums extend from tropical to temperate latitudes.

The datums at warm subtropical Site 208 represent changes in both tropical and temperate lineages. A greater number of datums are recorded in this site compared with tropical and temperate regions. However, several taxonomic first or last appearances are diachronous with regard to either tropical or temperate regions. Here, the Early Miocene and early Middle Miocene datums are generally formed by changes in tropical elements. In contrast, during the late Middle Miocene, Late Miocene and Pliocene the datums are more commonly formed by changes in temperate elements. This reflects global cooling related to the major development of Antarctic ice beginning in the Middle Miocene, concommitant with latitudinal shift of cool waters towards the equator.

Introduction

The central difficulty of biostratigraphic correlation within the Pacific basin is created by the diversity of biogeographic provinces, each of which has been marked by different biostratigraphic history and species groupings. This diversity has been clearly demonstrated by the extraordinary increase in the study of Cenozoic microfossil sequences associated with the Deep Sea Drilling Project. Hundreds of drilled sequences throughout the ocean basins have been studied and biostratigraphically subdivided by a large number of investigators. The faunal and floral provincialism during the Neogene has been largely related to distinct latitudinal zonality of surface-water masses and also to the cross-ocean paleooceanographic surface water gradients such as those that currently exist from west to east across the equatorial Pacific.

Within the planktonic foraminifera, strong provincial latitudinal differences in assemblages exist. Direct comparison between Neogene sequences of the tropics with those of temperate regions show little similarity. For example, the *Globoconella* group *(Globorotalia incognita, G. zealandica, G. miozea, G. conoidea, G. conomiozea, G. puncticulata* and *G. inflata)* is totally absent in tropical Site DSDP 289, whereas it is the dominant group in the more temperate Site 206. On the other hand, *Globigerinatella, Menardella (Globorotalia menardii, G. limbata, G. multicamerata)* and *Globorotalia (G. tumida)*, which are important elements in tropical Site 289, are almost absent in temperate

* *Graduate School of Oceanography, University of Rhode Island, Narragansett, R. I. 02882, U.S.A.*
** *Banaras Hindu University, Varanasi–221 005, India*

sites. Because of these and other major differences, it has become essential to understand the biostratigraphic succession at latitudes intermediate between the tropics and temperates—that is the warm subtropical and transitional water masses. DSDP Site 208 in the warm subtropical water mass of the South Pacific is such a site since it contains a mixture of both tropical and temperate planktonic foraminiferal assemblages.

The provincialism tied to different water masses has required the development of distinct biostratigraphic schemes which exhibit greater differences with increasingly different environmental conditions. At equatorial and tropical latitudes, the Neogene planktonic foraminiferal biostratigraphy of Blow (1969) has found wide acceptance, but for higher latitudes a number of zonal schemes are employed for different water mass regimes in different areas. Important zonal schemes established for the temperate areas are those of Jenkins (1967, 1971), based on New Zealand marine sequences, and Kennett (1973) and Srinivasan and Kennett (1981a, b), based on DSDP cores. Srinivasan and Kennett (1981a,b) have demonstrated the need for distinguishing at least 4 basically different biostratigraphic zonal schemes for the South Pacific associated with tropical, warm subtropical, temperate, transitional, and subantarctic water masses. Additional, but relatively minor latitudinal differences exist within each of these schemes.

Because of the latitudinal differences of planktonic foraminiferal assemblages, it is of particular importance to distinguish *datums* for correlation of Neogene deep sea sequences across the widest latitudes possible. Datums represent correlation planes joining levels in sediment sequences which, on paleontological or other grounds, appear to be isochronous (Hornibrook, 1966). This concept was well established before the beginning of the Deep Sea Drilling Project (Allan, 1956; Hornibrook, 1966; Jenkins, 1966a). In a practical sense, datums represent conspicuous evolutionary appearances or disappearances which are considered to be correlatable and isochronous over reasonable distances. However, the distances vary considerably over which individual datum levels can be correlated, ranging from within individual water masses to those applied in correlation between tropical and temperate areas. Datums, therefore, differ from zonal boundaries, although many are based on the same paleontological event.

The ultimate objective of stratigraphy is to place rocks and geological events within a radiometric time framework. Quantitative stratigraphy enables rates of change to be established and in turn provides a better frame of reference for understanding geological processes. The planktonic foraminiferal datums and zones have been integrated into a chronological framework largely by Berggren (1969, 1971, 1972), and Berggren and Van Couvering (1974, 1978). This has been largely accomplished by intercalibration of the available radiometric dates with the marine fossil sequence and by the direct application of paleomagnetic stratigraphy. Such dating of paleontological events has become known as biochronology (Berggren and van Couvering, 1974, 1978). In addition to Berggren's work, other contributions which deal with the dating of planktonic foraminiferal datum levels and zones include Glass *et al.* (1967), Hays *et al.* (1969), Ikebe *et al.* (1972), Saito (1972, 1977) and Ryan *et al.* (1974). Berggren and van Couvering (1974) provided a sequence of 13 radiometrically dated Neogene planktonic foraminiferal datums, whereas Saito (1977) increased the number of radiometrically dated datums to 37 in the tropical Pacific Neogene. Recently, Keller (1978, 1980, 1981a, b) identified a sequence of 45 "datum levels" for the Neogene equatorial and northern subtropical Pacific (DSDP Sites 77B, 71, 55, 319, 296, 173, and 310).

For the temperate region, Hornibrook and Edwards (1971) established 34 planktonic foraminiferal "datum levels" for the Neogene of New Zealand. These still require integration into the Neogene time scale. Three orders of reliability were established for the set of datums based on ease of identification, relative abundance, geographic distribution, and understanding of phylogenetic relations. Hornibrook and Edwards (1971) were also among the first workers to integrate datum levels based on two distinct paleontological groups (planktonic foraminifera and calcareous nannofossils). Since then, the Deep Sea Drilling Project has provided an important basis for the integration of different microfossil groups.

Southwest Pacific Sites

The southwest Pacific represents one of the few optimal areas of the world for the study of a traverse of high-quality planktonic foraminiferal biostratigraphic sequences ranging from the Equator to temperate regions. A north-south trending series of shallow pedestals (Macquarie Ridge, South Tasman Rise, New Zealand Plateau, Lord Howe Rise, and the Ontong-Java Plateau) have provided a sequence of high-quality, relatively continuously cored, carbonate Neogene sections, exhibiting minimum calcium carbonate dissolution. Srinivasan and Kennett (1981a,b) have examined a traverse of sections ranging from the Equator to subantarctic latitudes (Fig. 1) Sites include DSDP Sites 289, 208, 206, 207A, 284 and 281. Planktonic foraminiferal zonal schemes range from the subantarctic (Site 281) cool subtropical water mass to the south (Sites

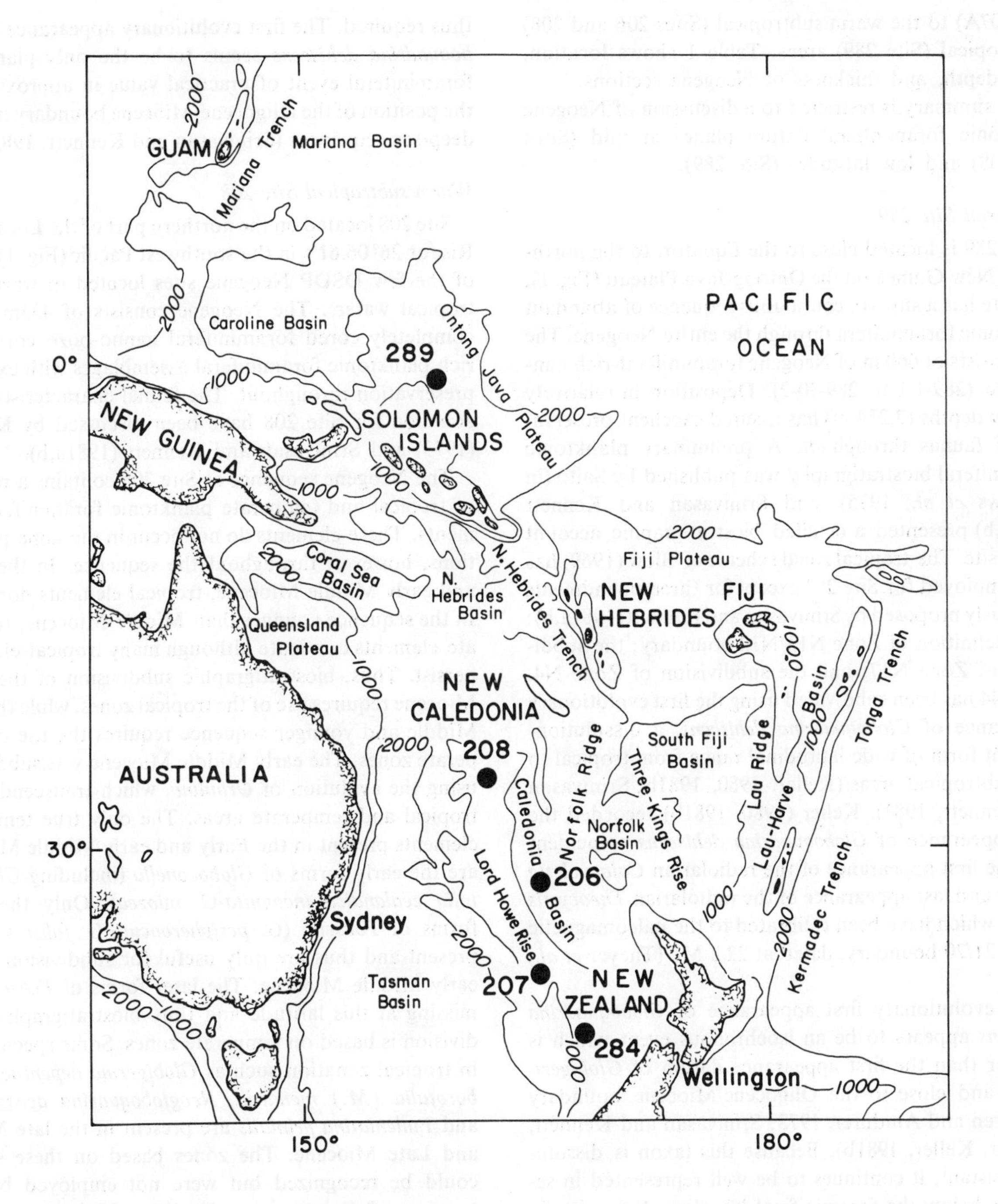

Fig. 1. Location of South Pacific DSDP drill sites examined. This paper reviews datums for Sites 289 (tropical), 208 (warm subtropical), and 206 (transitional).

Table 1. Location of DSDP Sites studied in the Equatorial and Southwest Pacific.

Site	Latitude	Longitude	Water Depth (m)	Thickness of Neogene
289	00°29.92′S	158°30.69′E	2,224	660 m
208	26°06.61′S	161°13.27′E	1,545	435 m
206	32°00.75′S	165°27.15′E	3,196	300 m
207A	36°57.75′S	165°26.06′E	1,389	120 m
284	40°30.48′S	167°40.81′E	1,068	208 m

284, 207A) to the warm subtropical (Sites 206 and 208) and tropical (Site 289) areas. Table 1 shows location, water depth, and thickness of Neogene sections.

This summary is restricted to a discussion of Neogene planktonic foraminiferal datum planes at mid (Sites 206, 208) and low latitudes (Site 289).

Equatorial Site 289

Site 289 is located close to the Equator, to the northeast of New Guinea on the Ontong-Java Plateau (Fig. 1). This site has a superb, continuous sequence of abundant planktonic foraminifera through the entire Neogene. The site consists of 660 m of Neogene foraminiferal-rich nanno-ooze (289-1-1 to 289-70-2). Deposition in relatively shallow depths (2,224 m) has assured excellent preservation of faunas throughout. A preliminary planktonic foraminiferal biostratigraphy was published by Saito (in Andrews *et al.*, 1975), and Srinivasan and Kennett (1981a,b) presented a detailed biostratigraphic account of this site. The tropical zonal scheme of Blow (1969) has been employed for Site 289 except for three amendments previously proposed by Srinivasan and Kennett (1981a,b): the redefinition of Zone N13/N12 boundary; the subdivision of Zone N17; and the subdivision of Zone N4. Zone N4 has been subdivided using the first evolutionary appearance of *Globoquadrina dehiscens*, a dissolution-resistant form of wide latitudinal range from tropical to cool subtropical areas (Keller, 1980, 1981b; Srinivasan and Kennett, 1981). Keller (1980, 1981b) recorded the first appearance of *Globoquadrina dehiscens* coincident with the first appearance of the radiolarian *Calocycletta virginis* and last appearance of the radiolarian *Theocyrtis annosa* which have been calibrated to the paleomagnetic Epoch 21/20 boundary, dated at 22.2 Ma (Theyer *et al.*, 1978).

The evolutionary first appearance of *Globoquadrina dehiscens* appears to be an isochronous event which is younger than the first appearance datum of *Globigerinoides* and close to the Oligocene/Miocene boundary (Berggren and Amdurer, 1973; Srinivasan and Kennett, 1981a,b; Keller, 1981b). Because this taxon is dissolution-resistant, it continues to be well represented in sequences below the foraminiferal lysocline. It persists in assemblages long after more fragile forms have dissolved. The Oligocene/Miocene boundary is usually placed at the first appearance of *Globigerinoides*. This is not considered to be an optimally useful datum because *Globigerinoides* is a dissolution-susceptible taxon and also appears to be a polyphyletic "genus." Furthermore, the first appearance of *Globigerinoides* is now placed within the latest Oligocene by some workers (Lamb and Stainforth, 1976; Van Convering and Berggren, 1977; Hardenbol and Berggren, 1978). A datum that may more closely approximate the Oligocene/Miocene boundary is thus required. The first evolutionary appearance of *Globoquadrina dehiscens* seems to be the only planktonic foraminiferal event of practical value in approximating the position of the Oligocene/Miocene boundary in many deep-sea sequences (Srinivasan and Kennett, 1983).

Warm subtropical Site 208

Site 208 located on the northern part of the Lord Howe Rise at 26° 06.61's in the southwest Pacific (Fig. 1) is one of the few DSDP Neogene sites located in warm subtropical waters. The Neogene consists of 435m of incompletely cored foraminiferal nanno-ooze containing rich planktonic foraminiferal assemblages with excellent preservation throughout. The faunal characteristics and zonation for Site 208 have been discussed by Kennett (1973) and Srinivasan and Kennett (1981a,b).

The Neogene sequence at Site 208 contains a mixture of tropical and temperate planktonic foraminiferal elements. These elements do not occur in the same proportions, however, throughout the sequence. In the Early and early Middle Miocene, tropical elements dominate. In the sequence younger than Middle Miocene, temperate elements dominate although many tropical elements persist. Thus, biostratigraphic subdivision of the Early Miocene requires use of the tropical zones, while the Late Middle and younger sequence requires the use of temperate zones. The early Middle Miocene was subdivided using the evolution of *Orbulina*, which transcends both tropical and temperate areas. The only true temperate elements present in the Early and early Middle Miocene are the early forms of *Globoconella* (including *Globorotalia zealandica incognita-G, miozea*). Only the early forms of *Fohsella* (*G. peripheroacuta-G. fohsi* s.l.) are present and thus are only useful for subdivision of the early Middle Miocene. The later forms of *Fohsella* are missing at this latitude and thus biostratigraphic subdivision is based on temperate zones. Some species used in tropical zonation such as *Globigerina nepenthes, Globorotalia (M.) menardii, Neogloboquarina acostaensis,* and *Pulleniatina primalis* are present in the late Middle and Late Miocene. The zones based on these species could be recognized but were not employed because temperate forms are more dominant. Major tropical elements that are largely absent in Site 208 include forms of *Globorotalia* (including *G. (G.) plesiotumida, G. (G.) tumida)* and *Sphaeroidinella.* Thus, the tropical zones based on these species are not recognized. The changeover of applicability of early Neogene tropical zones to late Neogene temperate zones results from global paleoclimatic change. The middle Middle Miocene represents the time of formation of the Antarctic ice cap accompanied by surface-water oceanic cooling, and thus the migration of temperate assemblages towards the tropical regions. Because Site 208 does represent a mixture of tropical and temperate elements, a greater number of faunal events is recorded compared with both tropical and temperate regions. However, faunal events at Site 208 do not necessarily represent datums, because the appearance or disappearance can be diachronous with either tropical or temperate regions. For example, the appearance of *Globigerinatella insueta* is much later than in tropical sequences, and thus this species is restricted to a narrow stratigraphic interval. Similarly, the appearance of *Globorotalia (G.) conomiozea* in Site 208 is a non-evolutionary event and is slightly later than its evolutionary appearance in temperate areas (Kennett, 1973).

Transitional Site 206

DSDP Site 206 is located in the New Caledonia Basin to the east of the central part of the Lord Howe Rise (Fig. 1). The site is strategically located in waters transitional between the warm subtropics and the temperate areas and is important for correlating between sequences of these areas. The Neogene at Site 206 is represented by 450 m of foraminiferal nanno-ooze containing abundant planktonic foraminifera. The sequence is dominated by temperate elements with tropical elements persistently present but rare. The sequence seems to be complete except for a hiatus centered in the latest Miocene. Core recovery is excellent to the base of the Late Miocene but poor to moderate in the earlier part of the Neogene. Despite this, Site 206 represents one of the highest-quality Neogene carbonate sequences in temperate to subtropical waters of the Southern Hemisphere. The relatively incomplete coring through the Early Miocene (160 m) at Site 206 is partly compensated for by relatively high sedimentation rates. Planktonic foraminiferal zonations have been developed by Kennett (1973) for the Late Miocene to Recent and by Srinivasan and Kennett (1981b) for the Early and Middle Miocene.

Neogene Datums

Early Miocene–latest Oligocene datums

Figure 2 summarizes datum levels and zones in three South Pacific DSDP Sites (DSDP 289, 208, and 206). Detailed discussions about intersite correlations have been provided earlier by Kennett (1973) and Kennett and Vella (1975) for the Pliocene and latest Miocene, and by Srinivasan and Kennett (1981a,b) for the entire Neogene. Datums identified in the South Pacific Neogene sequence are summarized below in sequential order beginning from the earliest Neogene. In the following section the abbreviations F.A. and L.A. stand for first appearance and last appearance, respectively. Some of the datum levels can be traced from tropical to temperate areas and are thus useful in correlating over a wide latitudinal range. These include the first appearances of *Globoquadrina dehiscens* and *Praeorbulina glomerosa* and the last appearances of *Globorotalia kugleri* and *Catapsydrax dissimilis*. The extent of others is more restricted.

Globigerinoides primordius F.A.

The first appearance of *G. primordius* is well marked in Site 289 (Core 70) and in Site 208 (Core 25), providing an excellent correlation datum. The *G. primordius* datum has been dated at 25.0 Ma by Van Couvering and Berggren (1977a) and at 22.5 Ma by Saito (1977). This datum is of restricted value because *G. primordius* is solution-susceptible, and thus absent in a large number of Pacific DSDP sequences. For instance, Keller (1980, 1981a) did not recognize this datum in DSDP Sites 71 and 77B in the equatorial north Pacific she examined. However, these sites exhibit much greater dissolution than Sites 289 and 208. The first evolutionary appearance of *G. primordius* has often been used to identify the base of the Miocene, but Srinivasan and Kennett (1983), Lamb and Stainforth (1976), and others have shown that its first appearance occurs within the latest Oligocene.

Globoquadrina dehiscens F.A.

The earliest distinctive planktonic foraminiferal datum in the Early Miocene is the first appearance of *G. dehiscens* which is clearly recorded in Sites 289, 208, and 206. This occurs within the upper part of the stratigraphic range of *Globorotalia kugleri* and well above the initial appearance of *Globigerinoides* spp. A similar sequence of biostratigraphic events has also been recorded by Keller (1980, 1981a) in equatorial North Pacific Sites 77B and 296 and Sites 292, 55, and 71. Keller has dafted the first appearance of *Globoquadrina dehiscens* at 22.2 Ma. The first appearance of *G. dehiscens* in New Zealand marks the base of the Waitakian Stage and the *G. dehiscens* Zone (Jenkins, 1971).

At DSDP Site 296, *G. dehiscens* first appears within the upper part of the nannofossil Subzone *C. abisectus* and radiolarian Zone *L. elongata* (Keller, 1980). The tops of these zones have estimated ages of 23.0 Ma and 22.2 Ma respectively (Bukry, 1975; Theyer *et al.*, 1978).

The ancestral form of *G. dehiscens* is *G. praedehiscens*, which seems to have derived from *G. tripartita* (Blow and Banner, 1962; Blow, 1969). The initial evolutionary appearance of *G. dehiscens*, in all sites and in the New Zealand sequences relative to the last appearance of *Globorotalia kugleri* and the first appearance of *Globigerinoides*, suggests that the datum is isochronous or near isochronous throughout the South Pacific. From a practical point of view, since the first appearance of *Globoquadrina dehiscens* is close to the Oligocene/Miocene boundary (Keller, 1981b), we recommend that the

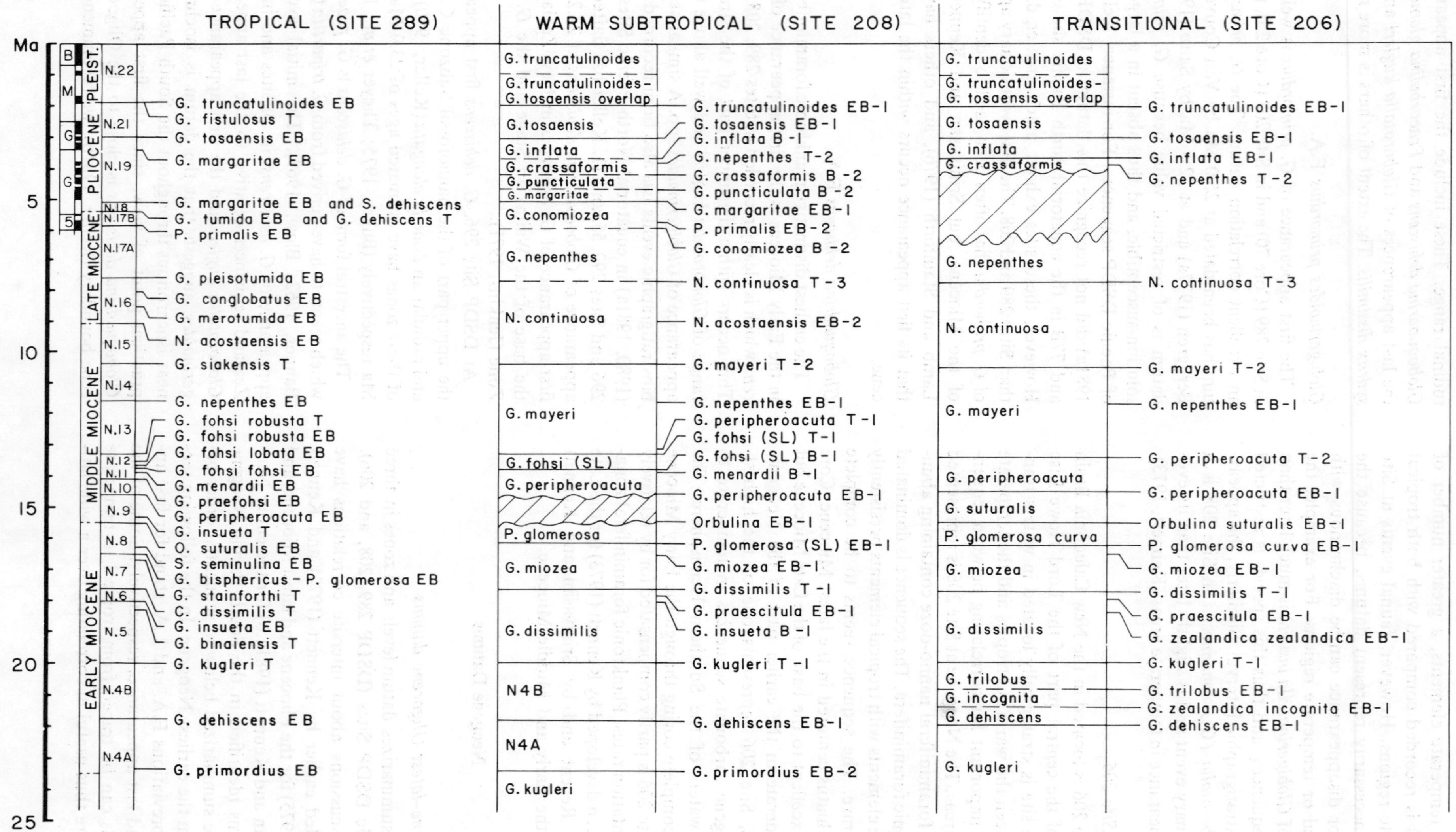

Ma 0 5 10 15 20 25

B M G G 5

PLEIST. PLIOCENE LATE MIOCENE MIDDLE MIOCENE EARLY MIOCENE

N.22 N.21 N.19 N.18 N.17B N.17A N.16 N.15 N.14 N.13 N.12 N.11 N.10 N.9 N.8 N.7 N.6 N.5 N.4B N.4A

TROPICAL (SITE 289)

G. truncatulinoides EB
G. fistulosus T
G. tosaensis EB
G. margaritae EB
G. margaritae EB and S. dehiscens
G. tumida EB and G. dehiscens T
P. primalis EB
G. pleisotumida EB
G. conglobatus EB
G. merotumida EB
N. acostaensis EB
G. siakensis T
G. nepenthes EB
G. fohsi robusta T
G. fohsi robusta EB
G. fohsi lobata EB
G. fohsi fohsi EB
G. menardii EB
G. praefohsi EB
G. peripheroacuta EB
G. insueta T
O. suturalis EB
S. seminulina EB
G. bisphericus - P. glomerosa EB
G. stainforthi T
C. dissimilis T
G. insueta EB
G. binaiensis T
G. kugleri T
G. dehiscens EB
G. primordius EB

WARM SUBTROPICAL (SITE 208)

G. truncatulinoides
G. truncatulinoides - G. tosaensis overlap
G. tosaensis
G. inflata
G. crassaformis
G. puncticulata
G. margaritae
G. conomiozea
G. nepenthes
N. continuosa
G. mayeri
G. fohsi (SL)
G. peripheroacuta
P. glomerosa
G. miozea
G. dissimilis
N4B
N4A
G. kugleri

G. truncatulinoides EB-I
G. tosaensis EB-I
G. inflata B-I
G. nepenthes T-2
G. crassaformis B-2
G. puncticulata B-2
G. margaritae EB-I
P. primalis EB-2
G. conomiozea B-2
N. continuosa T-3
N. acostaensis EB-2
G. mayeri T-2
G. nepenthes EB-I
G. peripheroacuta T-I
G. fohsi (SL) T-I
G. fohsi (SL) B-I
G. menardii B-I
G. peripheroacuta EB-I
Orbulina EB-I
P. glomerosa (SL) EB-I
G. miozea EB-I
G. dissimilis T-I
G. praescitula EB-I
G. insueta B-I
G. kugleri T-I
G. dehiscens EB-I
G. primordius EB-2

TRANSITIONAL (SITE 206)

G. truncatulinoides
G. truncatulinoides - G. tosaensis overlap
G. tosaensis
G. inflata
G. crassaformis
G. nepenthes
N. continuosa
G. mayeri
G. peripheroacuta
G. suturalis
P. glomerosa curva
G. miozea
G. dissimilis
G. trilobus
G. incognita
G. dehiscens
G. kugleri

G. truncatulinoides EB-I
G. tosaensis EB-I
G. inflata EB-I
G. nepenthes T-2
N. continuosa T-3
G. mayeri T-2
G. nepenthes EB-I
G. peripheroacuta T-2
G. peripheroacuta EB-I
Orbulina suturalis EB-I
P. glomerosa curva EB-I
G. miozea EB-I
G. dissimilis T-I
G. praescitula EB-I
G. zealandica zealandica EB-I
G. kugleri T-I
G. trilobus EB-I
G. zealandica incognita EB-I
G. dehiscens EB-I

Fig. 2. Planktonic foraminiferal datum levels and zones for DSDP Sites 289 (tropical), 208 (warm subtropical), and 206 (transitional) in the southwest Pacific. EB = Evolutionary Bottom; B = Bottom; T = Top; (SL) = *Sensu lato*. Value of datums in Sites 208 and 206 are ranked from 1 = highest reliability as datum; to 3 = lowest reliability as datum.

datum be employed to approximate the boundary in deep-sea carbonate sequences. The value of *G. dehiscens* is further enhanced because it is a dissolution-resistant form.

Globorotalia zealandica incognita F.A.

The evolution of *Globorotalia zealandica incognita* from *"Neogloboquadrina" continuosa* in Site 206 occurs within the upper range of *Globorotalia kugleri* and before the first appearance of *Globigerinoides trilobus.* Since the range of *Globorotalia kulgeri* is well known in tropical sequences, biostratigraphic relationships with this species in Site 206 indicate that the *G. zealandica incognita* first appearance datum is within the upper part of Zone N4. This concurs with the placement of the datum in North Pacific Sites 71, 77B, and 292 by Keller (1980, 1981c). At Site 208, *G. zealandica incognita* appears slightly later than the last appearance of *G. kugleri,* indicating placement within Zone N5 and thus diachroneity with the other sites. In New Zealand marine sequences, the first appearance of *G. zealandica incognita* is thus through evolution (Walters, 1965) and is judged to be isochronous with Sites 206 and 208.

Globigerinoides trilobus F.A.

The first appearance of *Globigerinoides trilobus* defines the base of the *G. trilobus* Zone in Site 206, and its first appearance in Sites 206 and 208 occurs well before the upper limit of *Globorotalia kugleri.* Its evolution is not clear in these sections, but the ancestor of *Globigerinoides trilobus* is considered to be *Globigerina woodi connecta,* as first shown by Jenkins (1965). The earliest forms of *Globigerinoides trilobus* are similar in all respects to *Globigerina woodi connecta* except for the addition of a small slit-like supplementary aperture. In tropical Site 289, its first appearance is slightly later than in the cool subtropics—only slightly earlier than the last appearance of *Globorotalia kugleri.* The ancestral form *Globigerina woodi connecta* is absent in the tropics and its first appearance is considered to reflect later migration following its evolution in the cool subtropics.

The evolution of *G. woodi connecta* to *Globigerinoides trilobus* was also recorded by Jenkins (1965) in temperate sites of New Zealand, defining the base of the *G. trilobus* Zone. The first evolutionary appearance of *G. trilobus* therefore seems synchronous throughout the temperate areas of the South Pacific. However, our correlation of the *G. trilobus* datum with the tropical zonation differs from that of Jenkins (1967). Since the first appearance of *G. trilobus* occurs below the upper limit of *Globorotalia kugleri* in all three sites (206, 208, and 289), the age of *Globigerinoides trilobus* datum is clearly within the upper part of Zone N4B and not correlatable with the middle part of the *Catapsydrax dissimilis* Zone (=Zone N6). Therefore, evidence in the South Pacific sequences suggests that the first evolutionary appearance of *Globigerinoides trilobus* forms a useful datum within temperate through subtropical areas. Its first appearance in the tropics, however, is diachronous but only slightly younger than in higher latitudes.

Globorotalia kugleri L.A.

One of the important datum levels in the South Pacific sequences is the last appearance of *Globorotalia kugleri,* an event well marked and apparently isochronous in Sites 289, 208, and 206. Keller (1980, 1981c) also recognizes its utility as a datum level in North Pacific sequences and dates its last appearance at 21.7 Ma based on indirect correlations between Site 77B using radiolarian datums with similar datums in paleomagnetically dated piston cores (Theyer and Hammond, 1974). The last appearance of *G. kugleri* marks the top of Zone N4 (Blow, 1969). The optimal biogeographic range of the species is in the tropics and subtropics. Toward high latitudes it decreases in abundance. Its last appearance in the temperate sites of New Zealand occurs earlier than in tropical areas, according to Jenkins (1967, 1975, 1978) who places it within the middle part of the *Globigerina woodi connecta* Zone.

Globorotalia zealandica zealandica F.A.

The evolutionary first appearance of *Globorotalia zealandica zealandica* from *G. zealandica incognita* defines a useful datum for correlating temperate through transitional sequences (Site 206 and New Zealand sequences including Sites 207A and 284). This evolution is discussed earlier and is clearly marked in Site 206. In New Zealand, Walters (1965) described the evolutionary first appearance of *G, zealandica zealandica* within the middle part of the Awamoan State. Its first appearance in warm subtropical Site 208 seems to be at the same stratigraphic level as in the higher latitude areas, but this is still not clear because the species is sparse. Nevertheless, the datum seems to be valuable for correlating temperate through transitional areas. Keller (1980, 1981 c) also recorded this datum in cool subtropical Sites 292 and 296 of the northwest Pacific. *Globorotalia zealandica zealandica* is widely distributed elsewhere, including the North Atlantic (Berggren and Amdurer, 1973; Poore and Berggren, 1975; Jenkins, 1977) and the South Atlantic (Tjalsma, 1976; Jenkins, 1978).

Globoquadrina binaiensis L.A.

G. binaiensis is a useful zonal marker for the Late Oligocene and Early Miocene of the tropical Indo-Pacific region. Its last appearance within Zone N5 at Site 289 (Core 60) and in the lower part of *C. dissimilis* Zone at Site 208 (Core 22) represents a useful datum for correlation within tropical and warm subtropical regions.

Globigerinatella insueta F.A.

The first appearance datum of *G. insueta* in Site 289 occurs in Core 59, defining the base of Zone N6. Because this taxon is highly dissolution-susceptible, its stratigraphic distribution is sporadic, especially in the lower part of its range at Site 208. Thus the first appearance of *G. insueta* may not represent the true first appearance datum at Site 208. Berggren and van Couvering (1974) estimated an age of 19.0 Ma for the *G. insueta* F.A. datum, whereas Saito (1977) provided an age at 18.6 Ma.

The last appearance datum of *G. insueta* at Site 289 occurs in Core 51 (upper part of Zone N9). This datum is dated at 15.8 Ma by Saito (1977). The exinction level of *G. insueta* is not recorded at Site 208 because of a disconformity. Keller (1980) records this species only in Site 319 where it is consistently present from Zone N8 to the middle of Zone N10.

Globorotalia praescitula F.A.

The evolutionary first appearance of *Globorotalia praescitula* from *G. zealandica zealandica* is clear in temperate and warm subtropical sites and provides another useful datum. Its appearance occurs only slightly later than its ancestor, *G. zealandica zealandica.* Since its appearance in Sites 208 and 206 occurs within the upper range of *Catapsydrax dissimilis,* it can be correlated with the upper part of Zone N6 of the tropical zonation. Walters (1965) recorded the first appearance of *Globorotalia praescitula* within the upper part of the Awamoan Stage.

Catapsydrax dissimilis L.A.

C. dissimilis is easily recognizable and is one of the most important early Neogene datum levels since it occurs in all sites examined from the tropics to the subantarctic and is a dissolution-resistant species. This species occurs abundantly at both Sites 289 and 208. It is an important Oligocene form but ranges upwards into the Early Miocene. Its last appearance in the Early Miocene together with *C. unicavus* marks a well-defined datum level at both Sites 289 and 208 delineating the Zone N6/N7 boundary (Blow, 1969). The last appearance of *C. dissimilis* has been recorded in New Zealand sequences above the first appearance of *G. trilobus* (Hornibrook and Edwards, 1971), which seems to be at the same stratigraphic level as in Sites 206 and 208. The extinction level of *C. dissimilis* has been paleomagnetically dated by Berggren and Van Couvering (1974) at 17.5 Ma, by Ryan *et al.* (1974) at 18.5 Ma, and by Saito (1977) at 18.0 Ma. However, Keller (1980) noted the *C. dissimilis* extinction datum nearly coincides with the base of the radiolarian datum *C. leptetrum* (Goll, 1972), dated at 21.6 Ma in the middle of magnetic Epoch 20 (Theyer *et al.*, 1978).

Catapsydrax stainforthi L.A.

The last appearance datum of *C. stainforthi* at Site 289 occurs in Core 55 in the upper part of Zone N7. The species is rare in Site 208. This datum has been dated at 17.3 by Saito (1977). The last appearance of *C. stainforthi* recorded by Keller (1980) in Sites 77B and 296 may be earlier than its genuine extinction in Site 289, because of dissolution at this level in these two sites.

Globorotalia miozea F.A.

The evolutionary first appearance of *Globorotalia miozea* from *G. praescitula* is well developed in Sites 206, 208, and 281 and forms a well-defined datum in temperate and warm subtropical areas. The datum is absent in tropical Site 289. The evolutionary first appearance of *G. miozea* immediately precedes that of *Praeorbulina,* and hence the age of this datum is within Upper Zone N7 to Lower Zone N8.

Globigerinoides bisphericus F.A.

At Site 289 the *G. bisphericus* (= *G. sicanus*) first appearance datum occurs above the extinction level of *C. stainforthi* and marks the base of foraminiferal Zone N8. In Site 208, this species first appears slightly later at approximately the same stratigraphic level as *Praeorbulina glomerosa curva.* Thus the delayed first appearance of *G. bisphericus* in Site 208 may be due to a later migration of warmer waters to the warm subtropics. There are a number of studies documenting similar diachronous first appearances of this species in DSDP sites (Bronnimann and Resig, 1971; Fleisher, 1974).

This datum has received a number of dates: Ikebe *et al.* (1972), 16.5–17.0 Ma; Ryan *et al.* (1974), 16.8 Ma; Keller (1980), 16.1; and Saito (1977), 17.2 Ma. These dates suggest a total range of about one million years.

Praeorbulina glomerosa F.A.

This valuable and well-known datum based on the evolutionary first appearance of *Praeorbulina glomerosa* from *Globigerinoides bisphericus* occurs within the lower part of Zone N8. It is well marked throughout the South Pacific and elsewhere.

Sphaeroidinellopsis seminulina F.A.

The first appearance of *S. seminulina* occurs simultaneously in Site 289 (lower Zone N8) and in Site 208 (*P. glomerosa curva* Zone) providing a useful datum level for intersite correlation. Jenkins and Orr (1972) recorded the first appearance of *S. seminulina* in the *P. glomerosa* Zone at Site 77B, in agreement with the stratigraphic range in the western Pacific.

Middle Miocene Datums

All the zonal diagnostic species established by Blow (1969) for the Middle Miocene were encountered at the equatorial Pacific Site 289. At Site 208, the *Globorotalia fohsi* lineage is represented only by *G. peripheroacuta and G. fohsi* (s.l.) because of cooler conditions.

Orbulina F.A.

The first evolutionary appearance of *Orbulina,* the "*Orbulina* datum," delineates the Early/Middle Miocene boundary (Zone N8/N9 boundary of Blow, 1969). *Orbulina* is one of the common solution-susceptible forms, and therefore in most DSDP sites its first evolutionary appearance is difficult to determine (Vincent, 1977; Keller, 1980). In Site 289 *Orbulina* is rare but can still be used to delineate the Early/Middle Miocene boundary. In Site 208, due to a disconformity, the first evolutionary appearance is not recorded and the species first appears accompanied by *G. peripheroacuta.* This datum is also widely recognized at temperate latitudes including New Zealand on-land sections. Both Berggren and Van Couvering (1974) and Saito (1977) estimated an age of 16.0 Ma for the first evolutionary appearance of *Orbulina.* Later Van Couvering and Berggren (1977) revised it to 15 Ma. Keller (1981a) also dated this datum at 15.0 Ma (Bukry, 1975).

Globorotalia fohsi lineage

The *Fohsella* lineage forms the basis for the standard tropical planktonic foraminiferal zonation in the Middle Miocene. The succession of the component species at Site 289 is similar to those established by Bolli (1957, 1967) and Banner and Blow (1965). At Site 208 only *G. peripheroacuta* and forms referable to *G. fohsi* (s.l.) are present because of the cooler conditions. At Site 206 and further south, even cooler conditions have eliminated *G. fohsi* leaving only *G. peripheroacuta.*

Globorotalia peripheroacuta F.A.

The first evolutionary appearance of *G. peripheroacuta* in Core 49 at Site 289 marks the base of Zone N10. At Site 208, the base of the *G. peripheroacuta* Zone is missing due to a disconformity, while at Site 206 the first and last appearance of *G. peripheroacuta* form clear datum levels. Saito (1977) estimated the first appearance of *G. peripheroacuta* to be 15.3 Ma, whereas Keller (1980) estimated its age at 14.6 Ma based on correlation with DSDP Site 77B.

Globorotalia praefohsi F.A.

The first apearance of *G. praefohsi,* which defines the Zone N10/N11 boundary, is (298 m) in Core 47 at Site 289 (446 m) and Core 22 cc in Site 208. Saito (1977) estimated an age of 14.7 Ma for the first appearance datum of *G. praefohsi.* Keller (1980) recorded the first appearance of *G. praefohsi* in Site 77B virtually coincident with the base of the *D. exilis* nannoplankton Zone dated at 14.0 Ma (Bukry, 1972, 1975) and the first appearance of the radiolarian *L. neotera* dated at 14.1 Ma (Theyer *et al.,* 1978).

Globorotalia fohsi fohsi F.A.

The first evolutionary appearance of *G. fohsi fohsi* from *G. praefohsi* at Site 289 occurs at 421 m delineating the N11/N12 boundary. At Site 208, on the other hand, *G. fohsi* s.l. develops from *G. peripheroacuta.* At this site, forms referable to *G. fohsi lobata* and *G. fohsi robusta* are absent, and the entire assemblage referable to the *G. fohsi* s.l. Zone broadly correlates with Zones N11 and N12 at Site 289.

Saito (1977) provided an estimated age of 13.9 Ma for the first appearance datum of *G. fohsi fohsi.* Keller (1980) estimated an age of 13.6 Ma for the *G. fohsi fohsi* first appearance at Site 77B, which is close to Saito's (1977) estimate, but considerably younger than the age of 15.5 Ma estimated by Ryan *et al.* (1974).

Globorotalia fohsi lobata F.A.

G. fohsi lobata is a distinct and short-ranging subspecies of *Fohsella,* and its first appearance level is a useful datum. In Site 289, *G. fohsi lobata* first appears shortly after *G. fohsi fohsi.* A similar stratigraphic range for *G. fohsi lobata* has been documented by Keller (1980) at Sites 77B and 319. *G. fohsi lobata* is absent in the cooler Site 208. Keller (1980) estimated an age of 13.1 Ma for the first appearance of *G. fohsi lobata* in Site 77B.

Globorotalia menardii F.A.

The first evolutionary appearance of *G. menardii* from *G. praemenardii* occurs in Site 289 at 425 m, in the upper part of Zone N11. In Site 208, *G. menardii* first appears in Core 20, approximately at the same level as the first appearance of *G. fohsi* s.l., providing a useful Middle Miocene datum.

Globorotalia fohsi robusta L.A.

The top of the *G. fohsi robusta* Zone as determined by the last appearance of *G. fohsi robusta* and of *Fohsella* provides an easily identifiable, useful Middle Miocene datum. Srinivasan and Kennett (1981a) recommended the use of this datum for defining the Zone N12/N13 boundary in preference to the first appearance of *S. subdehiscens* (Blow, 1969).

Berggren and Van Couvering (1974) estimated the

extinction level of *Fohsella* at 12.8 Ma. Saito (1977) estimated an age of 12.4 Ma for this datum based on Site 289 data.

Globigerina nepenthes F.A.

The first appearance of *G. nepenthes* defines the base of Zone N14 (Blow, 1969). This datum is well defined in Site 289, where *G. nepenthes* makes its initial appearance at 348 m. In Sites 208 and 206 this datum occurs the upper part of the *G. mayeri* Zone. Recently, Berggren and van Couvering (1978) discussed the chronology of *G. nepenthes* datum in detail and estimated an age of 12.7 Ma for this event.

Globorotalia mayeri and *G. siakensis* L.A.

The last appearances of *G. mayeri* and *G. siakensis* mark the top of Zone N14 (Blow, 1969) and form a useful datum in many Pacific DSDP sites (Saito, 1977; Keller, 1980, 1981b). This datum is easily recognized at Sites 289, 208, and 206. Saito (1977) estimated on age of 11.2 Ma for the *G. mayeri*/*G. siakensis* extinction datum.

Late Miocene Datums

By the beginning of the Late Miocene, the mid-latitude assemblages of the South Pacific had become much more temperate in character. Specifically, Zone N15 and most of the younger N zones of Blow (1969) recognized in Site 289 cannot be distinguished in Site 208 because of its predominantly temperate assemblages. Hence, a precise correlation between Site 289 and sites to the south is not possible for the Late Miocene and Pliocene. The Late Miocene and younger faunal assemblages of Site 208, instead, show closer affinities to the temperate assemblages in Sites 206, 207A, and 284. Thus intersite correlations are more easily invoked between Site 208 and sites in cooler waters to the south.

During the Late Miocene, the datums provided by changes in temperate forms have wider applicability. Well-known tropical Late Miocene datums that are well represented in Site 289 but not in other sites include *G. merotumida* F.A., *G. plesiotumida* F.A. *G. conglobatus* F.A., and *G. tumida* F.A.

Neogloboquadrina acostaensis F.A.

The first appearance of *N. acostaensis* defines the base of N16 (Blow, 1969). This datum is well defined in Site 289. In warm subtropical Site 208 its first appearance occurs within the *N. continuosa* Zone (Core 16). Saito (1977) estimated an age of 10 Ma for the *N. acostaensis* evolutionary appearance datum.

Neogloboquadrina continuosa L.A.

The last appearance of *N. continuosa* defines the top of the *N. continuosa* Zone (Kennett, 1973), which is a widespread datum at temperate and warm subtropical latitudes appearing in Sites 207A, 206, and 208.

Globorotalia conomiozea F.A.

Just as the *Fohsella* lineage provides useful datums in the Middle Miocene and the *Globorotalia* lineage (*G.* (*G.*) *merotumida, G.* (*G.*) *plesiotumida* and *G.* (*G.*) *tumida*) in the Late Miocene of tropical Site 289, species of *Globoconella* are of value in the Neogene of the cool subtropical to temperate regions (Kennett, 1966, 1967, 1970, 1973; Kennett and Vella, 1975; Collen and Vella, 1973; Jenkins, 1967, 1971).

In contrast to tropical regions, only a few radiometric and/or paleomagnetic age determinations are available for the subtropical to temperate datums of the South Pacific.

The first evolutionary appearance of *G. conomiozea* has been paleomagnetically dated at 6.3 Ma by Ryan *et al.* (1974) and Loutit and Kennett (1979). Within the middle of paleomagnetic Epoch 6, this datum is thus almost coeval with the *Pulleniatina primalis* datum in the tropics at 6.2 Ma (Saito, 1977).

D'Onofrio *et al.* (1975) and van Couvering *et al.* (1976) have considered the first appearance datum of *G. conomiozea* to be almost coincident with the base of the Messinian Stage and estimated an age of 6.5 Ma for the first appearance datum of this taxon, correlating it with the middle of Zone N17. In the South Pacific, the first evolutionary appearance of *G. conomiozea* is at the base of the Kapitean Stage (Kennett, 1967; Loutit and Kennett, 1979). The first appearance of *G. conomiozea* in Site 208 is non-evolutionary and is slightly after the first appearance of *P. primalis*. Thus, the *G. conomiozea* Zone of Site 208 broadly corresponds to Zone 17B in Site 289. A similar range was documented for *G. conomiozea* by Keller (1978) in the central North Pacific Site 310. The first appearance of *G. conomiozea* in Sites 207A and 284 are clear evolutionary gradations from *G. conoidea* (Kennett, 1973; Kennett and Vella, 1975).

Pulleniatina primalis F.A.

The first evolutionary appearance of *P. primalis* occurs within Zone N17 and was employed by Srinivasan and Kennett (1981b) to subdivide this zone into Zones 17A and N17B. The datum is clearly defined in Site 289 and in Site 208 but is missing in Site 206 because of an unconformity. This datum has been dated at 6.2 Ma by Saito (1977).

Pliocene Datums

Globorotalia margaritae F.A.

Blow (1969) considered the evolutionary first appearance of *Sphaeroidinella dehiscens* at the N18/N19 boundary to mark the Miocene/Pliocene boundary. Bolli (1970) considered the *Globorotalia margaritae* Zone to be entirely of Pliocene age based on DSDP cores from Leg 4 in the central Atlantic and Caribbean. He also considered the first appearance of *G. margaritae* to mark the Miocene/Pliocene boundary. Furthermore, Cita (1973) recorded the first appearance of *G. margaritae* to be slightly above the Miocene/Pliocene boundary in Mediterranean deep drilled sites, and found it to range to the end of the Early Pliocene. Berggren (1977) placed the approximate position of the Miocene/Pliocene boundary in DSDP Site 357 (Rio Grande Rise) at the initial appearance of *G. margaritae* and the extinction of *Globoquadrina dehiscens*. Krasheninnikov and Pflaumann (1978) regarded the first evolutionary appearance of *G. margaritae margaritae* from *G. margaritae primitiva* to mark the base of the Pliocene.

At Site 289, *Globorotalia tumida* first appears at 68 m (within Core 18), whereas *G. margaritae* (s.s.) and *S. dehiscens* both first appear 10 m higher at 58 m (within Core 17). The last appearance of *G. dehiscens* is in Core 18 at 65 m. Thus there is a stratigaphic overlap in the upper range of *G. dehiscens* and the lower range of *G. tumida*. Berggren (1973) regarded the initial appearance of *G. tumida* and the extinction of *G. dehiscens* (at 5.0 Ma) as the basis for recognition of the Miocene/Pliocene boundary. Srinivasan and Kennett (1981a) consider the first appearance of *G. margaritae* (s.s.) to represent the Miocene/Pliocene boundary rather than *G. tumida*, which overlaps the terminal part of the range of *G. dehiscens* at Site 289.

At Site 208, *G. tumida* is rare, and the first appearance of *S. dehiscens* is non-evolutionary, occurring near the base of Late Pliocene, thereby diminishing the value of these two species as datums in the warm subtropics. Nevertheless, the first appearance of *G. margaritae* marks the base of the Early Pliocene in Site 208 as well as in other South Pacific sites (Kennett, 1973) where *S. dehiscens* and *G. tumida* are not present. Thus, the *G. margaritae* first appearance datum seems to be a much more reliable indicator for the base of the Early Pliocene, and serves as a common datum for correlating Site's 289 and 208.

In recent years, there have been a number of papers dealing with the Miocene/Pliocene boundary with respect to the paleomagnetic stratigraphy (Ryan *et al.*, 1974; Kennett and Watkins, 1974; Berggren and Van Couvering, 1974; Cita, 1975; Van Couvering *et al.*, 1976; Burckle and Opdyke, 1977). It is now generally agreed that this boundary lies at approximately 5.0 to 5.1 Ma between the top of magnetic Epoch 5 and the lower part of the 'C' Event of the Gilbert. Berggren and Van Couvering (1974) estimated an age of 5.0 Ma for *S. dehiscens* first appearance datum, whereas Saito (1977) estimated an age of 5.0 Ma for the first appearance datum of *G. tumida* and 4.8 Ma for *S. dehiscens*.

Globorotalia puncticulata F.A.

In New Zealand, *G. puncticulata* first appears at the base of the Opoitian Stage, which has been paleomagnetically dated at 5.3 Ma (Loutit and Kennett, 1979). The first evolutionary appearance of *G. puncticulata* from its immediate ancestor *G. conomiozea* is clearly represented in Site 284 and 207A by very abundant specimens. In Site 284 *G. puncticulata* first appears at 142 m, while in Site 207A it appears at 76m (Core 3, Section 3). The Miocene/Pliocene boundary in these sites has been placed at the first appearance of *G. puncticulata* (Kennett, 1973), *G. margaritae* is absent in Site 284 because of cool conditions. Nevertheless, it first appears in Site 207A at approximately the same level as *G. puncticulata*.

In Site 206, *G. puncticulata*, *G. margaritae*, and *G. crassaformis* appear together because of a distinct unconformity representing a gap from latest Miocene to Early Pliocene. This site, thus, does not assist in intersite correlation of the Miocene/Pliocene boundary between these different water masses.

In Site 208, the first appearance of *G. puncticulata* is later than the last appearance of *G. conomiozea* and the first appearance of *G. margaritae*, reflecting later migration to warmer waters after its initial evolutionary appearance in cooler waters to the south. Therefore, the Miocene/Pliocene boundary is placed at the first appearance of *G. margaritae* rather than *G. puncticulata*. In Site 289 *G. puncticulata* is absent.

Globorotalia crassaformis F.A.

The first appearance of *G. crassaformis* in South Pacific Sites 284, 207A, 206, and 208 is recorded above the first evolutionary appearance of *G. puncticulata*, providing a useful datum within the Pliocene in temperate areas. However, in a number of instances these two events have been recorded at almost the same stratigraphic level (e.g. at Site 111 in the North Atlantic, Poore and Berggren, 1974; at Site 310 in the North Pacific, Vincent, 1975 and Keller, 1978). *G. crassaformis* is one of the most widely distributed Pliocene planktonic species which occurs both in tropical and temperate regions. In tropical areas, however, it first appears in the Middle Pliocene, later than in the temperate areas (Fleisher, 1974; Srinivasan, 1977; Vincent, 1977).

The first appearance of *G. crassaformis* predates the

Gilbert A Event in the Indian Ocean Core V20-163 (Saito *et al.*, 1975), indicating this datum to be at least 4.0 Ma. Berggren (1977b) suggested that the evolutionary first appearance of *G. crassaformis* from *G. puncticulata* occurred at 4.0 Ma.

Globorotalia inflata F.A.

The first evolutionary appearance of *G. inflata* at Sites 208, 206, 207A, and 284 provides a useful datum for correlation of DSDP sections in the South Pacific (Kennett, 1973; Kennett and Vella, 1975). Cita (1973) recorded the first evolutionary appearance of *G. inflata* in association with *G. tosaensis* in Mediterranean DSDP sites. In the South Pacific DSDP sites, however, the first appearance of *G. inflata* precedes the first occurrence of *G. tosaensis* as it does in most temperate sections (Vincent, 1977). Keller (1978) regarded the first appearance of *G. inflata* at North Pacific DSDP Site 310 to be coincident with the first appearance of *G. tosaensis* in the South Pacific. The base of the *G. puncticulata* Zone of Kennett (1973) was considered by Keller (1978) to be equivalent to the first appearances of *G. crassaformis* and *G. puncticulata* and to correspond to the base of Zone N19. The *G. crassaformis* Zone and the *G. tosaensis* Zone distinguished in the South Pacific (Kennett, 1973) were not distingushed by Keller (1978) in the North Pacific Site 310. Keller (1978) also pointed out the striking similarity in stratigraphic range and frequency distribution of *G. crassaformis* and the later forms of *Globoconella* in temperate regions of the North and South Pacific. Similar ranges have also been recorded in the North Atlantic (Poore and Berggren, 1974, 1975) and South Atlantic (Berggren, 1977a, b; Jenkins, 1978).

The first evolutionary appearance of *G. inflata* is at or just below the NN16/NN15 nannofossil zonal boundary. This datum at 3.0 Ma is slightly younger than the evolutionary appearance of *G. tosaensis* (Vincent, 1977).

Globorotalia tosaensis F.A.

The first evolutionary appearance of *G. tosaensis* defines the base of N21 in tropical Site 289 (Blow, 1969) and the *G. tosaensis* Zone in warm subtropical and transitional Sites 208 and 206. It is a valuable and widespread datum of 3.0 Ma age (Hays *et al.*, 1969). *G. tosaensis* is rare in temperate areas (e.g. Site 207A) and is unreliable as a datum (Kennett, 1973).

Quaternary Datum

Globorotalia truncatulinoides F.A.

The base of the Quaternary in all of the South Pacific sites examined (Sites 289, 208, 207A, 206, and 284) is taken to be at the first appearance of *Globorotalia truncatulinoides*. In these sites, the transition from *G. tosaensis* to *G. truncatulinoides* is not always simple. For instance, at Sites 284 and 207A (temperate) *G. truncatulinoides* appears suddenly and is continuously present throughout the Pleistocene. Whereas a fairly simple gradation is present between *G. tosaensis* and *G. truncatulinoides* in Site 206 (Kennett, 1973), in Site 208 the transition from *G. tosaensis* to *G. truncatulinoides* is not a simple gradational evolutionary change (Kennett, 1973). Instead, the relative abundance of the two forms fluctuates significantly within the interval of overlap, as in a South Pacific piston core described by Kennett and Geitzenauer (1969).

The first appearance of *G. truncatulinoides* has been dated 1.85 Ma (Olduvai Event) by Berggren and Van Couvering (1974) and 1.95 Ma by Saito (1977) based on magnetic stratigraphy.

Acknowledgements

We thank Dr. Ikebe, Chairman of the International Geological Correlation Programme, Project 114—Biostratigraphic Datum Planes of Pacific Neogene, for his leadership in this program during the last decade.

This research was supported by U.S. National Science Fundation grant OCE79-14594 (CENOP).

References

Allan, R. S., 1956: Report of the Standing Committee on datum planes in the geological history of the Pacific region. *8th Pacific Sci. Cong., Proc.*, v. **2**: p. 325–423.

Andrews, J. E., Packham, G., and others, 1975: *DSDP, Initial Reports,* U.S. Government Printing Office, Washington, D.C., v. **30**, p. 231–398.

Banner, F. T. and Blow, W. H., 1965: Progress in the planktonic foraminiferal biostratigraphy of the Neogene. *Nature*, v. **208**, p. 1164–1166.

Berggren, W. A., 1969: Cenozoic chronostratigraphy, planktonic foraminiferal zonation and the radiometric time scale. *Nature*, v. **224**, p. 1072–1075.

Berggren, W.A., 1971: Tertiary boundaries and correlations. *In* Funnell, B.M. and Riedel W. R. eds., *Micropaleontology of the Oceans*. Cambridge Univ. Press, p. 693–809.

Berggren, W. A., 1972: A Cenozoic time-scale—some implications for regional geology and paleobiology. *Lethaia*, v. **5**, p. 195–215.

Berggren, W. A., 1973: The Pliocene time scale; calibration of planktonic foraminiferal and calcareous nannoplankton zones. *Nature*, v. **243**, p. 391–397.

Berggren, W. A., 1977a: Late Neogene planktonic foraminiferal biostratigraphy of DSDP Site 357 (Rio Grande Rise). *DSDP, Initial Reports*, v. **39**, p. 591–614.

Berggren, W. A., 1977b: Late Neogene planktonic foraminiferal biostratigraphy of the Rio Grande Rise (South

Atlantic). *Mar. Micropal.*, v. **2**, p. 251–265.

Berggren, W. A. and Amdurer, M., 1973: Late Paleogene (Oligocene) and Neogene planktonic foraminiferal biostratigraphy of the Atlantic Ocean (lat. 30. N to lat. 30°S). *Ital. Paleontol. Riv.*, v. **79**, p. 337–392.

Berggren, W. A. and van Couvering, J. A., 1974: The Late Neogene: biostratigraphy, geochronology and paleoclimatology of the last 15 million years in marine and continental sequences. *Paleogeography Paleoclimatology Paleoecology*, v. **16**, p. 1–126.

Berggren, W. A. and van Couvering J.A., 1978: Biochronology. *In* G. V. Cohee, M. F. Glaessner and H.D. Hedberg, eds., *The geologic time scale*. Studies in geology, *Am. Assoc. Petrol. Geol.*, v. **6**.: p. 39–55.

Blow, W. H. and Banner, F. T., 1962: The Mid-Tertiary (Upper Eocene to Aquitanian) Globigerinaceae. *In* Eames, F. E. *et al.*, eds., *Fundamentals of Mid-Tertiary Stratigraphic Correlation*, Cambridge Univ. Press, Cambridge, 2. p. 61–151.

Blow. W. H., 1969: Late middle Eocene to Recent planktonic foraminiferal biostratigraphy. *In* Bronnimann, P. and Renz, H. H. eds., *First International Conference on Planktonic Microfossils, Geneva, 1967, Proc.*, p. 199–421.

Bolli, H. M., 1957: Planktonic foraminiefra from the Oligocene-Miocene Cipero and Lengua Formations of Trinidad, B. W. I. *U. S. Natl. Mus., Bull.*, v. **215**, p. 97–213.

Bolli, H. M., 1967: The subspecies of *Globorotalia fohsi* Cushman and Ellisor and the zones based on them. *Micropaleontology*, v. **13**, p. 502–512.

Bolli, H. M., 1970: The foraminifera of Sites 23–31, Leg 4. *DSDP, Initial Reports*, U.S. Government Printing Office, Washington, D. C., v. **4**, p. 577–643.

Bronnimann, P. and Resig, J., 1971: A Neogene globigerinacean biochronologic time scale of the south-western Pacific. *DSDP, Initial Reports*, U.S. Government Printing Office, Washington, D. C., v. **7**, p. 1235–1469.

Bukry, D., 1972: Coccolith stratigraphy, Leg 9 Deep Sea Drilling Project. *DSDP, Initial Reports*, U.S. Government Printing Office, Washington, D. C., v. **9**, p. 817–832.

Bukry, D., 1975: Coccolith and silicoflagellate stratigraphy, northwestern Pacific Ocean, Deep Sea Drilling Project, Leg 32. *DSDP, Initial Reports*, U. S. Government Printing Office, Washington, D. C., v. **32**, p. 677–692.

Burckle, L. H. and Opdyke, N. D., 1977: Late Neogene diatom correlations in the Circum-Pacific. *First International Congress on Pacific Neogene Stratigraphy, Tokyo, 1976, Proceedings*. p. 255–284.

Cita, M. B., 1973: Pliocene biostratigraphy and chronostratigraphy. *DSDP, Initial Reports*, U. S. Government Printing Office, Washington, D. C., v. **4**, p. 1343–1378.

Cita, M. B., 1975: The Miocene/Pliocene boundary: History and definition. *In* Saito, T. and Burckle, L. H. eds., *Late Neogene Epoch Boundaries*, Spec. Publ. no. 1., Micropaleontology Press, New York, p. 1–30.

Cita, M. B., 1976: Planktonic foraminiferal biostratigraphy of the Mediterranean Neogene. *In* Takayanagi, Y. and Saito, T. eds., *Progress in Micropaleontology, Selected Papers in Honor of Prof. K. Asano.* Micropaleontology Press, New York, p. 47–68.

Collen, J. D. and Vella, P., 1973: Pliocene planktonic foraminifera, southern North Island, New Zealand. *Jour. Foraminiferal Res.*, v. **3**, p. 13–29.

D'Onofrio, S., Gianelli, L., Iaccarino, S., Morlotti, E., Romeo, M., Salvatorini, G., Sampo, M., and Sprovieri, R., 1975: Planktonic foraminifera of the Upper Miocene from some Italian sections and the problem of the lower boundary of the Messinian. *Boll. Soc. Paleontol. Ital.*, v. **14**, p. 177–196.

Fleisher, R. L., 1974: Cenozoic planktonic foraminifera and biostratigraphy, Arabian Sea, Deep Sea Drilling Project, Leg. 23A. *DSDP, Initial Reports*, U. S. Government Printing Office, Washington, D. C., v. **23**, p. 1001–1072.

Glass, B., Ericson, D. B., Heezen, B. C., Opdyke, N. D., and Glass, J. A., 1967: Geomagnetic reversals and Pleistocene chronology. *Nature*, v. **216**, p. 437–442.

Goll, R. M., 1972: Section on radiolaria for synthesis chapter, Leg 9. *DSDP, Initial Reports*, U. S. Government Printing Office, Washington, DC, v. **9**, p. 947–1058.

Hays, J. D., Saito, T., Opdyke, N. D., and Burckle, L. H., 1969: Pliocene-Pleistocene sediments of the equatorial Pacific: Their paleomagnetic, biostratigraphic, and climatic record. *Geol. Soc. Amer., Bull.*, v. **80**, p. 1481–1513.

Hardenbol, J. and Berggren, W. A., 1978: A new Paleogene numerical time scale. *In* Cohee, G. V., Glaessner, M. F. and Hedberg, H. D. eds., *Contributions to the Geologic Time Scale*. Am. Assoc. Petrol. Geol., Stud. Geol., no. 6, p. 213–234.

Hornibrook, N. de B, 1966: New Zealand Tertiary microfossil zonation, correlation and climate. *In* Hatai, K. ed., *Tertiary correlations and climatic changes in the Pacific*. Symposium No. 25, 11th Pacific Sci. Congress, Tokyo, p. 29–39.

Hornibrook, N. de B. and Edwards, A. R., 1971: Integrated planktonic foraminiferal and calcareous nannoplankton datum levels in the New Zealand Cenozoic. *2nd Int. Conf. Planktonic Microfossils, Roma 1970, Proc.*, p. 649–657.

Ikebe, N., Takayanagi, Y., Chiji, M., and Chinzei, K., 1972: Neogene biostratigraphy and radiometric time scale of Japan—an attempt at intercontinental correlation. *Pacific Geology*, v. **4** p. 39–78.

Jenkins, D. G., 1965: The origin of the species *Globigerinoides trilobus* (Reuss) in New Zealand. *Cushman Found. Foram. Res., Contr.*, v. **16**, p. 116–120.

Jenkins, D. G., 1966a: Planktonic foraminiferal datum planes in the Pacific and Trinidad Tertiary: *New Zealand Jour. Geol. Geophys.*, v. **9**, p.424–427.

Jenkins, D. G., 1967; Planktonic foraminiferal zones and new taxa from the lower Miocene to the Pleistocene of New Zealand. *New Zealand Jour. Geol. Geophys.*, v. **10**, p. 1064–1078.

Jenkins, D. G., 1971: New Zealand Cenozoic planktonic foraminifera. *New Zealand Geol. Surv. Paleontol., Bull.*, no. 42, 278 pp.

Jenkins, D. G., 1975: Cenozoic planktonic foraminiferal biostratigraphy of the southwestern Pacific and Tasman

Sea—DSDP Leg 29. *In* Kennett, J. P., Houtz, R. E., *et al.*, eds., *DSDP, Initial Reports*, U. S. Government Printing Office, Washington, D.C., v. **29**. p. 449–467.

Jenkins, D. G., 1977: Lower Miocene planktonic foraminifera from a borehole in the English Channel. *Micropaleontology*, v. **23**, p. 297–318.

Jenkins, D. G., 1978: Neogene planktonic foraminifers from DSDP Leg 40, Sites 360 and 362 in the southeastern Atlantic. *In* Bolli, H. M., Ryan, W. B. F. *et al.*, eds., *DSDP, Initial Reports*, U. S. Government Printing Office, Washington, D.C., v. **40**, p. 723–739.

Jenkins, D. G. and Orr, W. N., 1972: Planktonic foraminiferal biostratigraphy of the eastern equatorial Pacific, Leg 9. *In* Hays, J. D. *et al.*, eds., *DSDP, Initial Reports*, U. S. Government Printing Office, Washington, D.C., v. **9**, p. 1059–1196.

Keller, G., 1978: Late Neogene biostratigraphy and paleoceanography of DSDP Site 310, Central North Pacific and correlation with the Southwest Pacific. *Mar. Micropal.*, v. **3**: p. 97–119.

Keller, G., 1980: Early to Middle Miocene planktonic foraminiferal datum levels of the equatorial and subtropical Pacific. *Micropaleontology*, v. **26**, p. 372–391.

Keller, G., 1981a: Planktonic foraminiferal faunas of the equatorial Pacific suggest Early Miocene origin of present oceanic circulation. *Mar. Micropal.*, v. **6**, p. 269–295.

Keller, G., 1981b: Miocene biochronology and paleoceanography of the North Pacific. *Mar. Micropal.*, v. **6**: p. 535–551.

Keller, G., 1981c: The genus *Globorotalia* in the Early Miocene of the equatorial and northwestern Pacific. *J. Foraminiferal Res.*, v. **11**, p. 118–132.

Kennett, J. P., 1966: The *Globorotalia crassaformis* bioseries in north Westland and Marlborough, New Zealand. *Micropaleontology*, v. **12**, p. 235–245.

Kennett, J. P., 1967: Recognition and correlation of the Kapitean Stage (Upper Miocene, New Zealand). *N. Z. Jour. Geol. Geophys.*, v. **10**, p. 1051–1063.

Kennett, J. P., 1970: Pleistocene paleoclimates and foraminiferal biostratigraphy in Subarctic deep-sea cores. *Deep-sea Research*, v. **17** p. 125–140.

Kennett, J. P., 1973: Middle and Late Cenozoic planktonic foraminiferal biostratigraphy of the southwest Pacific—DSDP Leg 21. *DSDP, Initial Reports*, v. **21**, p. 575–640.

Kennett, J. P., and Geitzenauer, K. R., 1969: The Pliocene-Pleistocene boundary in a South Pacific deep-sea core. *Nature*, v. **224**, p. 899.

Kennett, J. P., and Vella, P., 1975: Late Cenozoic planktonic foraminifera and paleoceanography at DSDP Site 284 in the cool subtropical South Pacific. *DSDP, Initial Reports*, v. **29**, p. 769–799.

Kennett, J. P. and Watkins, N. D., 1974: Late Miocene-Early Pliocene paleomagnetic stratigraphy, paleoclimatology and biostratigraphy in New Zealand. *Geol. Soc. Am., Bull.*, v. **85**, p. 1385–1398.

Krasheninnikov, V. A. and Pflaumann, U., 1978: Zonal stratigraphy of Neogene deposits of the eastern part of the Atlantic Ocean by means of planktonic foraminifers, Leg. 41 DSDP. *DSDP, Initial Reports*, v. **41**, 613–657.

Lamb, J. L. and Stainforth, R. M., 1976: Unreliability of *Globigerinoides* datum. *Am. Assoc. Petrol. Geol., Bull.*, v. **60**, p. 1564–1569.

Loutit, T. S. and Kennett, J. P., 1979: Application of carbon isotope stratigraphy of Late Miocene shallow marine sediments, New Zealand. *Science*, **204**, p. 1196–1199.

Poore, R. Z. and Berggren, W. A., 1974: Pliocene biostratigraphy of the Labrador Sea: calcareous plankton. *Jour. Foraminiferal Res.*, v. **4**, p. 91–108.

Poore, R. Z. and Berggren, W. A., 1975: Late Cenozoic planktonic foraminiferal biostratigraphy and paleoclimatology of Hatton-Rockall Basin: DSDP Site 116. *Jour. Foraminiferal Res.*, v. **5**, p. 270–293.

Ryan, W. B. F., Cita, M. B., Dreyfus, R. M., Burckle, L. H., and Saito, T., 1974: A paleomagnetic assignment of Neogene stage boundaries and the development of isochronous datum planes between the Mediterranean, the Pacific and Indian Oceans in order to investigate the response of the world ocean to the Mediterranean "salinity crisis." *Riv. Ital. Paleontol.*, v. **80**, p. 631–688.

Saito, T., 1972: Late Neogene epoch boundaries in deep sea sediments. *Internatl. Geol. Cong. 24th, Canada, Abstr.*, p. 540.

Saito, T., 1977: Late Cenozoic planktonic foraminiferal datum levels: the present state of knowledge toward accomplishing pan-Pacific correlation. *First International Congress on Pacific Neogene Stratigraphy, Tokyo, 1976, Proceed.* p. 61–80.

Srinivasan, M. S., 1977: Standard planktonic foraminiferal zones of the Andaman-Nicobar Late Cenozoic. *Recent Res. Geol.*, Hindustan Publ. Corp., Delhi, v. **3**, p. 23–29.

Srinivasan, M. S. and Kennett, J. P., 1981a: A review of Neogene planktonic foraminiferal biostratigraphy: applications in the Equatorial and South Pacific, *SEPM Spec. Publ.*, no. 32, p. 395–432.

Srinivasan, M. S. and Kennett, J. P., 1981b: Neogene planktonic foraminiferal biostratigraphy and evolution: Equatorial to Subantarctic, South Pacific, *Mar. Micropal.*, v. **6**, p. 499–533.

Srinivasan, M. S. and Kennett, J. P., 1983: The Oligocene/Miocene boundary in the South Pacific. *Geol. Soc. Am., Bull.*, in press.

Theyer, F. and Hammond, S. R., 1974: Paleomagnetic polarity sequence and radiolarian zones, Brunhes to polarity Epoch 20. *Earth Planet. Sci. Letts.*, v. **22**, p. 307–319.

Theyer, F., Mato, C. Y. and Hammond, S. R., 1978: Paleomagnetic and geochronologic calibration of latest Oligocene to Pliocene radiolarian events, Equatorial Pacific. *Mar. Micropal.*, v. **3**, p. 377–395.

Tjalsma, R. C., 1976: Cenozoic foraminifera from the south Atlantic, DSDP Leg. 36. *In* Barker, P. F., Dalziel, I. W. D., *et al.*, eds., *DSDP, Initial Reports*, U. S. Government Printing Office, Washington, D. C., v. **36**, p. 493–517.

Van Couvering, J. A. and Berggren, W. A., 1977: Biostratigraphical basis of the Neogene time scale. *In*: Kauffman and Hazel, eds., *Concepts and methods of biostratigraphy*, p· 283–306.

Van Couvering, J. A., Berggren, W.A., Drake, R. E., Aguirre, E. and Curtis, G. H., 1976: The terminal Miocene event. *Mar. Micropal.*, **1**, p. 263–286.

Vincent, E., 1975: Neogene planktonic foraminifera from the central North Pacific Leg 32, Deep Sea Drilling Project. *Initial Reports of the Deep Sea Drilling Project*, v. **32**, p. 165–801.

Vincent, E., 1977: Indian Ocean Neogene planktonic foraminiferal biostratigraphy and its paleoceanographic implications. *In* Heirtzler J. R. *et al.*, eds., *Indian Ocean Geology and Biostratigraphy*. Amer. Geophys. Union, p. 469–484.

Walters, R., 1965: The *Globorotalia zealandica* and *G. miozea* lineage. *N. Z. Jour. Geol. Geophys.*, v. **8**, p. 109–127.

I - 3

Neogene Calcareous Nannoplankton Datum Planes and Their Calibration to Magnetostratigraphy

Bilal U. HAQ* and Toshiaki TAKAYAMA**

Introduction

It has long been recognized that improvement of the biochronologic framework is a prerequisite for meaningful climatic, environmental, and "processes" studies of the geologic past. In exploration research an accurate biochronology can mean the difference between success and failure, and increasingly greater time accuracy is being demanded in the interpretations of seismic stratigraphy and sedimentary processes. At the present time, we are on the threshold of a breakthrough in developing the required accuracy and detail in biochronologic scale of the major plankton groups. This has become possible mainly through the recovery of the new Hydraulic Piston Cores (HPC) by the Deep Sea Drilling Project (DSDP) for much of the Cenozoic, and for the first time biostratigraphic studies can proceed alongside paleomagnetic studies of the same cores of pre-Pliocene sequences.

A prime example of the success of the Hydraulic Piston Corer in helping us establish accurate biochronology is provided by the results of the HPC studies obtained during Leg 73 in the eastern South Atlantic. During this leg a combination of HPC coring in soft sediments and rotary drilling in consolidated sediments (Sites 522, 523, and 524) recovered an almost complete Cenozoic sequence (with the exception of early Eocene) from which paleomagnetic studies and detailed biostratigraphic studies have yielded valuable new biochronologic data.

Similar results have also been obtained on cores recovered during Leg 72 (western South Atlantic, Rio Grande Rise) and subsequent DSDP Legs. Once the studies on these cores are completed we will be ushered into a new era in Cenozoic biochronology. In this report we present the preliminary results of such studies and also results from piston cores obtained in the equatorial Pacific that have been studied jointly for paleomagnetics and biostratigraphy.

Paleomagnetic Scale

In spite of the numerous paleomagnetic time scales suggested in recent years, controversy continues as to the accuracy of zonal and boundary assignments, within the correlation points for which radiometric determinations are available. The most widely accepted (and commonly used) Cenozoic polarity scale is that compiled by LaBrecque *et al.* (1977) (with modifications applied to absolute age values based on new decay constants). The subsequent compilations by Ness *et al.* (1980) and Lowrie and Alvarez (1981), although claiming to have more numerous calibration points, suffer from the confusion of using second-order biostratigraphically derived calibration points in the same manner as the radiometrically determined ages available for only a few spots on the stratigraphic column. This confusion, resulting from the use of two different types of data, has led to numerous erroneous calibrations, particularly for the Paleocene. Most current time scales of the Neogene, however, show little or no differences, and the differences that do exist are within the resolution limits of various radiometric methods.

Thus, the LaBrecque *et al.* (1977) scale, although based on fewer calibration points, with sea-floor spreading rates assumed to be relatively uniform within sets of two points, is still a good working model for the Neogene, especially if the age values are corrected, using new decay and abundance constants (Mankinen and Dalrymple, 1979). We have adopted the LaBrecque *et al.* (1977) scale here with the necessary age modifications.

Calibration of Neogene Datum Planes to Magnetostratigraphy

Late Neogene (Late Miocene-Pliocene) biochronology was developed over a decade ago, mainly because good

* *Woods Hole Oceanographic Institution, Woods Hole, Massachusetts 02543, U. S. A.*
** *College of Liberal Arts, Kanazawa University, Kanazawa 920, Japan*

piston-cores for that interval on which paleomagnetic work could be done have been commonly available from numerous ocean basins. Late Neogene land-based marine sections with both good paleomagnetic and biostratigraphic data have been studied both in Europe and Japan for sometime now. For example, Ryan *et al.* (1974) synthesized the available paleomagnetic/biostratigraphic data from numerous piston-cores of the equatorial Pacific and Neogene sections in Europe and New Zealand and presented a summary of datums for the Neogene planktonic Foraminifera, coccoliths and Radiolaria. Until recently these data have been used essentially unchanged as a good working model.

Gartner (1973) presented the first direct correlations of paleomagnetic data and Late Neogene nannofossil biostratigraphy in piston-cores from the equatorial Pacific. In spite of many improvements in detail, the overall correlations have held up in subsequent studies. For the early Miocene, however, most correlations were based on second-order comparisons with radiolarian stratigraphy that had been directly correlated to paleomagnetic stratigraphy by Theyer and Hammond (1974) and Theyer *et al.* (1977). The more recent data from DSDP Legs 72 and 73 in the South Atlantic and recent published data improve the Neogene calcareous nannofossil biochronology and extends it down to early Miocene; however, some gaps still remain in the middle Miocene, where good direct first-order correlations are not available at the time of this writing (1982).

Miocene–Pliocene Nannofossil Biochronology

The oldest Miocene datum levels have been recorded in paleomagnetically dated piston cores recovered from the equatorial Pacific by the Hawaii Institute of Geophysics. In core K72–42, the first appearance datum (FAD) of the late Oligocene-early Miocene species *Triquetrorhabdulus carinatus* was recorded within the normal polarity event of Anomaly 7 (see Fig. 1). A similar first occurrence of *T. carinatus* was also recorded in another equatorial Pacific core (K78–015), but closer to the bottom of Anomaly 7. In core K72–42, *Cyclicargolithus bisectus* has its last appearance datum (LAD) in Anomaly 6C. In another core from the same region, K78–016, *Discoaster druggi* and *Calcidiscus macintyrei* appear within Anomaly 6A.

Recently Stradner and Allram (1982) published results from DSDP Leg 66, Site 493 in the North Guatemala Basin (off Central Mexico) in the Pacific;, where early Miocene sequences were recovered and paleomagnetic stratigraphy was established by Niitsuma (1982). At Site 493, Stradner and Allram show that *Sphenolithus belemnos* and *Helicosphaera ampliaperta* first appear in an interval interpreted by us to be the upper part of magnetic epoch 18 (just above a probable hiatus). *Sphenolithus heteromorphus* appears within epoch 17 (magnetic Anomaly 5E).

Site 516 on the Rio Grande Rise cored during Leg 72 and studied here (see Fig. 2) reveals the occurrence of

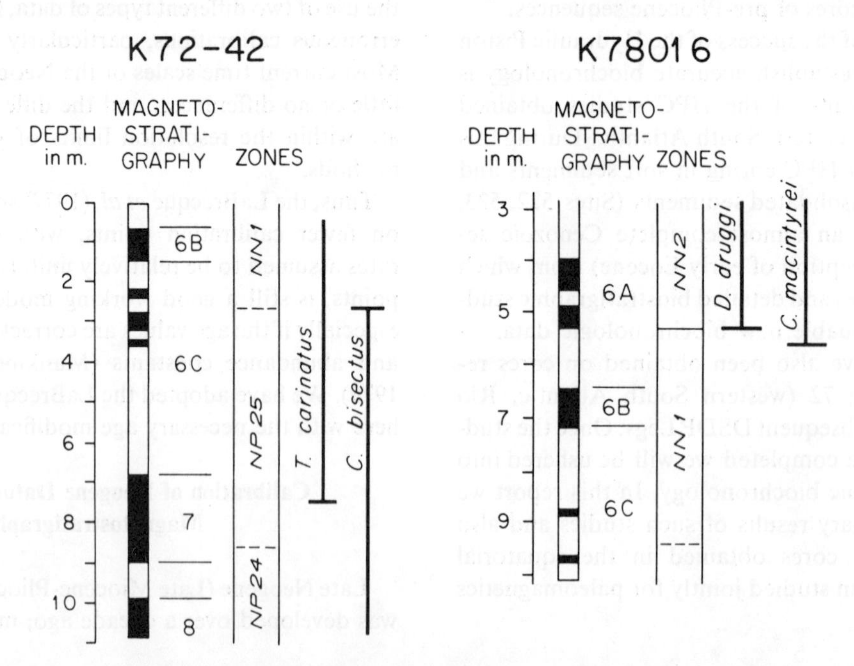

Fig. 1. Magneto- and biostratigraphic determinations in Oligocene-early Miocene piston cores from the equatorial Pacific collected by Hawaii Institute of Geophysics. Magnetostratigraphy by F. Theyer.

numerous important nannofossil datums of early Miocene: *Discoaster druggi* has its LAD within magnetic epoch 17 (Anomaly 5D); *Sphenolithus heteromorphus* and *S. grandis* both first appear at the top of a normal polarity event corresponding to Anomaly 5D within epoch 17; *S. belemnos* disappears within the reversed polarity interval at the bottom of epoch 16, and *Discoaster variabilis* makes its first appearance at about the same level.

Site 519 from the eastern South Atlantic (Leg 73), whose calcareous plankton were recently studied by Poore and Percival (personal communication, 1982), yields important information about numerous middle to late Miocene datum events. Magnetic epochs 9 through 7 were recorded in this section: a hiatus above epoch 7 is followed by the upper part of epoch 5 and an almost complete Gilbert and Gauss sequence. *Catinaster coalitus* both appears and disappears within epoch 9, and *C. calyculus* has a range that straddles the boundary between epochs 9 and 8, but there is some doubt about the reliability of these particular occurrences. *Discoaster neohamatus* has its FAD near the bottom of epoch 8. Numerous taxa appear suddenly at the inferred hiatus, and slightly higher *C. rugosus* appears within the lower normal polarity event of Gilbert. *Discoaster asymmetricus* FAD occurs near the bottom of the third (from top) normal polarity interval of Gilbert epoch. *D. tamalis* and *Pseudoemiliania lacunosa* come in just below and above the boundary between Gilbert and Gauss epochs, respectively.

Mazzei *et al.* (1979) studied the Miocene-Pliocene sequence from Site 397 on Cape Bojador, eastern Atlantic (Leg 47), for both nannofossil and planktonic foraminiferal biostratigraphy and paleomagnetics. A sequence ranging from magnetic epoch 7 through Gauss (with some gaps) is recorded. They show that *Discoaster neohamatus* has its last occurrence near the bottom of the normal polarity interval of magnetic epoch 7, and *D. berggrenii* and *D. quinqueramus* both disappear simultaneously within the upper of the two normal polarity intervals of epoch 5. They have also shown the first amauroliths, i.e. *Amaurolithus primus* and *A. delicatus*, to appear within an interval where they have no magnetic information, but which they interpret to be equivalent to magnetic epoch 6. *A. tricorniculatus* FAD occurs near the top of magnetic epoch 5, both *Discoaster asymmetricus* and *Ceratolithus rugosus* first appear within the middle normal polarity interval of epoch 5, and *Discoaster tamalis* appears in the upper part of epoch 5, similar to

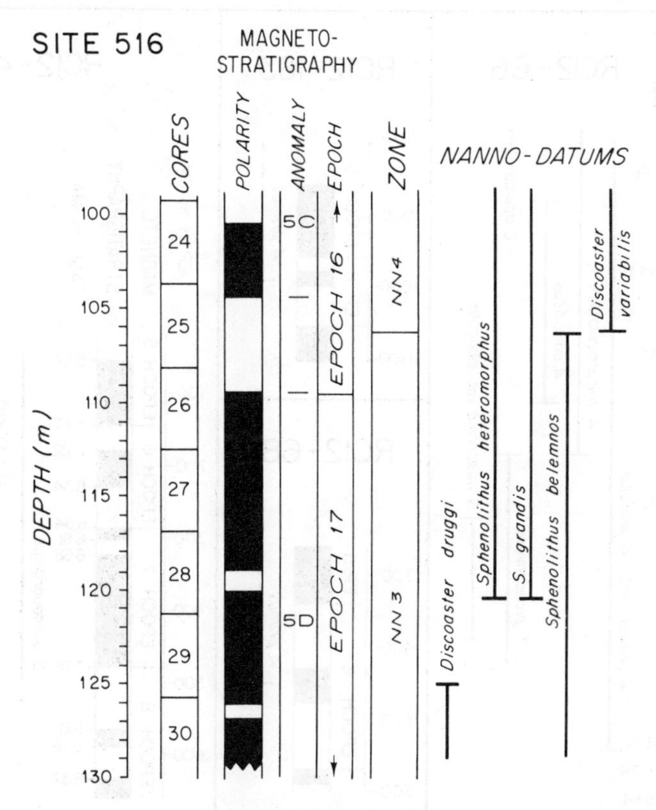

Fig. 2. Early Miocene magneto- and biostratigraphic relationships at DSDP Site 516 (Rio Grande Rise, western South Atlantic). Magnetostratigraphy by N. Hamilton.

its appearance at Site 519.

Haq *et al.* (1980) studied numerous late Miocene piston cores to assign an accurate age to an event characterized by a global decrease in benthic $\delta^{13}C$ and to identify bio-events associated with this isotopic event. These studies (see Fig. 3) showed that the FAD of *Discoaster quinquer-amus* occurs near the top of magnetic epoch 8 in core RC12–418 from the Emperor Seamount. In 1982, Strad-ner and Allram reported the similar occurrence of the FAD of *D. quinqueramus* from DSDP Site 493 near the top of magnetic epoch 8. *Discoaster surculus* first appears in the lower part of epoch 6, and *Amaurolithus primus* and *A. delicatus* both appear within the normal polarity interval of epoch 6 in core RC12–418. *A. tricorniculatus* and *A. amplificus* come in just above the top of epoch 6. Therefore, we believe that *A. amplificus* is erroneously reported to occur within epoch 6 by Mazzei *et al.* (1979). Similar occurrences of the FAD of *A. primus, A. deli-catus,* and *A. tricorniculatus* are also observed in cores RC12–66 (and RC12–68 for *A. primus* only) from the equatorial Pacific and RC12–163 from the Shikoku Basin (see Fig. 3). *Ceratolithus acutus,* an important marker of the Miocene/Pliocene boundary, first appears at the boundary of magnetic epoch 5 and the Gilbert in core RC12–66.

The Miocene-Pliocene nannofossil datum events have been summarized in Tables 1 and 2. As demonstrated by these tables, there seems to be a clear synchroniety of datum levels bteween the Pacific and the Atlantic Oce-ans. In the temperate latitudes, however, most datums do not apply, as many of the species are ecologically (climatically) excluded from the higher latitudes. A com-parison of late Neogene datum events compiled by Ikebe and Chiji (1980) from Japan shows that most Neogene nannofossil datums cannot be used in this area because of sporadic occurrences of asteroliths, ceratoliths, and amauroliths. Among the datum events tabulated in Tables 1 and 2, the following datum events can be recognized in the Neogene sections on the Pacific coast: *Catinaster coalitus* FAD, *C. calyculus* LAD, *Discoaster berggrenii* LAD, *D. asymmetricus* FAD, and *Reticulofenestra pseu-doumbilica* LAD (N. Honda, personal communication, 1982). In the sections on the coast of the Sea of Japan, only the extinction levels of *Sphenolithus heteromorphus* and *Cyclicargolithus floridanus* are useful in making an inter-regional correlation. However, precise correlations among these two extinction levels, paleomagnetic data, and radiometric ages have not yet been established.

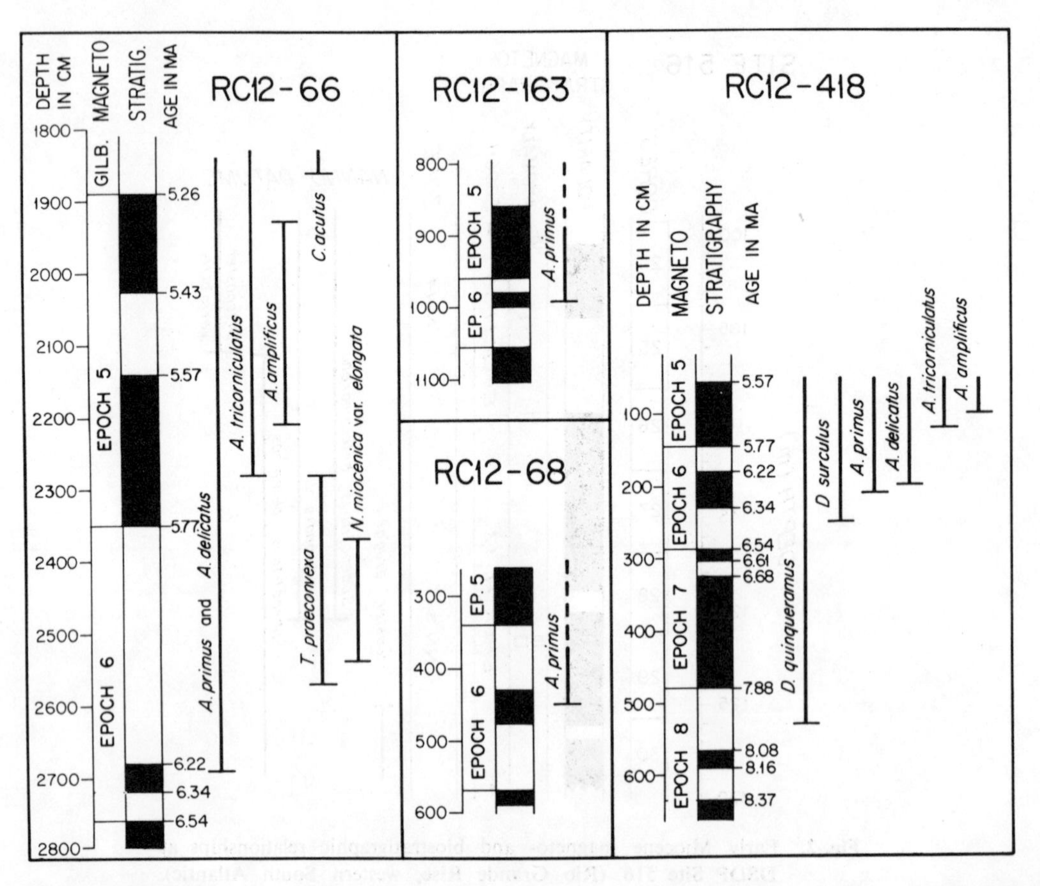

Fig. 3. Late Miocene magneto- and biostratigraphic relationships in piston cores from the Pacific Ocean as determined by Haq *et al.* (1980). Notice the consistent occurrence of *Amaurolithus primus* and *A. delicatus* within epoch 6.

Table 1. Miocene calcareous nannofossil datum levels, their occurrence in magnetic anomaly events and age in MYBP.

Datum	Normal Polarity Event in Anomaly	Age (MYBP)	
		Pacific	Atlantic
28. FAD *C. acutus*	Above 3A	5.1	—
27. LAD *D. bellus*	3A	5.3	—
26. LAD *D. berggrenii*	3A	5.4	—
25. LAD *D. quinqueramus*	3A	5.4	—
24. LAD *A. amplificus*	3A	5.4	—
23. FAD *A. tricorniculatus*	Base 3A	5.7	—
22. LAD *D. aulakos*	Between 4 and 3A	6.0	—
21. FAD *A. delicatus*	Between 4 and 3A	6.3	—
20. FAD *A. primus*	Between 4 and 3A	6.3	—
19. FAD *D. surculus*	Between 4 and 3A	6.4	—
18. LAD *D. neohamatus*	4	—	6.7
17. FAD *D. quinqueramus*	4A	8.0	—
16. FAD *D. neohamatus*	Top 5	—	8.5
15. LAD *C. calyculus*	Top 5?	—	8.4?
14. FAD *C. calyculus*	5?	—	8.8?
13. LAD *C. coalitus*	5?	—	9.0?
12. FAD *C. coalitus*	5?	—	9.6?
11. FAD *D. exilis*	Between 5B & 5C	—	15.5
10. FAD *D. variabilis*	Base 5C	—	16.9
9. LAD *S. belemnos*	Between 5C & 5D	—	17.2
8. FAD *S. grandis*	Mid 5D	—	18.0
7. FAD *S. heteromorphus*	Mid 5D	—	18.0
6. LAD *D. druggi*	5D	—	18.1
5. FAD *H. ampliaperta*	Base 5E	19.0	—
4. FAD *S. belemnos*	Base 5E	19.0	—
3. FAD *D. druggi*	6A	21.5	21.5
2. FAD *C. macintyrei*	6A	21.5	—
1. LAD *C. bisectus*	6C	23.6	23.6
Miocene/Pliocene Boundary	Above 3A	5.1	

Table 2. Pliocene-Pleistocene nannofossil datum levels, their occurrence in magnetic anomaly events and age in MYBP.

Datum	Normal Polarity Event in Anomaly	Age (MYBP)	
		Pacific	Atlantic
21. FAD *E. huxleyi*	Brunhes Epoch	0.27	0.27
20. LAD *P. lacunosa*	Brunhes Epoch	0.46	0.46
19. LAD *H. sellii*	Between 2 & Jaramillo Event	1.2	—
18. LAD *C. macintyrei*	Between 2 & Jaramillo Event	1.5	—
17. FAD *G. oceanica*	Just above 2	1.65	1.65
16. FAD *G. caribbeanica*	Top 2	1.68	1.68
15. LAD *D. brouweri*	Top 2	1.68	1.68
14. FAD *G. omega*	Base 2	—	1.8
13. FAD *G. aperta*	Below 2	2.0	2.0
12. LAD *D. pentaradiatus*	Between 2A and 2	2.1	—
11. LAD *D. asymmetricus*	Between 2A and 2	2.1	—
10. LAD *D. surculus*	Between 2A and 2	2.2	—
9. FAD *P. lacunosa*	Base 2A	3.4	3.4
8. LAD *R. pseudoumbilica*	Below 2A	3.5	3.5
7. FAD *D. tamalis*	Between 3 & 2A	3.6	3.6
6. LAD *C. acutus*	3	4.0	4.0
5. LAD *A. tricorniculatus*	3	4.4	—
4. FAD *D. asymmetricus*	3	—	4.3
3. FAD *D. pentaradiatus*	3	4.6	—
2. FAD *C. rugosus*	3	4.7	4.7
1. FAD *C. acutus*	Above 3A	5.1	—
Pliocene/Pleistocene Boundary	Top 2 (Olduvai Event)	1.65	

Pleistocene Biochronology

The Pleistocene nannofossil biochronology has been well studied and widely documented. Recently Gartner (1977) provided a summary of nannofossil biochronology and suggested a new zonation for the interval.

In spite of many studies, the placement of the Plio/Pleistocene boundary is still controversial. In 1977, Haq *et al.* used numerous calcareous nannoplankton datum events that referred to the Plio/Pleistocene boundary stratotype at Le Castella, the Calabrian stratotype at Santa Maria di Catanzaro in Calabria, Italy, and the Pleistocene nannofossil biochronology from numerous deep sea cores to arrive at an accurate age for the boundary based on magnetostratigraphy. The boundary was correlated by multiple overlapping criteria to a level equivalent to, or slightly younger than, the top of the Olduvai Event. It was shown that in deep sea cores *Discoaster brouweri* LAD occurs within or near the top of the Olduvai Event, and *Gephyrocapsa caribbeanica* and *G. oceanica* make their first appearance just below and above the top of the Olduvai Event; *G. omega* and *G. protohuxleyi* also appear within the Olduvai. The LADS of *Discoaster surculus, D. asymmetricus,* and *D. pentaradiatus* were also documented in this study as occurring slightly above the top of the Gauss, as does the FAD of *Gephyrocapsa aperta*. These relationships have since been reported from many other studies.

Recently Selli *et al.* (1977) proposed another section at Vrica in Calabria as a possible type section for the Neogene/Quaternary boundary. Although no paleomagnetics are available on this section, a nannofossil FAD and LAD sequence similar to that observed at Le Castella was recorded in the Vrica section as well (see Selli *et al.* 1977, Fig. 7). The Plio/Pleistocene boundary is proposed in a layer between the first occurrences of *Gephyrocapsa caribbeanica* and *G. oceanica*. However, a level (level *m*) within the critical interval, containing glass shreds, yielded an anomalously *old* fission-track age of 2.07 ± 0.33 m. y.; and a K-Ar date on the same shreds yielded an age of 2.2 ± 0.2 m. y. (as opposed to an age of about 1.65 to 1.6 m. y. based on paleomagnetic correlations with the top of the Olduvai Event).

Thierstein *et al.* (1977) used Oxygen-isotope stratigraphy to arrive at accurate ages for some late Pleistocene nannofossil datums and their synchroniety through latitudes. They established the global synchroniety of the extinction of *Pseudoemiliania lacunosa* and the FAD of *Emiliania huxleyi* by correlation with the oxygen-isotopic records in cores from tropical to subpolar regions. The *P. lacunosa* LAD was dated at 458 K yrs. as it occurred consistently within oxygen isotopic stage 12. The FAD of *E. huxleyi* was correlated with oxygen isotopic stage 8 and dated at 268 K. yrs. Table 2 includes a summary of all Quaternary nannofossil datums and their ages.

In temperate regions such as Japan it is difficult to ascertain the *Discoaster brouweri* LAD, as the youngest Cenozoic land section contains few discoasters. For this reason the first appearances of *Gephyrocapsa caribbeanica* and *G. oceanica* are more reliable criteria with which to identify the Plio/Pleistocene boundary. *Pseudoemiliania lacunosa,* one of the more useful time markers in the areas of the Pacific coast of central and southwestern Japan, is scarce in sediments on the coast of the Sea of Japan.

References

Gartner, S., 1973: Absolute chronology of the late Neogene calcareous nannofossil succession in the equatorial Pacific. *Geol. Soc. Am., Bull.*, v. **84**, p. 2021–2034.

Gartner, S., 1977: Calcareous nannofossil biostratigraphy and revised zonation of the Pleistocene. *Marine Micropaleont.*, v. **2**, p. 1–25.

Haq, B. U., Berggren, W. A. and van Couvering, J. A., 1977: Corrected age of the Pliocene/Pleistocene boundary. *Nature*, v. **269**, p. 483–488.

Haq, B. U., Worsley, T. R., Burckle, L. H., Douglas, R. G., Keigwin, L. D., Jr., Opdyke, N. D., Savin, S. M., Sommer, M. A., II, Vincent, E. and Woodruff, F., 1980: The late Miocene marine carbon-isotopic shift and the synchroneity of the some phytoplanktonic biostratigraphic events. *Geology*, v. **8**, p. 427–431.

Ikebe, N. and Chiji, M., 1980: Important datum-planes of the western Pacific Neogene (revised) with remarks on the Neogene stages in Japan. *In* Tsuchi, R., ed., *Neogene of Japan*. IGCP-114 Natl. W. G. Japan, p. 1–14.

LaBrecque, J., Kent, D. V. and Cande, S., 1977: Revised magnetic polarity time scale for Late Cretaceous and Cenozoic time. *Geology*, v. **5**, p. 330–335.

Lowrie, W. and Alvarez, W., 1981: One hundred million years of geomagnetic polarity history. *Geology*, v. **9**, p. 392–397.

Mankinen, E. A. and Dalrymple, G. B., 1979: Revised geomagnetic polarity time scale for the interval 0–5 m.y.B.P. *J. Geophy. Res.*, v. **84**, p. 615–626.

Mazzei, R., Raffi, I., Rio, D., Hamilton, N. and Cita, M. B., 1979: Calibration of late Neogene calcareous plankton datum planes with the paleomagnetic record of Site 397 and correlation with Moroccan and Mediterranean sections. *DSDP, Init. Repts.*, v. **47**, p. 375–389.

Ness, G., Levi, S. and Couch, R., 1980: Marine magnetic anomaly timescales for the Cenozoic and Late Cretaceous. A precis, critique and synthesis. *Geophy. and Space Phy. Rev.*, v. **18**, p. 753–770.

Niitsuma, N., 1982: Paleomagnetic results, Middle America Trench off Mexico, Deep Sea Drilling Project, Leg 66. *DSDP, Init. Repts.*, v. **66**, p. 737–768.

Ryan, W. B. F., Cita, M. B., Dreyfus Rawson, M., Burckle, L. H. and Saito, T., 1974: A paleomagnetic assignment of Neogene stage boundaries and the development of isochronous datum planes between the Mediterranean, the Pacific and the Indian Oceans in order to investigate the response of the World Ocean to the Mediterranean "Salinity Crisis". *Ital. Paleont. Riv.*, v. **80**, p. 631–688.

Selli, R., Accorsi, C. A., Bandini Mazzanti, M., Bertolani Marchetti, D., Bigazzi, G., Bonadonna, F. P., Borsetti, A. M., Cati, F., Colalongo, M. L., d'Onofrio, S., Landini, W., Menesini, E., Mezzetti, R., Pasini, G., Savelli, C. and Tampieri, R., 1977: The Vrica section (Calabria, Italy). A potential Neogene/Quaternary boundary stratotype. *Giornale di Geol.*, 2, v. **42** p. 181–204.

Stradner, H. and Allram, F., 1982: The nannofossil assemblage of Deep Sea Drilling Project Leg 66, Middle America Trench. *DSDP, Init, Repts.*, v. **66**, p. 589–639.

Theyer, F. and Hammond, S. R., 1974: Cenozoic magnetic time scale in deep-sea sediments. Completion of the Neogene. *Geology*, v. **2**, p. 487–492.

Theyer, F., Mato, C. Y. and Hammond, S. R., 1978: Paleomagnetic and geochronologic calibration of latest Oligocene to Pliocene radiolarian events, equatorial Pacific. *Marine Micropaleont.*, v. **3**, p. 377–395.

Thierstein, H. R., Geitzenauer, K. R., Molfino, B. and Shackleton, N. J., 1977: Global synchroneity of late Quaternary coccolith datum levels: validation by oxygen isotopes. *Geology*, v. **5**, p. 400–404.

I-4

Neogene Radiolarian Datum Planes of the Equatorial and Northern Pacific

Toyosaburo SAKAI

Introduction

Our knowledge of Neogene radiolarian biostratigraphy has increased significantly in recent years principally as a result of studies contributed to the Initial Reports of the Deep Sea Drilling Project. A low-latitude Neogene radiolarian zonation was developed during the early stage of the project, and this zonation has been satisfactorily employed in the later stages of the project's operation. This zonal scheme, with some modifications proposed by Riedel and Sanfilippo (1978), is generally accepted as the standard zonation for the tropical Neogene sequences.

Radiolarian Events of the Equatorial Pacific

Radiolarian events in their zonal scheme, defined by the bottom or top of each taxa's stratigraphic ranges, have been correlated to the magnetostratigraphy and chronological ages (Theyer *et al.*, 1978) (Table 1).

Although the range of uncertainty in the correlation of tropical sequences based on these events has been considered to fall between $\pm$ 0.4 Ma and $\pm$ 1 Ma (Westberg and Riedel, 1978), the 35 events which have been selected as the most reliable levels of correlation (Theyer *et al.*, 1978; and those indicated with an asterisk in Table 1) may help achieve considerable resolution in chronostratigraphy.

Radiolarian Events of the Northern Pacific

In the northern Pacific, only a few events effective in establishing "synchronous" correlation levels have been reported. However, many biostratigraphic events have been observed and some of them have been employed for regional correlation. Recently, many radiolarian events have been suggested as having a potential toward establishing accurate age assignment in middle to high latitudes (Sakai, 1980; Reynolds, 1980; Reynolds *et al.*, 1980). Some of them are given estimated ages on page 96 of "Neogene of Japan" (edited by R. Tsuchi and published in 1981) (Fig. 2). These estimations are made primarily on the basis of the biostratigraphy of DSDP Site 289 in the equatorial Pacific (Holdsworth, 1976) and of DSDP Site 436 in the northwestern Pacific (Sakai, 1980), by assuming constant sedimentainon between certain datum levels.

Those events which are considered to have a high reliability in correlation in both of these sequences include the morphotypical top of *Calocycletta costata* and transitional horizons from *Lithopera renzae* to *L. neotera*, *L. neotera* to *L. bacca*, *Cannartus petterssoni* to *Ommatartus hughesi*, and *Stichocorys delmontensis* to *S. peregrina*. Ages for these events are assigned temporarily by using 15.5 Ma date for the *Orbulina* datum (485 m in depth) and 5.8 Ma date for the *Pulleniatina* datum (202 m in depth) in Site 289. The age (t: Ma) versus depth (d:meter) relationship at Site 436 has been similarly calculated and the resulting equation is $t = 0.125d - 29.0$. Ages for those selected radiolarian events in the northwestern Pacific were obtained by using this equation (Fig. 1). The range of uncertainty in these estimated ages seems to be order of between $\pm$ 0.1 Ma and $\pm$ 0.4 Ma. Since these values appear to be slightly smaller than those estimated for the equatorial Pacific events (Theyer *et al.*, 1978), they indicate a remarkable degree of correspondence of these radiolarian events in spite of the uncertainties involved in the calculation of the estimated ages.

References

Holdsworth, B. K., 1976: Cenozoic Radiolaria biostratigraphy: Leg 30: Tropical and equatorial Pacific. *DSDP., Init. Rep.*, v. **30**, p. 499–538.

Reynolds, R. A., 1980: Radiolarians from the western North Pacific, Leg 57, Deep Sea Drilling Project. *In* Scientific Party, *DSDP., Init. Rep.*, Washington, U. S. Govt. Printing Office, v. **56**, **57**, p. 735–769.

Reynolds, R. A., Sakai, T., and Casey, R. E., 1980: Synthesis of radiolarian results from DSDP Legs 56 and 57 and

Department of Geology, Faculty of General Education, Utsunomiya University, Utsunomiya 320, Japan

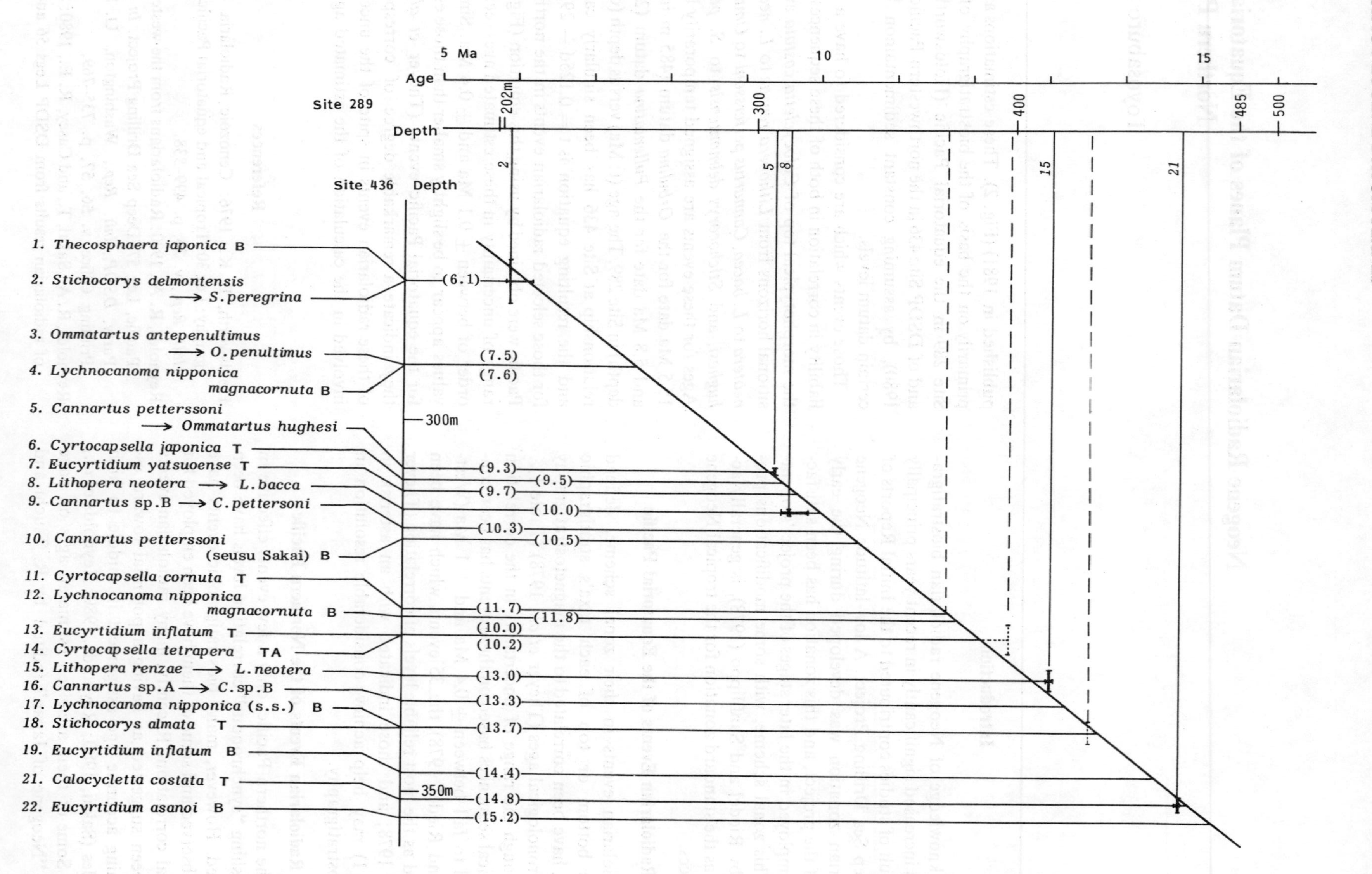

Fig. 1. Correlation of the radiolarian events between Site 289 (equatorial Pacific) and Site 436 (northwestern Pacific), and ages estimated for these events.

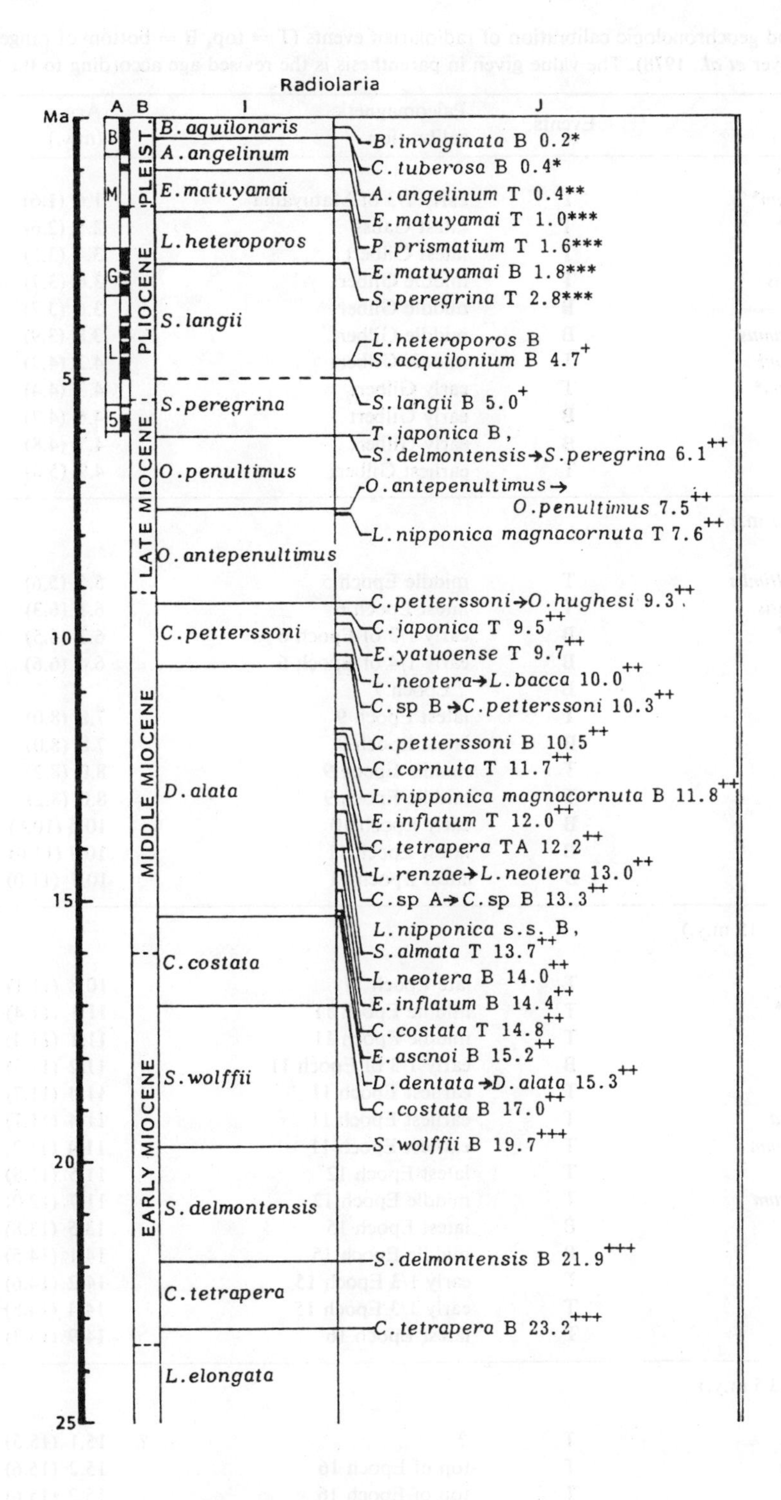

Fig. 2. Datum levels and zones applicable to the correlation of Neogene land-based sections in Japan. (Ref. p. 229)
A: Geomagnetic polarity time scale. B. Brunhes; G, Gauss; L, Gilbert; 5, Epoch 5.
B: Geologic age.
I: Radiolarian zones after Riedel and Sanfilippo (1978) and Sakai (1980).
J: Datum levels and their approximate ages (Ma). (after Tsuchi *et al.*, 1981)

Table 1. Paleomagnetic and geochronologic calibration of radiolarian events (T = top, B = bottom of range) observed in the equatorial Pacific (after Theyer *et al.*, 1978). The value given in parenthesis is the revised age according to the new decay constants.

Species	Events	Paleomagnetic calibration	Age (m.y.)
Pliocene (−1.6 to −5 m.y.)			
1. *Pterocanium prismatium**	T	early 1/3 of Matuyama	1.6 (1.6)
2. *Stichocorys peregrina**	T	latest Gauss	2.5 (2.6)
3. *Spongaster pentas*	T	latest Gilbert	3.4 (3.5)
4. *Ommatartus penultimus*	T	middle Gilbert	3.6 (3.7)
5. *Spongaster tetras**	B	middle Gilbert	3.6 (3.7)
6. *Ommatartus tetrathalamus*	B	middle Gilbert	3.8 (3.9)
7. *Spongaster berminghami*	T	middle Gilbert	4.2 (4.3)
8. *Solenosphaera omnituba**	T	early Gilbert	4.3 (4.4)
9. *S. pentas**	B	early Gilbert	4.6 (4.7)
10. *P. prismatium*	B	early Gilbert	4.7 (4.8)
11. *Acrobotrys tritubus**	T	earliest Gilbert	4.9 (5.0)
Late Miocene (−5 to −10.7 m.y.)			
12. *Ommatartus antepenultimus*	T	middle Epoch 5	5.5 (5.6)
13. *Stichocorys delmontensis*	T	latest Epoch 6	6.1 (6.3)
14. *Stichocorys peregrina**	B	early 1/3 of Epoch 6	6.3 (6.5)
15. *S. omnituba**	B	early 1/4 of Epoch 6	6.4 (6.6)
16. *S. berminghami*	B	? Epoch 7	
17. *Ommatartus hughesi**	T	latest Epoch 9	7.8 (8.0)
18. *O. penultimus**	B	latest Epoch 9	7.8 (8.0)
19. *Cannartus laticonus*	T	middle Epoch 9	8.0 (8.2)
20. *Cannartus petterssoni*	T	middle Epoch 9	8.0 (8.2)
21. *A. tritubus**	B	early Epoch 10	10.5 (10.8)
22. *O. antepenultimus**	B	latest Epoch 11	10.7 (11.0)
23. H. *hughesi**	B	latest Epoch 11	10.7 (11.0)
Middle Miocene (−10.7 to −15 m.y.)			
24. *Stichocorys wolffii*	T	late Epoch 11	10.8 (11.1)
25. *Cyrtocapsella cornuta**	T	middle Epoch 11	11.1 (11.4)
26. *Dorcadospyris alata**	T	middle Epoch 11	11.1 (11.4)
27. *C. petterssoni**	B	early 1/3 of Epoch 11	11.2 (11.5)
28. *Acrocubus octopylus*	T	earliest Epoch 11	11.4 (11.7)
29. *Cyrtocapsella tetrapera*	T	earliest Epoch 11	11.4 (11.7)
30. *Tympanidium binoctonum*	T	earliest Epoch 11	11.4 (11.7)
31. *Giraffospyris toxaria*	T	latest Epoch 12	11.5 (11.8)
32. *Cyclampterium leptetrum*	T	middle Epoch 12	11.7 (12.0)
33. *Cannartus laticonus**	B	latest Epoch 15	13.5 (13.8)
34. *Lithopera neotera**	B	middle Epoch 15	14.1 (14.5)
35. *Cannartus violina*	T	early 1/3 Epoch 15	14.2 (14.6)
36. *Calocycletta virginis**	T	early 1/3 Epoch 15	14.4 (14.8)
37. *Calocycletta costata**	T	latest Epoch 16	14.9 (15.3)
Early Miocene (−15 to −23.5 m.y.)			
38. *Liriospyris stauropora*	T	?	? 15.1 (15.5)
39. *Dorcadospyris dentata*	T	top of Epoch 16	15.2 (15.6)
40. *D. forcipata*	T	top of Epoch 16	15.2 (15.6)
41. *D. alata**	B	late Epoch 16	15.5 (15.9)
42. *Liriospyris parkerae*	B	middle Epoch 16	16.0 (16.4)
43. *Cannartus prismaticus*	T	middle Epoch 16	16.6 (17.0)
44. *G. toxaria*	B	early Epoch 16	16.9 (17.3)
45. *A. octopylus*	B	early Epoch 16	16.9 (17.3)
46. *Lychnocanoma elongata*	T	early Epoch 16	17.2 (17.6)
47. *Cannartus mammiferus**	B	bottom Epoch 16	17.5 (17.9)
48. *C. costata**	B	early Epoch 17	18.4 (18.9)
49. *D. dentata**	B	late Epoch 18	19.0 (19.5)

(*Continued*)

Species	Events	Paleomagnetic calibration		Age (m.y.)
50. *Liriospyris stauropora**	B	middle Epoch 18		19.1 (19.6)
51. *S. wolffii*	B	early Epoch 18		19.2 (19.7)
52. *Dorcadospyris praeforcipata*	T	earliest Epoch 18		19.4 (19.9)
53. *Dorcadospyris simplex*	T	late Epoch 19		19.6 (20.1)
54. *Cyclampterium pegetrum*	T	middle Epoch 19		20.0 (20.5)
55. *Dorcadospyris ateuchus**	T	middle Epoch 19		20.0 (20.5)
56. *Cannartus tubarius*	T	? early Epoch 19	?	20.7 (21.2)
57. *C. violina*	B	bottom Epoch 19		20.8 (21.3)
58. *Calocycletta serrata*	T	latest Epoch 20		21.0 (21.5)
59. *S. delmontensis**	B	late Epoch 20		21.3 (21.8)
60. *Atrophormis gracilis*	T	middle Epoch 20		21.6 (22.1)
61. *C. leptetrum**	B	middle Epoch 20		21.6 (22.1)
62. *Dorcadospyris papilio*	T	middle Epoch 20		21.6 (22.1)
63. *Calocycletta robusta*	T	middle Epoch 20		21.7 (22.2)
64. *C. virginis*	B	earliest Epoch 20		22.2 (22.8)
65. *Theocyrtis annosa*	T	earliest Epoch 20		22.2 (22.8)
66. *C. serrata**	B	latest Epoch 21		22.6 (23.2)
67. *C. tetrapera**	B	latest Epoch 21		22.6 (23.2)
68. *C. cornuta**	B	latest Epoch 21		22.6 (23.2)
69. *C. tubarius*	B	middle Epoch 21		23.2 (23.8)
Latest Oligocene (−23.5 to −27 m.y.)				
70. *Dorcadospyris circulus*	T	early Epoch 21		24.1 (24.7)
71. *D. simplex*	B	early Epoch 21		24.1 (24.7)
72. *L. elongata**	B	earliest Epoch 21		24.2 (24.8)
73. *Dorcadospyris riedeli*	T	late Epoch 22		25.0 (25.6)
74. *T. binoctonum*	B	middle Epoch 22		25.4 (26.0)
75. *C. robusta*	B	latest Epoch 23		26.2 (26.9)

their relation to other North Pacific sections. ***In*** Scientific Party, *DSDP., Init, Rep.,* Washington, U.S.Govt. Printing Office, v. **56**, **57**, p. 771–773.

Riedel, W. R. and Sanfillippo, A., 1978: Stratigraphy and evolution of tropical Cenozoic radiolarians. *Micropal.*, v. **24**, p. 61–96.

Sakai, T., 1980: Radiolarians from Sites 434, 435 and 436, North West Pacific, Leg 56. ***In*** Scientific Party, *DSDP., Init. Rep.*, Washington, U. S. Govt. Printing Office. v. **56**,**57**, p. 695–733.

Theyer, F., Mato, C. Y. and Hammond, S. R., 1978: Paleomagnetic and geochronologic calibration of latest Oligocene to Pliocene radiolarian events, equatorial Pacific. *Mar. Micropal.*, v. **3**, no. 4., p. 377–395.

Tsuchi, R., ed., 1981: *Neogene of Japan*. IGCP-114 National Working Group of Japan, Shizuoka, 140 p.

Westberg, M. J. and Riedel, W. R., 1978: Accuracy of radiolarian correlations in the Pacific Miocene. *Micropal.*, v. **24**, p. 1–23.

I - 5

Evaluation of Diatom Datum Planes of the Pacific Neogene

Itaru KOIZUMI* and Lloyd H. BURCKLE**

Introduction

There are two main problems attendant on the reliability of diatom datum planes: provincialism and accurate age assignment. By a detailed study of piston cores and DSDP drill sites the temporal and spatial distribution of subarctic, transitional, subtropical, and tropical diatom assemblages was determined (Koizumi, 1975a, 1975b, 1975c; Burckle and Opdyke, 1977). These regional subdivisions are based upon a study of the distribution of modern diatom assemblages (e.g., Kanaya and Koizumi, 1966). Figure 1 shows these regional subdivisions and the location of important diatomaceous sections in the North Pacific. First-order datum planes are those that are tied directly to the paleomagnetic reversal record (Donahue, 1970; Burckle and Opdyke, 1977), or radiometric dates (Koizumi, 1977; Barron, 1981). Datum planes extrapolated from sediment accumulation-rate curves based on these first-order age assignments are termed second-order datum planes. Burckle (1977, 1978) has ranked a number of diatom datum planes by order of reliability, using the criteria of Hornibrook and Edwards (1970). This ranking ranges from 1 (most reliable) to 3 (least reliable). Given these considerations, diatom datum planes are reviewed in the following regions:

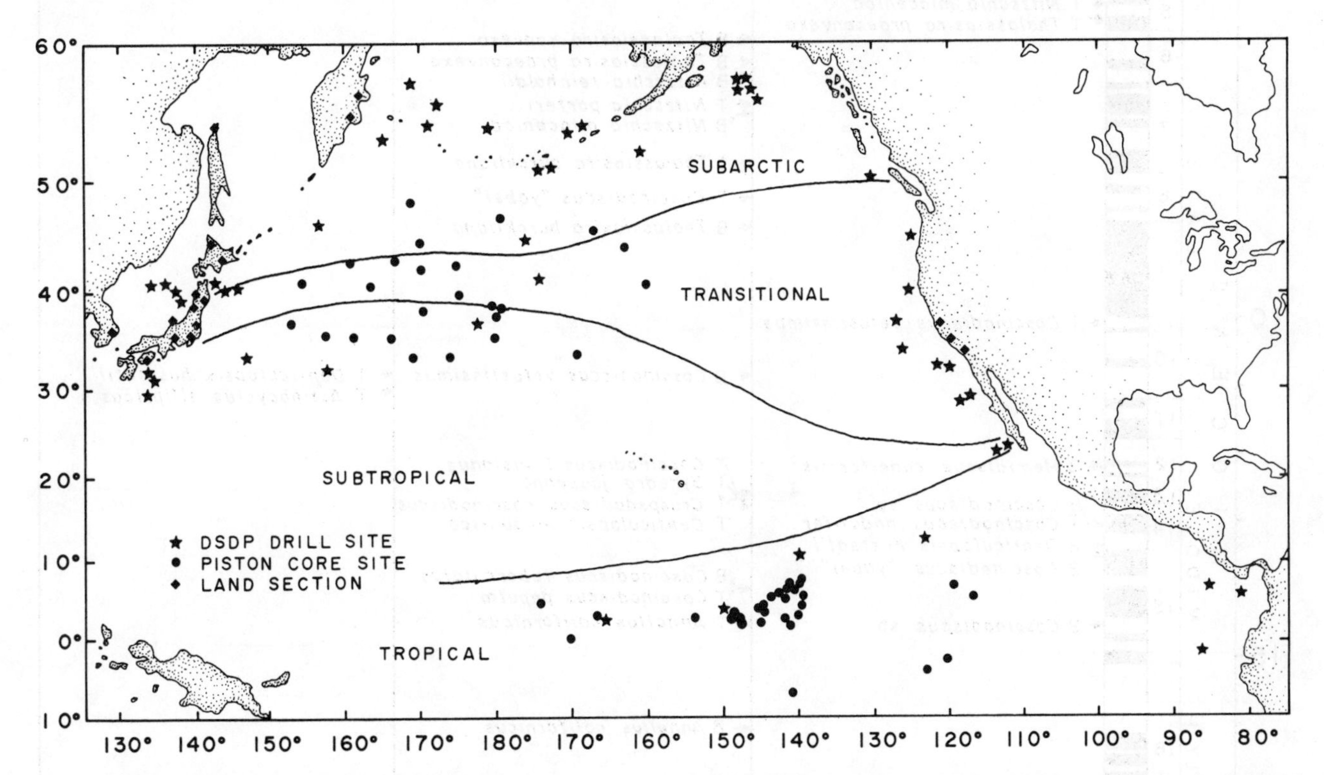

Fig. 1. Locations of cores and important diatomaceous sections and regional subdivisions of diatom assemblages through time in the North Pacific.

* *Institute of Geological Sciences, Osaka University, Osaka 560, Japan*
** *Lamont-Doherty Geological Observatory of Columbia University, Palisades, New York 10964, U. S. A.*

Tropical Region

More than 40 diatom datum planes are recognized and directly tied to the paleomagnetic reversal stratigraphy for the interval from the early Miocene to the Pleistocene (Burckle, 1972, 1977, 1978) (Fig. 2). Because these datum planes are directly tied to the paleomagnetics, they are first-order age assignments. Further, they are ranked by order of reliability using the criteria and format set down by Hornibrook and Edwards (1970).

Subtropical and Transitional Regions

In Japan and the Northwest Pacific, diatom datum planes for the interval from 0–5 m.y.b.p. are chronologically calibrated against the paleomagnetic stratigraphy (Donahue, 1970; Koizumi, 1975c; Koizumi and Kanaya, 1976; Burckle and Opdyke, 1977; Burckle *et al.*, 1978,

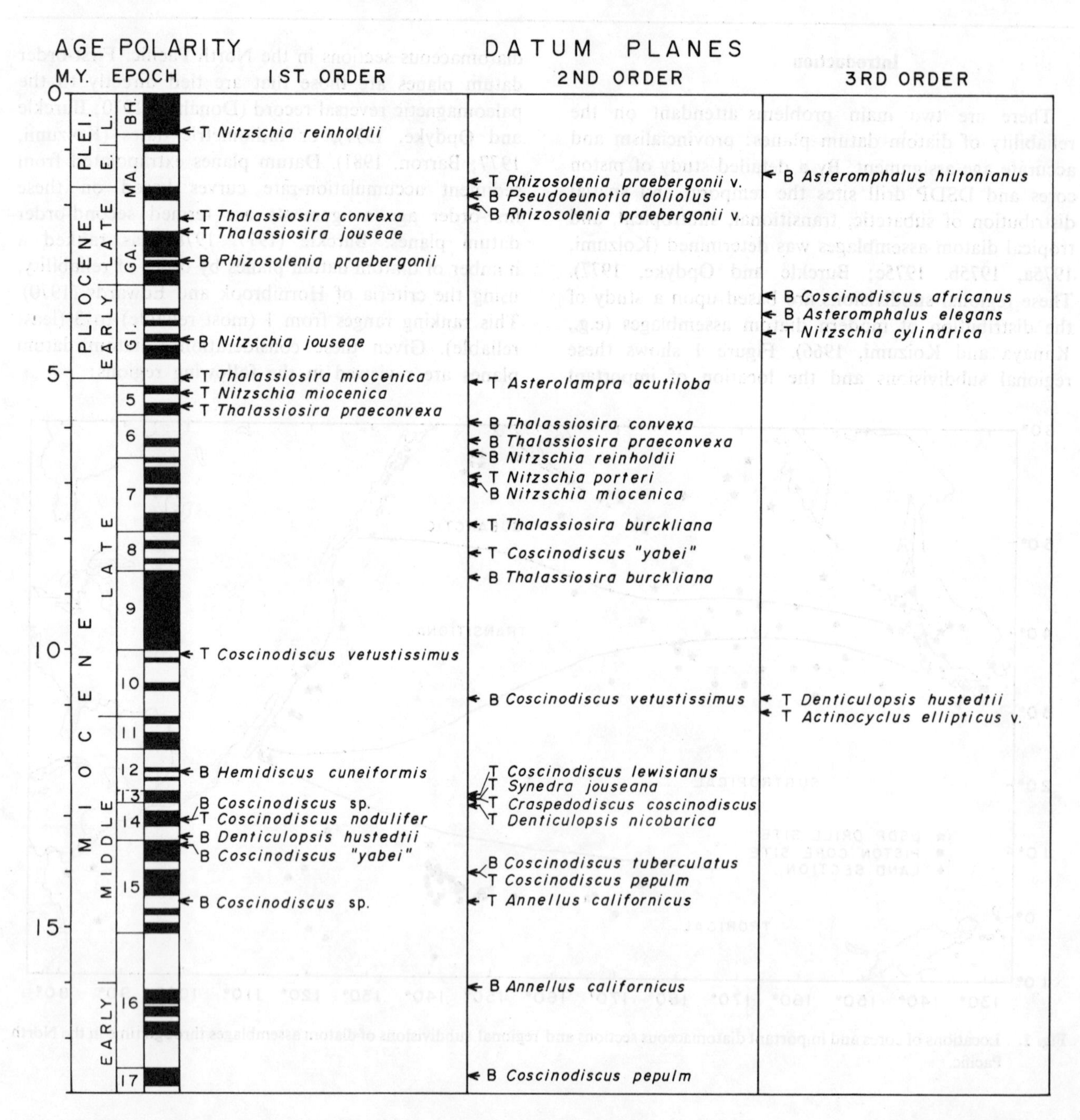

Fig. 2. Diatom datum planes for the Neogene and Quaternary in the tropical region. The paleomagnetic stratigraphy is that of Mankinen and Dalrymple (1979), which was modified from LaBrecque *et al.* (1977).

1980). For the interval before 5 m.y.b.p., datum planes are chronologically combined with K-Ar and FT dates (Koizumi, 1977). In the Northeastern Pacific and California, about 20 isochronous datum planes are identified in the interval from early Miocene to the Quaternary by Barron (1976, 1981; in Keller and Barron, 1981). Six datum planes are chronologically calibrated based on K-Ar dates on the Experimental Mohole cores (Barron, 1981; Burckle *et al.*, 1982). A number of datum planes are correlated to the paleomagnetic stratigraphy by second-order correlation to the lower latitude datum planes of Burckle (1972, 1977, 1978). A number of datum planes can be recognized throughout the transitional region with the exception of the interval from latest Miocene to early Pliocene in the Northeast Pacific.

Subarctic Region

Approximately 15 isochronous datum planes for the interval from 0–6.5 m.y.b.p. are tied to the paleo-

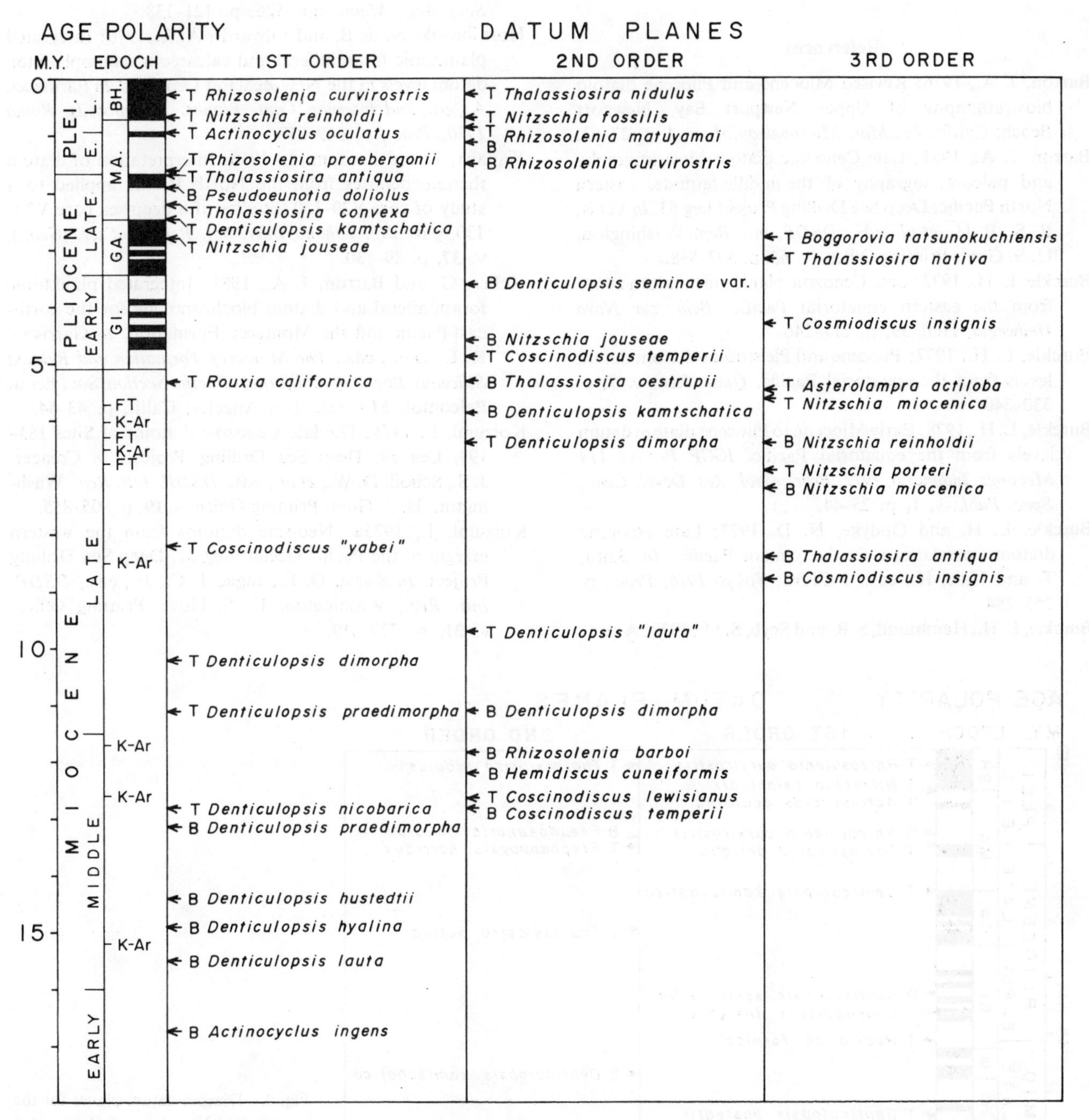

Fig. 3. Diatom datum planes for the Neogene and Quaternary in the subtropical-tropical region of the North Pacific.

magnetic stratigraphy along the west coast of Hokkaido and the high latitude North Pacific (Donahue, 1970; Koizumi, 1973, 1975c; Burckle and Opdyke, 1977; Ujiie, Saito *et al.*, 1977) (Fig. 4). In the interval prior to 6.5 m.y.b.p., however, there are no datum planes chronologically calibrated. It is in this region that we should collect chronologic dates for time control for diatom datum planes.

The first-order datum planes observed in the four separate areas are compiled in one column against the paleomagnetic stratigraphy in Fig. 5.

References

Barron, J. A., 1976: Revised Miocene and Pliocene diatom biostratigraphy of Upper Newport Bay, Newport Beach, California. *Mar. Micropaleontol.*, v. **1**, p. 27–63.

Barron, J. A., 1981: Late Cenozoic diatom biostratigraphy and paleoceanography of the middle-latitude, eastern North Pacific, Deep Sea Drilling Project Leg 63. *In* Yeats, R. S., B. U., *et al.*, eds., *DSDP, Init. Rep.*, Washington, U. S. Govt. Printing Office., v. **63**, p. 507–538.

Burckle, L. H., 1972: Late Cenozoic planktonic diatom zones from the eastern equatorial Pacific. *Beih. zur Nova Hedwegia*, Heft 39, p. 217–246.

Burckle, L. H., 1977: Pliocene and Pleistocene diatom datum levels from the equatorial Pacific. *Quat. Res.*, v. **7**, p. 330–340.

Burckle, L. H., 1978: Early Miocene to Pliocene diatom datum levels from the equatorial Pacific. *IGCP Project 114 Meetings, Indonesia, 1977, Proc., Geol. Res. Devel. Cent., Spec. Publ.*, v. **1**, p. 25–44.

Burckle, L. H. and Opdyke, N. D., 1977: Late Neogene diatom correlations in the circum-Pacific. *In* Saito, T. and Ujiie, H., eds., *1-CPNS, Tokyo 1976, Proc.*, p. 255–284.

Burckle, L. H., Hammond, S. R. and Seyb, S. M., 1978: A stratigraphically important new diatom from the Pleistocene of the North Pacific. *Pac. Sci.*, v. **32**, no. 2, p. 209–214.

Burckle, L. H., Dodd, J. R. and Stanton, R. J., Jr., 1980: Diatom biostratigraphy and its relationship to paleomagnetic stratigraphy and molluscan distribution in the Neogene Centerville Beach section, California. *Jour. Paleontol.*, v. **54**, no. 4, p. 664–674.

Burckle, L. H., Keigwin, L. D., Jr. and Opdyke, N. D., 1982: *Micropaleontology*, v. **28**, p. 329–334.

Donahue, J. G., 1970: Pleistocene diatom as climatic indicators in North Pacific sediments. *In* Hays, J. D., ed., *Geological Investigation of the North Pacific. Geol. Soc. Am., Mem.*, no. 126, p. 121–138.

Hornibrook, N. de B. and Edwards, A. R., 1970: Integrated planktonic foraminiferal and calcareous nannoplankton datum levels in the New Zealand Cenozoic. *In* Farinacci, A., ed., *2nd Internat. Conf. Plankt. Microfossils, Roma 1970, Proc.*, p. 649–657.

Kanaya, T. and Koizumi, I., 1966: Interpretation of diatom thanatocoenoses from the North Pacific applied to a study of core V20–130 (Studies of a deep-sea core V20–130, part IV). *Tohoku Univ. Sci. Rep., 2nd ser.* (*Geol.*), v. **37**, p. 89–130.

Keller, G. and Barron, J. A., 1981: Integrated planktonic foraminiferal and diatom biochronology for the northeast Pacific and the Monterey Formation. *In* Garrison, R. E., *et al.*, eds., *The Monterey Formation and Related Siliceous Rocks of California. Pacific Section*, Soc. Econ. Paleontol. Mineral., Los Angeles, Calif., p. 43–54.

Koizumi, I., 1973: The late Cenozoic diatoms of Sites 183–193, Leg 19, Deep Sea Drilling Project. *In* Creager, J. S., Scholl, D. W., *et al.*, eds., *DSDP, Init. Rep.*, Washington, U. S. Govt. Printing Office. v. **19**, p. 805–855.

Koizumi, I., 1975a: Neogene diatoms from the western margin of the Pacific Ocean, Leg 31, Deep Sea Drilling Project. *In* Karig, D. E., Ingle, J. C., Jr., eds., *DSDP, Init. Rep.*, Washington, U. S. Govt. Printing Office. v. **31**, p. 779–819.

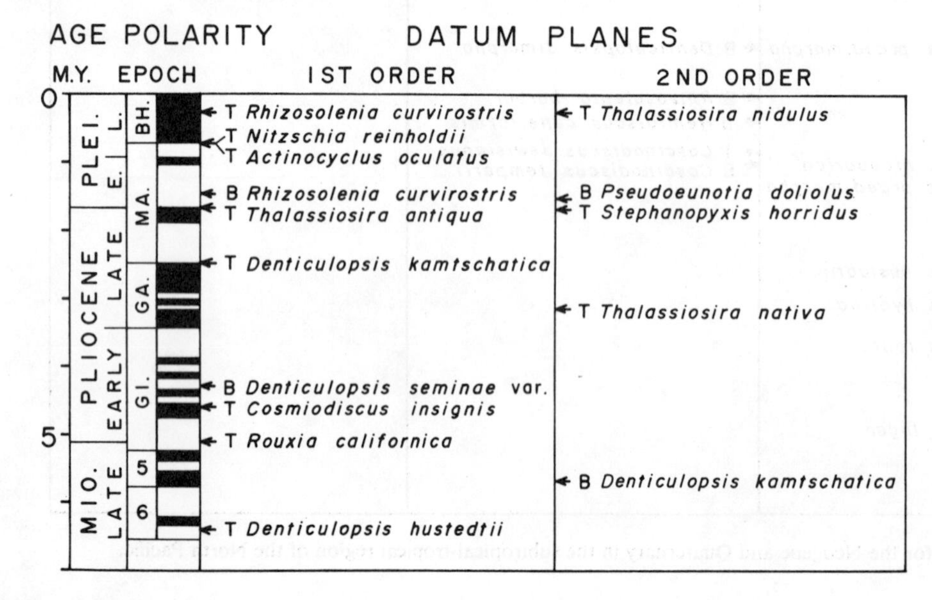

Fig. 4. Diatom datum planes for the late Neogene and Quaternary in the Arctic region.

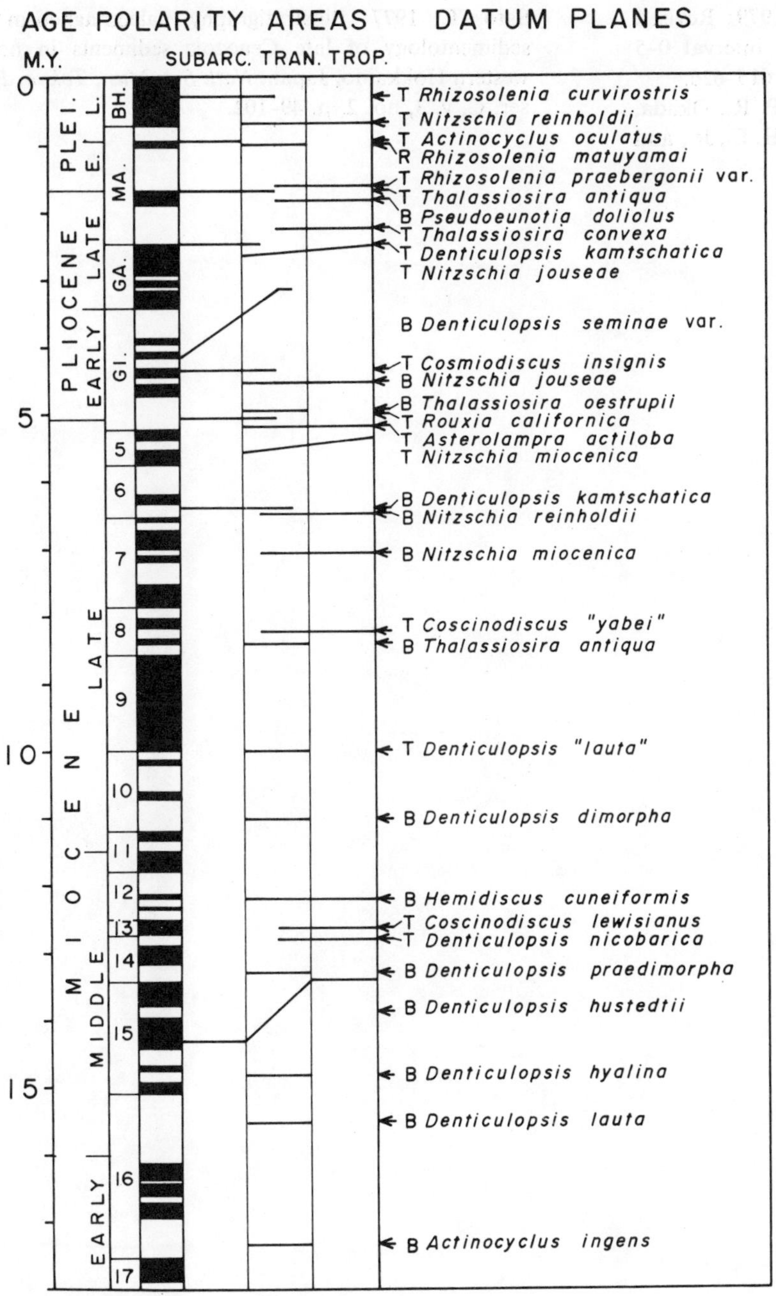

Fig. 5. Compilation of first-order diatom datum planes in the North Pacific and their geographic limit of application as datum planes.

Koizumi, I., 1975b: Neogene diatoms from the northwestern Pacific Ocean, Deep Sea Drilling Project. *In* Larson, R. L., Moberly, R., *et al.*, eds., *DSDP*, *Init. Rep.*, Washington, U. S. Govt. Printing Office, v. **32,** p. 865–889.

Koizumi, I., 1975c: Diatom events in late Cenozoic deep sea sequences in the North Pacific. *Geol. Soc. Japan, Jour.*, v. **81,** no. 10, p. 611–627.

Koizumi, I., 1977: Diatom biostratigraphy in the North Pacific region. *In* Saito, T. and Ujiie, H., eds., *1-CPNS, Tokyo 1976, Proc.*, p. 235–254.

Koizumi, I., and Kanaya, T., 1976: Late Cenozoic marine diatom sequence from the Choshi district, Pacific coast, central Japan. *In* Takayanagi, Y. and Saito, T., eds., *Progress in Micropaleontology*, Micropal. Press, Am. Mus. Nat. Hist., New York, p. 144–159.

LaBrecque, J. L., Kent, D. V. and Cande, S. C., 1977: Revised magnetic polarity time scale for Late Cretaceous and Cenozoic time. *Geology*, v. **5,** p. 330–335.

Mankinen, E. A. and Dalrymple, G. B., 1979: Revised geomagnetic polarity time scale for the interval 0–5 m.y.B.P. *Jour. Geophys. Res.*, no. 84, p. 615–626.

Ujiié, H., Saito, T., Kent, D. V., Thompson, P. R., Okada, H., Klein, G. de V., Koizumi, I., Harper, H. E., Jr., and Sato, T., 1977: Biostratigraphy, paleomagnetism and sedimentology of late Cenozoic sediments in northwestern Hokkaido, Japan. *Natl. Sci. Mus., Tokyo, Bull.*, ser. C, v. 3, no. 2, p. 49–102.

I - 6

Neogene Larger Foraminifera, Evolutionary and Geological Events in the Context of Datum Planes

Charles G. Adams

Abstract

A brief discussion of the Palaeogene/Neogene boundary problem is followed by a consideration of the datum plane concept as currently applied to larger foraminifera. Several kinds of first and last appearance datums are described, and are used to demonstrate that the evolutionary process known as punctuated equilibria may also be applicable to lineages of foraminifera hitherto thought to illustrate only phyletic gradualism. The two processes are not, therefore, believed to be mutually exclusive.

The "stage" boundaries (based essentially on datum planes) of the Neogene part of the East Indies Letter Classification are revised, and the ranges of stratigraphically important larger benthic taxa are plotted against Blow's planktonic zonation.

Dating and correlation problems caused by insufficient information, facies faunas, reworking (real and imagined), and simple misidentifications, are briefly discussed. The geographical distributions of *Flosculinella bontangensis* (Rutten) and *Borelis melo curdica* Reichel are used to demonstrate that the final Tertiary disconnection of the Mediterranean and Indian Ocean—the terminal Tethyan Event—had occurred by late Burdigalian times. Major eustatic changes are mentioned as possible stress factors leading to extinctions on a regional or global scale, and it is suggested that such events should be traceable in shallow-water shelf carbonates throughout the region. Finally, the first and last known occurrences of the Neogene genera shown on the range chart are listed and discussed.

Introduction

The primary purpose of this paper is to relate the ranges of some stratigraphically important Neogene larger foraminifera to the planktonic zonation and datum planes. This relatively simple objective has been attained by reviewing the literature and producing a chart showing the ranges of larger foraminifera and the levels at which there are independent age determinations based on plankton. However, in order to extract the maximum information from this chart and to understand its limitations, the palaeobiological events which determine stratigraphical ranges must be understood. The datum plane concept as applied to larger foraminifera is therefore reviewed, and consideration given to the nature of the events on which they are founded. Certain datum planes are informative in relation to the evolutionary processes known as phyletic gradualism and punctuated equilibria, and these too are briefly considered. The relevant part of the Letter Classification, which is based essentially on datum planes, is also reviewed. Finally, information drawn from the range chart is applied to the elucidation of two geological problems: dating the terminal Tethyan event, and testing recently postulated major Tertiary eustatic changes. However, since the majority of Neogene larger foraminiferal genera were already in existence during the late Palaeogene, it is first desirable to refer briefly to the problem of defining the Palaeogene/Neogene boundary.

The Oligocene/Miocene, and therefore the Palaeogene/Neogene boudary in the Indo-West Pacific region is usually regarded as being marked, in shallow water carbonate facies, by the first appearance of the genus *Miogypsina*. In other regions different criteria are employed, and an International Working Group is now trying to establish the best method of defining this boundary with the palaeontological and stratigraphical information currently available. This group's final recommendation is, however, most unlikely to enable biostratigraphers to employ *Miogypsina* as an Indo-West Pacific marker for the basal Miocene. Other considerations apart, its first appearances in different parts of the region are known not to be coeval (see p. 55). Since no other widely distributed genera are believed

British Museum (Natural History), Department of Palaeontology, Cromwell Road, London SW 7 5BD, U.K.

to have appeared or to have become extinct at or very near this boundary, and the majority range well down into the Oligocene, this account begins with the change from Palaeogene- to Neogene-type faunas in the Middle Oligocene.

Lyell (1833) introduced the first three Tertiary epoch terms, Eocene, Miocene and Pliocene; Beyrich (1854) later added the Oligocene, and the Chattian Stage (Fuchs, 1894) finally became generally accepted as its youngest division. Soon after the introduction of zonations based on planktonic foraminifera (Banner and Blow, 1965; Bolli, 1966; Blow, 1969), it was realized that the basal sediments in the type section of the succeeding Aquitanian stage could not be dated with the precision necessary for accurate regional correlation, and that the lowest datable beds were well within Blow's Zone N4. This might not have mattered if the zone fossil, *Globigerinoides primordius,* had not later been found to range down into N3, a discovery which left both the Aquitanian stage and Zone N4 without a good basal marker. For a review of the problem see Van Couvering and Berggren (1977).

Although larger foraminifera occur quite commonly in the Tertiary (including the Chattian and Aquitanian) of Europe, it has never been easy to correlate them with coeval faunas in the Indo-West Pacific. This is partly because tropical faunas tend to be richer and more varied, and partly because the Mediterranean region and southern Europe constituted a separate faunal province throughout much of Tertiary time (Adams, 1967; 1973) These difficulties had earlier led van der Vlerk and Umbgrove (1927) to erect the well-known Letter Classification for Tertiary strata in the East Indies.

Datum Planes

Datum planes or levels (the terms are synonymous and equally misleading) are stratigraphical horizons marked by two kinds of palaeobiological events: first appearances (stratigraphical bases), and extinctions (stratigraphical tops) of species or groups of species that can be recognized in sedimentary sections over a wide area. The application of the term "datum plane" to planktonic foraminifera was first suggested by Dr. N. de B. Hornibrook at a meeting reported on by Bolli (1969), and it has since been widely adopted and applied to a range of organisms from planktonic foraminifera to mammals. Although some first appearance datums are thought of as marking the first appearances of genera (e.g., the *Orbulina* datum), they all really mark the initial appearances of particular species (e.g., *Orbulina suturalis*).

First appearance datums based on larger foraminifera, in contrast to those based on plankton, tend to be of regional rather than circumglobal value. This reflects the tendency of larger foraminifera to be segregated into faunal provinces separated longitudinally by oceanic barriers rather than latitudinally by thermal gradients. Thus, although the *Orbulina* datum is at the base of the Middle Miocene throughout the circumtropical region, the *Lepidocyclina* datum would, if recognized, be within the Middle Eocene in the Americas but at the base of the Middle Oligocene in the Indo-West Pacific where it would be marked by a different species.

The first appearance datums (FADs) and last appearance datums (LADs) of Van Couvering and Berggren (1977) are each of two types (Fig. 1). FADs may be based either on evolutionary or migratory events. Amongst the Indo-West Pacific larger foraminifera, *Flosculinella* exemplifies an evolutionary FAD since it evolved in this region and is not known elsewhere. *Lepidocyclina*, on the other hand, typifies a migratory FAD in the Indo-West Pacific, since it almost certainly migrated there from the Americas, where it represented an evolutionary FAD in the Eocene. The value of a proposed regional datum plane does not, however, depend upon the nature (evolutionary or migratory) of the first appearance of the nominate taxon but on the rapidity of its dispersal throughout the region. This itself depends on many factors, but point of origin in relation to currents suitable for dispersal, and availability of suitable ecological niches in the areas to which individuals are transported are likely to be particularly important. Every newly evolved taxon and each newly arrived immigrant must have a first appearance datum, but in practice only a few can be used or even recognized, for the reasons given below.

Datum planes may also be based on the last appearances (extinctions) of taxa. LADs are usually based on the last species in a lineage (e.g., *Nummulites fichteli*), but at certain times, for example, the end of the Eocene, multiple coeval extinctions (e.g., *Pellatispira* spp., *Biplanispira* spp. and *Discocyclina* spp. are thought to have occurred. It is doubtful if similar events can be recognized in the Neogene. Theoretically at least it should be possible to recognize datum planes based upon the last appearances of intermediate species within a lineage, but in practice this cannot be done because each species seems to grade into its descendant. For the same reason, FADs based on intermediate species cannot usually be recognized. Although LADs based on extinctions are in general use, little or nothing has been written about them, and the nature of the events they are supposed to mark has never been discussed. It is therefore desirable to consider what information can be derived from a review of some widely employed extinction datums.

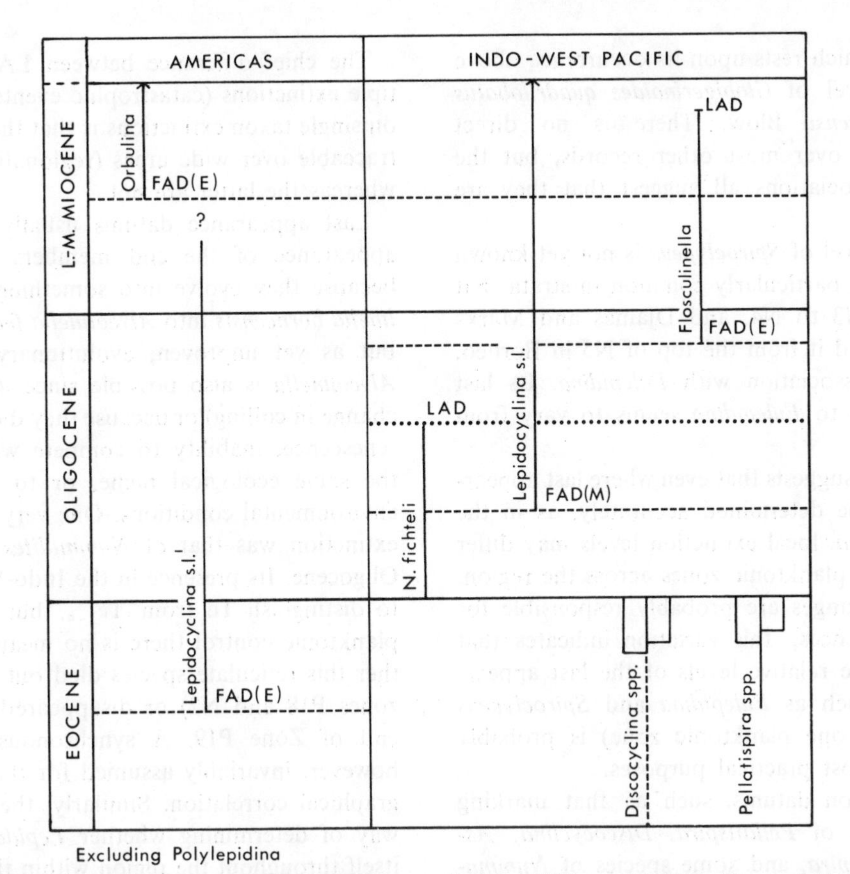

Fig. 1. Diagram showing examples of first and last appearance datums (FADs and LADs). E = evolutionary appearance; M = migratory appearance. The disappearance of *Discocyclina* and *Pellatispira* at the end of Eocene times constitutes a multiple extinction datum.

The genus *Austrotrillina* occurs commonly in shallow water carbonate assemblages of Oligocene and Early Miocene age, and although not one of the larger foraminifera, it was used by Van der Vlerk and Umbgrove (1927) in setting up the Letter Classification of the Tertiary and has remained in use ever since. The last species, *A. howchini* (Schlumberger), is known from East Africa to the Pacific Islands and southern Australia (Adams, 1968), the most advanced forms occurring in strata of N6 to N9 age. In southern Australia its last reported occurrence is in the Pata Limestone (Ludbrook, 1961), i.e., from above the *Orbulina* datum. It also occurs at about the same level in the Trealla Limestone of West Australia (Chaproniere, 1981). No other records from above the *Orbulina* datum are known to the writer from anywhere in the region, and most are no younger than N7 or N8. So although the LAD for *Austrotrillina* has to be drawn within N9, this does not seem to reflect its level of extinction over the region as a whole.

Lepidocyclina (Nephrolepidina) was even more widespread than *Austrotrillina* during the Miocene, and ranged from East Africa to Japan and New Zealand. It was widespread and abundant until Mid-Miocene times, after which its record becomes patchy and unsatisfactory. Haak and Postuma (1975) reported it from N16 in the Tonga Isles; Chang (1975) from N15–N16 in Taiwan; Matsumaru (1981) described it from beds of late N17 age in Japan which had previously been dated as N19 by Ibaraki and Tsuchi (1978); Van Vessem (1978) from N19 in East Borneo; and Adams *et al.* (1979) from early N19 in Fiji. There seems little doubt that the last two species were *L. radiata* and *L. rutteni*, but their local levels of extinction appear to vary widely. The LAD for *Lepidocyclina* is therefore even less widely applicable than that for *Austrotrillina.*

Eulepidina, the final disappearance of which marks the Te/Tf boundary, was widespread during Aquitanian times, but its extinction level in terms of the planktonic zonation has until recently been very difficult to determine. This is partly because it can be mistaken for microspheric forms of *Nephrolepidina* in random sections of limestone; partly because its abundance in older beds makes it particularly susceptible to reworking, and partly because it has been found only rarely with determinable plankton. Chaproniere (1981) has stated that its last known occurrence in the Mandu Limestone, Cape Range, West Australia, can be dated as early N6, a determination which rests upon its occurrence above the appearance level of *Globigerinoides quadrilobatus trilobus* (Reuss) *sensu* Blow. There is no direct planktonic control over most other records, but the benthic faunal associations all suggest that they are older.

The extinction level of *Spiroclypeus* is not yet known with certainty. It is particularly common in strata that can be dated as N3 to N4, and Djamas and Marks (1978) have reported it from the top of N5 in Borneo, although not in association with *Eulepidina.* Its last appearance relative to *Eulepidina* seems to vary from place to place.

Present evidence suggests that even where last appearance datums can be determined accurately, as in the case of *Austrotrillina,* local extinction levels may differ by as much as two planktonic zones across the region. Although facies changes are probably responsible for many major differences, this variation indicates that uncertainty over the relative levels of the last appearances of genera such as *Eulepidina* and *Spiroclypeus* (rarely more than one planktonic zone) is probably unimportant for most practical purposes.

Multiple extinction datums, such as that marking the disappearances of *Pellatispira, Discocyclina, Asterocyclina, Biplanispira,* and some species of *Nummulites* and *Spiroclypeus* at the end of Eocene times, have not been recognized in the Neogene. Multiple extinctions point to a major geological event, such as a large eustatic fall, having wiped out a number of genera—some on a global scale. However, the foraminiferal evidence for such sea-level falls (even in the Palaeogene) is tenuous, and no firm conclusions should be drawn until planktonic control over a number of last occurrences is available.

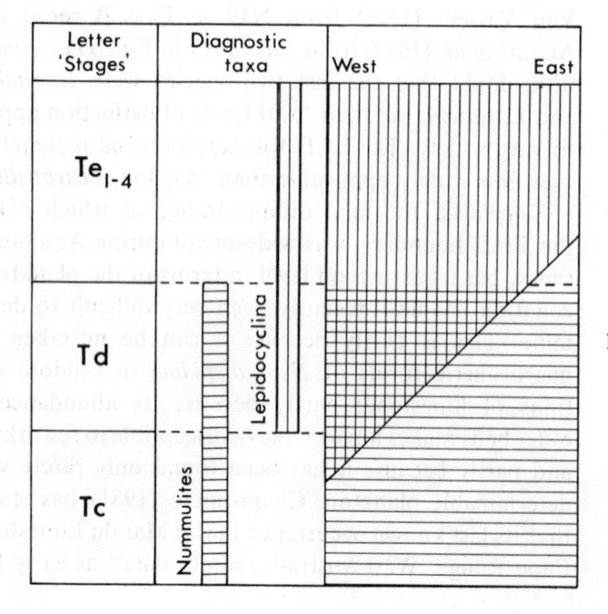

The chief difference between LADs based on multiple extinctions (catastrophic events) and LADs based on single taxon extinctions is that the former are readily traceable over wide areas (regionally or even globally) whereas the latter are not.

Last appearance datums usually represent the disappearance of the end members of lineages, either because they evolve into something new (e.g., *Flosculinella borneensis* into *Alveolinella fennemai* - a probable but as yet unproven, evolutionary event; *Borelis* to *Alveolinella* is also possible since it would involve no change in coiling) or because they die out through racial senescence, inability to compete with new arrivals in the same ecological niche, or to adapt to changing environmental conditions. One very important Tertiary extinction was that of *Nummulites fichteli* during the Oligocene. Its presence in the Indo-West Pacific is used to distinguish Td from Te_{1-4}, but in the absence of planktonic control there is no means of knowing whether this reticulate species died out gradually (between zones P18 and N2) or disappeared everywhere at the end of Zone P19. A synchronous disappearance is, however, invariably assumed for the purpose of stratigraphical correlation. Similarly, there is at present no way of determining whether *Lepidocyclina* established itself throughout the region within the time represented by one planktonic zone (P19) or whether it took considerably longer (Fig. 2). The possibility of correlative error in this part of the stratigraphical column is therefore considerable, and strata assigned to Td in one area could be coeval with strata dated as Tc or lower Te elsewhere. Errors of this kind, if not of this magnitude, can occur at any stratigraphic level, and it is probable that correlations based solely on single species of larger foraminifera are currently accurate only to $\pm$ one

Fig. 2. Diagram showing alternative correlations across the Mediterranean and Indo-West Pacific regions in the absence of planktonic control over the first and last appearances of two critical taxa. It is not known whether the dispersal of *Lepidocyclina* (*Eulepidina* and *Nephrolepidina*) was a geologically instantaneous event or whether it occupied a measurable amount of time. A west to east dispersal route is assumed (see Adams, 1967) but cannot be regarded as proven. If the extinction of *N. fichteli* was not everywhere coeval the possibility of correlative error is still greater.

planktonic zone. In general, extinction datums have only a limited value for regional correlation, and for purposes of interregional correlation they are frequently valueless.

Many systematists believe that when a new species arises, its immediate ancestor becomes extinct. This is implicit in the work of Drooger (1963, 1979) and is certainly believed by all cladists, although for different reasons. Some specialists see speciation as a process of gradual transformation, while others regard it as a sudden event. To the latter, gradualism in lineages such as *Miogypsinoides complanatus* to *M. dehaarti* merely indicates that the entire lineage represents a single species however different the end members may appear to be. However, both groups agree that primitive forms cannot exist concurrently with advanced forms in the same lineage unless geographic isolation has occurred; when this happens, evolution in one area will seem to be retarded relative to that in another. This has important implications for datum levels based on extinction and for stratigraphical correlation.

Datum Planes and Evolutionary Concepts

First appearance datums are almost invariably based on the first species in a new genus or lineage (e.g., *Orbulina suturalis, L. (N) isolepidinoides, Flosculinella reicheli,* etc.). These may be either evolutionary or migratory events. Reference is never made to an *O. universa* datum, an *N. japonica* datum, or an *F. bontangensis* datum, although these later species are equally well known and often more widely distributed than their predecessors. The reason that the first species in a lineage is stratigraphically more important than its descendents is usually that it shows no gradation to its immediate ancestor and is therefore easily recognizable. Similarly, the last species shows no gradation to any subsequent taxon. The levels of the first and last appearances of all intermediate species are difficult to determine, and there is the added problem that some species do not seem to have evolved everywhere at the same rate—a phenomenon known as retarded evolution (see Drooger, 1963, p.347).

Flosculinella, Miogypsinoides, and *Nephrolepidina* provide examples of lineages in which the various members have been shown to grade gradually into one another through time. The four species usually assigned to *Flosculinella (F. reicheli, F. globulosa, F. bontangensis,* and *F. borneensis)* are distinguished largely by their shape and size (see Mohler, 1950, for discussion). Members of the *Miogypsinoides* lineage have usually been defined according to the length of the initial spire and the thickness of the lateral walls, while species of *Nephrolepidina* are now mainly distinguished by the condition of the embryonic apparatus and shape of the equatorial chambers. The morphological differences between the species in these genera and subgenera are all gradational and are based on the secular modification of existing characters. When measured and expressed statistically, and the results plotted against time (Drooger, 1963; Van der Vlerk and Postuma, 1967; Chaproniere, 1980), they demonstrate that evolution proceeded slowly; in other words, they illustrate "phyletic gradualism." But when only the first species in each lineage is considered, a different picture emerges. The earliest *Flosculinella* (believed to be *F. reicheli*) is distinguished from its immediate ancestor, presumably a species of *Borelis,* by the possession of a row of secondary chamberlets in each whorl. The earliest *Miogypsinoides* differs from its immediate ancestor (thought to be *Pararotalia* by several authors) in possessing a layer of equatorial chambers, whereas the earliest *Austrotrillina* differs from its miliolacean ancestor in having alveoli in the wall (Fig. 3). All these are new characters (novelties), which seem to appear suddenly rather than gradually in the stratigraphical record and show no apparent gradation between ancestor and descendant. True gradation is, in fact, an impossibility since there can be no intermediate stage between the presence and absence of such morphological characters; only the degree to which they are developed can vary. These examples seem to provide support for the evolutionary process which Gould and Eldredge (1977) called "punctuated equilibria," and it can therefore be argued that phyletic gradualism and punctuated equilibria are not mutually exclusive since both are recognizable in the Foraminifera. However, it is possible that our perception of these "evolutionary processes" merely reflects current methods of discriminating between species (Adams, 1983). We tend to accord equal taxonomic status to most morphological characters at species level, although some are new and may properly be described as novelties, while others are merely modifications of previously developed structures. Some characters presumably reflect "important" mutations (i.e., those having major effects on test morphology) while others reflect either relatively minor mutations or the gradual effect of selection on characters with a high initial variability. It is unfortunate that our only indication of the importance of mutations is their effect on test morphology, since this may not be a true measure of their value to the organism.

Since some datum planes are based on the appearance of novelty, the presence of which determines whether punctuated equilibria can be recognized as an evolutionary process in the Foraminifera it is desirable that a definition should be formulated.

Novelty can be defined as the appearance of an en-

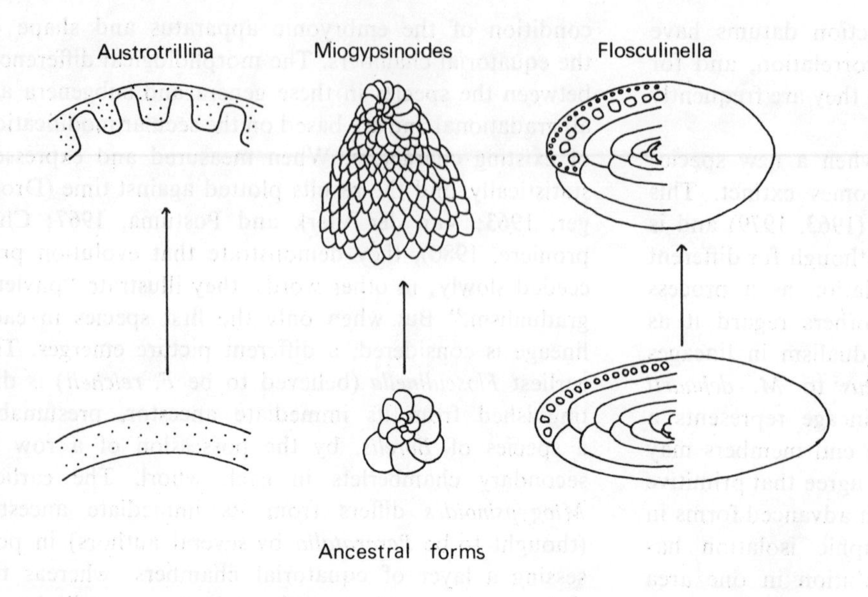

Fig. 3. Diagrammatic representation of the appearance of novelty in the genera *Austrotrillina* (the wall of the ancestral miliolacean lacks alveoli), *Miogypsinoides* (the ancestral genus, *Pararotalia*?, lacks equatorial chambers), and *Flosculinella* (the ancestral genus, *Borelis*, has but a single layer of chamberlets in each whorl).

tirely new test structure (e.g., equatorial chambers in *Miogypsinoides*, alveoli in *Austrotrillina*, and a second chamber layer in *Flosculinella*). Apertural changes should be so regarded only when they lead to the formation of new structures, e.g., the appearance of retrovert apertures in the initial spire of *Miogypsinoides* since they give rise to the equatorial chamber layer. The acquisition of additional sutural or areal apertures in members of the Globigerinacea cannot count as novelties unless new test structures are also produced. The evolution of *Orbulina* would thus be an example of phyletic gradualism as demonstrated by Blow (1956).

One important aspect of novelty is that it spreads so rapidly through populations that its appearance seems to constitute a virtually instantaneous event in geological terms. In reality, of course, novelties must require many generations to become widely distributed, but the geological time scale is too coarsely calibrated for them to be recorded (the finest division, ca. 1 m.y., probably represents at least one million generations). Deleterious mutations could also produce novelties, but these would not be perpetuated for obvious reasons, and fossil evidence for them should not normally be expected.

Planktonic Zones and the Letter Classification

The East Indies Letter Classification of the Tertiary is essentially a series of larger foraminiferal assemblage zones distinguished by datum planes based on the first and last appearances of a relatively small number of genera.

When first introduced by van der Vlerk and Umbgrove (1927) the Oligocene to Miocene part of this classification comprised only four divisions (c, d, e, and f) based on the ranges of nine genera and species (Fig. 4). In 1931, Leupold and van der Vlerk subdivided Tertiary e and f into eight parts using 30 taxa. They called the primary divisions "stages" (an unfortunate misnomer since they are really assemblage zones) and the secondary divisions zones (actually subzones). Two new "stages" Tg and Th were introduced for the latest Miocene and Pliocene, but the authors did not explain how they could be recognized.

With minor modifications, the Letter Classification was in use up to the time of the Geological Congress in 1948 when van der Vlerk (1950) stated that it had "more or less failed." However, he proceeded to redefine the e and f "stages" using 15 taxa but recognizing only two divisions of each unit (e_{1-4} and e_5; and f_1 and f_{2-3}) thus giving the classification a new lease of life. The Tg and Th divisions were quietly abandoned.

In his last revision, van der Vlerk (1955) placed the lower/upper e boundary at the extinction of *H. borneensis* and the first appearance of *Miogypsina* (including *Miolepidocyclina*), *Flosculinella, Trybliolepidina*, and *Multilepidina*. Between 1927 and 1955 van der Vlerk used 44 taxa to define the mid-Tertiary part of the Letter Classification, and of these only two (*Nummulites* and *Eulepidina*) have survived with their original ranges unchanged. Figure 4 shows how the classification altered as new information became available.

The writer (1970) attempted to up-date the Letter Classification and apply it to the whole Indo-West Pacific region. At that time there were few reliable records of planktonic foraminifera in association with larger foraminifera, but as the number increased, the

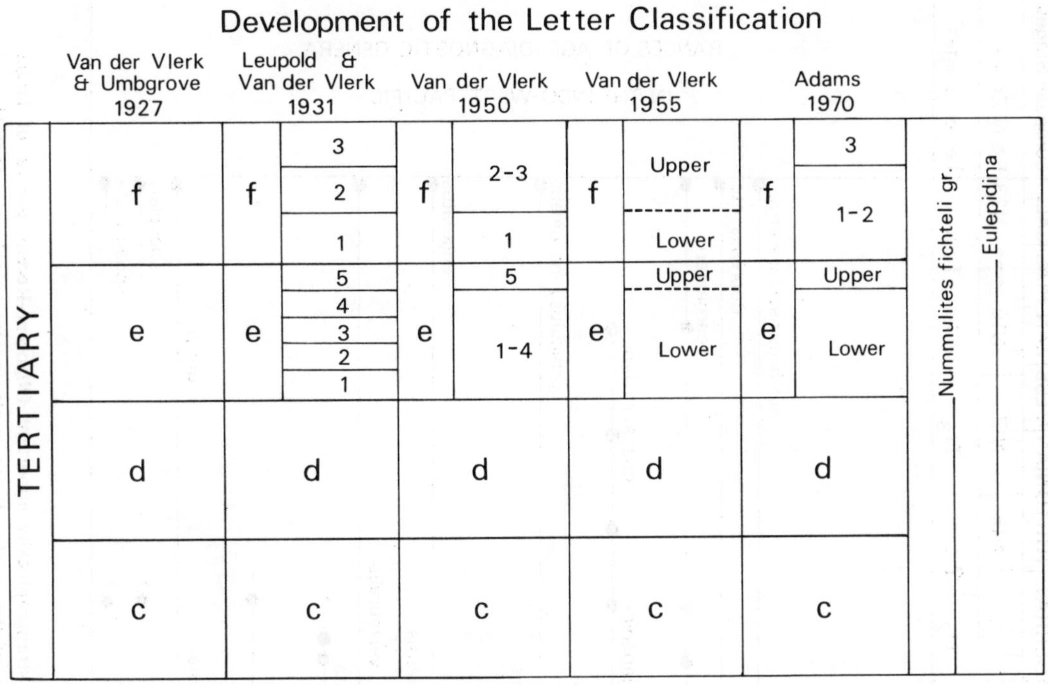

Development of the Letter Classification

	Van der Vlerk & Umbgrove 1927	Leupold & Van der Vlerk 1931		Van der Vlerk 1950		Van der Vlerk 1955		Adams 1970	
TERTIARY	f	f	3	f	2-3	f	Upper	f	3
			2						1-2
			1		1		Lower		
	e	e	5	e	5	e	Upper	e	Upper
			4		1-4		Lower		Lower
			3						
			2						
			1						
	d	d		d		d		d	
	c	c		c		c		c	

Nummulites fichteli gr.

Eulepidina

Fig. 4. Development of the mid-Tertiary part of the East Indies Letter Classification between 1927 and 1970. Only two taxa retained their original ranges during this time, and both are now under scrutiny.

need for a further revision became apparent. The last major application of the Letter Classification was by Hashimoto *et al.* (1977 b) who used it in an attempt to demonstrate the ranges of a large number of taxa in the Philippines.

Correlation of the Letter Classification with the planktonic zonation depends upon our ability to relate certain critical larger foraminiferal datum planes to the planktonic zones. Despite recent advances in knowledge, this remains a difficult exercise. Records of associated planktonic and larger foraminifera are not evenly distributed throughout the Neogene but tend to be concentrated at a few levels, and we therefore know much more about the larger foraminifera typical of some zones (e.g., N8) than others (e.g., N6), and little or nothing about those representing zones N3, N13, and N15. Moreover, only a few species of larger foraminifera are found with age-diagnostic plankton in any one locality, and numerous planktonic records would be required to determine accurately the ranges of even the commonest species. The ranges of age-diagnostic post-Eocene genera are shown in Fig. 5.

The Oligocene has conventionally been divided into three parts (Lower, Middle, and Upper) equivalent to the Lattorfian (old usage), Rupelian, and Chattian stages of Europe, and until recently thought to be more or less the same as the Tc, Td and Te_{1-4} divisions of the Letter Classification. Recently, however, the type Lattorfian has been shown to be late Eocene in age (Ritzkowski, 1981), and it was the possibility that this might soon be demonstrated, coupled with the realization that much of Oligocene time is represented by Chattian sediments, which led some authorities (e.g., Hardenbol and Berggren, 1978) to divide this epoch into two parts, Early (Rupelian) and Late (Chattian). However, in the Indo-West Pacific and Tethyan regions, three successive assemblages of larger foraminifera can be recognized within the Oligocene. The oldest, defining Tc, is characterized by the presence of *Nummulites* without *Lepidocyclina*, the second oldest includes both *Nummulites* and *Lepidocyclina*, and characterizes Td, while the youngest contains *Lepidocyclina* without *Nummulites*, and defines Te_{1-4}. Recent controversy over the definition of *Nummulites* now makes it necessary to specify that Td is characterized by the presence of reticulate species belonging to the *N. fichteli* group and/or *N. vascus* and *N. pengaronensis*, together with *Lepidocyclina* (*Eulepidina* or *Nephrolepidina*). This redefinition preserves the value of Tc, Td, and Te_{1-4}, while allowing systematists to use *Nummulites* in a broader sense than hitherto. The only evidence to suggest that a further change might be needed is the record of *N. fichteli* with a typical lower e fauna in float blocks of Bugton Limestone in the Mansala area, Philippines (Hashimoto *et al.*, 1977a). Confirmation that this important record is not of reworked specimens is required before the ranges

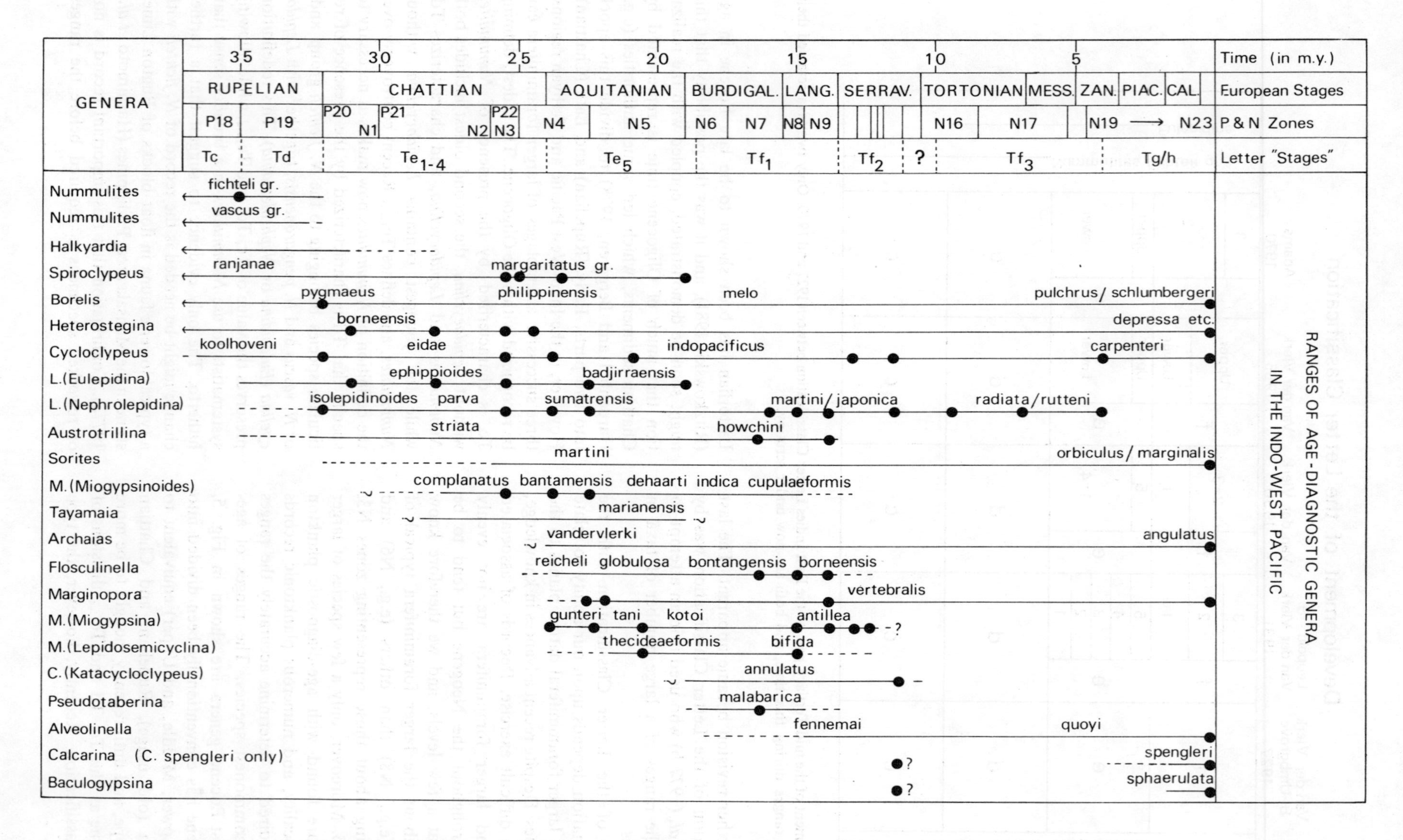

Fig. 5. Chart showing the levels at which there is planktonic control over the ranges of stratigraphically important Oligocene-Recent genera of larger foraminifera in the Indo-West Pacific region. Ranges may be different in other parts of the world. Black spots indicate planktonic control, and may represent more than one record per zone. Solid line = proven range; broken line = interval of uncertainty. A few commonly occurring Neogene species, including the first and last in each genus, are shown in their relative stratigraphical positions. They do not necessarily represent single lineages nor are they always controlled by plankton. Some species intergrade; others have overlapping ranges. The position of the f2/3 boundary is still uncertain. Chronometric scale, European stages, and P and N zones based on Van Couvering and Berggren (1977) and Hardenbol and Berggren (1978).

of several lower e species are extended downwards or that of *N. fichteli* upwards, since either course will necessitate a radical reappraisal of many Mid-Oligocene age determinations in the region. Throughout Indonesia, the lower part of Te_{1-4} is usually characterized by the association of *Heterostegina borneensis* and *L. (Eulepidina)* spp. (e.g., in the Melinau Limestone, Sarawak; Adams, 1965), and the upper part by the incoming of *Miogypsinoides* and the presence of *Spiroclypeus* in abundance. The Bugton samples with *N. fichteli* suggest that these differences may not be stratigraphically significant, as does the presence of *Miogypsinoides* at a very low level in the Chattian of Cutch, India (Raju, 1974). On the other hand, the interpretation of the Bugton fauna may be incorrect and the Indian sequence could either be condensed or conceal a disconformity.

There are as yet few reliable records of planktonic foraminifera in association with Tc and Td larger foraminifera. Eames *et al.* (1962) recorded *Nummulites intermedius-fichteli* from the *Globigerina oligocaenica* Zone (= P18/19) of the Lindi area, East Africa, and Mohan *et al.* (1978) reported *N. fichteli* from their *G. tapuriensis* Zone (= P18/19) of the Bombay offshore region.

Te_{1-4} larger foraminiferal faunas are especially important in the context of this paper because they are essentially Neogene in aspect. However, although common in the Indo-West Pacific region, they have been sampled systematically in only a few places, and knowledge of the order of appearance of their constituent taxa is heavily dependent on a relatively small number of publications: Adams (1965), Sarawak; Adams and Belford (1974), Christmas Island; Chaproniere (1981), West Australia; Cole (1954 and 1958), Marshall Islands; Cole (1969), Midway; Coleman (1963), Solomon Islands; Raju (1974), India. These publications can be supplemented by an abundance of data obtained from spot samples in places such as Saipan (Cole, 1957; Hanzawa, 1957; Matsumaru, 1976) and the Philippines (numerous publications; see Hashimoto *et al.* 1977b for summary of ranges), but in these and other papers the detail provided by close systematic sampling is lacking.

Records of associated larger foraminifera and plankton from Te_{1-4} are poor. Ujiié (1975) reported *Eulepidina* from Zone P21 (= N2) in a drill hole from the Philippine Sea. Djamas and Marks (1978) reported *H. borneensis* and *L. papuaensis* in N1, and *L. papuaensis* alone from N2. Both records were from Borneo. They also reported *Spiroclypeus* from N3, as did Natori (1978 b). Several species of *L. (Nephrolepidina)* and *L. (Eulepidina)* are known from N3, as are *H. borneensis* and *Cycloclypeus*. It should be noted that *H. borneensis*, which is almost always the only species of the genus reported from strata of late Oligocene age in the region, is not easy to recognize and that many of the published records may therefore be incorrect.

The Miocene epoch saw a considerable number of first appearances and extinctions, some of which can be related to the planktonic zonation. Adams (1970), in attempting to follow van der Vlerk's concept of the Letter Classification, defined the earliest Miocene faunas as those including either *Miogypsina* or *Flosculinella*, thus equating the boundary of Te_5 with the base of the Miocene and therefore of the Aquitanian. Unfortunately, *Flosculinella* is now thought to range down into Te_{1-4} (Adams and Belford, 1974), and there is no particular reason to suppose that the first appearance of *Miogypsina* corresponds exactly to the Chattian/Aquitanian boundary in Europe, or even to the base of N4. The oldest species from the drill holes in the Marshall Isles (Cole, 1954, 1958) are said to be *M. borneensis* and *M. musperi*. In the Melinau Limestone, Sarawak, the earliest *Miogypsina* is thought to be *M. thecideaeformis* although it is impossible to be certain. On Christmas Island (Adams and Belford, 1974), it is *M. neodispansa* (Jones and Chapman), probably a senior synonym of *M. kotoi* (or of *N. droogeri;* Belford, pers. comm); in the Cauvery Basin it is *M. gunteri* Cole; and in Kutch, *M. tani* Drooger (Raju, 1974). In the Philippines several species, including such disparate forms as *M. gunteri* and *indonesiensis*, appear together according to Hashimoto *et al.* (1977b), but this seems to be unlikely for evolutionary reasons. The oldest *Miogypsina* in the New Hebrides is *M. thecideaeformis* according to Coleman (1963). Bock (1976) has drawn attention to the possible evolution of *M. borneensis* from *Miogypsinoides dehaartii*, a species which crosses the lower/upper e boundary and is also known from Tf_1. In the Americas, *Miogypsina* presumably evolved either from *Miogypsinoides bermudezi* Drooger or *Miogypsinoides complanatus* since no other species of *Miogypsinoides* is known from the region. *Miogypsina* may therefore be polyphyletic, but even if it is not, its first appearances in the Indo-West Pacific region must indicate slightly different levels in the early Miocene.

Records of N4 and N5 larger foraminifera shown on the range chart are taken from Binnekamp (1973), Chaproniere (1975, 1981), Djamas and Marks (1978), Haak and Postuma (1975), Mohan *et al.* (1978), Natori (1978 a and b), O'Herne and van der Vlerk (1971), Palmieri (1974), van der Vlerk and Postuma (1967), and Van Vessem (1978).

The Te_5/f boundary must be placed at the extinction of *L. (Eulepidina)* and/or *Spiroclypeus* in accordance with van der Vlerk's original proposal. Unfortunately, these two events were not everywhere coeval, and the exact position of the boundary in terms of the planktonic zonation is therefore uncertain. A recent survey of the literature has revealed that neither *Eulepidina* nor *Spiroclypeus* ranges far up into the Lower Miocene. Both occur near its base in the Solomons (Coleman, 1963); Hashimoto *et al.* (1977 a and b) have reported similar associations from a few isolated samples in the Philippines, and Cole (1954, 1958) has reported them from drill holes on Bikini and Eniwetok. Adams (1965) thought that *Eulepidina* and *Spiroclypeus* occurred throughout the whole of that part of the Melinau Limestone which he referred to Te_5, although he was unable to determine their ranges precisely. Subsequent collecting (C.G.A., 1966) along the supposed Te_5 section exposed on the banks of the Terikan River failed to reveal these taxa, and a recent reappraisal of the earlier work has shown that the only high Te_5 occurrence of *Eulepidina* and *Spiroclypeus* was probably based on wrongly labelled slides since new sections cut from the original sample have failed to reveal either genus. Present evidence therefore suggests that the extinction level of *Eulepidina* is just above the N5/N6 boundary as suggested by Chaproniere (1975, 1981). Unfortunately, *Spiroclypeus* does not occur with *Eulepidina* in the Mandu Limestone of western Australia, and its last known occurrence seems to be at the top of N5 in the Kutei Basin, Kalimantan (Djamas and Marks, 1978), where it also occurs without *Eulepidina*.

Tf_1 includes the *Orbulina* datum and is usually characterized by the association of *Miogypsinoides dehaarti, Miogypsinoides cupulaeformis, Miogypsina* spp, *Austrotrillina howchini,* and *Pseudotaberina malabarica,* together with a number of species of *Nephrolepidina* including *N. japonica, N. martini, N. sumatrensis,* and *N. howchini.* Its upper boundary is marked by the extinction of *Austrotrillina* and *Miogypsinoides.* The last occurrence of *Austrotrillina* is in N9 (Adams, 1968; Chaproniere, 1981), but it is currently impossible to establish the extinction level of *Miogypsinoides* although it seems to be below the *Orbulina* datum in most places (see p. 63). Most known Tf_1 faunal assemblages are pre-*Orbulina* datum; only a few are known to be younger. Tf_1 is therefore almost exactly equivalent to the combined Burdigalian and Langhian stages of Europe, and certainly straddles the Lower/Middle Miocene boundary.

Records of plankton associated with Tf_1 larger foraminiferal faunas are quite numerous, although there are few unequivocal records from N6 (Chaproniere, 1981, is an exception). Rasheed and Ramachandran (1978) have described the important fauna of the Quilon Beds of Southern India and dated it as N7. Tsuchi *et al.* (1979) have summarized the considerable amount of information on the planktonic and larger foraminifera of zones N7–N9 in Japan, and there are several N8 records including those of Cole (1975), Java; Van Vessem (1978), Java; Matsumaru (1971), New Zealand; and Adams and Whittaker (unpublished report, 1981), East Kalimantan.

It now seems possible to recognize Tf_2 by the occurrence of *C. (Katacycloclypeus) annulatus* Tan and/or *Miogypsina* following the extinction of *Austrotrillina* and *Miogypsinoides. Lepidocyclina rutteni, L. martini, L. japonica, Cycloclypeus indopacificus, Alveolinella quoyi* and *Marginopora vertebralis* are commonly recorded from strata of this age.

Records of Middle Miocene plankton and larger foraminifera include those of Adams and Frame (1979) from Fiji; Baumann (1975) from Java and Irian Jaya; Chang (1975) from Taiwan; Chaproniere (1975, 1981) from West Australia; Clarke and Blow (1969) from Sumatra; Ikebe and Chiji (1971), and Matsumaru (1980) from Japan; Palmieri (1973, 1974) from eastern Australia; Todd (1977) from the New Hebrides; van der Vlerk and Postuma (1967) from Java; Ujiié (1973) from Japan; and Van Vessem (1978) from Java.

The Tf_2/f_3 boundary is difficult to relate to the planktonic zonation since the extinction levels of *C. (K.) annulatus* and *Miogypsina* are not well established. Adams and Frame (1979) concluded that *C. (K.) annulatus* ranged up to zone N14 in Fiji, but there are unverifiable reports of it at higher levels. *Miogypsina* has long been considered to die out at about N12–N14 (Clarke and Blow, 1969) although proof has been lacking. Matsumaru (1971) reported it from N14 in the New Zealand but it seems likely that these specimens are reworked. McTavish (1967) reported *Miogypsina* from his *Sphaeroidinellopsis seminulina* assemblage in the New Hebrides. This certainly post-dates the FAD of *G. nepenthes,* and is therefore either N15 or younger if the specimens are not reworked. It seems likely that the extinction level of *Miogypsina* varied across the region. The main reason for the present uncertainty is that few carbonates of N12–N16 age have so far been described.

Tertiary f_3 is defined, as before, by the presence of *Lepidocyclina* after the extinction of all taxa used to delimit the earlier letter divisions. Its upper boundary is determined by the extinction of the last species (*L. (N.) radiata* or *L. (N.) rutteni*), both of which have now been reported from the early Pliocene (see p. 62).

Tertiary g may be defined as that part of the Pliocene following the extinction of *Lepidocyclina.* It is characterized by the presence of several long-ranging species such as *A. quoyi, M. vertebralis,* and *Borelis pulchrus.* Tertiary h is an unrecognizable and unnecessary division. The relatively few records of associated planktonic and larger foraminifera in the Upper Miocene include those of Baumann (1972), Java; Billman and Kartaadipura (1975), Kalimantan; Haak and Postuma (1975), Tonga; and Matsumaru (1981), Japan. Plio-Pleistocene records are also few, but include those of Adams, Rodda and Kiteley (1979), Fiji; Huang (1975), Taiwan; Matsumaru (1976b), Ryuku Islands; Todd (1960), Western Pacific; and Van Vessem (1978), Borneo.

Of the seven boundaries mentioned above, two (Tc/d and lower/upper e) are defined by FADs, and five by LADs. These divisions work quite well in practice but do not lend themselves to accurate correlation with the planktonic zonation because LADs, as discussed earlier, are rather unreliable for long-distance corelation. It would obviously be preferable to define all boundaries on FADs but this is not possible at present since new Miocene taxa such as *Alveolinella* and *Marginopora* were neither particularly common nor widely distributed during the first few million years of their existence.

The main changes in the Neogene part of the Letter Classification since the last complete revision (Adams, 1970) are the lowering of the upper e/f boundary from the base of N9 to near the base of N6, a change begun by Haak and Postuma (1975) and completed by Chaproniere (1981), and the recognition of f_2 as a distinct unit. The lower and upper boundaries of Tf_3 have both been raised.

Stratigraphical and Palaeontological Problems

Stratigraphical correlation by Tertiary larger foraminifera is still hindered by insufficient knowledge and technical problems, and is often further complicated by difficulties of our own making. It may not be generally appreciated that there is as yet very little published information on Middle and Late Miocene carbonates, and their faunas are therefore relatively poorly known. It is possible that shallow water limestones of latest Miocene (Messinian) age are almost entirely absent throughout the region, as noted by Adams *et al.* (1977); this seems to be confirmed by the correlation chart for the Philippines published by Hashimoto *et al.* (1977 b). Regional correlation could, however, be greatly improved if carbonate successions such as the Yalam Limestone (Te_5–$Tf_{2\ or\ 3}$) of New Britain, the New Guinea Limestone, the Kennon and Uling Limestones (Tf_{1-2}) in the Philippines, and the Karren and Parigi Limestones (Mid-Late Miocene) of Java were to be fully described. These, or their age equivalents elsewhere, should yield sufficient information on larger foraminifera to effect a lasting improvement in regional correlation. It is possible that species of *Archaias, Borelis, Heterostegina, Operculina,* etc., currently regarded as of little stratigraphical value, would prove to be useful and certainly more would be learned about the FADs and LADs of *Alveolinella, Baculogypsina, Calcarina, Flosculinella, Lepidocyclina,* and *Miogypsina.*

Facies faunas were once thought likely to pose serious problems. It is well known that species of larger foraminifera are restricted to particular ecological niches. For example, in the Red Sea, *Sorites orbiculus* Ehrenberg lives mainly epiphytically on large plants between the surface and 40 m, whereas *Heterostegina depressa* prefers a hard bottom substrate and a water depth of 40 m or more (Hottinger, 1977). Fortunately, despite the diversity of shelf habitats open to larger foraminifera, facies restriction has not proved to be the impediment to correlation that was once feared, since elements of the various assemblages tend to become mixed after death, and changes in the relative levels of land and sea ensure that faunas characteristic of different environments alternate in stratigraphical position. However, some assemblages have proved difficult to date. As mentioned earlier, conformation of the age and stratigraphical value, if any, of the *Eulepidina/H. borneensis* assemblage (thought by some authors, including the writer, to be typical of the early part of lower e) of Indonesia is required since it might prove to be coeval with the *Miogypsinoides, Eulepidina, Spiroclypeus* assemblage of India. The *Borelis melo curdica* fauna of the Middle East is still difficult to correlate directly with Indo-Pacific faunas, although it is thought to straddle the Lower/Middle Miocene boundary (Adams *et al.*, in press). Now that the Quilon fauna (assemblage of *A. howchini, P. malabarica,* and *Miogypsina* sp.) has been dated (Rasheed and Ramachandran, 1978), very few other important assemblages pose serious problems.

In the writer's opinion, some of the most serious correlation problems are posed by reworking and by the attitude this induces in palaeontologists. Reworking itself is a common enough phenomenon everywhere and is particularly evident in the Earth's mobile belts. It can, however, be difficult to recognize in shallow-water carbonates, where it gives special cause for concern. Equally serious, however, is the assumption of reworking which seems, psychologically, to be an occupational hazard amongst palaeontologists. Far too often, reworking is assumed to have occurred simply because it explains inconvenient faunal associations; this assumption is a measure of the faith palaeontologists have in datum planes based on extinctions, and of their unwillingness to admit that it might sometimes be misplaced. Eames (in Adams, 1965) was reluctant to allow that *Dictyoconus* could range up into the Oligocene, largely because it was thought to be a reliable Paleocene-Middle Eocene marker in S.E. Asia. However, it had already been discovered in Tc beds elsewhere in Sarawak (Adams 1964) and was subsequently found in the Tb part of the Chimbu Limestone, New Guinea (Bain and Binnekamp, 1973).

Another interesting example of Oligocene "reworking" is provided by Hashimoto *et al.* (1978) who found *Discocyclina* sp. with *N. fichteli* and *H. borneensis* in the Lutak Hill Limestone, and assumed that the discocyclinids were reworked. This may well be correct, but if the range of *Discocyclina* is locally longer than usual, its occurrence in Tc limestones with *Nummulites* (its usual associate in the Eocene) should be expected. Records of the reworking of single species need to be carefully evaluated since this process tends to be unselective amongst foraminifera and usually involves more than one taxon.

An anomalous association of *Nummulites fichteli* and three Te species in the Bugton Limestone, Mindanoro (Hashimoto *et al.* 1977a), was, in the absence of evidence for reworking, explained by extending the ranges of the Te species downwards—the maintenance of one well-known LAD being preferred to that of three seemingly less important FADs. This decision was probably correct, but such changes require careful consideration.

The difficulty of identifying species of larger foraminifera in random thin sections of limestone poses problems since only a few diagnostic characteristics are ever visible in one specimen. But names assigned to individuals that can, at best, only doubtfully be determined create rather than solve stratigraphical problems. It is hard for a taxonomist to admit that he cannot identify commonly occurring species, but it is better to provide good figures and give reasons for expressing doubt about individual determinations than to create confusion by assigning incorrect specific names. Unfortunately, the literature on the Tertiary foraminifera of Southeast Asia is burdened with unverifiable identifications, many of which are clearly incorrect.

The Mediterranean/Indian Ocean Disconnection

The terminal Tethyan event may seem to be an irrelevance in the context of Indo-West Pacific datum planes, but consideration of the distribution of larger foraminifera in the region shows that this is not so. It is self-evident that species evolving in the Mediterranean or Indo-Pacific provinces after they became separated should have remained restricted to their regions of origin, but this simple fact has not always been reflected by taxonomic usage.

Palaeobiogeographers agree that a broad marine connection between the Mediterranean and Indian Ocean existed across what is now the Middle East in Palaeogene times, and although there are differences at specific level between the larger foraminiferal faunas of the Mediterranean and Indo-West Pacific, the genera were broadly similar in each region (Adams, 1967, 1973). Specific names erected in Europe and the Mediterranean region for such Palaeogene genera as *Alveolina* and *Nummulites* have been employed for taxa occurring as far east as New Guinea (Bain and Binnekamp, 1973). However, many early and mid-Miocene Mediterranean species such as *Lepidocyclina morgani* and *Borelis melo curdica* are less certainly present in the Indo-Pacific, while such typical Indo-West Pacific taxa as *Flosculinella bontangensis* and *Miogypsina polymorpha* are unknown from the Mediterranean region. Determination of the date of separation of the two regions would therefore help to prevent the misuse of names and the confusion which inevitably follows.

Some progress towards a solution of this problem has been made in the last few years. Drooger (1979) argued persuasively for separation during Late Oligocene (Chattian) times using evidence drawn largely from the distribution of *Miogypsina* and *Lepidocyclina* (*Nephrolepidina*). He concluded that any similarities between the faunas of the two regions in Miocene times were a consequence of parallel evolution. Adams *et al.* (in press) have shown that a land barrier across the Persian Gulf in late Burdigalian times can be inferred from the distribution of *Borelis melo curdica* and *Flosculinella bontangensis*. The problem is to decide whether this represents the first appearance of the barrier or whether it was there earlier, perhaps even in late Chattian times. The apparent absence of marine Aquitanian sediments in the central and southern parts of the Persian Gulf area (possible a significant fact in itself) and the paucity of studies of matrix-free material, makes the solution of this problem difficult.

A review of the literature has indicated that *B. melo curdica* is not known from the Indo-West Pacific (*contra* Adams, 1970); that *B. melo* occurs in the early Burdigalian of the Indian Ocean region but is usually absent east of India; that *Flosculinella* is not known north of the Persian Gulf (Mediterranean records are believed to be misidentifications); that the most advanced forms of *A. howchini* are not known from the Mediterranean region; and that all records of *A. howchini* from the Middle East are probably incorrect and refer to *A. paucialveolata* Grimsdale or *A. asmariensis* Adams. The absence of *A. howchini* from strata in the Middle East which seem to be perfectly suited to this genus argues strongly for a separation by early Burdigalian times.

Direct correlation between the Indo-Pacific and the Mediterranean region using larger foraminifera remains difficult, and it is desirable that species names based on post-Oligocene Mediterranean taxa should not be used in the Indo-Pacific region unless it can be clearly demonstrated that the forms concerned are morphologically indistinguishable in the two areas.

Eustatic Movements

The probable effects of the major eustatic changes postulated by Vail *et al.* (1977) on the larger foraminiferal faunas of the Cenozoic have been discussed in some detail elsewhere (Adams, 1983), and here it is necessary only to draw attention to the possibility of using the datum levels observable in shallow-water carbonate successions to test the validity of claims for Neogene sea level highstands and lowstands. The proposed test depends upon the fact that any marked sea level fall would expose shallow-water carbonates to weathering and would force the larger forminifera to migrate towards the shelf edge where they would be subjected to intense competition for space in the greatly narrowed euphotic zone. This would inevitably result in a diminution of their numbers and would be likely to cause extinctions. The fall and subsequent rise of sea level would later be reflected in the sedimentary succession. Sediments resting on weathered limestone surfaces should show signs of having been deposited in deepening water, and even in wholly carbonate sequences, a faunal change should indicate the horizon marking the eustatic event.

Evidence (extinctions, disconformities, lithological changes, etc.) consistent with a global eustatic fall is found widely at the Eocene/Oligocene boundary. Unfortunately, it has not yet been possible to demonstrate similar changes in carbonate sequences of early Chattian age (30 m.y. B.P.), at which time the greatest Cenozoic sea level fall is supposed to have occured. Nevertheless, these and other Tertiary carbonates, including those of Neogene age, should be re-examined for indications of changes at the relevant levels. These events would not, of course, be recognizable in deep water sediments or in shallow-water successions where the rate of subsidence locally kept pace with falling sea level.

FADs and LADs of Some Important Taxa

The taxa listed below are mainly those currently used to define the divisions of the Letter Classification. The first- and last-known Neogene occurrences of each genus are cited wherever possible (FADs are not always given for taxa ranging up from the Oligocene), and some problems and points of special interest are discussed.

Alveolinella Douvillé, 1906

First appearance. Within Tf_1, exact level uncertain. Fairly common from Tf_2 onwards. First species *A. fennemai* (Checchia-Rispoli). Earliest verifiable record seems to be that of *A.* sp. (Eames *et al.* 1962, pl. VI, D) from beds said to be of Tf_{1-2} age (Tf_1 since they contain *Austrotrillina*) from the Darai Limestone, Papua. This particular specimen appears to be intermediate between *Flosculinella* and *Alveolinella.*

Last occurrence. Extant. The genus seems to be represented only by *A. quoii* (d'Orbigny) from Tf_2 times onwards and is often found associated with *Marginopora vertebralis.*

Archaias de Montfort, 1808

First Neogene appearance. Base of Miocene (ranges up from the Paleogene). One of the earliest records must be of *A.* cf. *vandervlerki* in Haile and Wong (1965) from the Gomantong Limestone (Te5) of Borneo. Although occurring commonly in the Middle East, this genus has so far been found too infrequently in the Indo-West Pacific for its full stratigraphical potential to be realized.

Last occurrence. Extant. *A. angulatus* (Fichtel and Moll) is a commonly occurring Recent species. See also *Pseudotaberina malabarica.*

Austrotrillina Parr, 1942

First Neogene appearance. Basal Miocene, but known from late Td in parts of the region (Melinau Limestone, Sarawak: Adams, 1965). *A. striata* Todd and Low in the first known Indo-Pacific species; *A. asmariensis* Adams has a similar range in the Middle East (see Adams, 1968).

Last occurrence. *A. howchini* (Schlumberger); Pata Limestone, southern Australia, where it occurs above the *Orbulina* datum. See Ludbrook (1961, 1963). It also occurs with *Orbulina* in the Trealla Limestone of western Australia (see Chaproniere, 1981).

Remarks. Rahaghi's record (1980) of this genus from the Middle Eocene of Iran is suspect since the beds concerned are immediately overlain by *Austrotrillina-bearing* Oligocene limestones, and the genus has never been reported from beds of late Eocene age.

Baculogypsina Sacco, 1893

First appearance. Kleinpell (1954) reported this genus from the type section (Tf_2 as defined here) of the Futuna Limestone, Lau Islands, Fiji, and the writer has found it in the same locality. All other records are believed to be Pleistocene or younger (see Todd, 1960).

Last occurrence. Extant. *B. sphaerulata* (Parker and Jones) is the only known species according to Todd (1960).

Remarks. The anomalous occurrence of *B. sphaerulata* (identical in all respects with the types) in the Middle Miocene, Futuna Limestone, suggests that these specimens may be Recent forms occupying an infilled fissure in the limestone. A further field examination is required. See *Calcarina spengleri* below.

Borelis de Montfort, 1808

First Neogene appearance. Basal Miocene (ranges up from the Palaeogene). First Neogene species probably *B. pygmaeus* Hanzawa.

Last occurrence. Extant. *B. pulchrus* (d'Orbigny) and *B. schlumbergeri* (Reichel) live in the region today.

Remarks. Numerous species names have been applied to Neogene forms, but it is unlikely that many can be justified. *B. melo melo* (Fichtel and Moll) occurs in the Miocene of East Africa and could well be found elsewhere, although most records from Indonesia and the western Pacific (that of Cole, 1969, from Midway seems to be an exception) seem to be erroneous. Matsumaru (1978; 1980, Text-fig. 1) gave the range of *B. melo* as upper part of Te_{1-4} and Te_5, based on its supposed occurrence in the Minimakazi Limestone. *B. melo melo* is, however, not fusiform in the Mediterranean region, and until gradation to fusiform variants is clearly established in populations of Miocene age, the name should be restricted to globular individuals without secondary axial thickening. *B. melo curdica* Reichel is not known in the Indo-West Pacific region (*contra* Adams, 1970).

Calcarina d'Orbingy, 1826

A relatively long-ranging genus, only one species of which, *C. spengleri* (Gmelin), is important in the present context.

First appearance. Difficult to determine. Although usually considered a Plio-Pleistocene species, it has been found in the Mid-Miocene type section of the Futuna Limestone, Fiji, and recorded (Kleinpell, 1954), as *Siderolites mbalavuensis* sp. nov. It is probable that these specimens are Recent contaminants but they have been found independently by the writer in the same outcrop. See *Baculogypsina* above.

Last occurrence. Extant *C. spengleri* (Gmelin) and *C. hispida* Brady: widespread in Plio-Pleistocene limestones.

Cycloclypeus (Cycloclypeus) Carpenter, 1856

First Neogene appearance. Basal Miocene. Ranges up from the Early Oligocene (Tc). Binnekamp's record (1973, p. 5) from the Eocene of New Britain has been checked by the writer and found to be based on a misidentification of *Heterostegina*. First species, *C. (C.) koolhoveni* Tan; first Miocene species, *C. (C.) eidae* Tan, which ranges up from the Chattian.

Last occurrence. Extant. One living species, *C. (C.) carpenteri* Brady, *C. (C.) guembelianus* Brady is a subjective junior synonym (Adams and Frame, 1979).

At least eight species have been named from Miocene strata, but it is doubtful if more than a few are useful. Some authors (e.g., Chaproniere, 1980) recognize only two Miocene species, *C. eidae* and *C. carpenteri*, but it seems desirable on both taxonomic and stratigraphical grounds to recognize a third, *C indopacificus* Tan.

Cycloclypeus (Katacycloclypeus) Tan, 1932

First appearance. ?Te_5 (late), Angat Formation, Philippines, *K. transiens* Tan (Hashimoto *et al.*, 1977b); common in Tf_{1-2}. There are many records from random thin sections of limestone of Tf_3 age, but all are probably incorrect.

Last occurrence. N14-N15, Futuna Limestone, Lau Islands, Fiji (Adams and Frame, 1979). One species only, *C. (K.) annulatus* K. Martin, is currently recognizable (Adams and Frame, 1979), although several other names are in use.

Flosculinella Schubert, 1910

First Neogene appearance. Usually basal Miocene (Te_5), but there is one record of *Flosculinella* sp. (Adams and Belford, 1974) from limestones believed to be of late Te_{1-4} age on Christmas Island, and one of *F. globulosa* (unfigured) from the top of Te_{1-4} in the Philippines (Hashimoto *et al.*, 1977b). *F. globulosa* occurs with *Spiroclypeus* on Tinian (Hanzawa, 1957) and in a drill hole on Midway (Cole, 1969).

Last occurrence. Extinction level currently difficult to establish but the last known occurrence is probably that of *F. borneensis* in the f_2 part of the Yalam Limestone, New Britain (Binnekamp, 1973).

Remarks. *F. reicheli* Mohler occurs in Te_5 (and possibly Te_{1-4}) strata. *F. globulosa* (Rutten) seems to be restricted to Te_5 although Hashimoto *et al.*, (1977b) report it from the top of Te_{1-4} (Binangonan fauna) in the Philippines. *F. bontangensis* (Rutten) seems to range from late Te_5 through Tf_1 where it is common. It occurs above the *Orbulina* datum in western Australia (Chaproniere, 1975, 1981). *F. borneensis* (Tan) is supposedly the last species and is common in beds of Tf_{1-2} age. It often occurs in association with *F. bontangensis*. The record of *F. globulosa* from the Butong Limestone (e_4, Binangonan fauna) of Rizal Province, Luzon, requires confirmation. *F. fusiformis* Hashimoto and Matsumaru has so far been reported only from the Philippines.

Halkyardia Heron-Allen and Earland, 1918

Not a Neogene genus but known to range up from the Eocene into beds of Oligocene (Tc-Te_{1-4}) age (Adams, 1970; Hashimoto *et al.*, 1978) and therefore shown on Fig. 5. Ranges of individual Oligocene species difficult to determine.

Heterostegina d'Orbigny, 1826

First Neogene appearance. Basal Miocene (Te_5); ranges up from the Palaeocene.

Last occurrence. Extant *H. depressa* d'Orbigny is a common living species.

Remarks. *H. borneensis* van der Vlerk probably ranges up into Te_5 from Te_{1-4} where it is common. However, misidentifications are likely to be frequent since it has usually been determined from random thin sections in which all the essential characters are rarely visible. Hashimoto *et al.*, (1977b) gave the range of this species as Tc (Lutak Hill Limestone) to Tf_1 (Uhling Limestone) in the Philippines, but confirmation is required. Chaproniere (1980) has reported it from the Middle Miocene (N10-N12) in New Zealand, but noted that it could possibly be reworked.

Lepidocyclina Gümbel, 1870

This is probably the most important genus of Neogene larger foraminifera, since it is widespread, abundant, and long-ranging. Its morphological diversity is such that it has been given a large number of specific names, most of which are unfortunately of little value.

Since 1960 Cole has maintained that the subgenera *Eulepidina* H. Douvillé and *Nephrolepidina* H. Douvillé should be combined. His grounds were that the first two chambers showed complete gradation in shape in certain species, *e.g. L. radiata*, and that *L. augusticamera* combined embryonic and other features of *Nephrolepidina* and *Eulepidina*. Various authors have attempted to refute this, and Coleman (1963) has drawn attention to the probable importance of the periembryonic chambers which had been disregarded by Cole. The two subgenera may, in the Indo-West Pacific at least, be distinguished as follows:

Eulepidina

Protoconch and deuteroconch always unequal in size. Periembryonic chambers small, and often indistinct when seen in thin section. Adauxiliary chambers numerous and not necessarily confined to the equatorial plane; when visible, appearing to be no larger than adjacent chambers. Equatorial chambers always either spatulate or hexagonal in shape.

Nephrolepidina

Protoconch and deuteroconch subequal or unequal in size. Embryonic apparatus surrounded by periembryonic chambers of several kinds. Primary auxiliary and adauxiliary chambers distinct; never more than 10 in number (usually 4–8 depending in part on the size of the deuteroconch) and usually larger than adjacent chambers and the first cycle of equatorial chambers. Equatorial chambers rhombic, ogival, spatulate or hexagonal in shape.

Figure 6 shows how specimens of *Lepidocyclina* in which the first two chambers are neither clearly eulepidine nor nephrolepidine in form can be distinguished. It is quite possible that *Eulepidina* and *Nephrolepidina* merit generic status but this decision should await a full investigation of the American species and of the stolon systems. Fortuin (1970) proved that the auxiliary chambers in *L. (E.) dilatata*, and by implication in other species of this subgenus, are neither regularly spaced nor confined to the equatorial plane.

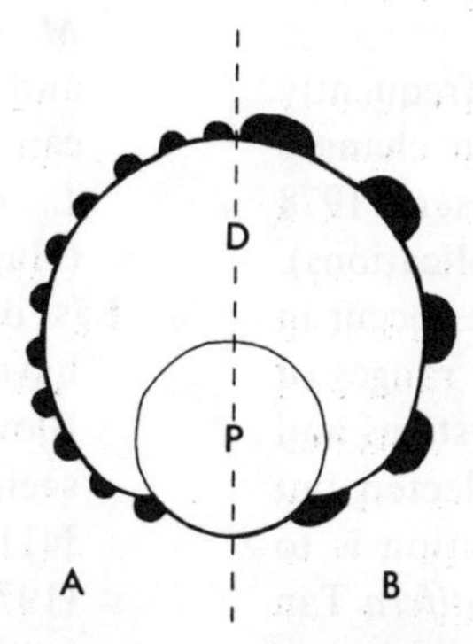

Fig. 6. Schematic diagram showing how *Eulepidina* (A) may be distinguished from *Nephrolepidina* (B) when the protoconch (P) and deuteroconch (D) are neither clearly eulepidine nor nephrolepidine in shape. The auxiliary and adauxiliary chambers (black) are quite different in size.

L. (Eulepidina)

First Neogene appearance. Basal Miocene (Te_5) at which level it is widespread, Ranges up from the Middle Oligocene (Td) in the Indo-West Pacific region. Associated with *Miogypsina* in Saipan (Hanzawa, 1957) and elsewhere.

Last occurrence. Believed to be *L. (E.) badjirraensis* Crespin from just within Zone N6, western Australia (Chaproniere, 1975, 1981).

Remarks. Numerous names are available for the late Oligocene and early Miocene species, but a thorough taxonomic revision will be necessary before proper synonymies can be established and stratigraphic ranges determined for the valid species.

L. (Nephrolepidina)

First Neogene appearance. Basal Miocene (Te_5) by which time it is widespread. Ranges up from the Middle Oligocene (Td) in the Indo-West Pacific region. Several species reported from the earliest Miocene; *N. sumatrensis* (Brady) is probably the best known.

Last occurrence. *L. (N.) rutteni* van der Vlerk has been reported from beds of early Pliocene age (N19) in Borneo by Van Vessem (1978), and *L. (N.) radiata* van der Vlerk (= *L. suvaensis* Whipple) from beds of the same age in Fiji (Adams and Frame, 1979). There is also an N19 record of *L. (N.) japonica* from Japan (Ibaraki and Tsuchi, 1978), but Matsumaru (1981) has referred these specimens to *L. (N.) rutteni* and assigned the beds concerned to Zone N17. Ibaraki (1981) has summarized the stratigraphical distribution of *Lepidocyclina* in Izu Peninsula, Japan, and concluded that it occurs in beds of N14, N17, N18, and N19 age, but the writer doubts whether these specimens are all *in situ*. The extinction levels of late Miocene species of *Lepidocyclina* are difficult to determine owing to the paucity of described carbonate sequences of this age.

Remarks. Strata of N3–N8 age are now frequently dated using biometric methods based on changes in the embryonic apparatus (see Van Vessem, 1978 and Chaproniere, 1980 for recent applications). We do not as yet know how many lineages occur in the Miocene, nor do we know the exact ranges of any species. The investigaton of stolon systems and lateral chamberlets has been rather neglected but must be undertaken if a sound classification is to be achieved. One striking species, *L. stratifera* Tan (*L. omphalus* Tan is probably a synonym), although known only from the microspheric form, seems to be restricted to a narrow interval within Zones N8 and N9.

Marginopora Quoy and Gaimard, 1830

First appearance. Early Miocene (within Te_5). Matsumaru's (1978) record from the Minamizaki Limestone, Chichi-Jima, could be the first. These beds were dated as Te_5, but the evidence is slim and they rest on strata of Te_{1-4} age. Slightly younger records, but still within Te_5, are those of Matsumaru (1976a) from the upper e limestones of Saipan; Binnekamp (1973) from the lower part (Te_5) of the Yalam Limestone, New Britain; and Hashimoto *et al.* (1977b) from the Angat Formation (Te_5) in the Philippines. Hashimoto *et al.* (1981) reported *M. vertebralis* with a fauna dated as Te_{1-4} from Cagraray Point, Batan Island Group, Philippines, but provided no illustration. It is common in beds of Tf_1 age in Australia (see Adams, 1968, 1970; Chaproniere, 1981).

Last occurrence. Extant. *M. vertebralis* Quoy and Gaimard is widely distributed.

Remarks. Not common until Tf times when it is found frequently with *Flosculinella bontangensis* and later with *Alveolinella quoyi*. Appears always to have been restricted to very shallow, inshore environments.

Miogypsina Sacco, 1893

Miogypsina includes three subgenera in the Indo-West Pacific region. The most primitive, *M. (Miogypsinoides)*, is the only one to possess a canal system in the dorsal wall; the other two possess lateral chamberlets. For a review see Bock (1976).

Miogypsina (Miogypsina)

First appearance. Basal Miocene (defines base of Te_5) in the Indo-West Pacific region. Planktonic control over most of the earlier occurrences is lacking, and it is obvious from the evolutionary grades attained by the embryonic apparatus in these early forms that the oldest Miocene sediments are not everywhere coeval. Three of the oldest occurrences must certainly be M. *gunteri* Cole from the Cauvery Basin, *M. tani* Drooger from Kutch, India (Raju, 1974), and *M. tani* from Sabah (Van Vessem, 1978). None can be younger than N4. *Miogypsina* occurs with *L. (Eulepidina)* early in Te_5 in the Philippines (Hashimoto *et al.*, 1977b, p. 116).

Last occurrence. Very difficult to establish. Thought to have become extinct in Zone N14 by Clarke and Blow (1969), but the youngest verifiable records seem to be from N12 (Baumann, 1975, p.19) and N11 (Van Vessem, 1978, p.20) Hashimoto *et al.* (1977b) show the genus ranging to the top of Tf_3 in their distribution table, but this extended range is not substantiated in their text. The longest-ranging species are given as *M. polymorpha* (actually a *Lepidosemicyclina*) and *M. globulina*, both of which are shown as extending to the top of Tf_2. The former occurs in the Kennon Limestone, for which Hanzawa and Hashimoto (1970) were able to establish only a general Tf_{1-2} age, while the latter is recorded as a member of the Sibul Fauna and therefore dated as N12.

M. (Lepidosemicyclina) Rutten, 1911

First appearance. Within Te_5; numerous records based largely on specimens seen in random thin sections of limestone. *M. (L.) thecideaeformis* is the first species of *Miogypsina* recorded in many sequences, but the determinations may not always be correct, and in the absence of planktonic control its presence should not necessarily be taken to indicate the base of Te_5. Its first appearance in the Cauvery Basin, India (Raju, 1974), seems to be in N5, but absolute proof is lacking.

Last occurrence. Van Vessem (1977) has reported *M. (L.) bifida* from strata of N7/8 age in Borneo, Chaproniere (1981) recorded *M.* cf *thecideaeformis* from the Trealla Limestone (N9) of Western Australia, and there are other uncorroborated reports from strata as young as N12 in Java (e.g., Clarke and Blow, 1969).

M. (Miogypsinoides) Yabe and Hanzawa, 1928

First Neogene appearance. Basal Miocene (Te_5); ranges up from the late Oligocene (Te_{1-4}). The first known appearance in the Indo-West Pacific is from beds of Chattian age in Cutch (Raju, 1974), where it almost immediately follows the extinction of *Nummulites fichteli*. Elsewhere in the region a considerable thickness of strata usually separates the disappearance of these nummulites from the incoming of *Miogypsinoides*, e.g., in Sarawak (Adams, 1965) and Christmas Island (Adams and Belford, 1974) where it appears well above the base of lower e. A stratigraphic hiatus or condensed sequence may therefore be present in Cutch. Hashimoto *et al.* (1977a) reported *Miogypsinoides* as possibly occurring in strata of Td age in the Philippines, but since no evidence was provided and all their other records are cited as Te_4, confirmation is required.

Last occurrence. Undoubtedly ranges up into Tf_1 in the Marshall Isles (Cole, 1954; 1958) and into beds of equivalent age in Cutch (Raju, 1974), but has not to the writer's knowledge been found above the *Orbulina* datum. *M. cupulaeformis* has been reported (Cole, 1963) with *F. bontangensis, M. vertebralis, L. (N.) japonica* and *L. (N.) rutteni* from the Bonya Limestone of Guam. This strongly suggests, but does not prove, that the genus ranges above the *Orbulina* datum in some places.

Pseudotaberina Eames (in Davies, 1971)

First appearance. Quilon limestones, India (Rasheed and Ramachandran, 1978). Tf_1. This monotypic genus has been reported from the Persian Gulf (Henson, 1950) to New Guinea (Eames *et al.*, 1962) and is usually associated with *A. howchini* and *Flosculinella bontangensis*. *P. malabarica* may eventually prove to be a species of *Archaias,* but pending a thorough taxonomic revision these genera should not be placed in synonymy. It may grade into *A. vandervlerki* de Neve. Kureshy (1970) listed a Gaj fauna from Mangopir which is consistent with an N6 or N7 age.

Last occurrence. Uncertain. Not known above the level of the Gaj Series in India and beds assigned to Tf_{1-2} in Indonesia.

Sorites Ehrenberg, 1839

First Neogene appearance. Basal Miocene (Te_5). Ranges up from the Palaeogene (Te_{1-4}). Two of the earliest known appearances must be from Christmas Island (Adams and Belford, 1974) where it occurs with *Miogypsinoides* spp. in Te_{1-4}, and from Chichi-jima where it is also associated with *Miogypsinoides* in beds of the same age (Matsumaru, 1978).

Last occurrence. Extant. *S. marginalis* (Lamarck) lives in the region today.

Remarks. It is probable that *Sorites* will eventually be found in rocks older than Te_{1-4} since it is known from the Early Oligocene of the Middle East (Henson, 1950, p. 56). The early species are difficult to identify in random thin sections.

Spiroclypeus Douvillé, 1905

First Neogene appearance. Basal Miocene (Te_5). Ranges up from the Palaeogene. Until relatively recently there was no known occurrence in rocks of Early Oligocene age, but Kaever (1970) reported *Spiroclypeus* sp. from the Lower Oligocene of Afghanistan, and Rasheed and Ramachandran (1978) have recorded *S. ranjanae* Tewari from the Lower and Middle Oligocene of Cutch. There is at present no satisfactory way of discriminating between the eight species reported from Te strata in the region.

Last occurrence. Within Te_5, at about the same level as *Eulepidina*. In some places it may be earlier, in others later. Almost certainly does not occur in strata younger than N5. Known in association with *Flosculinella globulosa* in the Philippines (Hanzawa and Hashimoto, 1970) and also with *Miogypsina* in the same area (see Hashimoto *et al.*, 1977b for refs.). It was also found with *Miogypsina* in Sydney's Dale, Christmas Island (Adams and Belford, 1974), although it usually occurs with *Miogypsinoides.*

Tayamaia Hanzawa, 1967

First Neogene appearance. Basal Miocene (Te_5) but known to range up from the Palaeogene (Chattian; Te_{1-4}). Reported from both lower and upper e of Christmas Island (Adams and Belford, 1974).

Last occurrence. Believed to be within Te_5. Known in association with *Miogypsinoides dehaarti* and *Miogypsina* cf. *neodispansa* on Christmas Island; also known from the Tagpochau Limestone of Saipan (Hanzawa, 1957).

Acknowledgements

Drs. D.J. Belford and G.C.H. Chaproniere kindly read and commented helpfully on the typescript at very short notice. Miss Lois Cody drafted the text-figures.

References

Adams, C. G., 1964: On the age and foraminiferal fauna of the Bukit Sarang Limestone, Sarawak, Malaysia. *Malaysia Geol. Surv., Ann. Rept., Borneo Region,* p. 156–162.

Adams, C. G., 1965: The foraminifera and stratigraphy of the Melinau Limestone, Sarawak, and its importance in Tertiary correlation. *Q. Jl. Geol. Soc. Lond.*, v. **121**, p. 283–338.

Adams, C. G., 1967: Tertiary Foraminifera in the Tethyan, American and Indo-Pacific Provinces. *In* Adams, C.G. and Ager, D.V., eds., *Aspects of Tethyan Biogeography.* Syst. Assoc. Publ. London, no. 7, p. 195–217.

Adams, C. G., 1968: A revision of the foraminiferal genus *Austrotrillina* Parr. *Br. Mus. Nat. Hist., Bull.* (Geol.), v. **16**, no. 2, p. 73–97.

Adams, C. G., 1970: A reconsideration of the East Indian letter classification of the Tertiary. *Br. Mus. Nat. Hist., Bull.*(Geol.), v. **19**, no. 3, p. 87–137.

Adams, C. G., 1973: Some Tertiary foraminifera. *In* Hallam, ed., *Atlas of Palaeobiogeography*, Elsevier Scientific Publishing Co., Amsterdam, p. 455–468.

Adams, C. G., 1983: Speciation, phylogenesis, tectonism and eustasy: factors in the evolution of Cenozoic larger foraminiferal bioprovinces. *In* Simms, R. W., Price, J.H., and Whalley, P. E., eds. *Evolution, Time and Space: The Emergence of the Biosphere.* Academic Press, London. p. 255–289.

Adams, C. G., and Belford, D. J., 1974: Foraminiferal biostratigraphy of the Oligocene-Miocene limestones of Christmas Island (Indian Ocean). *Palaeontology*, v. **17**, no. 3, p. 475–506.

Adams, C. G., Benson, R.H., Kidd, R.B., Ryan, W.B.F., and Wright, R.C., 1977: The Messinian Salinity Crisis and evidence of late Miocene eustatic changes in the world ocean. *Nature,* v. **269**, p. 383–386.

Adams, C. G., Gentry, A. and Whybrow, P., 1983: Dating the terminal Tethyan event. *Utrecht Micropal. Bull.*, v. **30** (in press).

Adams, C. G., Rodda, P. and Kiteley, R.J., 1979: The extinction of the foraminiferal genus *Lepidocyclina* and the Miocene/Pliocene boundary problem in Fiji. *Marine Micropal.*, v. **4**, p. 319–339.

Adams, C. G. and Frame, P., 1979: Observations on *Cycloclypeus* (*Cycloclypeus*) Carpenter and *Cycloclypeus* (*Katacycloclypeus*) Tan (Foraminiferida). *Br. Mus. Nat. Hist., Bull.*(Geol), v. **32**, p. 3–17.

Bain, J.H.C. and Binnekamp, J.G., 1973: The foraminifera and stratigraphy of the Chimbu Limestone, New Guinea. *Bur. Miner. Min. Res. Geol. Geophys, Bull.*, v. **139**, p. 1–12.

Banner, F.T. and Blow, W.H., 1965: Progress in the planktonic foraminiferal biostratigraphy of the Neogene. *Nature,* v. **208**, no. 1, p. 164–6.

Barker, W. and Grimsdale, T.F., 1963: A contribution to the phylogeny of the orbitoidal foraminifera, with descriptions of new forms from the Eocene of Mexico. *Jour. Pal.*, v. **10**, p. 231–247.

Baumann, P., 1972: Les faunes de Foraminifères de l'Éocène supérieur la base du Miocène dans le Bassin de Pasir, sud de Kalimantan (Bornéo). *Revue Inst. fr. Pétrole,* v. **27**, p. 817–829.

Baumann, P., 1975: The Middle Miocene diastrophism, its influence to the sedimentary and faunal distribution of Java and the Java Sea Basin. *Natl. Inst. Geol. & Mining, Bull.* Bandung, v. **5**, p. 13–28.

Belford, D.J., 1981: Co-occurrence of Middle Miocene larger and planktic smaller foraminifera, New Ireland, Papua New Guinea. *Bur. Miner. Resour. Geol. Geophys. Aust. Bull.*, 209, p. 1–21.

Beyrich, E., 1854: Uber die Stellung der Hessischen Tertiärbildungen. *Ber. Verh. kgl. preuss. Akad. Wiss. Berlin,* p. 640–66.

Billmann, H.G. and Kartaadipura, L.W., 1975: Late Tertiary biostratigraphic zonation, Kutei Basin, offshore East Kalimantan, Indonesia. *3rd. Ann Conv. Indones. Petrol. Ass.*, Jakarta, *Proc.* p. 301–310.

Binnekamp, J.G., 1973: Tertiary larger foraminifera from New Britain PNG. *Bur. Miner. Resour. Geol. Geophys. Aust., Bull.*, v. **140**, p. 1–26.

Blow, W.H., 1956: Origin and development of the foraminiferal genus *Orbulina* d'Orbigny. *Micropaleontology,* v. **2**, p. 57–70.

Blow, W.H., 1969: Late Middle Eocene to Recent planktonic foraminiferal biostratigraphy. *In* Brönnimann, P. & Renz, H.H. eds., *Ist. Conf. on planctonic microfossils, Proc.* (Geneva, 1967). E.J. Brill, Leiden, p. 200–421.

Bock, J.F., de, 1976: Studies on some *Miogypsinoides-Miogypsina* s.s. associations with special reference to morphological features. *Scripta Geologica,* v. **36**, p. 1–135.

Bolli, H. M., 1966: Zonation of Cretaceous to Pliocene marine sediments based on planktonic foraminifera. *Inform. Assoc. Venez. Geol. Min. Petrol. Bol.*, v. **9**, p. 3–32.

Bolli, H. M., 1969: Report on a meeting of the Working Group for a biostratigraphic zonation of the Cretaceous and the Cenozoic as a basis for the correlation in marine geology. *Geol. Newsletter,* no. 3, p. 199–207.

Chang, L.A., 1975: Biostratigraphy of Taiwan. *Geol &*

Palaeont. Southeast Asia, **15**, p. 337–361.

Chaproniere, G.C.H., 1975: Palaeoecology of Oligo-Miocene larger Foraminiferida, Australia, *Alcheringa*, v. **1**, p. 37–58.

Chaproniere, G.C.H., 1980: Biometrical studies of early Neogene larger Foraminiferida from Australia and New Zealand. *Alcheringa*, v. **4**, p. 153–181.

Chaproniere, G.C.H., 1981: Australasian mid-Tertiary larger foraminiferal associations and their bearing on the East Indian Letter Classification. *Miner. Resour. Geol. Geophys. Aust. Bur. Bull.*, v. **6**, p. 145–151.

Clarke, W.J. and Blow, W.H., 1969: The inter-relationships of some late Eocene, Oligocene and Miocene larger foraminifera and planktonic biostratigraphic indices. *lst. Int. conf. on Planktonic Microfossils, Proc.* Leiden, E.J. Brill. **2**, p. 82–96.

Cole, W.S., 1954: Larger foraminifera and smaller diagnostic foraminifera from Bikini drill holes. *U.S. Geol. Surv., Prof. Pap.*, 260–O, p. 569–608.

Cole, W.S., 1957: Larger foraminifera of Saipan Island. *U.S. Geol. Surv., Prof. Pap.*, 280–1, p. 321–360.

Cole, W.S., 1958: Larger Foraminifera from Eniwetok Drill Holes. *U.S. Geol. Surv., Prof. Pap.*, 260–V, p. 741–784.

Cole, W.S., 1960: Variability in embryonic chambers of *Lepidocyclina. Micropaleontology*, **6**, p. 133–140.

Cole, W.S. 1963: Tertiary larger foraminilera from Guam. *U.S. Geol. Surv. Prof. Pap.*, 403–E, p. 1–28.

Cole, W.S., 1969: Larger foraminifera from deep drill holes on Midway Atoll. *U.S. Geol. Surv., Prof. Pap.*, 680–C, p. 1–15.

Cole, W.S., 1975: Concordant age determinations by larger and planktonic foraminifera in the Tertiary of the Indo-Pacific region. *Jour. Foram. Research*, v. **5**, p. 21–39.

Coleman, P.J., 1963: Tertiary larger foraminifera of the British Solomon Islands, southern Pacific. *Micropaleontology*, v. **9**, p. 1–38.

Dasgupta, A., 1977: Two species of *Austrotrillina* from western Cutch. *Geol. Soc. India, Jl.*, v. **18**, p. 65–71.

Davies, A.M., 1971: *Tertiary Faunas* Vol. 1. Revised by F.E. Eames, George Allen & Unwin Ltd., London.

Djamas, Y.S. and Marks, E., 1978: Early Neogene foraminiferal biohorizons in E. Kalimantan, Indonesia. *Geol. Res. & Develop. Centre, Bandung, Spec. Pub.*, **1**, p. 111–124.

Drooger, C.W., 1963: Evolutionary trends in the Miogypsinidae. *In* Koenigswald, G.A.R. von *et al.*, eds., *Evolutionary Trends in Foraminifera*, Elsevier, Amsterdam, p. 314–349.

Drooger, C.W., 1979: Marine connections of the Neogene Mediterranean, deduced from the evolution and distribution of larger foraminifera. *Ann. Geol. Pays Hellen. VIIIth Int. Cong. Med. Neogene, Athens*, Fasc. I, p. 361–369.

Eames, F.E., Banner, F.T., Blow, W. H., and Clarke, W.J., 1962: *Fundamentals of mid-Tertiary stratigraphical correlation.* Camb. Univ. Press., 163 p.

Fortuin, A.R., 1970: The early ontogenetic stages in *Eulepidina dilatata. Kon. Ned. Akad. Wetensch., Proc., Ser. B.*, v. **73**, no. 3, p. 196–208.

Fuchs, Th., 1892–1894: Tertiärfossilien aus dem Kohlenführended miozänablagerungen der Umgebung von Krapina und Radoboj und über die Stellung der sogenannten 'Aquitanischen' Stufe. *A.M. k Földt. Int. Evkonyue.*, v. **10**, no. 5.

Gould, S.J. and Eldredge, N., 1977: Punctuated equilibria: the tempo and mode of evloution reconsidered. *Paleobiology*, v. **3**, p. 115–151.

Haak, R. and Postuma, J.A., 1975: The relation between the tropical planktonic foraminiferal zonation and the Tertiary Far East Letter Classification. *Geol. en Mijnb.*, v. **54**, p. 195–198.

Haile, N. H. and Wong, N.P.Y., 1965: The geology and mineral resources of the Dent Peninsula, Sabah. *Geol. Surv. Borneo Region. Malaysia, Mem.*, **16**, p. 1–199.

Hanzawa, S., 1957: Cenozoic foraminifera of Micronesia. *Geol. Soc. Am. Mem.*, v. **66**, 63 pp.

Hanzawa, S. and Hashimoto, W., 1970: Larger foraminifera from the Philippines. Pt. 1. *Geol. & Palaeont. Southeast Asia*, v. **8** p. 187–230.

Hardenbol, J. and Berggren, W.A., 1978: A new Paleogene numerical time scale. *Studies in Geology*, A.A.P.G., Tulsa, no. 6, p. 213–234.

Hashimoto, W. and Matsumaru, K., 1975: On the *Lepidocyclina*-bearing limestone exposed at the Southern Cross Mountain Highway, Taiwan. *Geol. & Palaeont. Southeast Asia*, v. **16**, p. 103–116.

Hashimoto, W., Matsumaru, K. and Kurihara, K., 1977a: Larger foraminifera from the Philippines., V. Larger foraminifera from Cenozoic limestones in the Mansalay Vicinity, Oriental Mindoro, with an appendix "An Orbitoid-bearing limestone from Barahid, Bongabong. *Geol. & Palaeont. Southeast Asia*, v. **18**, p. 59–76.

Hashimoto, W., Matsumaru, K., Kurihara, K., David, P.P., and Balce, R., 1977b: Larger foraminiferal assemblages useful for the correlation of the Cenozoic marine sediments in the mobile belt of the Philippines. *Ibid.*, v. **18**, 103–124.

Hashimoto, W., Matsumaru, K. and Kurihara, K., 1978: Larger foraminifera from the Philippines, VII. Larger foraminifera from the Lutak Hill Limestone, Pandan Valley, Central Cebu. *Ibid.* v. **19**, p. 73–80.

Hashimoto, W., Matsumaru, K. and Sugaya, M., 1981: Larger foraminifera from the Philippines, Xl. On the Coal Harbor Limestone, Cagraray Island, Batan Island Group, Albay Province. *Ibid.*, v. **22**, p. 55–62.

Henson, F.R.S., 1950: Middle Eastern Tertiary Peneroplidae (Foraminifera), with remarks on the phylogeny and taxonomy of the family. Wakefield, England, 70 p., 10 pls.

Hottinger, L., 1977: Distribution of larger Peneroplidae, *Borelis* and Nummulitidae in the Gulf of Elat, Red Sea. *Utrecht Micropal. Bull.*, v. **15**: 35–109.

Huang, T., 1975: Late Neogene foraminiferal zonation of southwestern Taiwan. *In* Saito, T. and Burckle, L., eds., *Late Neogene Epoch Boundaries*. Micropaleontology, Spec. Pub. no. 1, p. 106–114.

Ibaraki, M., 1981: Geologic ages of "*Lepidocyclina*" and *Miogypsina* horizons in Japan as determined by planktonic foraminifera. *Int. Workshop on Pacific Neogene*

Biost., Proc. (IGCP-114), Osaka, Japan, p. 118–119.

Ibaraki, M. and Tsuchi, R., 1978: Planktonic foraminifera from *Lepidocyclina* horizons at Namegawa in the southern Izu Peninsula Central Japan. *Shizuoka Univ. Fac. Sci., Rep.*, v. **12**, p. 115–130.

Ikebe, N. and Chiji, M., 1971: Notes on Top-datum of *Lepidocyclina* sensu lato in reference to planktonic foraminiferal datum. *Osaka City Univ., Geosc. Jour.*, v. **14**, p. 19–52.

Kaever, M., 1970: Die alttertiärn Grossforaminiferen sudost-Afghanistans unter besonderer Berucksichtigung der Nummulitiden-Morphologie, Taxonomie und Biostratigraphie. *Münst. Forsch. Geol. Paläont.*, H16/17, 450 p.

Kleinpell, R.M., 1954: Neogene smaller foraminifera from Lau, Fiji. *Bernice P. Bishop Mus., Bull.*, v. **211**, p. 1–96.

Kureshy, A.A., 1970: The larger and pelagic foraminifera of Mangopir, West Pakistan. *Cush. Contr. Foram. Res.*, v. **21**, no. 2, p. 78–80.

Leupold, W. and Vlerk, I.M. van der, 1931: The Tertiary. *Leid. Geol. Meded.*, v. **5**, p. 611–648.

Ludbrook, N.H., 1961: Stratigraphy of the Murray Basin in South Australia. *Geol. Surv. S. Aust., Bull.*, v. **36**, p. 1–96.

Ludbrook, N.H., 1963: Correlation of the Tertiary rocks of South Australia. *Roy. Soc. S. Aust., Trans.*, v. **87**, p. 5–15.

Lyell, C., 1833: *Principles of Geology* (lst ed.), John Murray, London, v. **3**, p. 1–398.

Matsumaru, K., 1971: The genera *Nephrolepidina* and *Eulepidina* from New Zealand. *Trans. Proc. Palaeont. Soc. Japan*, N.S., no. 84, p. 179–189.

Matsumaru, K., 1974: Larger foraminifera from East Mindanao, the Philippines. *Geol. Palaeont. Southeast Asia*, v. **14**, p. 101–115.

Matsumaru, K., 1976a: Larger foraminifera from the islands of Saipan and Guam, Micronesia. *Progress in Micropaleontology*, Micropal. Press, N.Y., p. 190–213.

Matsumaru, K., 1976b: Larger foraminifera from the Ryuku-Group, Nansei Shoto Islands, Japan. *1st Int. Symp. on Benthonic Foraminifera of Continental Margins.* Part A. Paleoecology and Biostratigraphy Maritime Sediments. Spec. Publ. 1, p. 401–424.

Matsumaru, K., 1978: Biostratigraphy and paleoecological transition of larger Foraminifera from the Minamizaki Limestone, Chichi-Jima, Japan. *Geol. Res. Devel. Centre (Indonesia). Spec. Pub.* I, p. 63–88.

Matsumaru, K., 1980: Cenozoic larger foraminiferal assemblages of Japan, Pt. 1. A comparison with Southeast Asia. *Geol. Paleont. Southeast Asia*, **21**, p. 211–224.

Matsumaru, K., 1981: On *Lepidocyclina* (*Trybliolepidina*) *rutteni* van der Vlerk from Zone N17 at Mitsugane, Izu Peninsula, Shizuoka Prefecture, Japan. *Imp. Acad., Proc.*, v. **57**, ser. B, p. 115–118.

McTavish, R.A., 1967: Tertiary faunas of smaller foraminifera from the British Solomon Islands. *Geol. Rec. Brit. Solomon Islands*, III Rept., no. 81, p. 66–71.

Mohan, M., Narayanan, V. and Kumar, P., 1978: Paleocene and Early Neogene biostratigraphy of Bombay offshore region. *In* Rasheed, D.A., ed., *VII Ind. Coll. Micropal. Strat., Proc.*, p. 180–194c.

Mohler, W.A., 1950: *Flosculinella reicheli* sp. nov. aus dem Tertiar von Borneo. *Eclog. Geol. Helv.*, v. **42**, p. 521–527.

Natori, H., 1978a: Foraminifera from West Java. *In* Untung, M. and Sato, Y., eds., *Gravity and Geological Studies in Jawa, Indonesia. Spec. Pub.*, no. 6, p. 82–89.

Natori, H., 1978b: Biostratigraphic relationships between larger foraminifera, 'Letter stages' and planktonic foraminiferal zones. *In* Untung, M. and Sato, Y., eds., *Gravity and Geological studies in Jawa, Indonesia, Spec. Pub.*, no. 6, p. 130–132.

O'Herne, L. and Vlerk, I.M., van der, 1971: Geological age determinations on a biometrical basis (comparison of eight parameters). *Soc. Paleont. Ital., Bol.*, v. **10**, p. 3–18.

Palmieri, V., 1973: Comparison of correlation methods for planktonic and larger foraminifera in the Capricorn Basin, Queensland. *Queensland Gov. Mining Jl.*, v. **74**, no. 863, p. 312–317.

Palmieri, V., 1974: Correlation and environmental trends of the subsurface Tertiary Capricorn Basin. *Geol. Surv. Queensland, Rept.*, no. 86, 14 p.

Rahaghi, A., 1980: *Tertiary faunal assemblage of Qum-Kashan, Sabzewar and Jahrum areas.* Publ., no. 8, National Iranian Oil Co. Geological Laboratories, p. 1–64. 30 pls.

Raju, D.S.N., 1971: Upper Eocene to Early Miocene planktonic foraminifera from the subsurface sediments in Cauvery Basin, South India. *Jb. Geol. B.A.*, 17, p. 7–68.

Raju, D.S.N., 1974: Study of Indian Miogypsinidae. *Utrecht Micropal. Bull.*, v. **9**, 148 pp.

Rasheed, D.A. and Ramachandrand, K.K., 1978: Foraminiferal bio-stratigraphy of the Quilon Beds of Kerala State, India. *In* Rasheed, D.A., ed., *VII Indian Colloq. Micropal. & Strat. Proc.*, p. 299–320.

Ritzkowski, S., 1981: Latdorfian. *In* Pomerol, C., ed., *Stratotypes of Paleogene stages.* Bull. d'Information des Géologues du Bassin de Paris, *Mém. Hors Serie* 2, Paris, p. 149–166.

Sen Gupta, B.K., 1964: Tertiary biostratigraphy of a part of North-Western Kutch. *Geol. Soc. India, Jl.*, v. **5**, p. 138–158.

Todd, R., 1960: Some observations on the distribution of *Calcarina* and *Baculogypsina* in the Pacific. *Tôhoku Univ., Sci. Rept., 2nd Ser. (geol)*, v. **4**, p. 100–107.

Todd, R., 1977: *In* Mallick, D.I.J. and Greenbaum, D., eds., Geology of southern Santo. *Reg. Rept. New Hebrides Condominium. Geol. Surv.*, p. 1–84.

Tsuchi, R. *et al.*, 1979: Correlation of Japanese Neogene sequences (1): A synthesis based upon biostratigraphic datum levels, zones, stages and radiometric ages. *In* Tsuchi, R., ed., *Fundamental data on Japanese Neogene bio- and chronostratigraphy*. Shizuoka Univ., p. 143–150.

Ujiié, H., 1970: Miocene foraminiferal faunas from the Sandakan Formation, North Borneo. *Geol. Palaeont. Southeast Asia*, v. **8**, p. 165–185.

Ujiié, H., 1973: Distribution of the Japanese *Miogypsina*, with description of a new species—Restudy of the

Japanese Miogypsinids, Part 3. *Natl. Sci. Mus., Bull.*, v. **16**, no. 1, p. 99–114.

Ujiié, H., 1975: An Early Miocene planktonic foraminiferal fauna from the Megami Formation, Shizuoka Prefecture. *Natl. Sci., Mus., Bull., ser C. (geol)*, v. **1**, no. 3, p. 83–92.

Ujiié, H., 1975b: Planktonic foraminiferal biostratigraphy in the western Philippine Sea, DSDP Leg 31. *DSDP, Initial Repts.*, v. **31**, p. 677–691.

Vail, P.R., Mitchum, R.M. Y. Jr., and Thompson, S., 1977: Seismic stratigraphy and global changes of sea level, Pt. 4: Global cycles of relative changes of sea level. *Am. Assoc. Petrol. Geol. Mem.*, v. **26**, p. 83–97.

Van Couvering, J.A. and Berggren, W.A., 1977: Biostratigraphic basis of the Neogene time scale. *In* Kauffman, E.G. & Hazel, J.E., eds., *Concepts and methods in biostratigraphy*, Dowden, Hutchinson & Ross, Stroudsberg, Penn, p. 283–306.

Van Vessem, E.J., 1977: The internal structure of *Miogypsina polymorpha* and *M. bifida*. *Kon. Ned. Akad. Wetensch., Proc. Ser. B.*, v. **80**, no. 5, p. 421–428.

Van Vessem, E.J. 1978: Study of Lepidocyclinidae from South East Asia, particularly from Java and Borneo. *Utrecht Micropal. Bull.*, **19**, 163 pp.

Vlerk, I.M. van der, 1950: Stratigraphy of the Caenozoic of the East Indies based on foraminifera. Rep. 18th Int. Geol. Congr. Pt. XV, *Int. Pal. Union. Proc.* p. 61–63.

Vlerk, I.M. van der, 1955: Correlation of the Tertiary of the Far East and Europe. *Micropaleontology*, v. **1**, p. 72–75.

Vlerk. I.M. van der and Postuma, J.A., 1967: Oligo-Miocene lepidocyclinas and planktonic foraminifera from East Java and Madura, Indonesia. *Kon. Ned. Akad. Weten., Proc.* Amsterdam, Ser. B,, **70**, p. 391–398.

Vlerk, I.M. van der and Umbgrove, J.H., 1927: Tertiary Gidsforaminiferen van Nederlandisch Oost-Indie. *Wet. Meded. Dienst. Mijnb. Ned. O-Ind.*, v. **6**, p. 3–35.

I - 7

Eastern Pacific Molluscan Bio-Events and Their Relation to Neogene Planktonic Datum Planes

Louie MARINCOVICH, Jr.

Introduction

Few direct ties are known between Neogene molluscan faunas of the eastern Pacific and planktonic microfossils. Planktonic foraminifers and diatoms are principally found in deep-water deposits that generally lack molluscan fossils, especially age-diagnostic ones, and abundantly fossiliferous Neogene molluscan faunas almost always represent shallow-water environments in which planktonic foraminifers and diatoms are few or absent. In strata where planktonic microfossils have been found with shallow-water molluscan faunas, the microfossils usually have been transported major distances and are too few in number and too worn and broken to be of use in age determinations. Thus, correlations between molluscan bio-events and planktonic datum planes are, in most cases, indirectly inferred in two ways: 1) the use of benthic foraminifers, which may range from shallow to deep water, as intermediaries between molluscan and planktonic faunas and floras; and 2) the reliance on radiometric techniques for independent dating of molluscan faunas and microfossil assemblages.

The use of either benthic foraminifers or radiometric dates for correlation of molluscan and planktonic assemblages involves limitations inherent in each method. It is well known that many eastern Pacific benthic foraminiferal stages and zones are time-transgressive when compared to planktonic foraminiferal datum planes (Ingle, 1967; Bandy, 1972), so that molluscan-to-planktonic correlations are not precise and are accurate only within a certain range of error. It has also long been recognized that the accuracy of radiometric dates is highly dependent not only upon the material analyzed, but also upon the analytical techniques employed and the judgement and skill of the analyst. Thus, in instances where such indirect methods of correlation are used, there will always remain a margin for error in equating molluscan bio-events and planktonic datum planes.

In the eastern Pacific, molluscan bio-events may be placed in two classes: those related to climatic changes that result in the shifting of faunal-province boundaries, and those caused by the opening or closing of migration routes, such as Bering Strait or the Isthmus of Panama, between adjacent oceanic regions. In either case, changed oceanographic and biological conditions result not only in the geographic movement of faunas, but in the origin of some species and the extinction of others. Changes in faunal composition are, in turn, used to identify and interpret the nature and extent of the bio-events.

Neogene molluscan faunas are relatively well known in the northern part of the eastern Pacific, especially in California, Oregon, and Washington, and much more poorly known in the southern part. Therefore, the present discussion will largely center on the northeastern Pacific, with references where possible to more southerly parts of the eastern Pacific.

Climatic Events

Northeastern Pacific Neogene paleoclimatic trends inferred from mollusks have been documented in several studies, notably those of Addicott (1969, 1970a, b). The most extensive work has been done on faunas of the San Joaquin basin of central and southern California, where periods of relatively warm climate have been identified in about the middle of the "Temblor" Stage, the middle of the "Margaritan" Stage, and in the upper "Jacalitos" to middle "San Joaquin" Stages (Fig. 1). Of these three the Temblor warm-event is inferred to have had by far the highest shallow-water temperatures (Addicott, 1969, 1970b), and was the most significant post-Eocene interval of climatic warming in this portion of the Pacific. The largest documented warm-water molluscan fauna of the Temblor event consists of over 200 species from the uppermost part of the Olcese Sand and the lower part of the Round Mountain Silt of the Kern River area, southern San Joaquin basin (Addicott, 1970b, c). Tropical taxa such

U.S. Geological Survey, 345, Middlefield Rd., MS15, Menlo Park, California 94025, U.S.A.

as *Ficus, Anadara s.s. Dosinia, Natica s.s.* and *Natica (Naticarius)* achieved their northernmost post-Eocene distribution at this time (Addicott, 1969; Marincovich, 1977), never afterward appearing in the middle-to high-latitude northeastern Pacific. Besides occurring in middle Temblor faunas of the San Joaquin basin, these tropical taxa are also found in coeval faunas of the Astoria Formation in Oregon (latitude 46° N) (Moore, 1964), which is assigned to the Newportian molluscan Stage of the Pacific Northwest provincial chronology (Addicott, 1976). In fact, *Anadara s.s. Dosinia,* and *Ficus* even occur as far north as 57° N latitude, in the stratotype of the Narrow Cape Formation on Kodiak Island, western Gulf of Alaska, which contains a Newportian fauna (Allison, 1978). Benthic foraminifers associated with this Temblor warm-event in the San Joaquin basin are assigned to the upper part of the Relizian and lower part of the Luisian Stages (Addicott, 1972) and are correlated approximately with diatoms of the *Actinocyclus ingens* Zone and Subzone a of the *Denticulopsis lauta* Zone, and with North Pacific planktonic foraminiferal zones ranging from the upper half of N7 through N9 (Poore and others, 1981). Based on the somewhat tentative correlation of molluscan and microfossil biostratigraphic schemes by Poore and others (1981), the "Temblor" warm-event would have occurred during the interval of approximately 15–17 Ma, and thus straddled the early Miocene to middle Miocene boundary. Previously, a middle Miocene age had been inferred for this Neogene thermal maximum based on shallow-water molluscan faunas of the southern San Joaquin basin (Addicott, 1969, 1970 a, b). It should be re-emphasized here that the correlation of "Temblor" molluscan faunas and planktonic microfossils is imprecise and is made by utilizing the sometimes unreliable intermediaries of benthic foraminifers. Correlation of the early or middle Miocene warm-water fauna inferred for the Kodiak Island Narrow Cape Formation with faunas in Oregon, Washington, and California is based entirely on mollusks. The only microfossils so far known from the Narrow Cape Formation are poorly preserved ostracodes of Miocene age (Elisabeth Brouwers, personal communication, 1982).

The next youngest influx of relatively warm-water molluscan taxa into middle latitudes of the northeastern Pacific occurred in the middle part of the "Margaritan" molluscan Stage of central and southern California. Some of the molluscan genera that help to characterize the "Temblor" warm-event also occur in "Margaritan" faunas, but are represented by much smaller numbers of species. The principal "Margaritan" molluscan faunas occur in parts of the Monterey Formation and Santa Margarita Sandstone of the central California Coast Ranges (Anderson and Martin, 1914; Addicott others, 1978) Warm-water "Margaritan" mollusks include species of *Lima, Olivella, Forreria,* and *Cerithium,* which are less tropical in character than mollusks of the "Temblor" warm-event. Benthic foraminifers from the Santa Margarita Sandstone are of extremely shallow-water aspect and represent the Mohnian Stage (Kleinpell, 1938; Smith, 1968; Addicott and others, 1978) It is possible that the "Margaritan" warm-event is also coeval with the type Delmontian benthic foraminiferal Stage, which planktonic microfossil studies have shown to be entirely coeval with a lower part of the Mohnian Stage (Barron, 1976; Poore and others, 1981). The type Delmontian Stage was shown by Barron (1976) to be equivalent to at least Subzone c and possibly in part to Subzone d of the *Denticulopsis hustedtii-Denticulopsis lauta* Zone. In terms of planktonic foraminifers, the type Delmontian ranges from about the middle of Zone N13 to the middle of N15 and consists of terminal middle Miocene deposits ranging in age from 11 to 12.5 Ma (Poore and others, 1981). The "Margaritan" warm-event is inferred to have occurred at approximately the same time. North of California there is no evidence of warm-water mollusks in coeval faunas

The final northeastern Pacific Neogene warm-event took place within the "Etchegoin" and "San Joaquin" molluscan Stages of central and southern California (Fig. 1). Both faunal and geochemical analysis of molluscan faunas of the Etchegoin and San Joaquin Formations, in the Kettleman Hills area of the southern San Joaquin basin, suggest temperatures somewhat higher than those at the same latitude along the modern Californian coastline (Stanton and Dodd, 1970). A warm-temperate paleoclimate inferred from molluscan faunal composition was reinforced by a mean paleotemperature of 14° C derived from strontium concentration in the outer shell layer of *Mytilus* specimens from the Etchegoin (late Miocene and Pliocene) and San Joaquin (early Pliocene) Formations (Stanton and Dodd, 1970). This evidence for a warm-temperate marine paleoclimate contrasts with the mild-temperate conditions that now prevail at the same latitude in the northeastern Pacific (Hall, 1964). Stanton and Dodd (1970) also inferred warm-temperate paleoclimates for late Miocene molluscan faunas of the Jacalitos Formation of Arnold and Anderson (1908) and Pancho Rico Formation, which are referred to the "Jacalitos" Stage (Fig. 1). The sum of their inferred faunal paleotemperature estimates indicate relatively warm intervals in the upper "Jacalitos" and lower "Etchegoin" Stages, and in the middle "San Joaquin" Stage (Stanton and Dodd, 1970).

The upper "Jacalitos" through lower "Etchegoin" warm interval corresponds generally to the upper part of the Mohnian Stage and lower part of the *Bolivina*

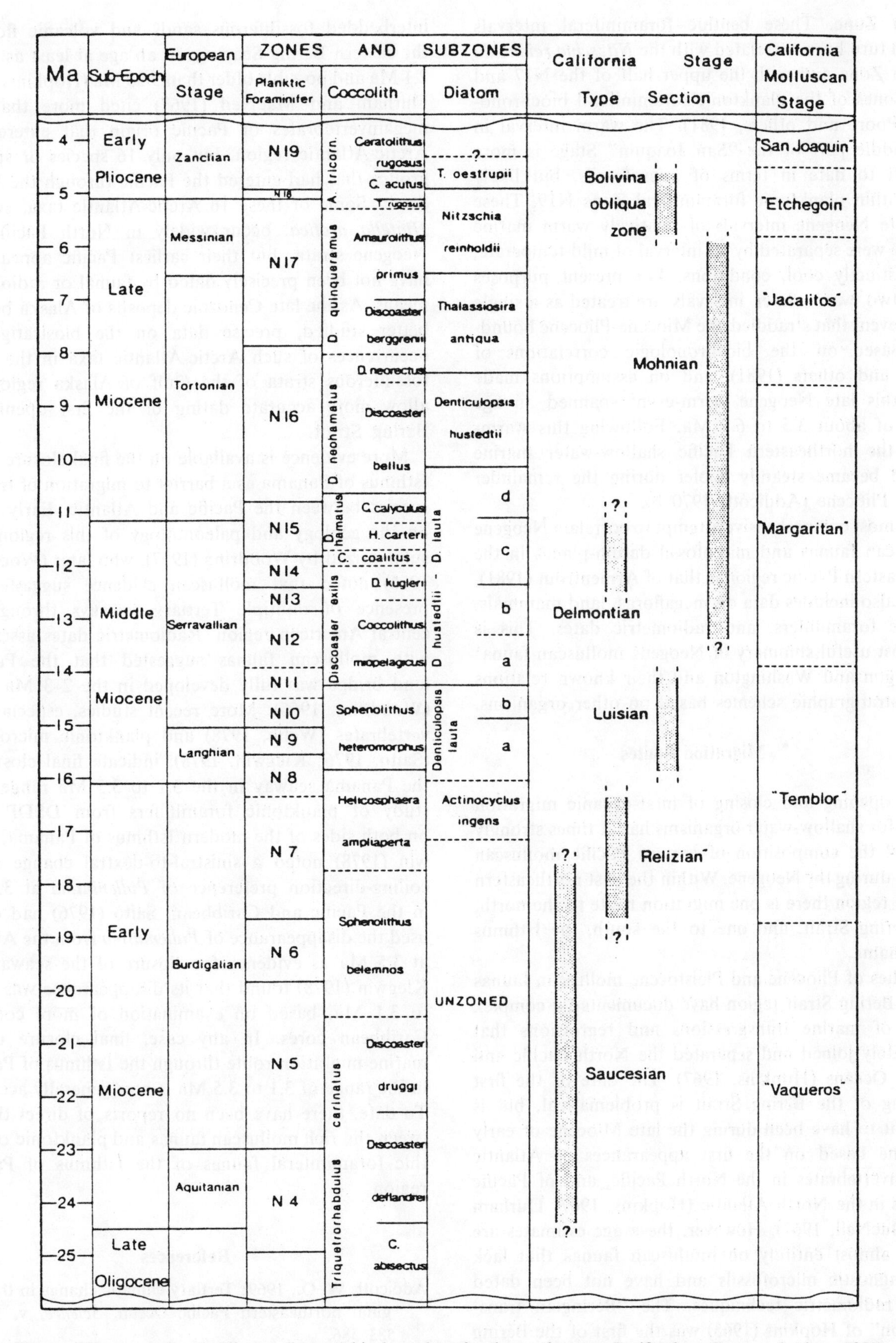

Fig. 1. Miocene and early Pliocene (in part) biochronology for the northeastern Pacific (from Poore and others, 1981).

obliqua Zone. These benthic foraminiferal intervals have in turn been correlated with the *Nitzschia reinholdii* diatom Zone and with the upper half of the N17 and N18 Zones of the planktonic foraminiferal biochronology (Poore and others, 1981). The warm interval in the middle part of the "San Joaquin" Stage is more difficult to date in terms of microfossils, but likely falls within planktonic foraminiferal Zone N19. These two late Neogene intervals of relatively warm marine climate were separated by an interval of mild-temperate, not distinctly cool, conditions. For present purposes these two warm-water intervals are treated as a single warm-event that straddled the Miocene-Pliocene boundary. Based on the biochronologic correlations of Poore and others (1981), and on assumptions made here, this late Neogene warm-event spanned an age range of about 3.5 to 6.5 Ma. Following this warm-event the northeastern Pacific shallow-water marine climate became steadily cooler during the remainder of the Pliocene (Addicott, 1970 b).

The most comprehensive attempt to correlate Neogene molluscan faunas and microfossil datum-planes in the northeastern Pacific region is that of Armentrout (1981), which also includes data on megafloras, land mammals, benthic foraminifers, and radiometric dates. This is the most useful summary of Neogene molluscan faunas in Oregon and Washington and their known relations to biostratigraphic schemes based on other organisms.

Migration Routes

The opening and closing of inter-oceanic migration routes for shallow-water organisms has at times strongly affected the composition of eastern Pacific molluscan faunas during the Neogene. Within the vast northeastern Pacific region there is one migration route to the north, the Bering Strait, and one to the south, the Isthmus of Panama.

Studies of Pliocene and Pleistocene molluscan faunas in the Bering Strait region have documented a complex series of marine transgressions and regressions that alternately joined and separated the North Pacific and Arctic Oceans (Hopkins, 1967). The date of the first opening of the Bering Strait is problematical, but is thought to have been during the late Miocene or early Pliocene based on the first appearances of Atlantic megainvertebrates in the North Pacific, and of Pacific species in the North Atlantic (Hopkins, 1967; Durham and MacNeil, 1967). However, these age estimates are based almost entirely on molluscan faunas that lack age-diagnostic microfossils and have not been dated using radiometric techniques. The "Beringian transgression" of Hopkins (1965) was the first of the Bering Strait transgressions and potassium-argon dates from interbedded fossiliferous sands and volcanic flows in the eastern Bering Sea suggest an age at least as old as 2.1 Ma and possible older than 3.35 Ma (Hopkins, 1967). Durham and MacNeil (1967) cited more than 125 megainvertebrates of Pacific origin that entered the Arctic-Atlantic region, but only 16 species or species-groups that had entered the Pacific through the Bering Strait. Some of these 16 Arctic-Atlantic taxa, such as *Hiatella arctica*, occur widely in North Pacific late Neogene strata, but their earliest Pacific appearances have not been precisely dated by faunal or radiometric means. As the late Cenozoic deposits of Alaska become better studied, precise data on the biostratigraphic occurrences of such Arctic-Atlantic taxa in the thick, fossiliferous strata of the Gulf of Alaska region will allow more accurate dating of the first opening of Bering Strait.

More evidence is available on the final closure of the Isthmus of Panama as a barrier to migration of tropical species between the Pacific and Atlantic. Early works on the geology and paleontology of this region were summarized by Woodring (1957), who later (Woodring, 1966) noted that molluscan evidence suggested the presence of multiple Tertiary seaways through the central American region. Radiometric dates associated with molluscan faunas suggested that the Panama land bridge was fully developed in the 2–3 Ma range (Woodring, 1966). More recent studies, especially on vertebrates (Webb, 1978) and planktonic microfossils (Saito, 1976; Kiegwin, 1978), indicate final closure of the Panama seaway in the 3.1 to 3.5 Ma range. In a study of planktonic foraminifers from DSDP cores on both sides of the modern Isthmus of Panama, Kiegwin (1978) noted a sinistral-to-dextral change in the coiling-direction preference of *Pulleniatina* at 3.5 Ma in the Pacific and Caribbean. Saito (1976) had earlier used the disappearance of *Pulleniatina* from the Atlantic at 3.5 Ma as evidence for closure of the seaway, but Kiegwin (1978) found that its disappearance was closer to 3.1 Ma, based on examination of more complete Caribbean cores. In any case, final closure of the marine migration route through the Isthmus of Panama in the range of 3.1 to 3.5 Ma is now generally accepted. To date, there have been no reports of direct ties between the rich molluscan faunas and planktonic or benthic foraminiferal faunas of the Isthmus of Panama region.

References

Addicott, W. O., 1969: Tertiary climatic change in the marginal northeastern Pacific Ocean. *Science*, v. **165**, p. 583–586.

Addicott, W. O., 1970a: Latitudinal gradients in Tertiary

molluscan faunas of the Pacific coast. *Palaegeography, Palaeoclimatology, Palaeoecology*, v. **8**, p. 287–312.

Addicott, W. O., 1970b: Tertiary paleoclimatic trends in the San Joaquin basin, California. *U. S. Geol. Survey, Prof. Paper*, 644–D, p. D1-D19.

Addicott, W. O., 1970c: Miocene gastropods and biostratigraphy of the Kern River area, California. *U.S. Geol. Survey, Prof. Paper*, 642, p. 1–174, pls. 1–21.

Addicott, W. O., 1972: Provincial middle and late Tertiary molluscan stages, Temblor Range, California. *Pacific, Coast Miocene Biostrat. Symposium, 47th Ann. Mt. Proc.* Soc. Economic Paleontologists and Mineralogists, Pacific Section, Bakersfield, Calif., March 9–10, 1972, p. 1–21, pls. 1–4.

Addicott, W. O., 1976: Neogene molluscan stages of Oregon and Washington. *Neogene Symposium*, Soc. Economic Paleontologists and Mineralogists, Pacific Section Ann. Mt., San Francisco, Calif., April 21–24, 1976, p. 95–115, pls. 1–5.

Addicott, W. O., Poore, R. Z., Barron, J. A., Gower, H. D., and McDougall, K., 1978: Neogene biostratigraphy of the Indian Creek–Shell Creek area, northern La Panza Range, California. *U. S. Geol. Survey, Open-file Rept.* 78-446, p. 49–83.

Allison, R. C., 1978: Late Oligocene through Pleistocene molluscan faunas in the Gulf of Alaska region. *Veliger*, v. **21**, no. 2, p. 171–188.

Anderson, F. M. and Martin, B., 1914: Neogene record in the Temblor Basin, California, and Neogene deposits of the San Juan district, San Luis Obispo County. *Calif. Acad. Sci., Proc.*, ser. 4, v. **4**, p. 15–112, pls. 1–10.

Armentrout, J. M., 1981: Correlation and ages of Cenozoic chronostratigraphic units in Oregon and Washington. *Geol. Soc. America., Spec. Paper*, 184, p. 137–148.

Arnold, R. and Anderson, R., 1908: Preliminary report on the Coalinga oil district, Fresno and Kings counties, California. *U. S. Geol. Survey, Bull.*, 357, p. 1–142.

Bandy, O. L., 1972: Neogene planktonic foraminiferal zones, California, and some geologic implications. *Palaeogeography, Palaeoclimatology, Palaeoecology*, v. **12**, p. 131–150.

Barron, J. A., 1976: Marine diatom and silicoflagellate biostratigraphy of the type Delmontian Stage and the type *Bolivina obliqua* Zone, California. *U.S. Geol. Survey, Jour. Res.*, v. **4**, p. 339–351.

Durham, J. W. and MacNeil, F. S., 1967: Cenozoic migrations of marine invertebrates through the Bering Strait region. *In* D. M. Hopkins, ed., *The Bering land bridge*, Stanford Univ. Press, Stanford, Calif., p. 326–349.

Hopkins, D. M., 1965: Chetvertichnye morskie transgressii na Alyaske (Quaternary marine transgressions in Alaska). *In* Antropogenovye Period v Arktike i Subarktike (Anthropogene Period in the Arctic and Subarctic). *Nauchno-Issled. Inst. Geol. Arktiki Trudy*, v. **143**, p. 131–154.

Hopkins, D. M., 1967: Quaternary marine transgressions in Alaska. *In* D. M. Hopkins, ed., *The Bering land bridge*, Stanford Univ., Stanford, Calif., p. 47–90.

Ingle, J. C., Jr., 1967: Foraminiferal biofacies variation and the Miocene-Pliocene boundary in southern California. *Amer. Paleontology, Bull.*, v. **52**, no. 236, p. 217–394.

Keigwin, L. D., Jr., 1978: Pliocene closing of the Isthmus of Panama, based on biostratigraphic evidence from nearby Pacific Ocean and Caribbean Sea cores. *Geology*, v. **6**, p. 630–634.

Kleinpell, R. M., 1938: Miocene stratigraphy of California. *Amer. Assoc. Petrol. Geologists, Bull.*, Tulsa, Oklahoma, p. 1–450.

Marincovich, L., Jr., 1977: Cenozoic Naticidae (Mollusca: Gastropoda) of the northeastern Pacific. *Amer. Paleontology, Bull.*, v. **70**, no. 294, p. 169–494, pls. 17–42.

Moore, E. J., 1963 [1964]: Miocene marine mollusks from the Astoria Formation in Oregon. *U.S. Geol. Survey, Prof. Paper*, 419, p. 1–109, pls. 1–32.

Poore, R. Z., Barron, J. A. and Addicott, W. O., 1981, Biochronology of the northern Pacific Miocene. *IGCP-114 International Workshop on Pacific Neogene Biostratigraphy, Proc.*, Osaka, Japan, Nov. 25–29, 1981,p. 91–97.

Saito, T., 1976: Geologic significance of coiling direction in the planktonic foraminifera *Pulleniatina*. *Geology*, v. **4**, p. 305–309.

Smith, P. B., 1968: Paleoenvironment of phosphate-bearing Monterey Shale in Salinas Valley, California. *Amer. Assoc. Petrol. Geologists, Bull.*, v. **52**, no. 9, p. 1758–1791.

Stanton, R. J., Jr. and Dodd, J. R., 1970: Paleoecologic techniques—comparison of faunal and geochemical analyses of Pliocene paleoenvironments, Kettleman Hills, California. *Jour. Paleo.*, v. **44**, no. 6, p. 1092–1121.

Webb, S. D., 1978: A history of savanna vertebrates in the new world. Part 2. South America and the great interchange. *Ann. Reviens Ecol. Syst.*, v. **9**, p. 393–426.

Woodring, W. P., 1957: Geology and paleontology of Canal Zone and adjoining parts of Panama. *U.S. Geol. Survey, Prof. Paper*, 306-A, p. 1–145, pls. 1-23.

Woodring, W. P., 1966: The Panama land bridge as a sea barrier. *Amer. Philosoph. Soc., Proc.*, v. **110**, no. 6, p. 425–433.

I - 8

Western Pacific Molluscan Bio-Events and Their Relation to Neogene Planktonic Datum Planes

Ryuichi TSUCHI* and Tsugio SHUTO**

Introduction

Recent progress in defining the Neogene chronologic framework of the western Pacific has been made by incorporating detailed biostratigraphy with new data on planktonic microfossils, radiometric dating, and magnetostratigraphy. This has provided a basis for precise correlation of mollusc-bearing horizons and has also provided a more precise definition of the chronologic succession of benthic marine molluscan faunas.

In the western Pacific, including the Japanese Islands, marine molluscs have played an important role in the correlation and geologic age assignment of Neogene marine sediments. In Japan, the precise biochronologic position of many of the mollusc-bearing horizons and formations has been disclosed by associated planktonic microfossils and, also, those from juxtaposed horizons. As a result, it has become necessary to refine in time and space the distribution of molluscan species and faunas with reference to biohorizons of planktonic microfossils.

Chronological Successions of Molluscan Faunas in Japan

Neogene molluscan faunas in Japan can be grouped into four successive phases of different ages. Faunas in each phase are composed of two different water systems, namely, warm and cool, as seen in the Recent faunas of the warm Kuroshio and the cold Oyashio currents. The chronologic division of Tertiary faunas in Japan was first outlined by Otuka (1939). Recent ecologic and zoogeographic analyses of the Neogene faunas have been made by Chinzei (1978).

Molluscan faunas of the first phase are called Ashiya and Asagai-Poronai. The Ashiya fauna, taking molluscs of the Ashiya Group in north Kyushu as its type, is characterized by such shallow-water inhabitants as *Turritella infralirata, Glycymeris cisshuensis,* and *Venericardia vestitoides.* The fauna can be regarded as a warm-water one as it includes such genera as *Crassatella, Dosinia,* and *Pitar.* The fauna also includes some Asagai-Poronai elements, e.g., *Yoldia laudabilis* and *Venericardia subnipponica.* According to recent planktonic foraminiferal studies, the Ashiya Group is mostly included in Zone P. 21 of late Oligocene age (Tsuchi *et al.*, 1983). Leading members of the Ashiya fauna are also found in molluscs of the upper part of the Nichinan Group in the eastern part of Kyushu and the Hikokubo Group in the Chichibu Basin which are assignable, respectively, to Zone N. 4 and N. 6–7 (Shuto, 1963; Ujiie, 1959) of latest Oligocene and early Miocene age. The Ashiya fauna can, therefore, be regarded as ranging from late Oligocene to early Miocene.

The Asagai-Poronai fauna was named from molluscs of the Asagai Sandstone of the Joban coal field in central Honshu and from of the Poronai Group in central Hokkaido. The fauna is characterized by such shallow mud and sand dwellers as *Portlandia watasei, Yoldia asagaiensis,* and *Periploma besshoense.* As some elements of the fauna are found mixed in the Ashiya fauna, the former can be regarded as a cool-water equivalent of the latter, although the vertical range of the Asagai-Poronai fauna may extend down to horizons lower than the Ashiya (Kaiho, 1981; Ibaraki, 1983).

Molluscan faunas of the second phase are called Kadonosawa and Chikubetsu. The Kadonosawa fauna was originally named from molluscs of the lower part of the Kadonosawa Formation in northeastern Honshu. The fauna is dominated by tropical and subtropical species such as *Vicarya callosa japonica, Anadara ninohensis, Dosinia nomurai,* and *Siratoria siratoriensis.* The fauna of the Kurosedani Formation of the Yatsuo Group on the Sea of Japan side in central Honshu, an equivalent of the Kadonosawa fauna, contains such mangrove swamp dwellers as *Telescopium* and *Geloina,* as-

* *Geoscience Institute, Faculty of Science, Shizuoka University, Shizuoka 422, Japan*
** *Department of Geology, Faculty of Science, Kyushu University, Fukuoka 812, Japan*

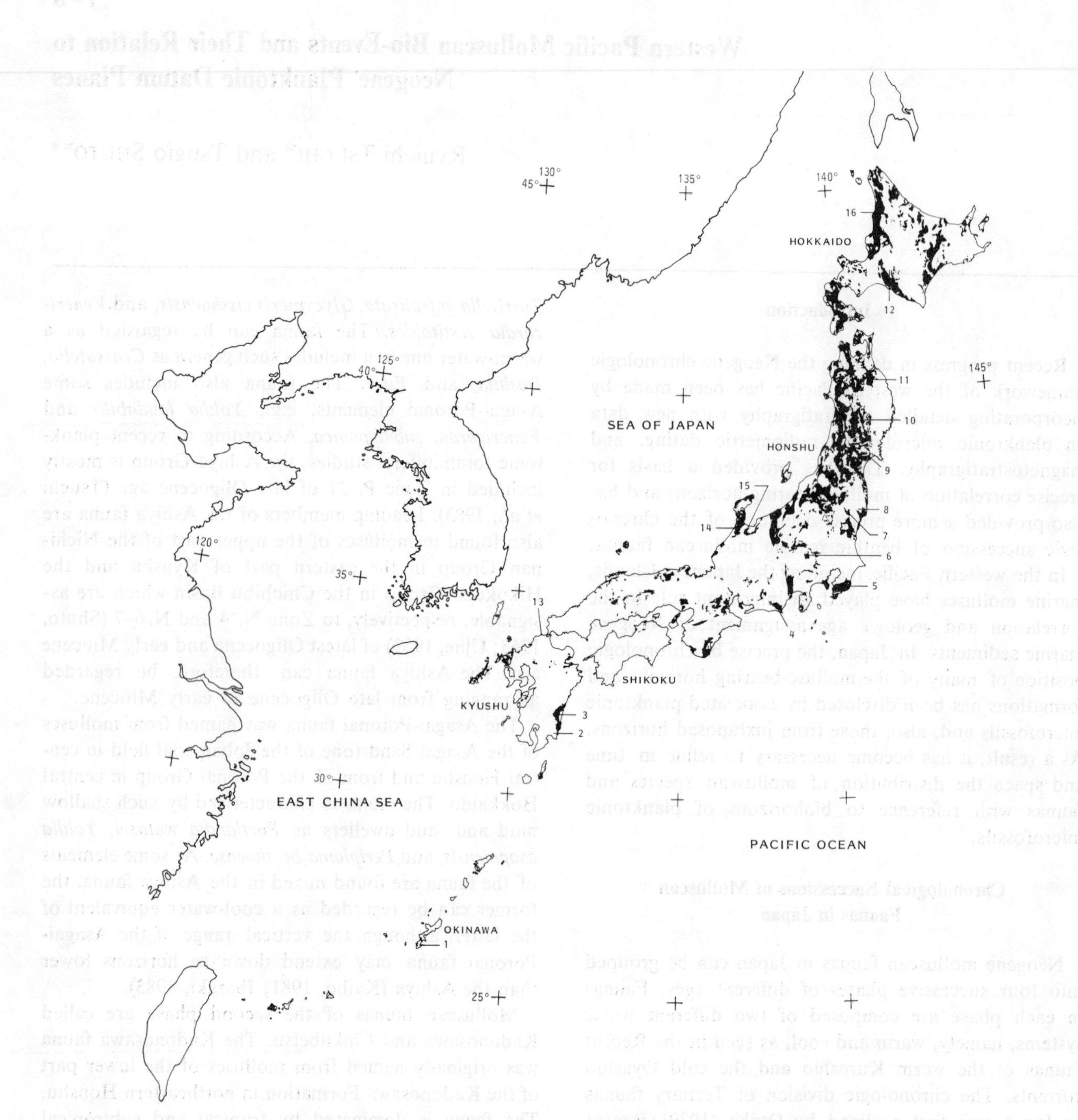

Fig. 1. Index map and distribution of marine Neogene strata in Japan, compiled from Geological map of Japan, Geol. Survey of Japan(1978). 1. Shimajiri, 2. Nichinan, 3. Miyazaki, 4. Kakegawa, 5. Boso Peninsula, 6. Chichibu, 7. Tanakura, 8. Asagai, 9. Sendai, 10. Kurosawa, 11. Kadonosawa, 12. Poronai, 13. Ashiya, 14. Omma, 15. Yatsuo, 16. Haboro.

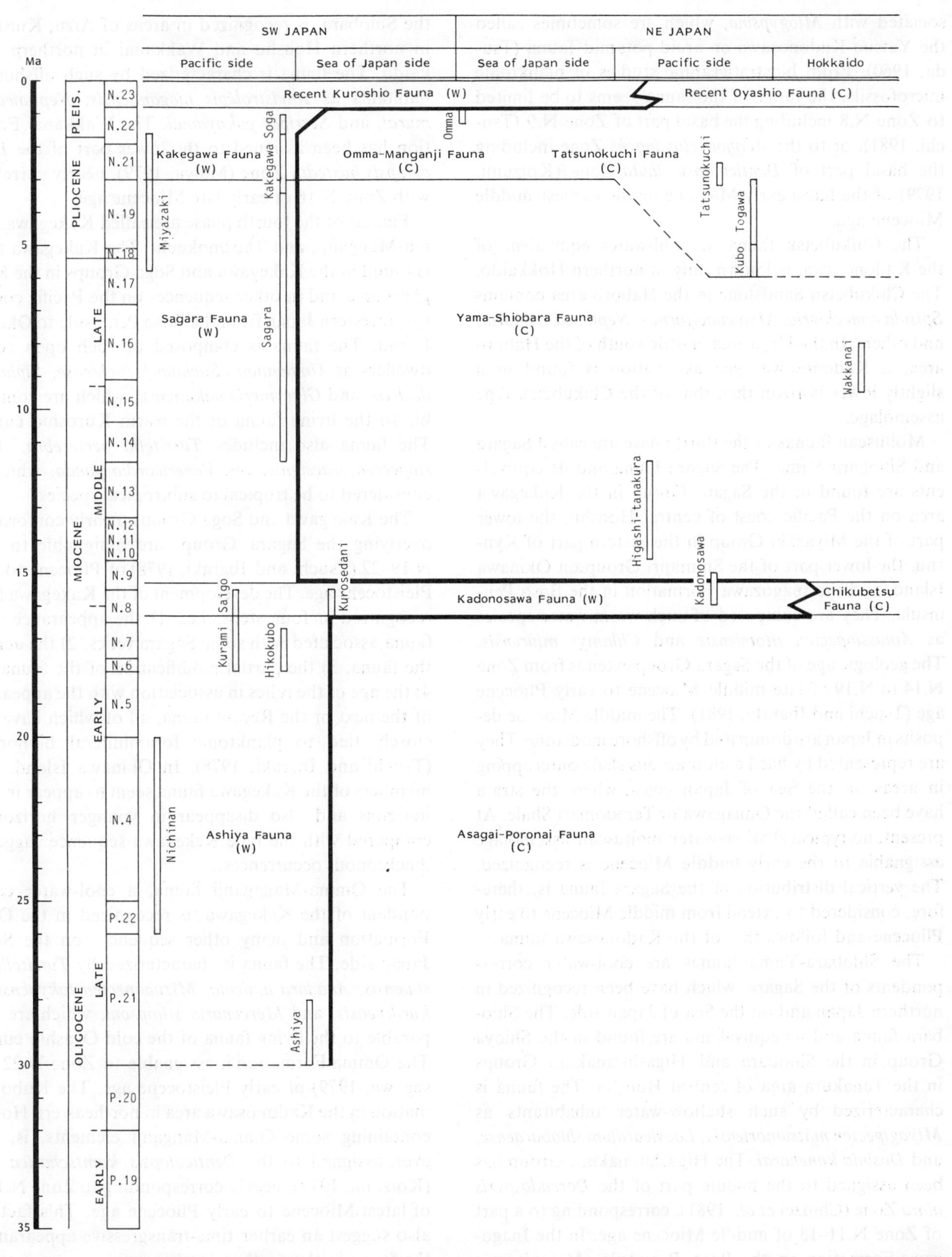

Fig. 2. Schematic diagram for the chronologic and geographic distribution of marine Neogene molluscan faunas in Japan. Chronologic distribution of selected mollusc-bearing formations is also indicated.

The Neogene time scale is that given by Tsuchi *et al.* (1981) and Tsuchi (1983), and that of the Paleogene is given by Hardenbol and Berggren (1978). (W) (C) indicate warm and cool water faunas, respectively.

sociated with *Miogypsina*, which are sometimes called the Yatsuo-Kadonosawa or arcid-potamid fauna (Tsuda, 1960). From biostratigraphic studies of planktonic microfossils, the range of the fauna seems to be limited to Zone N.8 including the basal part of Zone N.9 (Tsuchi, 1981), or to the *Actinocyclus ingens* Zone including the basal part of *Denticulopsis lauta* Zone (Koizumi, 1979) of the latest early Miocene or the earliest middle Miocene age.

The Chikubetsu fauna, a cool-water equivalent of the Kadonosawa, is known only in northern Hokkaido. The Chikubetsu Sandstone in the Haboro area contains *Spisula onnechiuria, Mya cuneiformis, Neptunea nomurai*, and others. In the Uryu area, a little south of the Haboro area, a Kadonosawa type association is found in a slightly lower horizon than that of the Chikubetsu type assemblage.

Molluscan faunas of the third phase are called Sagara and Shiobara-Yama. The Sagara fauna and its equivalents are found in the Sagara Group in the Kakegawa area on the Pacific coast of central Honshu, the lower part of the Miyazaki Group in the eastern part of Kyushu, the lower part of the Shimajiri Group on Okinawa Island, and the Inagozawa Formation in the Boso Peninsula. They are composed of such warm-water species as *Amussiopecten iitomiensis* and *Chlamys miurensis*. The geologic age of the Sagara Group extends from Zone N.14 to N.19 of late middle Miocene to early Pliocene age (Tsuchi and Ibaraki, 1981). The middle Miocene deposits in Japan are dominated by offshore mudstone. They are represented by hard diatomaceous shale outcropping in areas on the Sea of Japan coast, where the strata have been called the Onnagawa or Teradomari Shale. At present, no typical shallow-water molluscan assemblage assignable to the early middle Miocene is recognized. The vertical distribution of the Sagara fauna is, therefore, considered to extend from middle Miocene to early Pliocene and follows that of the Kadonosawa fauna.

The Shiobara-Yama faunas are cool-water correspondents of the Sagara, which have been recognized in northern Japan and on the Sea of Japan side. The Shiobara fauna and its equivalents are found in the Shioya Group in the Shiobara and Higashitanakura Groups in the Tanakura area of central Honshu. The fauna is characterized by such shallow-water inhabitants as *Miyagipecten matsumoriensis, Laevicardium shiobaraense*, and *Dosinia kaneharai*. The Higashitanakura Group has been assigned to the middle part of the *Dorcadospyris alata* Zone (Chinzei *et al.*, 1981), corresponding to a part of Zone N.11–13 of middle Miocene age. In the Inagozawa Formation on the Boso Peninsula, *Miyagipecten matsumoriensis* occurs in association with *Amussiopecten iitomiensis*, a common constituent of the Sagara fauna. The Yama fauna, an equivalent offshore assemblage of the Shiobara, is recognized in areas of Aizu, Kurosawa in northern Honshu and Wakkanai in northern Hokkaido. The fauna is characterized by such offshore inhabitants as *Ancistrolepis mogamiensis, Neptunea nomurai*, and *Serripes yokoyamai*. The Wakkanai Formation has been assigned to the lower part of the *Denticulopsis hustedtii* Zone (Maiya, 1979), nearly correlative with Zone N.16 of early late Miocene age.

Faunas of the fourth phase are called Kakegawa, Omma-Manganji, and Tatsunokuchi. The Kakegawa fauna is found in the Kakegawa and Soga Groups in the Kakegawa area and in other sequences on the Pacific coast of southwestern Japan from the Boso Peninsula to Okinawa Island. The fauna is composed of such open coastal dwellers as *Umbonium (Suchium) suchiense, Siphonalia declivis*, and *Glycymeris nakamurai*, which are comparable to the living fauna of the warm Kuroshio current. The fauna also includes *Turritella perterebra, Amussiopecten praesignis*, and *Venericardia panda*, which are considered to be tropical to subtropical species.

The Kakegawa and Soga Groups, nearly conformably overlying the Sagara Group, are assignable to Zone N.19–22 (Tsuchi and Ibaraki, 1978) of Pliocene to early Pleistocene age. The development of the Kakegawa fauna recognized in four steps, i.e., 1) the appearance of the fauna associated with some Sagara relics, 2) the acme of the fauna, 3) the partial modification of the fauna, and 4) the age of the relics in association with the appearance of the next or the Recent fauna, all of which have been closely tied to planktonic foraminiferal biohorizons (Tsuchi and Ibaraki, 1978). In Okinawa Island, some members of the Kakegawa fauna seem to appear in older horizons and also disappear in younger horizons as compared with the type Kekegawa sequence suggesting diachronous occurrences.

The Omma-Managanji fauna, a cool-water correspondent of the Kakegawa, is recognized in the Omma Formation and many other sequences on the Sea of Japan side. The fauna is characterized by *Turritella saishuensis, Anadara amicula, Mizuhopecten tokyoensis hokurikuensis*, and *Mercenaria stimpsoni*, which are comparable to the living fauna of the cold Oyashio current. The Omma Formation is assignable to Zone N.22 (Hasegawa, 1979) of early Pleistocene age. The Kubo Formation in the Kadonosawa area in northeastern Honshu, containing some Omma-Manganji elements, is, however, assigned to the *Denticulopsis kamtschatica* Zone (Koizumi, 1979), nearly corresponding to Zone N.17–18 of latest Miocene to early Pliocene age. This fact may also suggest an earlier time-transgressive appearance of the fauna in the northern region.

The Tatsunokuchi fauna, another cool-water correspondent of the Kakegawa fauna, is typically found in the Tatsunokuchi Formation near Sendai, in northern

Honshu. It is also known from northeastern Honshu and Hokkaido, sometimes in association with the Omma-Manganji elements. The Tatsunokuchi fauna is characterized by *Fortipecten takahashii, Anadara tatunokutiensis, Dosinia tatunokutiensis,* and *Pseudoamiantes sendaicus.* The Tatsunokuchi Formation has been assigned to the upper part of *Denticulopsis seminae* var./ *Denticulopsis kamtschatica* Zone (Tsuchi *et al.*, 1981), nearly corresponding to Zone N.19–21 of Pliocene age. The Omma-Manganji and the Tatsunokuchi faunas are thought to be separated zoogeographically from each other.

The distribution of the above-mentioned Neogene molluscan faunas in Japan in time and space is schematically summarized in Fig. 2. The microplanktonic time scale of the Neogene used here is that given by Tsuchi *et al.* (1981) and Tsuchi (1983). The most pronounced bio-event of the Neogene is the predominance of warm-water molluscs in the latest early Miocene or the earliest middle Miocene at about 16 Ma in a limited duration of about 1 Ma, when tropical-subtropical shallow-water species represented by the Kadonosawa fauna prevailed throughout almost all of the Japanese Islands, except northern Hokkaido. A paleogeographic map at that time has been drawn by the IGCP-114 National working group of Japan (1981). Another invasion of warmwater molluscs is represented by the appearance of the Kakegawa fauna during the middle Pliocene, at 3–2 Ma, when species of tropical-subtropical aspect flourished on the Pacific coast of southwestern Japan.

With the precise age assignment of mollusc-bearing horizons by means of related microplanktonic bio-events, it is possible to estimate evolutionary rates of some species successions. The *Suchium suchiense* bio-series, a Japonic endemic form-group typically found in the Kakegawa area on the Pacific coast of central Japan, includes three taxa: *Suchium suchiense paleosuchiense,* which appeared in the Pliocene at about 3 Ma, evolved successively into *S. suchiense* s.s. at 2.4 Ma, and *S. suchiense subsuchiense* at 1.9 Ma, which can be found up to nearly 1.0 Ma (Tsuchi *et al.*, 1981).

Western Pacific Molluscan Bio-Events

As already expressed in the preceding chapter, the predominance of warm-water species represented by the Kadonosawa fauna in Japan seems to be traceable as a major circum-North Pacific molluscan event that occurred during the *Actinocyclus ingens* Zone or near the base of the *Denticulopsis lauta* Zone of latest early Miocene or earliest middle Miocene age of about 16 Ma. This event is recorded from the Chengogsa fauna in South Korea (Yoon, 1976), Uglegorskian fauna in South Sachalin (Gladenkov, 1983), Ilynian fauna in Kamchatka (Menner and Gladenkov, 1979), Newportian and the middle Temblor faunas in the Northeast Pacific (Addicott, 1976; Marincovich, 1983), etc. Such climatic events provide a basis for evaluating the chronostratigraphic relationship of the molluscan units of different geographic areas.

Remarkable differentiation and flourishing of turritellid, pectinid, dosiniid, and anadaran species during the Neogene in Japan have been recognized in two different ages of "early Miocene" and "early Pliocene" accompanied by marine transgressions. The former is nearly correlative with the age of the Kadonosawa fauna and the latter is roughly correlative with the age of the appearance and the acme of the Kakegawa fauna.

Anadara (Hataiarca) kakehataensis and / or *A. (Anadara) makiyamai,* found in association with *Vicarya* in the Kadonosawa fauna and its equivalents in Japan, are widely recognized in Korea. Such arcid-potamid associations are also known from early Miocene sequences of Indonesia and the Philippines. *Anadara suzukii,* described from the Pliocene Tonohama Group in south Shikoku, occurs in the Kakegawa area southward to Formosa, and *Anadara sedanensis* from the Pliocene of Indonesia is widely distributed and is found also in south Shikoku, Japan (Noda, 1978).

Fortipecten takahashii, a leading member of the Tatsunokuchi fauna of Japan and described originally from south Sakhalin, has a wide geographic distribution from northern Honshu to Hokkaido, Sakhalin, and Kamchatka (Masuda, 1962). Formations in Sendai and Haboro containing this species have been assigned to the upper part of *Denticulopsis seminae* var. / *Denticulopsis kamtschatica* Zone (Koizumi, 1978; Tsuchi *et al.*, 1981) nearly corresponding to planktonic foraminiferal zones N.19–21 of Pliocene age.

Late Neogene sequences of Indonesia and the Philippines have a number of gastropods in common, while those of Japan and the Philippines share a few common species which are generally useless for correlation because of their long vertical ranges. However, a number of Pliocene species of Japanese mainland and the Philippines occur in the Ryukyu Islands. Included are such gastropods as *Turris panaiensis* and *Makiyamaia subdeclivis.* These species are expected to serve as the tie between the faunas of Japan and those of the Philippines (Shuto, 1981).

Some Turritellid bio-series in different regions exhibit progressive increases in density and strength of secondary and tertiary spiral sculpture, suggesting a parallel evolution; sudden and considerable change in sculpture in each bio-series may be synchronous (Kotaka, 1978). Such Turritellid bio-series also have the possibility of aiding in inter-regional correlation.

As mentioned above, some of the characteristic species

having extensive geographic distributions would be useful for inter-regional correlation, and the horizons in which they occur have a good chance of being correlative. In any case, they would have more reliability if their precise chronologic positions were reexamined in the light of planktonic microfossils.

Acknowledgements

Dr. Warren O. Addicott of U.S. Geological Survey reviewed the manuscript and also gave us valuable suggestions on molluscan bio-events. Our deep appreciation is due to him for his kind help.

References

Addicott, W.O., 1969: Tertiary climatic change in the marginal northeastern Pacific Ocean. *Science*, v. **165**, p. 583–586.

Chinzei, K., 1978: Neogene molluscan faunas in the Japanese Islands: An ecologic and zoogeographic synthesis. *The Veliger*, v. **21**, no. 2, p. 155–170.

Chinzei, K., Iwasaki, Y. and Matsui, S., 1981: Neogene of the Tanakura area, Fukushima Prefecture: Its stratigraphy and faunas. *Guide book for excursion, Ann. Meet. Geol. Soc. Japan 1981 Tokyo*, p. 87–102.

Hardenbol, J. and Berggren, W.A., 1978: A new Paleogene numerical time scale. *In* Cohee, G.V. *et al.*, eds., *Contr. Geologic Time Scale, Studies in Geology*, AAPG., no. 6, p. 213–234.

Hasegawa, S., 1979: Foraminifera of the Himi Group, Hokuriku Province, central Japan. *Tohoku Univ., Sci. Rep. 2nd. ser. (Geol.)*, v. **49**, no. 2, p. 89–163.

Ibaraki, M., 1983: Occurrences of Middle Eocene planktonic foraminifers in a mollusca-bearing horizon and limestone of the Takisawa Formation, the Setogawa Group. *Geol. Soc. Japan, Jour.*, v. **89**, p. 57–59.

IGCP-114 National Working Group of Japan, Paleogeographic map of the Japanese Islands during 16–15Ma, the earliest Middle Miocene. *In* Tsuchi, R., ed, *Neogene of Japan,* IGCP-114 Natl. Working Group of Japan, p. 105–109.

Kaiho, K., 1981: Planktonic foraminifera from Paleogene of Hokkaido (Poronai, Utsunai and Nuibetsu Formations) *Ann. Meet. Geol. Soc. Japan 1981 Tokyo, Abst.* p. 263.

Koizumi, I., 1973: The stratigraphic ranges of marine planktonic diatom and diatom biostratigraphy in Japan. *Geol. Soc. Japan, Mem.*, no. 8, p. 35–44

Koizumi, I., 1979: Kadonosawa-Sannohe area (2) *In* Tsuchi, R. ed., *Fundamental data on Japanese Neogene Bio- and Chronostratigraphy*, IGCP-114 Natl. W. G. Japan, p. 53–55.

Kotaka, T., 1978: World-wide correlation based on Turritellid phylogeny. *The Veliger*, v. **21**, no. 2, p. 189–196.

Maiya, S., 1979: Embetsu area, Hokkaido. *In* Tsuchi, R. ed. *Fundamental Data on Japanese Neogene Bio- and Chronostratigraphy*, IGCP-114 Natl. W.G. Japan, p. 57–58.

Marincovich, L., Jr., 1983: Eastern Pacific molluscan bio-events and the relation to Neogene planktonic datum planes. *In* Ikebe, N. and Tsuchi, R., eds., *Pacific Neogene datum planes, Univ. Tokyo Press*, p. 69–73.

Masuda, K., 1962: Notes on the Tertiary Pectinidae of Japan *Tohoku Univ., Sci. Rep., Spec. Vol.*, v. **5**, p. 159–193.

Masuda, K., 1978: Neogene Pectinidae of the Northern Pacific. *The Veliger*, v. **21**, no. 2, p. 197–202.

Menner, V.V. and Gladenkov, Y.B., 1979: On the creation of a correlation scheme for the Neogene of the northern part of the Circumpacific belt. *16th Pac. Sci. Cong., Abst., Sec.B-3*, v. **2**, p. 93–94.

Noda, H., 1978: Neogene anadaran distribution in Japan and Southeast Asia. *Univ. Tsukuba. Inst. Geosci., Ann. Rep.*, no. 4, p. 33–37.

Otuka,Y., 1939: Tertiary crustal deformation in Japan, with short remarks on Tertiary palaeogeography. *Jubilee Pub. Commem. Prof. H.Yabe, M.I.A. 60th Birthday*, v. **1**, p. 481–520.

Shuto,T., 1963: Geology of the Nichinan area, with the special reference to the Takachiho disturbance. *Kyushu Univ. Fac. Sci., Mem., ser.D*, v. **6**, no. 2, p. 135–166.

Shuto,T., 1981: Biogeography and correlation of late Neogene gastropod faunas of Southeast and East Asia. *IGCP-114 Int. Workshop Pac. Neog. Biostrat. Osaka, 1981,Proc.* p. 141–142.

Tsuchi,R., 1981: A topic of marine biogeography of Japan in the Early/Middle Miocene. *Kaseki(Fossils)*, Pal. Soc. Japan, no. 30, p. 1–5.

Tsuchi,R., 1983: Neogene bio- and chronostratigraphy in Japan. *Jap. Assoc. Petrol. Tech.*, v. **48**, no. 1, p. 35–48.

Tsuchi,R. and Ibaraki,M., 1978: Late Neogene succession of molluscan fauna on the Pacific coast of southwestern Japan, with reference to planktonic foraminiferal sequence. *The Veliger*, v. **21**, no. 2, p. 217–222.

Tsuchi,R. and Ibaraki,M., 1981: Kakegawa area. *In* Tsuchi, R. ed., *Neogene of Japan–Its biostratigraphy and chronology-*. IGCP-114 Natl. W.G. Japan, p. 37–41.

Tsuchi,R. and IGCP-114 National Working Group of Japan, 1981: Bio- and chronostratigraphic correlation of Neogene sequences in the Japanese Islands. *In* Tsuchi,R., ed., *Neogene of Japan*, IGCP-114 Natl. Working Group of Japan, p. 91–104.

Tsuchi,R., Shuto,T. and Ibaraki,M., 1983: Geologic ages of the Ashiya Group and the Ashiya Fauna as determined by planktonic foraminifera. *Ann. Meet. Geol. Soc. Japan. Kagoshima, 1983, Abst.*, p. 125.

Tsuchi,R., Takayanagi,Y. and Shibata,K., 1981: Neogene bio-events in the Japanese Islands. *In* Tsuchi,R., *Neogene of Japan,* IGCP-114 Natl. W. G. of Japan, p. 15–32.

Tsuda,K., 1960: Paleo-ecology of the Kurosedani Fauna. *Niigata Univ. Fac. Sci., Jour.*, v. **2**, no. 3, p. 171–203.

Ujiie,H. and Iijima,H., 1959: Miocene foraminifera from the Akahira Group, Saitama Pref., Japan. *Chichibu Mus. Nat. Hist., Bull.*, no. 9, p. 69–94.

Ujiie,H., Saito,T., Kent,D., Thompson.P.R., Okada,H.,

Klein, G.de V., Koizumi, I., Harper, H.E., Jr., and Sato, T., 1977: Biostratigraphy, paleomagnetism and sedimentology of Late Cenozoic sediments in northwestern Hokkaido, Japan. *Natl. Sci. Mus., Bull. ser. C(Geol.)*, v. **3**, no.2, p.49–102.

Yoon, S., ed., 1981: Neogene geology of southern Korea, *Guide book for excursion, IGCP-114 lst. Int. Workshop Pac. Neog. Biostrat. Osaka 1981*, p. 1–26.

I-9

Neogene Magnetostratigraphy and Chronostratigraphy

William A. BERGGREN,* Hisao NAKAGAWA,[2*] Tsunemasa SAITO,[3*]
N. J. SHACKLETON,[4*] and Ken SHIBATA[5*]

For the Neogene, there is relatively little difference between the various current magnetic anomaly time-scales. For instance, the oldest current estimate for the base of Anomaly 6 is 24.76 Ma, and the youngest estimate 23.75 Ma. Radiometric dating is not sufficiently accurate to distinguish between these extremes. We have adopted the time scale of Ness *et al.* (1980) because it is up-to-date, clearly explained in their publication, and geologically acceptable for the Paleogene as well as for the Neogene.#

In every piston core and DSDP site in which magnetic measurements have been made, an element of interpretation is necessary to identify particular anomalies in the imperfectly recovered sedimentary record. At present this represents an uncertainty which is relatively larger than uncertainties in the details of the chronology of Ness *et al.* (1980). We believe that these interpretational uncertainties, together with inevitable biostratigraphic uncertainties, are small enough that our time scale still represents the best available in 1981.

We have attempted to use all the relevant data available, from the Atlantic and from Italian sections as well as from the Pacific. At present there is no sufficient data available to cover the whole Neogene using only Pacific data; there is almost no direct magnetostratigraphic correlation for Pacific calcareous microfossils older than latest Miocene. Figure 1 shows a comparison between the marine magnetic anomaly time scale (Ness *et al.*, 1980), the magnetostratigraphy of Pacific deep-sea sediments (Theyer *et al.*, 1978) and of the Mediterranean Neogene stage stratotype sections (Nakagawa *et al.*, 1974, unpublished results).

References

Nakagawa, H., Kitamura, N., Takayanagi, Y., Sakai, T., Oda, M., Asano, K., Niitsuma, N., Takayama, T., Matoba, Y., and Kitazato, H., 1974: Magnetostratigraphic correlation of Neogene and Pleistocene between the Japanese Islands, central Pacific and Mediterranean regions. *1st Internat. Congr. Pacific Neogene Stratigr., Proc.* Tokyo 1976, p. 285–310.

Ness, G., Levi, S. and Couch, R., 1980: Marine magnetic anomaly timescale for the Cenozoic and Late Cretaceous: A precis, critique, and synthesis. *Geophys. Space Phys., Rev.*, v. **18**, p. 753–770.

Theyer, F., Mato, C. Y. and Hammond, S. R., 1978: Paleomagnetic and geochronologic calibration of latest Oligocene to Pliocene radiolarian events, Equatorial Pacific. *Mar. Micropaleontol.*, v. **3**, p. 377–395.

Postscript: But see the critique by Berggren, this volume, concerning the methodology employed by Ness *et al.* (1980) in the formulation of their time scale.

* *Woods Hole Oceanographic Institution, Woods Hole, Massachusetts 02543, U.S.A.*
2* *Institute of Geology and Paleontology, Tohoku University, Sendai 980, Japan*
3* *Department of Earth Sciences, Yamagata University, Yamagata 990, Japan*
4* *Sub-Department of Quaternary Research, The Godwin Laboratory, Free School Lane, Cambridge, England CB2 3RS, U.K.*
5* *Geological Survey of Japan, Yatabe-machi, Tsukuba-gun, Ibaraki 305, Japan*

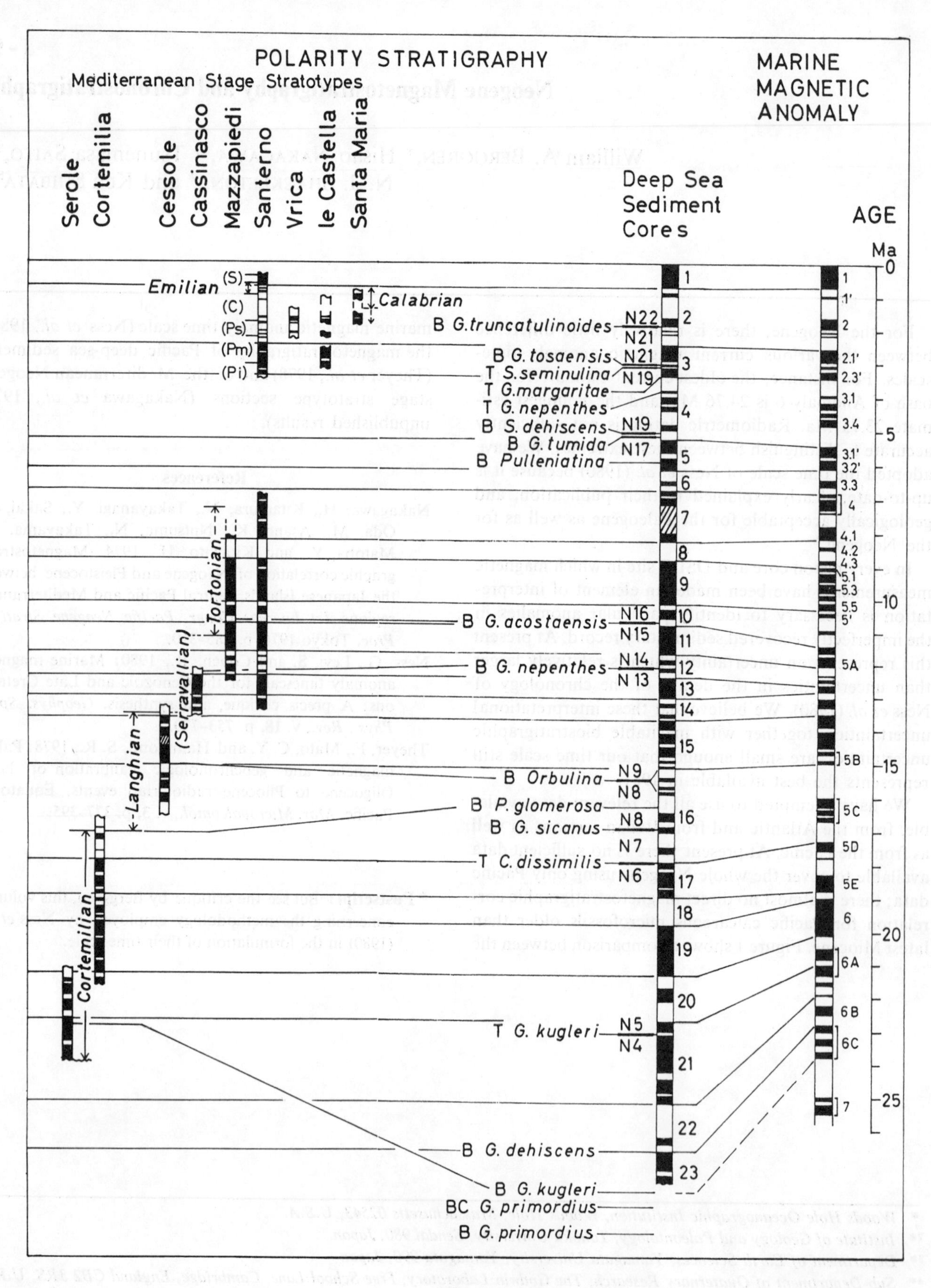
POLARITY STRATIGRAPHY
Mediterranean Stage Stratotypes
Serole
Cortemilia
Cessole
Cassinasco
Mazzapiedi
Santerno
Vrica
le Castella
Santa Maria
MARINE MAGNETIC ANOMALY
Deep Sea Sediment Cores
AGE
Ma
Emilian
Calabrian
Tortonian
Serravallian
Langhian
Cortemilian
(S)
(C)
(Ps)
(Pm)
(Pi)
B G.truncatulinoides
B G. tosaensis
T S. seminulina
T G.margaritae
T G. nepenthes
B S. dehiscens
B G. tumida
B Pulleniatina
B G. acostaensis
B G. nepenthes
B Orbulina
B P. glomerosa
B G. sicanus
T C. dissimilis
T G. kugleri
B G. dehiscens
B G. kugleri
BC G. primordius
B G. primordius
N22
N21
N19
N17
N16
N15
N14
N13
N9
N8
N7
N6
N5
N4
1
1'
2
2.1'
2.3'
3.1
3.4
3.1'
3.2'
3.3
4
4.1
4.2
4.3
5.1
5.3
5.5
5'
5A
5B
5C
5D
5E
6
6A
6B
6C
7
0
5
10
15
20
25

Fig. 1

Radiometric Dating Related to Pacific Neogene Planktonic Datum Planes

Ken SHIBATA,* Susumu NISHIMURA[2*] and Kiyotaka CHINZEI[3*]

Precise biostratigraphic positioning and reliable dating are the basic requirements for a sample used for calibrating a time scale. To obtain a reliable age, a careful petrographic examination is indispensable. Degree of weathering and alteration should always be examined and the freshest possible sample should be selected, although this is sometimes difficult.

Dating results for Japanese Neogene volcanic rocks indicate that whole-rock samples which contain zeolite as an alteration product give ages younger than those obtained from fresh samples. Whole rock samples composed mainly of glass tend to be younger, and those of loose tuff also give ages much younger than expected. On the contrary, biotite separated from tuff ocasionally contains grains of detrital origin, giving an older age than expected. Most of K-Ar ages on glauconite from the Japanese Neogene are younger.

Taking account of these lines of evidence, we carefully selected K-Ar ages well controlled by biostratigraphic data, for the Japanese and central and southern Pacific Neogene including Australia and New Zealand. In addition, fission-track ages for the Japanese Neogene were selected on the same basis. The K-Ar and fission-track ages are listed in Tables 1 and 2, respectively. Ages are recalculated, if necessary, by the new decay constants (Steiger and Jäger, 1977), and $\lambda = 7.03 \times 10^{-17}$/y for fission-track age.

The number of ages thus selected is not many, and some of them are not so well controlled. Nevertheless, these ages are correlated to the reference scale that is made by combining the magnetic anomaly scale of Ness *et al.* (1980) with the important datum levels (Fig. 1). K-Ar ages are generally concordant with biostratigraphic data, whereas some of fission-track ages are very discordant. Ages much older than estimated are probably caused by detrital zircons. Further collaboration of geochronologists and biostratigraphers is needed to refine the Neogene timescale.

Biostratigraphic and lithostratigraphic positions of the Japanese Neogene rocks dated are summarized below.

17–1 Hosoya Tuff (Kakegawa area) 1.9 ± 0.4 (Ma, omitted hereafter): The tuff located at 30m above the first appearance level of *Globorotalia truncatulinoides*, and above the top level of *Globoquadrina asanoi* in the Kakegawa area (Ibaraki and Tsuchi, 1974).

17–2 Iozumi Tuff (Kakegawa area) 2.3 ± 0.5: Upper part of the *Globoquadrina asanoi* occurrence range in the Kakegawa area, about 100m below the horizon of change in coiling direction from dextral to sinistral in *Plleniatina* spp. (*Pulleniatina* DS horizon) (Ibaraki and Tsuchi, 1976).

16–1 A tuff in the middle part of the Wakimoto Formation (Oga Peninsula) 2.8 ± 0.6: Near the top occurrences of *Globorotalia inflata inflata* and *Ommatartus tetrathalamus* in the Oga section, within the Brunhes Normal Epoch. (Kitazato, 1975).

17–3 Arigaya Tuff (Kakegawa area) 3.2 ± 0.6: Approximately 70m below the *Globoquadrina asanoi* base level, 120m above the lowest occurrence of *Globorotalia tosaensis*. (Ibaraki and Tsuchi, 1976).

16–2 A tuff in the middle part of the Kitaura Formation (Oga Peninsula) 3.5 ± 0.7: The tuff is named Km-4 by Kitazato (1975), 300m above the top occurrence of *Globoquadrina asanoi*, within the zone of dextral *Neogloboquadrina pachyderma*. According to Koizumi and Kanaya (1977) the tuff is located at about the middle part of their *Actinocyclus oculatus* zone.

18 Hirugaya Tuff (Kakegawa area) 4.1 ± 0.2: Approximately 50m below the top datum of *Globorotalia margaritae* in the Kakegawa area, and above the level of *Pulleniatina* SD. (Ibaraki, personal communication, Ibaraki and Tsuchi, 1976).

19 HK-Tuff (Miura Peninsula) 3.7 ± 1.1: The upper part of the Zushi Formation. The tuff is located at about the midpoint between the base of

* *Geological Survey of Japan, Yatabe-machi, Tsukuba-gun, Ibaraki 305, Japan*
[2*] *Department of Geology and Mineralogy, Faculty of Science, Kyoto University, Kitashirakawa, Sakyo-ku, Kyoto 606, Japan*
[3*] *Geological Institute, Faculty of Science, University of Tokyo, 7-3-1 Hongo, Bunkyo-ku, Tokyo 113, Japan*

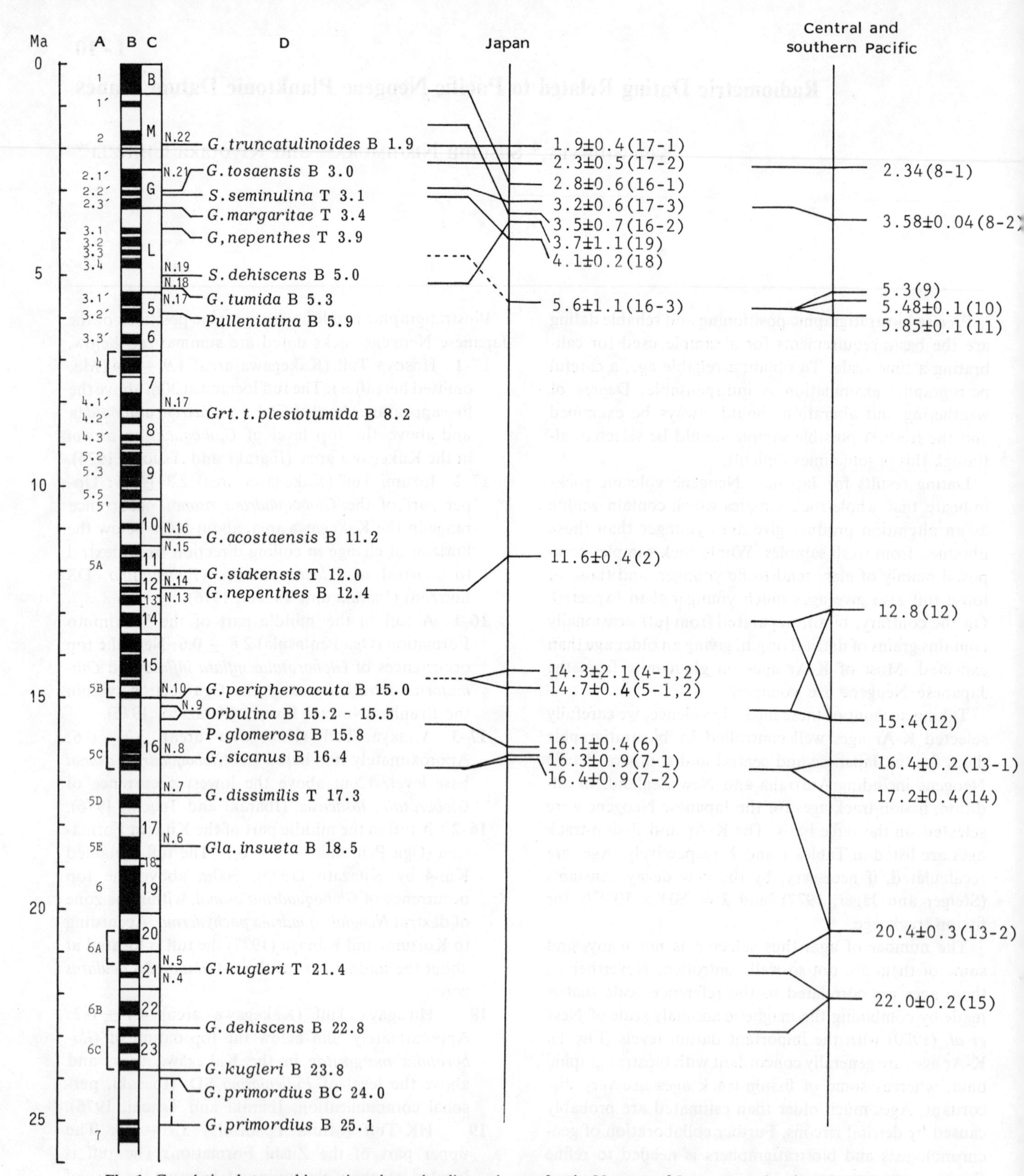

Fig. 1. Correlation between biostratigraphy and radiometric ages for the Neogene of Japan, central and southern Pacific.

A: Magnetic anomaly number (Talwani *et al.*, 1971 and others)

B: Marine magnetic anomaly timescale of NLC-80 (Ness *et al.*, 1980)

C: Magnetic polarity epoch (Foster and Opdyke, 1970 and others)

D: Datum levels of selected planktonic foraminifers tied to magnetic polarities and their estimated ages based upon NLC-80. B: Base datum; T: Top datum; BC: Base of common occurrence.

Globorotalia tumida and the base of *Sphaeroidinella dehiscens* (Yoshida, 1979).

16–3 Anzenji Tuff in Oga Peninsula (upper part of Kitazato's Mm) 5.6 ± 1.1: The lower part of the Funakawa Formation. Around the boundary horizon between the *Denticulopsis kamtchatica* zone and the *D. kamtchatica* / *D. seminae fossilis* zone of Koizumi and Kanaya (1977), below the base of *D. seminae fossilis*).

2 Fuiiki Tuff (Tomioka area) 11.6 ± 0.4: Middle part of the Tomioka Group. The tuff is located a few meters below the lowest occurrence of *Globigerina nepenthes* in the Tomioka area (Chiji and Konda, 1978; Takayanagi *et al.*, 1978).

4–1,2; 5–1,2 Kumano Acidic Rocks, 14.3 ± 2.1, 14.7 ± 0.4: Rhyolite intruded in the sediments containing *Orbulina universa.* (Ikebe and Chiji, 1971, Ikebe *et al.*, 1972)

6 Iwakurayama Rhyolito (Noto Peninsula) 16.1 ± 0.4: A rhyolite lava flow considered to be interbedded between the Hojuji Diatomite (lower) and the Iida Diatomite (Ishida, 1959, Kaseno, 1965). It is located above the basal occurrences of *Globigerinoides sicanus* and *Sphenolithus heteromorphus,* probably close to the top of *S. heteromorphus* (Takayama, 1977; Kami *et al.*, 1981).

7–1,2 Iwaine Andesite (Toyama area) 16.3 ± 0.9, 16.4 ± 0.9: 300 to 400m stratigraphically below the first appearance of *Orbulina universa,* and below the horizon of *Globigerinoides sicanus* in the Yatsuo section (Hasegawa, 1979)

Abbreviation for Tables 1 and 2

F: Formation
G: Group
FAD: First appearance datum
LAD: Last appearance datum
DS, SD: Horizon of coiling change in the test of planktonic foraminifera from dextral to sinistral and sinistral to dextral

Table 1. K-Ar ages well controlled by biostratigraphic data for the Pacific Neogene.

Locality, Formation	Material dated	Biostratigraphy	Age (Ma)	Reference	Data No.
Japan					
Aomori, Isoyama F.	Sandstone, glauconite	Upper part of *D.h.* Zone	6.6	Ueda and Suzuki (1973)	1–1
Akita, Nadakayama F.	Sandstone, glauconite	*D.h.* Zone	8.5	Ueda and Suzuki (1973)	1–2
Akita, Nishikurosawa F.	Sandstone, glauconite	Upper part of *D.h-D.l.* Zone	8.6	Ueda and Suzuki (1973)	1–3
Akita, Onnagawa F.	Sandstone, glauconite	Upper part of *D.h-D.l.* Zone	9.6	Ueda and Suzuki (1973)	1–4
Gunma, Fujiki Tuff	Tuff, biot.	*Gna. nepenthes*	11.6 ± 0.4	Shibata *et al.* (1979)	2
Hokkaido, Tokushibetsu And.	Andesite, whole rock	*Desmostylus*	13.7 ± 0.7	Shibata *et al.* (1981b)	3–1
〃	〃 , ess. lens		13.8 ± 0.9	〃	3–2
Nara, Ominesan	Granodiorite, biot.		14.3 ± 2.1	Shibata and Nozawa (1967)	4–1
Wakayama, Kumano Acid Rock	Granite porphyry, biot.	Upper limit of *Orbulina* datum	14.3 ± 2.1	〃	4–2
〃	〃 , sandine		14.7 ± 0.4	Kawai and Hirooka (1966)	5–1
〃	〃 , biot.		14.7 ± 0.4	〃	5–2
Noto, Najimi Mudstone	Sandstone, glauconite	Upper part of *D.1.* Zone	15.2	Ueda and Suzuki (1973)	1–5
〃 , Iwakurayama Rhy.	Rhyolite, biot.	Below *Orbulina* datum (N.8)	16.1 ± 0.4	Shibata *et al.* (1981a)	6
Toyama, Iwaine F.	Andesite, whole rock	300–400m below top level of *Gds. sicanus*	16.3 ± 0.9	Shibata (1973)	7–1
〃	〃		16.4 ± 0.9	〃	7–2
Central and southern Pacific					
Australia, Glenelg River	Basalts, whole rock	Underlying N22	2.34	Singleton *et al.* (1976)	8–1
〃 , Portland	Basalts, whole rock	Overlying N19	3.58 ± 0.04	〃	8–2
New Zealand, Chatham Island	Basalts, hornb.	Betw. LAD of *Grt. conomiozea* (sensu lato) and FAD of *Grt. crassaformis* (unkeeled)	5.3	Grindley *et al.* (1977)	9
Fiji, Sambeto volcanics	Andesites, whole rock, biot.	Pre-N18	5.48 ± 0.1	McDougall (1963)	10
〃 , Namosi And.	Andesites, whole rock, hornb.	Pre-N18 (poss, N17)	5.85 ± 0.1	Gill and McDougall (1973)	11
New Guinea, Lower Tf	Lavas, whole rock, min.	Tf_{1-2} (N9–N12)	12.8 – 15.4	Page and McDougall (1970)	12
New Zealand, Muriwai And.	Andesite, whole rock	Immed. below FAD of *Gds. sicanus*	16.4 ± 0.2	Adams (1975)	13–1
〃	Andesites, whole rock plag.		17.2 ± 0.4	Bandy *et al.* (1970)	14
〃 , Gee Greensand,	Sandstone, glauconite	FAD of *Gds. trilobus* and *Grt. praescitula*	20.4 ± 0.3	Adams (1975)	13–2
Australia, Maude Basalt	Basalts, whole rock	*Gqa. dehischens* (N.4)	22.0 ± 0.2	Abele and Page (1974)	15

Table 2. Fission-track ages well controlled by biostratigraphic data for the Japanese Neogene.

Locality, Formation	Material dated	Biostratigraphy	Age (Ma)	Reference	Data No.
Oga Peninsula					
Wakimoto middle Tuff	Zircon	Top of *Grt. inflata inflata*	2.8 ± 0.6	Sasajima *et al.*, 1978	16–1
Kitaura Tuff, Funakawa F.	〃	Above top of *Gqa. asanoi*	3.5 ± 0.7	〃	16–2
Anzenji Tuff, Funakawa F.	〃	Boundary of *D. kamtchatica and D.kamtchatica/D.seminae fossilis*	5.6 ± 1.1	〃	16–3
Kakegawa area					
Hosoya Tuff, Kakegawa G.	〃	Above base of *Grt. turncatulinoides*	1.9 ± 0.4	Nishimura, 1975	17–1
Iozumi Tuff, 〃	〃	Below *Pulleniatina* DS	2.3 ± 0.5	〃	17–2
Arigaya Tuff, 〃	〃	Above base of *Gqa. asanoi*	3.2 ± 0.6	〃	17–3
Hirugawa Tuff, 〃	〃	Below top of *Sps. seminulina*	4.1 ± 0.2	Nishimura, 1981	18
Miura area					
HK-Tuff, Zushi F.	〃	Between *Grt. tumida* and *Sps. dehiscens*	3.7 ± 1.1	Yoshida, 1982	19

Abbreviation for generic and specific names
- *D.h.: Denticulopsis hustedtii*
- *D.l : Denticulopsis lauta*
- *Gds.: Globigerinoides*
- *Gna.: Globigerina*
- *Gqa.: Globoquadrina*
- *Grt.: Globorotalia*
- *Sps.: Sphaeroidinellopsis*

Acknowledgments

We thank Dr. N. de B. Hornibrook of New Zealand Geological Survey and Dr. I. Mc-Dougall of Australian National University for valuable information on age data used in this compilation.

References

Abele, C. and Page, R.W., 1974: Stratigraphic and isotopic ages of basalts at Maude and Aireys Inlet, Victoria, Australia. *Roy. Soc. Victoria, Proc.*, p. 86.

Adams, C.J.D., 1975: New Zealand potassium-argon age list-2. *N. Z. Jour. Geol. Geophys.*, v. **18**, p.443–467.

Bandy, O.L., Hornibrook, N. de B. and Schofield, J.C., 1970: Age relationships of the *Globigerinoides trilobus* zone and the andesite at Muriwai Quarry, New Zealand. *N. Z. Jour. Geol. Geophys.*, v. **13**, p. 980–995.

Chiji, M. and Konda, I., 1978: Planktonic foraminiferal biostratigraphy of the Tomioka Group and the Nishiya-tsushiro and Shizukawa Groups, Central Japan, with some considerations of the Kaburan Stage (Middle Miocene). *Cenozoic Geology of Japan,* Osaka, p. 73–92.

Foster, J.H. and Opdyke,N.D., 1970: Upper Miocene to Recent magnetic stratigraphy in deep sea sediments. *Jour. Geophys. Res.*, v. **75**, p. 4465–4473.

Gill, J.B. and McDougall, I., 1973: Biostratigraphic and geological significance of Miocene-Pliocene volcanism in Fiji. *Nature*, v. **241**, p. 176–180.

Grindley, G.W., Adams, C.J.D., Lumb, J.T., and Watters, W.A., 1977: Paleomagnetism, K-Ar dating and tectonic interpretation of upper Cretaceous and Cenozoic volcanic rocks of the Chatham Islands, New Zealand. *N. Z. Jour. Geol. Geophys.*, v. **20**, p. 425–467.

Hasegawa, S., 1979: Yatsuo Area, Toyama Pref., 2. *In* Tsuchi, R., ed., *Fundam. Data Japan. Neogene Bio- and Chronostratigraphy*, IGCP 114 Natl. W.G. of Japan, p. 83–84.

Ibaraki, M. and Tsuchi,R., 1974: Planktonic foraminifera from the upper part of the Kakegawa Group and the Soga Group, Shizuoka Prefecture, Japan. *Shizuoka Univ. Fac. Sci., Rept.*, v. **9**, p. 115–130.

Ibaraki, M. and Tsuchi, R., 1976: Planktonic foraminifera from the lower part of the Kakegawa Group, Shizuoka Prefecture, Japan. *Shizuoka Univ. Fac. Sci., Rept.*, v. **11**, p. 161–178.

Ikebe, N. and Chiji, M., 1971: Notes on top-datum of *Lepidocyclina* sensu lato in reference to planktonic foraminiferal datum. *Osaka City Univ. Geosci., Jour.*, v. **14**, p. 19–52.

Ikebe, N., Takayanagi, Y., Chiji, M., and Chinzei, K., 1972: Neogene biostratigraphy and radiometric time scale of Japan–An attempt at intercontinental correlation. *Pacific Geology*, no.4, p. 39–78.

Ishida, S., 1959: The Cenozoic strata of Noto, Japan. *Univ. Kyoto Coll. Sci., Mem., Ser. B*, v. **26**, p. 83–101.

Kami, S., Kato, M., Kuchida, K., and Takayama, T., 1981: Geological ages of calcareous sandstones on the Noto Peninsula. *Kanazawa Univ. Coll. Sci. Lib. Arts, Ann. Rep.*, v. **18**, p. 47–63.

Kaseno, Y., 1965: Geology of Noto Peninsula. Rept. of the Sci. Res. in the Noto Peninsula. Ishikawa Pref. Office, p. 1–84.

Kawai, N. and Hirooka, K., 1966: Some age results on Cenozoic igneous rocks from Southwest Japan. *Abstr. Symp. "Age of formation for Japanese acid rocks by dating results"*, Geol. Soc. Japan, p.5.

Kitazato, H., 1975: Geology and geochronology of the Younger Cenozoic of Oga Peninsula. *Tohoku Univ., Inst. Geol. Pal., Contr.*, no. 75, p. 17–49.

Koizumi, I. and Kanaya, T., 1977: Correlation of Late Neogene sections on the Oga Peninsula and Akita City,

Northeast Japan. *Prof. K. Huzioka Mem. Vol., Akita,* p. 401–412.

McDougall, I., 1963: Potassium-argon ages of some rocks from Viti Levu, Fiji. *Nature,* v. **198**, p. 677.

Ness,G., Levi,S. and Couch,R., 1980: Marine magnetic anomaly time-scales for the Cenozoic and Late Cretaceous: A precis, critique and synthesis. *Geophys. Sp. Phys., Rev.*, v. **18**, p. 753-770.

Nishimura, S., 1975: On the Problem of the application of fission-track dating for tephrochronology. *Kyoto Univ. Tsukumo Earth Science, Rep.*, v. **10**, p. 1–8.

Nishimura, S., 1981: On the Neogene fission-track dating of tuffs. *IGCP-114 International Workshop on Pacific Neogene Biostratigraphy, Proc.*, p. 132–136.

Page, R.W. and McDougall, I., 1970: Potassium-argon dating of the Tertiary f_{1-2} stages in New Guinea and its bearing on the geological time-scale. *Am. Jour. Sci.*, v. **269**, p. 321–342.

Sasajima, S., Nishimura, S. and Ishida, S., 1978: Geomagnetic chronology and radiometric age, and their relevance to some Neogene series in Japan. *Cenozoic Geology of Japan* (*Prof. N.Ikebe Memorial Volume*), p. 135–154.

Shibata, K., 1973: K-Ar ages of volcanic rocks from the Hokuriku Group. *Geol. Soc. Japan, Mem.*, v. **8**, p. 143–149.

Shibata, K. and Nozawa, T., 1967: K-Ar ages of granitic rocks from the Outer Zone of Southwest Japan. *Geochem. Jour.*, v. **1**, p. 131–137.

Shibata, K., Sato, H. and Nakagawa, M., 1981a: K-Ar ages of Neogene volcanic rocks from the Noto Peninsula. *Japan. Assoc. Petrol. Min. Econ. Geol., Jour.*, v. **76**, p. 248–252.

Shibata, K., Uchiumi, S. and Nakagawa, T., 1979: K-Ar age results-1. *Geol. Surv. Japan., Bull.*, v. **30**, p. 675–686.

Shibata, K., Yamaguchi, S., Ishida, M., and Nemoto, T., 1981b: Geochronology of the *Desmostylus*-bearing formation from Utanobori, Hokkaido. *Geol. Surv. Japan., Bull.*, v. **32**, p. 545–549.

Singleton, O.P., McDougall, I. and Mallett, C.W., 1976: The Pliocene-Pleistocene boundary in southern Australia. *Geoi. Soc. Aust., Jour*, v. **23**, p. 229–311.

Steiger, R.H. and Jäger, E., 1977: Subcommisson on Geochronology: Convention on the use of decay constants in geo- and cosmochronology. *Earth Planet. Sci. Letters,* v. **36**, p. 359-362.

Takayama,T., 1977: On the geological age of the "Hojuji diatomaceous mudstone", Noto Peninsula, based on the calcareous nannofossils. *Kanazawa Univ. Coll. Sci. Lib. Arts, Ann. Rep.*, v. **14**, p. 71–78.

Takayanagi, Y., Sakai, T., Oda, M., Takayama, T., Oriyama, J., and Kaneko,M., 1978: Problems relating to the Kaburan Stage. *Cenozoic Geology of Japan*, p. 93–111.

Talwani, M., Windisch, C.C. and Langseth, M.G., Jr., 1971: Reykjanes Ridge Crest: A detailed geophysical study. *J. Geophys. Res.*, v. **76**, p. 473–517.

Ueda, Y. and Suzuki, T., 1973: K-Ar ages of the glauconite and celadonite from Northeast Japan. *Geol. Soc. Japan., Mem.*, v. **8**, p. 151–159.

Yoshida, S., 1979: Miura Peninsula. *In* Tsuchi, R., ed., *Fundam. Data Japan. Neogene Bio- and Chronostratigraphy*, IGCP 114 Natl. W.G. Japan, p. 20–21.

II

Relation of Pacific Neogene Datum Planes to Those of the Atlantic and Mediterranean

II - 1

Correlation of Atlantic, Mediterranean, and Indo - Pacific Neogene Stratigraphies: Geochronology and Chronostratigraphy*

William A. BERGGREN

Abstract

The Neogene includes the Miocene (T = 23.5–5.4 Ma), Pliocene (T = 5.4–1.7 Ma), and Pleistocene (T = 1.7–0 Ma) epochs. A revised Neogene time scale is presented based upon a review and integration of recent data in the fields of biostratigraphy, magnetostratigraphy, and radiochronology. Of the various magnetobiochronologies currently in vogue, that of Ryan *et al.* (1974) is found to be most consistent with biostratigraphically controlled radiochronology. At the same time, the correlation of radiolarian biostratigraphy to core magnetostratigraphy by Theyer and Hammond (1974a,b; Theyer *et al.* 1978) has been used as a standard for correlation and calibration and results in some minor, but important, revisions to the chronology. The resulting "magnetobiochronologic" scale does not differ significantly, however, from those previously proposed by some authors (Ryan *et al.*, 1974; Berggren and Van Couvering, 1974; Van Couvering and Berggren, 1977).

The revised Neogene time scale provides a chronologic framework for regional correlation of the standard Mediterranean Neogene Age/Stage sequence with those developed in geographically nearby (Paratethys) and distant (Indo-Pacific, West Coast North America) regions.

Introduction

This is the first of two papers which deal with interregional correlation of Neogene stratigraphies. In this paper I present a review of data in the fields of bio- and radiochronology and magnetostratigraphy insofar as they pertain to the construction of a revised, updated Neogene time scale which will provide, in turn, the framework for regional chronostratigraphic correlations.

In the second paper I have presented a summary of the biostratigraphic and paleobiogeographic distribution of Neogene planktonic foraminifera from data available in the Initial Reports of the Deep Sea Drilling Project (Legs 1–50) and supplementary, ancillary studies.

Biomagnetoradiochronology

The relationship between geochronologic systems based on biologic evolution (biochronology) and on decay rates of stable isotopes (radiochronology) to the stratigraphic record of fossils (biostratigraphy) and geomagnetic polarity reversals (magnetostratigraphy) has been discussed at length (Berggren and van Courvering, 1978). Basically biochronology organizes geologic time according to the irreversible process of evolution. It represents an ordinal framework which can measure time (with the exception of the last 4–5 m.y.) with a greater resolution, if with less accuracy, than radio-chronology. Bio- and magnetostratigraphy are iterative and consist of observed superpositional sequences of fossils and magnetic reversals, respectively. The arrangement and correlation in time of bio- and magnetostratigraphies and their calibration to well-chosen radiochronologically determined "instants in time" constitute the basic data from which a "magnetobiochronologic" time scale can be constructed.

There are various methods that can be used in constructing a time scale:

1. Assigning absolute ages to describe lithostratigraphic levels (with or without biostratigraphic control). The inherent limitation in the resolution of the radiochronologic method is well known. The

Department of Geology and Geophysics, Woods Hole Oceanographic Institution, Woods Hole, Massachusetts 02543, and Department of Geology, Brown University, Providence, Rhode Island 02912, U.S.A.

* *Reprinted from Proceedings of International Workshop on Pacific Neogene Biostratigraphy, Osaka Japan (1981) by the courtesy of the author. [Editor's note]*

assignment of precise ages to biostratigraphic datum levels or magnetostratigraphic boundaries is impossible beyond *ca.* 4 Ma because the experimental error in the dating methods exceeds the discrimination of the biologic event and the duration of magnetic reversal-chrons and/or events.

2. Integrating biostratigraphic data to magnetostratigraphy and/or radiochronology, correlation of the pattern of sea-floor magnetic anomalies to magnetostratigraphy in deep sea cores, and interpolation between carefully selected radiochronologically dated calibration points assuming linear rates of either sedimentation and/or sea-floor spreading. The resolution and accuracy of age estimates of discrete levels and the time interval between them is *greater* using this essentially biochronologic or magnetobiochronologic method than that obtainable by radiochronology alone (see Odin, 1978: 254 for a dissenting opinion). This is the method adopted by Berggren (1972), Berggren and van Couvering (1974, 1978), and van Couvering and Berggren (1977).

Ryan *et al.* (1974) have essentially taken this method a step further by estimating paleomagnetic reversal boundaries according to a linear-regression "fit," extrapolated between selected radiometrically dated calibration points with the (approximately located) calibration points back-fitted to give the straightest possible time-magnetostratigraphy curve.

A comparison of some of the age estimates assigned to Miocene planktonic foraminiferal datum levels and/or boundaries based on the biochronologic method is shown in Table 1. Absolute age daterminations assigned to various planktonic foraminiferal datum levels, zonal boundaries, or zonal intervals in the Paratethys and

Table 1. Estimated biochronologic age of late-Oligocene/Miocene planktonic foraminiferal zonal boundaries. Calibration points used in calculation shown by*.

Zonal boundary	Criterion	Berggren (1972)	Age Estimate in Ma Berggren and van Couvering (1974); van Couvering and Berggren (1977)	Ryan *et al.* (1974)	Saito (1977)
N17/18	⊥ tumida	5.0*	5.0*	—	5.0
	⊥ Pul. primalis			5.8	6.2*
N16/17	⊥ plesiotumida	6.5	8.6		7.7
N15/16	⊥ acostensis	10.0	10.4	10.8	10.0
N14/15	⊤ siakensis	12.0	11.5	12.0	11.2
N13/14	⊥ nepenthes	13.5	11.7	12.7	11.5
N12/13	⊤ fohsi lobata	14.0	12.2	—	12.4
N11/12	⊥ fohsi		12.7	15.5	13.9
N10/11	⊥ praefohsi		13.2		14.7
N 9/10	⊥ peripheroacuta	15.5	14.0	—	15.3
N 8/9	⊥ Orbulina	16.0*	15.0*	16.0*	16.0*
N 7/8	⊥ sicanus	16.5	15.6	16.8*	17.2
N 6/7	⊤ dissimilis	17.0	17.5	18.5	18.0
N 5/6	⊥ insueta	19.0	19.0	19.4	18.6
N 4/5	⊤ kugleri	20.5	22.5		20.5
N 3/4	⊥ Globigerinoides	22.5*	25.0	22.7	22.5*
	⊥ kugleri				25.5
P21/P22	⊤ opima	26.0	26.0	—	26.0

Table 2. Radiometric dates and estimated biochronologic ages of Miocene planktonic foraminiferal zonal boundaries or datum levels (compiled from Ikebe, 1977).

		Japan	
Zonal boundary	Criterion or Datum level	Radiometric date (Ma)	Estimated Biochronologic Age (Ma)
N17/18	—S. tumida	5.3	
—	—Pull. primalis		7.0 – 7.5
N15/16	—acostaensis	9.5 – 10.0	
N13/14	—nepenthes	11.3 ± 0.4	
N9/10	—peripheroacuta	—	14.0
N8/9	—Orbulina	15.0 ± 0.5	—
(upper N8)	—Praeorb. glomerosa curva	—	16.0
N 7/8	—sicanus	—	16.5 – 17.0

Table 3. Radiometric dates on (predominantly) marine Miocene levels of the Paratethys region (from Steininger *et al.*, 1977). Additional data may be found in Vass and Bagdisarjan (1978) and Vass (1979). (Radiometric dates not corrected for new ^{40}K decay constant.)

Paratethys

Age/State	Radiometric Date (Ma)	Biostratigraphic control	Biostratigraphic and Chronostratigraphic Correlation
Sarmatian	13.2 ± 2.1 12.7 ± 2.1 12.9 ± 0.5 (2 dates)		Sarmatian is non-marine and equivalent to upper Serravallian essentially.
Badenian	17.5 ± 0.9 15.35 ± 0.7	intercalated with *Praeorbulina* and *Orbulina*	Badenian corresponds to NN5-NN7 and to the Langhian-Serravallian
	16.8 ± 0.75 16.0 ± 0.3 16.3 ± 0.9 15.3 ± 0.1	overlie *Praeorbulina* and intercalated with lower-middle Badenian marine beds.	
	15.2 ± 0.9 14.6 (prelim. date) 14.9 ± 0.9	No Biostrat. date	
Karpatian	20.7 ± 1.2	below *G. sicanus* intercalated with "schlier" type mollusc fauna	The Karpatian corresponds to zones NN4 (*partim*) and lower part of NN5 and N7 (upper part) and N8 (lower part = ⊥of *G. sicanus*)
	17.3 ± 2.2, 18.6 ± 1.0, 18.8 ± 2.5, 18.9 ± 1.5, 16.5 ± 2	apparently below *G. sicanus* level but in zone NN4.	
Ottnangian	22.0 ± 2.0	discordantly overlies fossiliferous Eggenburgian, underlain by Ottnangian "Rzehakia" beds	Ottnangian corresponds to NN3 (*partim*)—NN4 (partim) and the upper part of the Burdigalian
	7 dates from 19.9 ± 3.5— 22.3 ± 4.3 (aver. 21.2 ± 1.2)	Ottnangian (? Eggenburgian)	
	20.7 ± 1.6, 23.5 ± 2.7, 19.1 ± 1.2, 19.5 ± 3.5, (aver. 20.5 ± 0.8)		
Eggenburgian	18.0 ± 1.1 (17.6 ± 1.6) poor quality	Typical Eggenburgian benthic foram fauna directly overlying rich Eggenburgian mollusc fauna	Eggenburgian corresponds to zones NN1-NN3 and thus to late Aquitanian-Burdigalian (*partim*)
Egerian	20.2 ± 1.2 21.5 ± 1.1	With *Miogypsina septemtrionalis* and *M. formosensis*	Egerian corresponds to zones NP24/25-NN1 (*partim*) and to Chattian and Early Aquitanian.

Japan are shown in Tables 2 and 3, respectively.

A comparison of the data in these three tables leads to the following observations.

1. Construction of a Miocene time scale based solely on radiometric dates results in a significant departure from linearity in sea-floor spreading rates and sedimentation rates (assumptions which can be demonstrated to be reliable over discrete intervals of time).
2. The biochronologic time scales shown on Table 1 possess a relatively high degree of internal consistency. Major differences arise because of 1) the difference in the estimated age of the *Orbulina* (calibration) level, 2) adjustment of the estimated age of the paleomagnetic reversal boundaries according to a linear-regression "fit," extrapolated between selected radiometrically dated points with the (approximately located) calibration points back-fitted to give the straightest possible time-magnetostratigraphy curve (Ryan *et al.*, 1974). The main result of this method is a revised (expanded) Middle Miocene chronology.
3. More recent radiometric dates between 15–16 Ma associated with *Praeorbulina-Orbulina* levels in the Badenian of the Paratethys (cited in Vass, 1979) are more in line with those in Japan centering on 15 ± 0.5 Ma for the *Orbulina* datum and a biochronologic estimate of the *Praeorbulina* datum of approximately 16 Ma (Ikebe, 1978) and biochronologic estimates made by biostratigraphers (see Table 1).
4. Radiometric dates of 20–22 Ma on Ottnangian (= NN3–NN4 (*partim*) = N6–N7 = Burdigalian) in the Paratethys region are considerably offset (out of phase) with biochronologic age estimates of 16–18 Ma on late early Miocene zones N7–8 (Burdigalian-early Langhian) in the Mediterranean and in Pacific Ocean deep sea cores. The radiometric calibration of Early Miocene biochronology in the Paratethys

Table 4. Data base for revised Neogene time scale.

Data Base	Author (S)	Geographic Area
Radiochronology	McDougall and Page (1975)	Australasia
	Vass (1979)	
	Vass and Bagdasarjen (1978)	Paratethys
	Steininger *et al.* (1976)	
	Ikebe (1977)	Japan
	Odin (1978)	
	Kreuzer *et al.* (1980)	Europe
Paleomagnetic stratigraphy (with integrated paleontology = biochronology)	Nakagawa *et al.* (1975)	Mediterranean stratotypes
	Nakagawa *et al.* (1977)	Japan
	Theyer and Hammond (1974a,b)	deep sea cores
	Opdyke *et al.* (1974)	
	Ryan *et al.* (1974)	Sea-floor spreading in ocean basins
Sea-floor magnetic anomaly stratigraphy	Ness *et al.* (1980)	Sea-floor spreading in ocean basins

Table 5. Comparison of Early Miocene magnetic polarity time scales.

	Theyer and Hammond (1974)	Ryan *et al.* (1975) (= Blakey, 1974)	La Brecque *et al.* (1977)	Ness *et al.* (1980)
epoch 16	14.8 – 17.6 (2.8)	15.6 – 18.1 (2.5)	14.7 – 17.1 (2.4)	15.0 – 17.3 (2.3)
epoch 17	17.6 – 18.7 (1.1)	18.1 – 19.8 (1.7)	17.1 – 18.7 (1.6)	17.3 – 18.8 (1.5)
epoch 18	18.7 – 19.5 (0.8)	19.8 – 20.1 (0.3)	18.7 – 19.0 (0.3)	18.8 – 19.0 (0.2)
epoch 19	19.5 21.0 (1.5)	20.1 – 21.3 (1.2)	19.0 – 20.1 (1.1)	19.0 – 20.1 (1.1)
epoch 20	21.0 – 22.4 (1.4)	21.3 – 22.3 (1.0)	20.1 – 21.0 (0.9)	20.1 – 21.0 (0.9)
epoch 21	22.4 – 24.7 (2.3)	22.3 – 23.4 (1.1)	21.0 – 22.0 (1.0)	21.0 – 21.9 (0.9)
epoch 22	24.7 – 26.1 (1.4)	23.4 – 24.4 (1.0)	22.0 – 23.0 (1.0)	21.9 – 22.8 (0.9)

Note: Chronology of Theyer and Hammond (1974b) derived by linear extrapolation of sedimentation rates between microfossil datum levels and ages proposed by Berggren and van Couvering (1974). Ryan *et al.* (1975) use Blakely (1974) for chronology from 8.71 Ma—22.69 Ma (mid-Epoch 21) and Heirtzler *et al.* (1968) below that. La Brecque *et al.* (1977) use calibration ages of 3.32 Ma (anomaly 2.3′ (0)), 7.39 Ma (anomaly 4.1 (y)), and 54.9 ma (anomaly 29 (o) at the Cretaceous/Tertiary boundary). Ness *et al.* (1980) use calibration points which fit anomaly 5.5 (0) at 10.30 Ma and anomaly 24 at 55 Ma. The result of these latter two chronologies is a pronounced foreshortening (compression) of the time-scale in the middle part of the scale (particularly in the Early Miocene) which is inconsistent with other geologic evidence (see text for discussion). Different methodologic approach also probably explains major discrepancy in estimating duration of Epoch 21 (0.9 to 2.3 m.y.) by these four groups.

remains in irreconcilable conflict with the time scale based on a synthesis of biostratigraphy, magnetostratigraphy, and dating in western Europe, Africa, North America, and Pacific Ocean deep sea cores, as pointed out by van Couvering and Berggren (1977: 303; cf. Vass, 1979: 1437).

In an attempt to construct a revised, updated Neogene time scale there are several basic components that must be carefully considered and evaluated (Table 4).

Neogene biostratigraphic zones, zonal boundaries, and datum events have been directly correlated with a magnetic polarity reversal scale and/or indirectly cross-correlated from piston-cores, DSDP drill sites (e.g., site in the equatorial Pacific), outcrops, etc., to the magnetic polarity scale by means of biochronologic methods, i.e., assuming linear sedimentation rates (Theyer and Hammond, 1974 a,b; Opdyke *et al.*, 1974) or magnetostratigraphic methods, i.e., assuming linear sea-floor spreading rates (Ryan *et al.*, 1974). (The integration of biostratigraphy and paleomagnetic stratigraphy in classic type) Neogene marine stratigraphic sections of the Mediterranean (Nakagawa *et al.*, 1977) has further provided the framework for regional, and, indeed, global correlation of standard, regional, and local chronostratigraphic schemes. Finally, the continued addition of radiometric dates (see Table 1) provides the basis for a geochronologic scale to which the other schemes must, ultimately, be calibrated.

For example, a magnetobiostratigraphic framework exists for the Miocene. The Miocene/Pliocene boundary, as stratotypified in the Mediterranean region, has been biostratigraphically linked with a level close to, if not essentially contemporaneous with, the Gilbert/Epoch 5 magnetic polarity boundary (Saito, in Hays *et al.*, 1969; Berggren, 1973; Cita and Gartner, 1973). Raidometric age determinations on this level have centered on *ca.* 5 Ma (McDougall and Page, 1975). Age estimates have been progressively lowered as a result of the newly proposed ^{40}K decay constants applied to biostratigraphically controlled radiometric dates on marine sections (McDougall and Page, 1975) or to mangetostratigraphic sequences of lavas on Iceland (McDougall *et al.*, 1976a, b; 1977; Harrison *et al.*, 1979). Thus the current estimate of the Gilbert/Epoch 5 boundary is at 5.44 Ma (McDougall, 1977).

In a similar manner the Oligocene/Miocene boundary (=Aquitanian/ Chattian boundary as stratotypified in the Aquitaine Basin, in France) has been biostratigraphically correlated to a level close to, if not coincident with, the magnetic Epoch 21/22 boundary (Theyer and Hammond, 1974b) for which the best estimated age is *ca.* 23–24 Ma (Berggren and van Couvering, 1974; 1977; Kreuzer *et al.*, 1980).

But there remain skeltons in the closet which continue to bedevil attempts at choosing between alternative Neogene chronologies. Four magnetic polarity chronologies are shown in Table 5 for the interval of time spanning *ca.* 15–23 Ma. The chronologic scale of Theyer and Hammond (1974b) and Ryan *et al.* (1974) utilize biostratigraphic correlations and radio-chronologic data in their construction.

La Brecque *et al.* (1977) have revised the chronology of the magnetic reversal scale of Heritzler *et al.* (1968) by choosing calibration ages of 3.32 Ma (anomaly 2.3′), 7.39 Ma (anomaly 4.1′ (y)) and 64.9 Ma (anomaly 29 (o) at the Cretaceous/Tertiary boundary) and making interpolations based on assumed linear spreading rates. The revised magnetochronology was then used to derive revised age estimates on various Neogene chronostratigraphic studies of Ryan *et al.* (1974). This scale was, in turn, revised for the interval 0–13 Ma (Harrison *et al.*, 1979) to reflect the new decay constant(s) (Steiger and Jager, 1977, 1978). The changes are of very minor nature in the context of this study. More to the point is the fact this revised scale, although it correctly integrates biostratigraphic and magnetostratigraphic data (see discussion of Ness *et al.*, 1980, below), does not agree well with the radiochronologic calibration of the biostratigraphy in the early Miocene. The Oligocene/Micoene boundary is shown in the magnetochronology of La Breque *et al.* (1977, Fig. 1) to lie within Epoch 22 at about 22.5 Ma (it should be more properly located near the Epoch 21/22 boundary; Theyer and Hammond, 1974b), which in the chronology of La Breque *et al.* (1977) is at about 22 Ma, or slightly younger. Either age is considerably younger than the radiometric dates which suggest an age of 23–24 Ma for this boundary (Odin, 1978; Kreuzer, 1980).

The "time scale" of Ness *et al.* (1980) is a *pure magnetochronology* in that it does not employ correlations with the geologic record. Ness *et al.* (1980) have redetermined the age of magnetic anomalies (and, by extension) magnetic polarity epoch boundaries by using newly estimated ages (based, in turn, on newly proposed decay constants) for several calibration points. These points are:

1) anomaly 2.3' (3.4 Ma)
2) anomaly 5.5 (0) (10.30 Ma)
3) anomaly 24 (55 Ma)
4) anomaly 29/30 at the Cretaceous/Tertiary boundary (66.7 Ma)

Several problems arise with the resulting chronology, not the least of which, in the context of this paper is a considerable foreshortening (compression) of the chronology in the middle part of the sequence—the Early Miocene. Comparison of the chronologies in Table 5 show an offset of almost 3 m.y. between the four at the base of epoch 22 (ranging from 22.8 Ma to 26.1 Ma). The interpretations of Theyer and Hammond (1974b) and Ryan *et al.* (1974) of the duration of individual epochs between 16 and 22 are relatively consistent with the notable exception of epoch 21 where the values are estimated at 2.3 m.y. and 1.1 m.y., respectively. Presumably this discrepancy may be attributed to the use by the former of (assumed) linear sedimentation rates in the calcluation of epoch boundary ages and of (assumed) linear spreading rates by the latter in their calculations.

Of greater concern here is the related problem of the position of several geologic boundaries in a magnetostratigraphic context on the one hand, and in a geochronologic context on the other.

The stratotype of the Burdigalian Stage is associated with magnetic polarity epochs 19 (and anomaly 6) and 20 (Ryan *et al.*, 1974). The *Calocycletta virginis* (radiolarian) datum occurs just below the magnetic polarity 20/21 boundary (at a level estimated at 22.6 Ma) and is closely correlative with the planktonic foraminiferal N4/ N5 boundary (Theyer and Hammond, 1974b). The Burdigalian stratotype has a planktonic foraminiferal fauna indicative of Zone N5 (Anglada, 1971) and NN2 (*druggi*) (Muller *in* Bizon and Muller, 1979) or NN2-NN3 (*druggi-belemnos*) zones (Schmidt *in* Benda *et al.*, 1977) (see also van Couvering and Berggren, 1979). Bukry and Crouch (1978) indicate that the oldest stratigraphic levels in California assigned to the Saucesian stage belong to the *S. belemnos* (NN3) Zone (which would have an age younger than 20 Ma (van Couvering and Berggren, 1977), whereas radiometric dates (Turner, 1970) indicate an age of *ca.* 22.5 Ma for the base of the Saucesian. On the other hand, Warren (1980) suggests that the Saucesian stage includes (in its lower part) strata that can be assigned to the *druggi* Subzone and perhaps the *deflandrei* Subzone of the *Triquetrorhabdulus carinatus* Zone of Bukry (1973). Biostratigraphic correlation of the level(s) dated by Turner (1970) is not well documented, so it is not clear whether the dated levels are equivalent to Zone NN2 or NN3. Radiometric dates of 22–23 Ma are cited for the Burdigalian of Europe (Odin, 1973). Further to the point, DSDP site 15 was drilled on anomaly 6 in the South Atlantic and the oldest sediments above basement belong to Zone N5 (Blow,

1970). Thus Zone N5 (and the Burdigalian) appears to be associated with epochs 19 and 20 at least.

In the "time scale" of Ness *et al.* (1980) the base of epoch 20 is, at 21.0 Ma, considerably younger than the estimate of 22.3 Ma by Ryan *et al.* (1974), which is concordant with the radiometric determinations on Burdigalian levels and biostratigraphic cross-correlations with magnetostratigraphy.

Moving downward we come to the problem of the Aquitanian Stage and, by extension, the Oligocene/Miocene boundary. As is the case with all chronostratigraphic boundaries discussed here, there are two aspects of the problem: the position of the boundary in terms of magnetostratigraphy and the radiometric age of the boundary.

The Oligocene/Miocene boundary is approximately coincident with the calcareous nannoplankton NP25/NN1 boundary, within planktonic foraminiferal Zone N4, and lies between the FADs of *Globoquadrina dehiscens* + *Globigerinoides primorius* (below) and *Globigerinoides trilobus* + *G. altiapertúrus* (above; Bizon and Muller, 1979; this work). The *Lychnocanoma elongata* (radiolarian) datum occurs in the earliest part of epoch 21 at an estimated age of 24.3 Ma (Theyer and Hammond, 1974b; in the chronology of Ness *et al.* (1980) this level is at about 21.7 Ma, a 2.5 m.y. difference, and in the chronology of Ryan *et al.* (1974) at about 23+ Ma, a 1.3 + m.y. difference) and is approximately correlative with the NP25/NN1 boundary. Thus the Oligocene/Miocene boundary would be biochronologically dated at about 24 Ma (within the *L. elongata* Zone = NP25/NN1 boundary) in the chronology of Theyer and Hammond (1974b). This agrees closely with radiometric dates available at that time (Odin, 1973) on the Oligocene/Miocene boundary centered on 23–24 Ma which led Berggren and van Couvering (1974) and van Couvering and Berggren (1977) to suggest an age of 23.5 Ma for this boundary. These estimates have recently been reinforced by additional glauconite dates on the Chattian (late Oligocene) and Vierlandian (early Miocene) stages of NW Germany which suggest an age of about 23 Ma for the Oligocene/Miocene boundary (Kreuzer *et al.*, 1980). Ryan *et al.* (1974, Fig. 6) show the Aquitanian Stage extending down into the *Dorcadospyris papilio* (radiolarian) Zone (=NP25) in epoch 22 with no explanation (although this may be based upon the conviction that the base of the Aquitanian = base Miocene = FAD of *Globigerinoides primoridus*). The FAD of *Globigerinoides primordius* does occur in stratigraphic levels belonging to Zone NP25 in deep sea cores which are pre-Aquitanian in age of that the base Aquitanian = Oligocene/Miocene boundary is, in fact, above this datum level and probably very close to the epoch 21/22 boundary.

In the chronology of Ness *et al.* (1980) the epoch 20/21 boundary is at 21.9 Ma. The important point here is the fact that the biostratigraphically correlated Oligocene/Miocene boundary has an age of *ca.* 21.5–22 Ma on this "time-scale" which is considered too young in view of the evidence cited above and elsewhere (Odin, 1978; Kreuzer, 1976).

It can be seen from this discussion that in calculating their "time scale" Ness *et al.* (1980) have placed their magnetic polarity epochs (and boundaries) in a chronologic framework which ignores the biostratigraphically determined position of chronostratigraphic boundaries in terms of magnetostratigraphy and the radiometric data base which serves to provide calibration points. In a word, they have ingnored the geology (cf. Ryan *et al.*, 1974).

To compound the problem further Ness *et al.* (1980) have recalculated the ages of chronostratigraphic boundaries (in this instance for the Neogene) by applying the recently suggested decay constants to the age estimates made by Berggren and van Couvering (1974). This procedure is viewed as quite inappropriate at best, misleading at worst. The age estimates of Berggren and van Couvering (1974) are just that, age estimates and not radiometric dates. They are the result of a "best estimate" of the relationship between the biostratigraphic position of a chronostratigraphic boundary and its relationship to a careful assessement of current radiochronology. The methodology is deliberately selective and certain dates are omitted from consideration (see Vass, 1979; Vass and Bagdasarjan, 1978; for an honest *mea culpa* from radiochronologists faced with inconsistent data) just as inconsistent biostratigraphic data are omitted where justified. The point is that it is possible to recalculate a radiometric date, but not an estimated age. But more to the point is the fact that in the case of the Oligocene/Miocene boundary this boundary has been biostratigraphically positioned to a level within Epoch 21 in the magnetostratigraphic scale. Any change in the estimate of the age of this (or any other boundary in the Neogene) *must* be made within the framework of magnetobiostratigraphically correlated limits of the chronostratigraphic units. In the example cited here the "manipulations" by Ness *et al.* (1980) result in the following *reductio ad absurdum* (Fig. 1):

1. The magnetobiostratigraphically correlated Oligocene/ Miocene boundary (in Epoch 21) has an estimated age of 21.5–22 Ma.
2. The estimated (radiometrically based; biostratigraphically correlated) age of the Oligocene/Miocene boundary by Berggren and van Couvering (1974; ca. 24 Ma) is recalculated to 24.6 Ma and lies within a long reversed polarity interval between anomalies 6c and 7 which can be biostratigraphically demon-

strated to lie comfortably within (and out of harms way from the geophysicists) the late Oligocene. These two Oligocene/Miocene "boundaries" are about 3 m.y. apart!

Carrying this analysis a step further reveals that the dichotomy between the radiometric limits of the Miocene and the magnetobiostratigraphically correlated position of the chronostratigraphic boundaries vary from 5.1 Ma and 5.44 Ma at the top to 24.6 Ma and 21.9 Ma, respectively, a difference of 0.3 m.y. to 2.6 m.y.

Ness *et al.* (1980) have ignored the fact that the magnetobiochronologic scale is a relative (sliding) scale and that if the anomaly and epoch boundaries are moved up or down according to a newly derived calibration chronology, the biostratigraphically correlated fossil datum events, zones, and zonal boundaries, and the chronostratigraphic limits which are based on these correlations, must also be adjusted accordingly. In summary, the chronolgic framework of the recently proposed magnetostratigraphic scheme of Ness *et al.* (1980) is found to be inappropriate for the purpose of a revised and updated Neogene time scale.

The most recent revision of the Neogene magnetic polarity time scale (save for that by Ness *et al.*, 1980) is that of Hailwood *et al.* (1979) and involves the following steps:

1. Derivation of a Late Neogene magnetic polarity time scale based on a correlation of magnetobiostratigraphic data from DSDP Site 400A (Bay of Biscay) with the "standard" magnetic polarity time scale for the past 10 m.y. The resulting age vs. depth calibration was used to assign numerical ages to the principal zonal boundaries.
2. Magnetobiostratigraphic studies on DSDP Sites 403–405 (Rockall Plateau) were combined with similar studies on the Paleocene part of the Gubbio Section in Italy (Premoli-Silva *et al.*, 1974) and information on the biochronologic age of sediments overlying identifiable marine magnetic anomalies at the DSDP sites. Synthesis of this data produced significant changes to the current Paleogene and early Neogene magnetic polarity time scale (Hailwood *et al.*, 1979, Fig. 5). The numerical ages assigned to the biochronologic scale were derived from three sources: a) 0–10 Ma: ages of zone boundaries proposed for North Atlantic region based on results at Site 400A; b) 10–24 Ma: Ryan *et al.* (1974); c) 24–65 Ma: Hardenbol and Berggren (1978).

The ages of anomalies from 0–23 Ma were modified directly from the "standard" of the Heirtzler *et al.* (1968) scale, following Talwani *et al.* (1971) and Blakely (1974). Correlation lines connecting the various biochronologically "dated" anomalies do not form a single straight line as might be expected if spreading rates have been linear, as assumed in the original model proposed by Heirtzler *et al.* (1968). Rather three straight line segments are formed with gradient changes occurring at anomalies 24B and 7A, with estimated numerical ages of 52 Ma (early Eocene) and 28 Ma (late Oligocene), respectively. (These ages were not corrected for new decay constants, but approximately 1.0 and 0.5 m.y. may be added to each age, respectively, to conform with recently proposed versions.)

The geochronometric implications are clear, however. The newly proposed time scale of Hailwood *et al.* (1979) indicate's significant changes in the rate of sea floor spreading in the North Atlantic in the early and late Paleogene.

The Neogene portion of the time sclae of Hailwood *et al.* (1979) is thus essentially the same as that of Ryan *et al.* (1974), which is, essentially, the one used here with minor changes in biostratigraphic correlation and epoch boundary (and anomaly) age corrections (see below).

In preparing a revised Neogene time scale (Fig. 2) have expressed the philosophy and methodlogy eloquently expressed by Theyer *et al.* (1978: 380),

> ". . . the complexity and diversity of the data employed to compile the geochronology of the paleomagnetic "time scale" (the definition of the latter is an altogether separate issue from the simple elucidation of the sedimentary magnetostratigraphic sequence), is expressed in the revisions (for example, Ryan *et al.*, 1974; LaBrecque *et al.*, 1977) which have appeared since the original data were published. Because most of these revisions result from indirect correlations with DSDP cores, evidence from marine magnetic anomalies or involved correlations with land-based sections, the present authors decided to adhere to the original chronology (Theyer and Hammond, 1974a, b) instead of incorporating each proposed change. Furthermore, at the present stage, the most important facts is not so much the "absolute" chronology, but the *direct* (that is, within the same material) calibration of the biostratigraphic events against an independent system of discrete events of global character, such as magnetic reversals. Once this direct calibration is accomplished for the Tertiary, the geochronological estimates will acquire a much more solid basis than is their present foundation."

Three steps have been followed in the construction of this current version of a Neogene time scale.

1. I have essentially utilized the magnetostratigraphic and magnetobiochronologic framework established by Ryan *et al.* (1974) with minor chronometric adjustments as suggested by Dalrymple (1979) and Harrison *et al.* (1979) to reflect the recently proposed new ^{40}K decay constant(s) (Steiger and Jager, 1977, 1978). This results in a slight downward (older) expan-

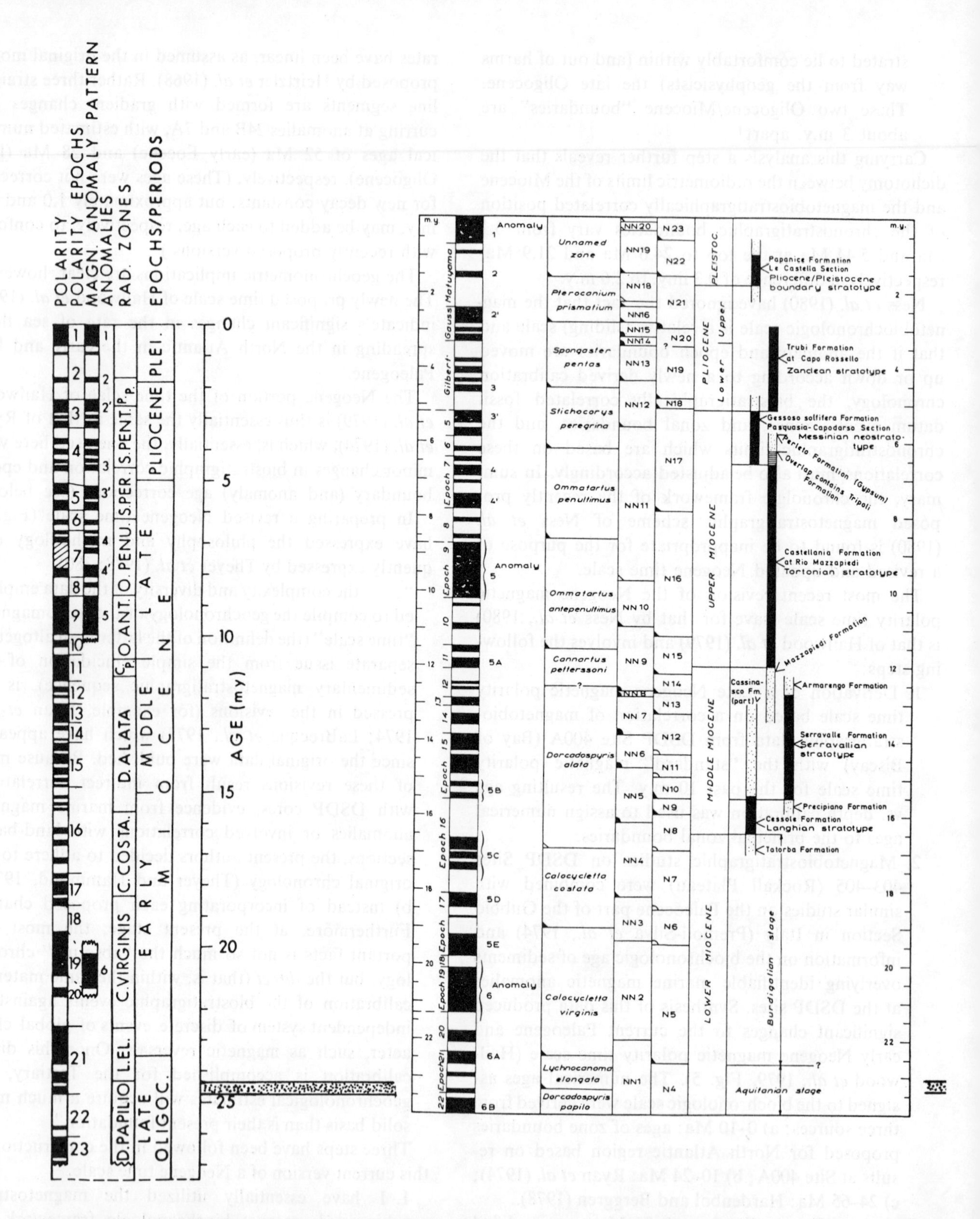

Fig. 1. Comparative magnetobiochronology of the Neogene. The studies by Theyer and Hammond (1974b) and Opdyke *et al.* (1974) and Ryan *et al.* (1974) have integrated biostratigraphic, paleomagnetic, and magnetostratigraphic and radiochronologic data into a chronologic framework which essentially limits the Miocene to the interval between the Gilbert/Epoch 5 boundary (top) and the Epoch 21/22 boundary (bottom) with estimated ages of *ca.* 5 and 23–24 Ma, respectively. Only that part of the revised magnetochronology of Ness *et al.* (1980) spanning the interval between anomalies 5e and 10 is shown on the right in Fig. 1. Using revised calibration points of 10.30 (anomaly 5.5) and 55 Ma (anomaly 24), a revised magnetochronology is presented which shows considerable compression in the middle part of the intervening sequence (note the approximately 1.5–2.5 m.y. offset in the

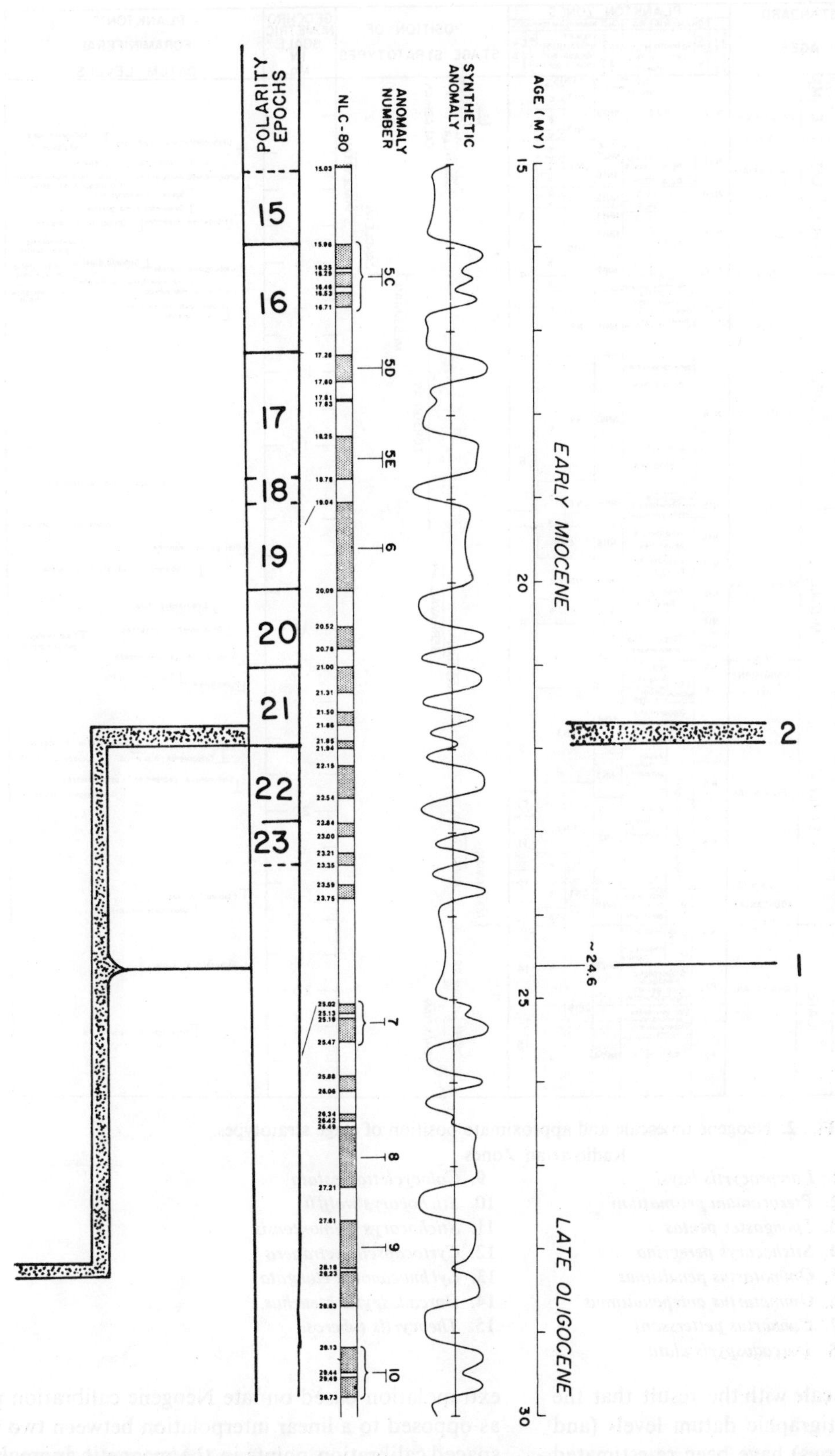

position of the Epoch 21/22 boundary relative to the other two chronologies). Ness *et al.* (1980) have recalculated the age of the Oligocene/Miocene boundary at 24.6 Ma (by applying a new ^{40}K decay constant) to an *estimated* age (not radiometric date) of 24 Ma suggested by Berggren and van Couvering (1974). As a result there are two Oligocene/Miocene boundaries, shown in positions 1 and 2, to the right of the figure: number 1 is the radiochronologic one at 24.6 Ma, and number 2 is the magnetobiostratigraphic one at *ca.* 22 Ma, associated with the Epoch 21/22 boundary, over 2.5 m.y. apart. The discrepancy results from procedural flaws in the methodlogy of Ness *et al.* (1980), i.e., ignoring the geologic record (see text for further discussion).

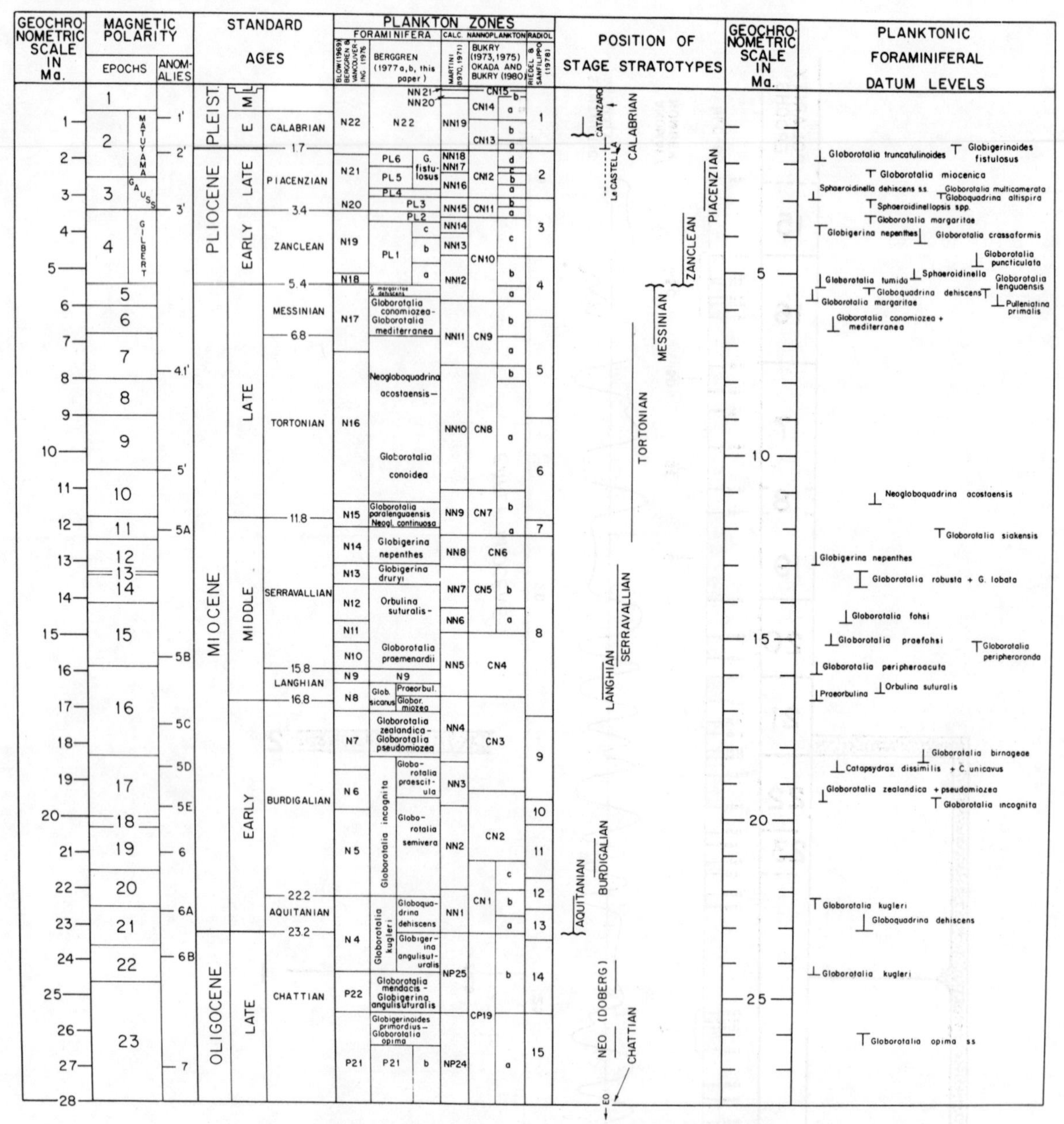

Fig. 2. Neogene timescale and approximate position of stage stratotypes.

Radiolarian Zones

1. *Lamprocyrtis haysi*
2. *Pterocanium prismatium*
3. *Spongaster pentas*
4. *Stichocorys peregrina*
5. *Ommatartus penultimus*
6. *Ommatartus antepenultimus*
7. *Cannartus petterssoni*
8. *Dorcadospyris alata*
9. *Calocycletta costata*
10. *Stichocorys wolffii*
11. *Stichocorys delmontensis*
12. *Cyrtocapsella tetrapera*
13. *Lychnocanoma elongata*
14. *Dorcadospyris ateuchus*
15. *Theocyrtis tuberosa*

sion of the Neogene time scale with the result that the ages of the various biostratigraphic datum levels (and chronostratigraphic boundaries) have been re-estimated and found to be roughly 0.1 (in the Pliocene) to 0.3 m.y. (early Miocene) older.

Since the methodology of Ryan *et al.* (1974), following Blakely (1974), consists essentially of a downward linear extrapolation based on late Neogene calibration points, as opposed to a linear interpolation between two widely spaced calibration points in the magnetic anomaly scale used by Ness *et al.* (1980), we may expect significant differences in the resulting chronology. As I have pointed out above, the magnetobiochronologic framework of Ryan *et al.* (1974) is more consistent with available

Table 6. Estimated age (in Ma) of standard Neogene chronostratigraphic units.

Chronostratigraphic units			T in Ma
Pleistocene (1.7–0)	Late		0.125 – 0
	Middle		0.7 – 0.125
	Early		1.7 – 0.7
Pliocene (5.4 – 1.7)	Late (Piacenzian)		3.4 – 1.7
	Early (Zanclean)		5.4 – 3.4
Miocene (23.2 – 5.4)	Late	Messinian	6.8 – 5.4
	(11.8 – 5.4)	Tortonian	11.8 – 6.8
	Middle	Serravallian	15.8 – 11.8
	(16.8 – 11.8)	Langhian	16.8 – 15.8
	Early	Burdigalian	22.2 – 16.8
	(23.2 – 16.8)	Aquitanian	23.2 – 22.2

geologic information. I have chosen the "Ryan-Heirtzler" chronology in favor of those developed by Theyer and Hammond (1974a,b) and Theyer *et al.* (1978) because the former is based on the assumption of linear spreading rates whereas the latter is based upon the assumption of linear sedimentation rates. As a result, the magnetobiostratigraphically correlated portion of various chronostratigraphic boundaries is more consistent with current radiochronologic data in the case of the "Ryan-Heirtzler" chronology than is the case with that of the "Theyer-Hammond."

2. The correlation of radiolarian biostratigraphy to core mangetostratigraphy (Theyer and Hammond, 1974a,b; Theyer *et al.*, 1978) has been used as a standard for correlation and calibration.

3. Interzonal correlation has been made to this standard following the data of Berggren and van Couvering (1974; 1978); van Couvering and Berggren (1977); Ryan *et al.* (1974); Keller (1980, 1981) and various publications in the Initial Reports of the Deep Sea Drilling Project, and current records in our laboratory at Woods Hole (unpublished).

The resulting magnetobiochronologic scale does not differ significantly from that previosuly proposed by Berggren and van Couvering (1974), van Couvering and Berggren (1977), and Ryan *et al.* (1974). Several points may be noted, however:

1. The Oligocene/Micoene boundary (= base Aquitanian) is closely linked with the NP25/NN1 boundary which lies within the lower part of the *Lychnocanoma elongata* Zone at an estimated age of 23.2 Ma. This agrees well with recent estimates of an age of 23 Ma based on geochronometry for this boundary (Kreuzer *et al.*, 1980).
2. The stratotype Burdigalian is biostratigraphically linked with Zone N5 (although the relationship of its base to the N4/N5 or NN1/NN2 boundaries is not clear). It is considered correlative with the N4/N5 boundary here with an estimated age of 22.2 Ma.
3. The position of the base of the Tortonian remains perplexing. Berggren and van Couvering (1974; 168) and van Couvering and Berggren (1977; 287) observe that the Serravallian/Tortonian (= Middle/Late Miocene) boundary occurs within Zone 15 and NN9 and approximately at the *Cannartus petterssoni*/*Ommatartus antepenultimus* boundary, which coincided with the magnetic polarity epoch 10/11 boundary.

Councurrent studies by Ryan *et al.* (1974) suggested correlation of the base Tortonian with a level near the base of epoch 11 with an estimated age of 12 Ma, and that was followed by Berggren and van Couvering (1978) in their revisions to Neogene chronology. However, there remain problems with this correlation. Berggren and van Couvering (1978) have shown that the FAD of *Globigerina nepenthes* is linked with Zone NN8 and that the base of the *Cannartus petterssoni* Zone (radiolarian Zone 7 in Figure 2) was closely linked with Zone NN9 and the LAD of *Globorotalia fohsi lobata*.

The relationship of the base of the *Cannartus petterssoni* Zone to the magnetic polarity stratigraphy is critical to positioning the base of the Tortonian and estimating the age of the Middle/Late Miocene boundary. Ryan *et al.* (1974: Fig. 6) position the base of the *C. petterssoni* Zone questionably with the mid-part of epoch 12 and near the base of Zones NN9 and N14. Cross-correlation between the Tortonian stratotype and deep sea stratigraphy suggested that the base of the Tortonian was situated near the base of epoch 11 with an estimated age of 12 Ma (Ryan *et al.*, 1974). However, Theyer *et al.* (1978) have shown that the *C. petterssoni* Zone is bracketed by the limits of magnetic polarity epoch 11. The correlation linking the base of the *C. petterssoni* Zone and the *Discoaster hamatus* (NN9) Zones with a level within Zone N14 and within epoch 11, suggests that the base of the Tortonian, which is within Zone N15 and near the top of the *C. petterssoni* Zone, may lie close to, if not coincident with, the epoch 10/11 boundary, as originally

Fig. 3. Regional Neogene correlation.

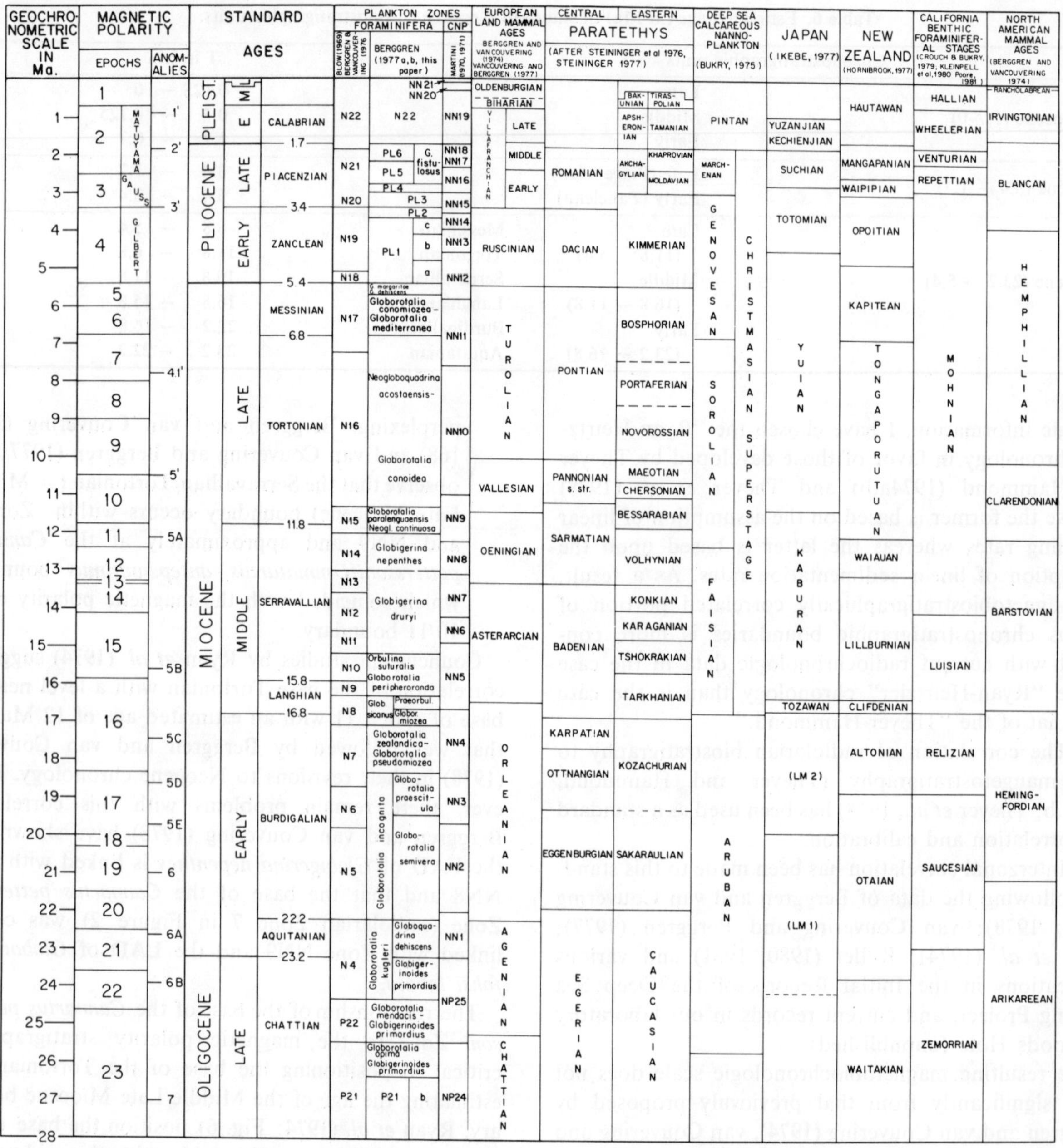

suggested by Berggren and van Couvering (1974). A revised age of 11.8 Ma is estimated for the Middle/Late Miocene boundary here.

Revised estimated ages of standard Neogene chronostratigraphic units are shown in Table 6.

Regional Chronostratigraphic Correlations

The Neogene time scale shown in Fig. 2 provides the framework for the regional chronostratigraphic correlations shown in Fig. 3. These correlations, in turn, are based on an assesment of a large amount of biostratigraphic and paleomagnetic and rediometric data which are tabulated in Table 7. The syntheses presented by Berggren and van Couvering (1974) and van Couvering and Berggren (1977) remain the primary sources of the correlations shown in Fig. 3. These have been supplemented and updated by an analysis of data that has appeared in the past several years, particularly in the Paratethys region and in Japan. Several recent studies dealing with deep sea biostratigraphy have also proved useful in regional biochronologic correlation.

The reader should be aware of the problems inherent in attempting regional correlations on the scale shown here. There is a basic problem in correlating chronostratigraphic units with well defined and biostratigraphically controlled (boundary) stratotpye sections (as are the European units, particularly the late Neogene ones), "open-ended" units with no clearly defined stratotypes or boundary stratotypes (as is the case in

Table 7. Data base for revised Neogene time scale (Fig. 2) and regional chronostratigraphic correlations (Fig. 3).

Region	Reference	Biostratigraphy Biochronology	Paleomagnetic Stratigraphy	Geochronology
Mediterranean-Europe	Colalongo *et al.* (1974)	×		
	d'Onofrio *et al.* (1974)	×		
	Cita (1975, 1976)	×		
	Ryan *et al.* (1974)	×	×	
	Nakagawa *et al.* (1975)		×	
	Odin (1975, 1978)			×
	Kreuzer (1976)			×
	Kreuzer *et al.* (1980)			×
	Odin *et al.* (1978)			×
	Bizon, Muller *et al.* (1979)	×		
	Cavelier & Roger (1980)	×		
Paratethys	Senes (ed., 1977–1978)	×		
	Steininger *et al.* (1976)	×		
	Steininger (1977)	×		
	Steininger & Rögl (1979)			
	Cicha *et al.* (1971, 1975)	×		
	Odin & Hunziker (1975)			×
	Vass *et al.* (1971)			×
	Vass (1975)			×
	Vass *et al.* (1975)			×
	Vass & Bagdasarijan (1978)			×
	Vass (1979)			×
California	Bukry & Crouch (1978)	×		
	Kleinpell (1980)			
Japan and North Pacific	Ikebe *et al.* (1972)	×		×
	Ikebe (1973)	×		×
	Ikebe & Asano (1973)	×		×
	Asano *et al.* (1975)	×		
	Ikebe (1977)	×		×
	Nakagawa *et al.* (1977)	×		×
Equatorial-Temporate Pacific Ocean	Saito *et al.* (1975)	×	×	
	Saito (1977)	×		
	Keller (1978, 1980, 1981)	×		
	Theyer & Hammond (1974a,b)	×	×	
	Theyer *et al.* (1978)	×	×	
Equatorial-Temperate Atlantic Ocean	Berggren (1977a,b)	×		
	Haq & Berggren (1978)	×		
Indian Ocean	Vincent (1977)	×		
Regional (interocenaic)	Thunell (1981)	×		
General Studies	Berggren & van Couvering (1974)	×		
	van Couvering & Berggren (1977)	×		
	McDougall & Page (1975)			×
	McDougall (1977)		×	
	Heirtzler *et al.* (1968)		×	
	Blakely (1974)		×	
	Talwani *et al.* (1971)		×	
	La Brecque *et al.* (1977)		×	
	Ness *et al.* (1980)		×	
	Hailwood *et al.* (1979)		×	

Also papers in Saito and Burckle (eds.) (1975); Takayanagi and Saito (eds.) (1976); Saito and Ujiie (eds.) (1977); Proceedings VI International Congress Mediterranean Neogene Stratigraphy, Bratislava (1975); Proceedings VII International Congress Mediterranean Neogene Stratigraphy, Athens 1979, Ann. Geol. Pays Helleniques Fasc. I-III (1979)

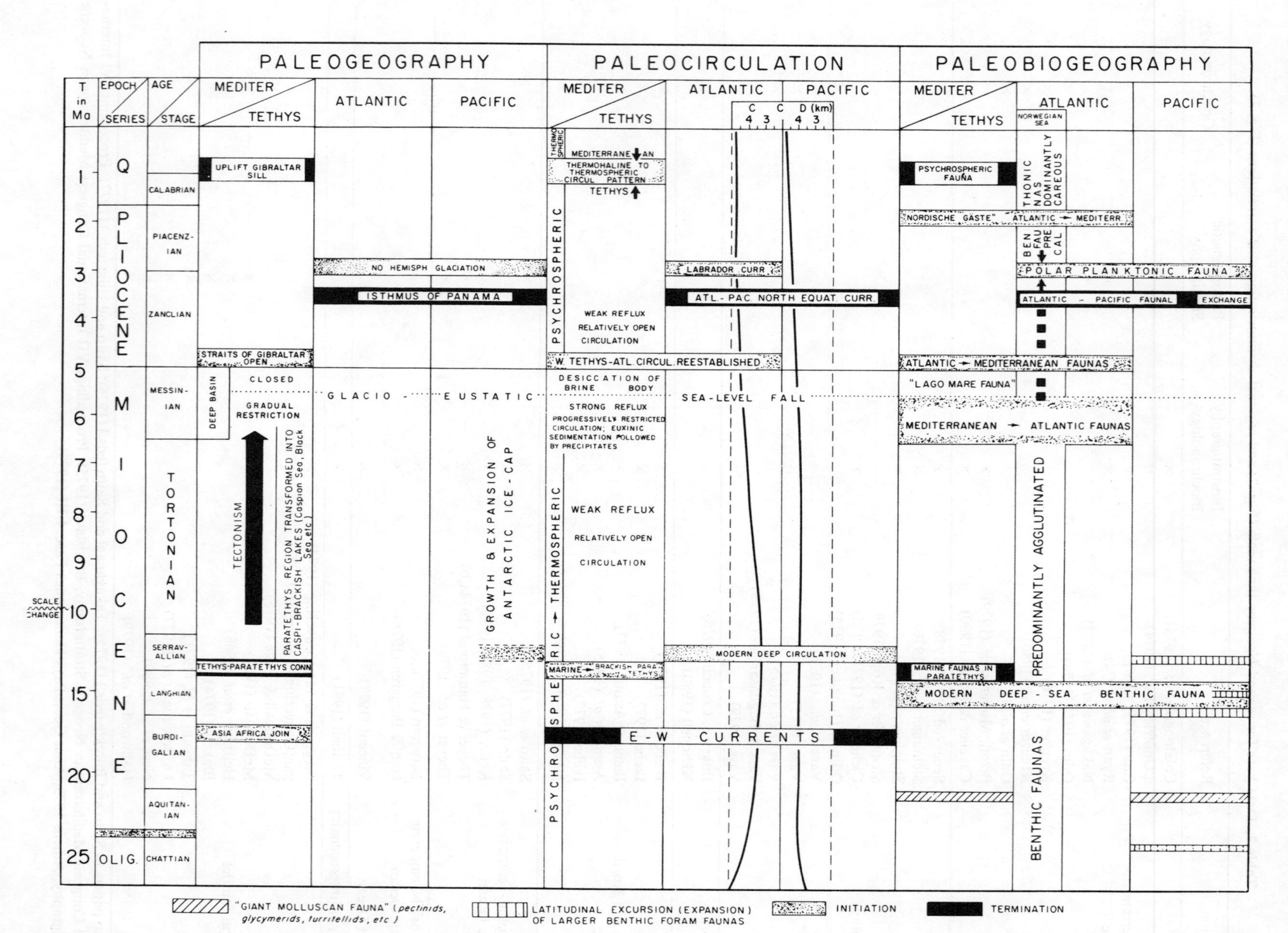

Fig. 4. Chronology and correlation of Neogene paleogeographic, paleocirculation, and paleobiographic events in the Atlantic, Pacific, and Mediterranean-Tethyan regions.

the Paratethys region), units which are based (defined) upon benthic foraminiferal assemblage criteria and which are demonstrably time-transgressive (as is the case with the California Neogene; see Bukry and Crouch, 1978; Kleinpell, 1980), and units based upon essentially biochronologic rather than lithostratigraphic criteria (as is the case with the mammalian "ages"; Berggren and van Couvering, 1974). The result is that the correlations shown in Fig. 3 represent the integration of a large amount of data via first, second, and third order correlation. Nevertheless, I believe that the correlations shown in Fig. 3 accurately represent the current "state of the art" and provide a suitably accurate framework for interpreting the related synchrony or diachrony of regional and global biogeographic and tectonic patterns (Fig. 4; McGowran, 1979). Finally, they provide additional evidence, if such be needed, that "biogeomagnetostratigraphic correlation" is at least the art of the possible and the likely, if not the science of the probable and the precise that we like to credit it with being.

Acknowledgements

This is the first of two papers prepared for presentation at the invitation of Dr. N. Ikebe, leader of IGCP Project 114 (Biostratigraphic Datum Planes of Pacific Neogene) at the International Workshop on Pacific Neogene Biostratigraphy (Osaka and Kobe, 24–29 November 1981). I would like to thank Dr. Ikebe and Dr. M. Chiji, Chairman and Secretary, respectively, of the IWPNB organizing committee, for the invitation to participate at this meeting.

Discussions with, and a critical review of an early draft of, the text of this paper by Drs. Richard Z. Poore (Menlo Park) Edith Vincent (La Jolla), and J.A. Van Couvering (New York) are gratefully acknowledged. I am also grateful to Dr. Poore who generously provided unpublished data on California Neogene biostratigraphy for use in compiling Fig. 3.

This research has been supported by grants OCE 78–19769 and OCE 80–23728 from the Submarine Geology and Geophysics Branch of the National Science Foundation. This is W.H.O.I. Contribution Number 4907.

Postscript

Following the completion of this paper Dr. R.Z. Poore (Menlo Park) informed me of the following:

1. Coccoliths in strata directly and conformably below the type section of the Saucesian stage of California are referable to the *Sphenolithus ciperoensis* (NP 25) to? *Cyclicargolithus abisectus* Subzone of the *Triquetrorhabdulus carinatus* (NN 1) Zone.
2. Three potassium-argon dates averaging 23.8 Ma are available from the Iversen Basalt near Iversen Point in northern California. Interbedded sediments are considered to be either Zemorrian or Saucesian on the basis of benthic foraminifera. Calcareous nannoplankton from these same samples include *C. abisectus*, *Dictyococcites bisectus* and *D. scrippsae* and are thus very close to the Oligocene/Miocene boundary.

References

Anglada, R., 1971: Sur la limite Aquitanian-Burdigalien. Sa place dans la Sud-Est de la France. *C.R. Acad. Sci. Paris*, v. **272**, p. 1948–1951.

Asano, K., Takayanagi, Y. and Takayama, T., 1975: Late Neogene epoch boundaries in Japan. *In*. Saito, T. and Burckle, L. H. eds., *Late Neogene Epoch Boundaries*. Micropaleontology Press, p. 115–123.

Benda, L., Meulenkamp, J. E., Schmidt, R.R., Steffens, P., and Zachariasse, J., 1977: Biostratigraphic correlations in the Eastern Mediterranean Neogene. 2. Correlation between sporomorph associations and marine microfossils from the Upper Oligocene-Lower Miocene of Turkey. *Newsl. Stratigr.*, v. **6**, no.1, p. 1–22.

Berggren, W.A., 1972: A Cenozoic time scale: implications for regional paleogeography and paleobiogeography. *Lethaia*, v. **5**, p. 195–215.

Berggren, W.A., 1973: The Pliocene time scale: calibration of planktonic foraminiferal and calcareous nannoplankton zones. *Nature*, v. **243**, no. 5407, p. 391–397.

Berggren, W.A., 1977: Late Neogene planktonic foraminiferal biostratigraphy of the Rio Grande Rise (South Atlantic). *Mar. Micropaleontol.*, v. **2**, p. 265–313.

Berggren, W.A., 1977b: Late Neogene planktonic foraminiferal biostratigraphy of DSDP Site 357 (Rio Grande Rise). *In* Supko, P.R., Perch-Nielsen, K., *et al.*, eds., *DSDP*, *Initial Reports*, U. S. Government Printing Office, Washington, D.C., v. **39**, p. 591–614.

Berggren, W.A. and van Couvering, J.A., 1974: The Late Neogene; biostratigraphy, geochronology, and paleoclimatology of the last 15 million years in marine and continental sequences. *Paleogeog.*, *Paleoclimatol.*, *Paleoecol.*, v. **16**. no. 1–2, p. 1–216.

Berggren, W.A. and van Couvering, J.A., 1978: Biochronology. *Am. Assoc. Petrol. Geol.*, *Studies in Geol.*, no. 6, p. 39–55.

Berggren, W.A., Kent, D.V. and van Couvering, J.A. (in press). Neogene geochronology and chronostratigraphy. *In* Snelling, N.J., ed., *Geochronology and the Geological Record*. *Geological Society of London*, *Special Paper*.

Bizon, G., Muller, C., *et al.*, 1979: Report of the Working Group on Micropaleontology (*sic*!). *Ann. Geol. Pays. Hellen.*, *Tome hors Serie*, *1979*, fasc., v. **3**, p. 1335–1364.

Blakeley, R.J., 1974: Geomagnetic reversals and crustal spreading rates during the Miocene. *Jour. Geophys. Res.*, v. **79**, p. 2979–2985.

Bukry, D., 1973: Low latitude coccolith biostratigraphic zonation. *In* Edgar, N.T., Saunders, J.B. *et al.*, eds., 1973: *DSDP*, *Initial Reports,* U.S. Government Printing Office, Washington, D.C., v. **15**, p. 685–703.

Bukry, D., 1975a: Coccolith and silicoflagellate stratigraphy, northwestern Pacific Ocean, Deep Sea Drilling Project 32. *In* R.L. Larson, R. Moberly *et al.*, eds., *DSDP, Initial Reports,* U.S. Government Printing Office, Washington, D.C., v. **32**, p. 677–701.

Bukry, D., 1975b: New Miocene to Holocene stages in the ocean basins based on calcareous nannoplankton zones. *In* T. Saito and L.H. Burckle eds., *Late Neogene Epoch Boundaries.* Micropaleontology Press, New York, p. 162–166.

Cavalier, C. and Roger, J., eds., 1980: Les etages francais et leurs stratotypes. *Bur. Rech. Geol. Min., Mem.*, no. 109, p. 3–295.

Cicha, I., Hagn, H. and Martini, E., 1971: Das Oligozan und Miozan des Alpen und des Karpaten. Ein Vergleich mit Hilfe planktonischer organismen. *Mitt. Bayer, Staatssaml. Paleont. Hist. Geol.*, **11**, p. 279–293.

Cicha, I., Marinescu, F., Senec, J., *et al.*, 1975: Correlation des Neogene de la Paratethys Centrale. *Geol. Surv. Prag*, p. 1–33.

Cita, M.B., 1975: The Miocene/Pliocene boundary: history and definition. *In* Saito, T. and Burckle, L.H. eds., *Late Neogene Epoch Boundaries,* Micropaleontology Press. New York, p. 1–30.

Cita, M.B., 1976: Planktonic foraminiferal biostratigraphy of the Mediterranean Neogene. *In* Takayanagi, Y. and Saito, T., eds., *Progress in Micropaleontology.* Micropaleontology Press, New York. p. 47–68.

Cita, M.B. and Gartner, S., 1973: The Stratotype Zanclean foraminiferal and nannofossil stratigraphy. *Riv. Ital. Pal. Strat.*, v. **79**, no.4, p. 503–558.

Crouch, J.K. and Bukry, D., 1979: Comparison of Miocene provincial foraminiferal stages to coccolity zones in the California continental borderland. *Geology,* v. **7**, p. 211–215.

Dalrymple, G.B., 1979: Critical tables for conversion of K-Ar ages from old to new constants. *Geology,* v. **7**, p. 558–560.

Hailwood, E.A., Bock, W., Costa, L., Dupeuble, P.A., Muller, C., and Schnitker, D., 1978: Chronology and biostratigraphy of northeast Atlantic sediments, DSDP Leg 48. *In* Montadert, L., Roberts, D.G., *et al.*, eds., *DSDP, Initial Reports,* U. S. Government Printing Office, Washington, D.C., v. **48**, p. 1119–1141.

Haq, B. and Berggren, W.A., 1978: Late Neogene calcareous plankton biochronology of the Rio Grande Rise (South Atlantic Ocean). *Jour. Paleontol.*, v. **52** no.6, p. 1167–1194.

Hardenbol, J. and Berggren, W.A., 1978: A new Paleogene numerical time scale. *Am. Assoc. Petrol. Geol., Studies in Geology*, no. 6, p. 213–234.

Harrison, C.G.A., McDougall, I. and Watkins, N.D., 1979: A geomagnetic field reversal time scale back to 13.0 million years before present. *Earth and Planetary Science Letters*, **42**, p. 143–152.

Hays, J.D., Saito, T., Opdyke, N.D., and Burckle, L.H., 1969: Pliocene-Pleistocene sediments of the equatorial Pacific—their paleomagnetic biostratigraphic and climatic record. *Geol. Soc. Amer., Bull.*, v. **80**, p. 1481–1514.

Heirtzler, J.R., Dickson, G.O., Herron, E.M., Pitman, W.C. III, and Le Pichon, X., 1968: Marine magnetic anomalies, geomagnetic field reversals, and motions of the ocean floor and continents. *Jour. Geophys. Res.*, v. **73**, p. 2119–2136.

Hornibrook, N. de B., 1977: The Neogene (Miocene-Pliocene) of New Zealand. *In* Saito, T. and Ujiie, H., eds., *First Intern. Congress Pacific Neogene Stratigraphy, Tokyo, 1976, Proc.*, p. 145–150.

Ikebe, N., 1973: Neogene biostratigraphy and radiometric time scale. *Osaka City Univ., Jour. Geosci.*, v. **16**, no.4, p. 51–67.

Ikebe, N. and Asano, K., 1973: Interregional correlation of the Neogene based on planktonic datum-levels and radiometric datings. *Geol. Soc. Japan, Mem.*, no. 8, p. 203–214 (in Japanese).

Ikebe, N., Takayanagi, Y., Chiji, M., and Chinzei, K. eds., 1972: Neogene biostratigraphy and radiometric timescale of Japan—an attempt at intercontinental correlation. *Pacific Geology*, v. **4**, p. 39–78.

Ikebe, N., *et al.*, 1977: Summary of bio- and chronostratigraphy of the Japanese Neogene. *In* Saito, T. and Ujiie, H. *First Intern. Congress Pacific Neogene Stratigraphy, Tokyo, 1976, Proc.* p. 93–114.

Keller, G., 1978: Late Neogene biostratigraphy and paleooceanography of DSDP Site 310 Central North Pacific and correlation with the Southwest Pacific. *Mar. Micropaleont.*, v. **3**, p. 97–119.

Keller, G., 1980: Middle to Late Miocene planktonic foraminiferal datum levels and paleoceanography of the North and Southeastern Pacific Ocean. *Mar. Micropaleont.*, v. **5**, p. 249–281.

Keller, G., 1981: Early to Middle Miocene planktonic foraminiferal datum levels of the equatorial and subtropical Pacific. *Mar. Micropaleont.*, v. **26** no.4, p. 372–391.

Kleinpell, R.M. *et al.*, 1980: The Miocene stratigraphy of California revisited. *Am. Assoc. Petrol. Geol., Studies in Geology*, no.11, 182 p.

Kreuzer, H., 1976: Problems of dating glauconite and some results of dating Tertiary glauconite from N.W. Germany. *In* Tobien, H. ed., *The N.W. European Tertiary Basin, Rep.*, 1: Ad Hoc Meeting Mainz (FRG), Dec. 11–12, 1975, p. 40–42.

Kreuzer, H., Daniels, C.H. von, Gramann, F., Harre, W., and Mattrat, B., 1973: K/Ar dates of the N.W. German Tertiary Basin (abstract), *Fortschr. Mineral. 50., Beih.* 3, p. 94–95.

Kreuzer, H., Kuster, H., Daniels, C.H. von, Hinsch, W., Spiegler, D., and Harre, W., 1980: K-Ar dates for Late Oligocene glauconites from N.E. Lower Saxony (N.W. Germany), *Geol. Jb.*, A54, p. 61–74.

LaBrecque, J., Kent, D.V. and Cande, S.C., 1977: Revised magnetic polarity time scale or Late Cretaceous and Cenozoic time. *Geology*, v. **5**, p. 330–335.

Martini, E., 1971: Standard Tertiary and Quaternary calcareous nannoplankton zonation. *2nd Conf. Planktonic*

Microfossils, Proc., v. **2**, p. 739–785.

McDougall, I., 1977: The present status of the geomagnetic polarity time scale. *In* McElhinny, M.W., ed., *The Earth: its origin, structure and evolution*. Academic Press, London.

McDougall, I., Watkins, N.D. and Kristjansson, L., 1976a: Geochronology and paleomagnetism of a Miocene-Pliocene lava sequence at Bessastadaa, eastern Iceland. *Am. J. Sci.*, v. **276**, p. 1078–1095.

McDougall, I., Watkins, N.D., Walter, G.P.L., and Kristjansson, L., 1976b: Potassium-argon and paleomagnetic analysis of Icelandic lava flows: Limits on the age of anomaly 5. *Jour. Geophys. Res.*, v. **81**, p. 1505-1512.

McDougall, I., Saemundsson, K., Johannesson, H., Watkins, N.D., and Kristjansson, L., 1977: Extension of the geomagnetic polarity time scale to 6.5 my: K-Ar dating, geological and paleomagnetic study of a 3,500-m lava succession in western Iceland. *Geol. Soc. Am., Bull.*, v. **88**, p. 1–15.

McGowran, B., 1979: Some Miocene configurations from an Australian standpoint. *Ann. Geol. Pays Hellen., Tome hors series, 1979*, fasc. II, p. 767–77.

Nakagawa, H., Niitsuma, N., Kimura, K., and Sakai, T., 1975: Magnetic stratigraphy of Late Cenozoic stages in Italy and their correlation in Japan. *In* Saito, T. and Burckle, L.H., eds., *Late Neogene Epoch Boundaries*. Micropaleontology Press, New York, p. 64–70.

Nakagawa, H., Kitamura, N., Takayanagi, T., Sakai, T., Oda, M., Asano, K., Niitsuma, N., Takayama, T., Matoba, Y., and Kitazato, H., 1978: Magnetostratigraphic correlation of Neogene and Pleistocene between the Japanese Islands, Central Pacific, and Mediterranean regions. *In* Saito, T. and Ujiie, H., eds., *First Internat. Congress Pacific Neogene Stratigraphy, Tokyo, 1976, Proc.*, p. 285–310.

Ness, G., Levi, S. and Couch, R., 1980: Marine magnetic anomaly time scales for the Cenozoic and Late Cretaceous: a precis, critique, and synthesis. *Geophys. and Space Physics, Rev.*, v. **18**, no. 4, p. 753–770.

Odin, G.S., 1973: Resultats de datations radiometriques dans les series sedimentaires du Tertiare de l'Europe occidentale. *Geographie Phys. Geologie Dynam., Rev.*, v. **15**, no. 3, p. 317–330.

Odin, G.S., 1975: De glauconiarum, constitutione, origine, aetateque. *These Doct. Etat, Paris*, 280 p.

Odin, G.S., 1978: Isotopic dates or a Paleogene time scale. *Am. Assoc. Petrol. Geol., Studies in Geology*, no. 6, p. 247–257.

Odin, G.S., Curry, D., and Hunziker, J.C., 1978: Radiometric dates from N.W. European glauconites and the Paleogene time scale. *Geol. Soc. Lond., Jour.*, v. **135**, p. 481–497.

Odin, G.S., Hunziker, J.C. and Lorenz, C.R., 1975: Age radiometrique du Miocene inferieur en Europe occidentale et centrale: *Geol. Rundschau*, v. **64**, no. 2, p. 570–592.

Okada, H. and Bukry, D., 1980: Supplementary modification and introduction of code numbers to the low-latitude coccolith biostratigraphic zonation (Bukry, 1973; 1975). *Marine Micropaleontology*, v. **5**, p. 321–325.

Onofrio, S.d', Giannelli, L., Iaccarino, S., Marlotti, E., Romeo, M., Salvatorini, G., Sampo, M., and Sprovieri, R., 1975: Planktonic foraminifera of the Upper Miocene from some Italian sections and the problem of the lower boundary of the Messinian. *Soc. Paleont. Ital., Bull.*, v. **14** no. 2, p. 177–196, pls. 1–5.

Opdyke, N.D., Burckle, L.H. and Todd, A., 1974: The extension of the magnetic time scale in sediments of the Central Pacific Ocean. *Earth and Planetary Sci. Letters*, v. **22**, p. 300–306.

Ryan, W.B.F., Cita, M.B., Dreyfus Rawson, M., Burckle, L.H., and Saito, T., 1974: A paleomagnetic assignment of Neogene stage boundaries and the development of isochronous datum planes between the Mediterranean, the Pacific and Indian Oceans in order to investigate the response of the world ocean to the Mediterranean "salinity crisis". *Ital. Paleont. Rev.*, v. **80** no.4, p. 631–688.

Saito, T., 1977: Late Cenozoic planktonic foraminiferal datum levels: the present state of knowledge toward accomplishing pan-Pacific stratigraphic correlation. *First Intern. Congress Pacific Neogene Stratigraphy, Tokyo, 1976, Proc.*, p. 61–80.

Saito, T., Burckle, L.H., and Hays, J.D., 1975: Late Miocene to Pleistocene biostratigraphy of equatorial Pacific sediments. *In* Saito, T. and Burckle, L.H. eds., *Late Neogene Epoch Boundaries*, Micropaleontology Press, New York, p. 226–264.

Senes, J., ed., 1967–1978: Chronostratigraphie und Neostratotypen-Miozan der Zentralen Paratethys.-1 (Karpatien), 1967; 2 (Eggenburgien), 1971; 3 (Ottnangien); 4 (Sarmatien), 1974; 5 (Egerien), 1975; 6 (Badenien), 1978. Slovak Akad. Vied., Bratislava.

Steininger, F., 1977: Integrated assemblage-zone biostratigraphy at marine-nonmarine boundaries: examples from the Neogene of Central Europe. *In* Kauffman, E.G. and Hazel. J.E. eds., Concepts and Methods in Biostratigraphy. Dowden, Hutchinson and Ross, Inc., Stroudsburg, p. 235–256.

Steininger, F.F., Rögl, F. and Martini, E., 1976: Current Oligocene/Miocene biostratigraphic concept of the Central Paratethys (Middle Europe). *Newsl. Stratigr.*, v. **4**, no. 3, p. 174–202.

Steininger, F., and Rögl, F. 1979: The Paratethys history–a contribution towards the Neogene dynamics of the Alpine Orogene (an abstract). *Ann. Geol. Pays Hellen., Tome hors Serie, 1979*, fasc. III, p. 1153–1165.

Talwani, M., Windisch, C.C., Langseth, M.G., Jr., 1971: Reykyanes Ridge Crest: a detailed geophysical study. *Jour. Geophys. Res.*, v. **76**, p. 473–517.

Theyer, F. and Hammond, S.R., 1974a: Paleomagnetic polarity sequence and radiolarian zones, Brunhes to polarity epoch 20. *Earth and Planetary Sci. Lett.*, v. **22**, p. 307–319.

Theyer, F. and Hammond, S.R., 1974b: Cenozoic magnetic time scale in deep-sea cores: completion of the Neogene. *Geology*, v. **2**, p. 487–492.

Theyer, F., Mato, C.Y. and Hammond, S.R. 1978: Paleomagnetic and geochronologic calibration of latest Oligocene to Pliocene radiolarian events, Equatorial Pacific.

Marine Micropaleontlogy, v. **3**, p. 377–395.

Thunell, R.C., 1981: Late Miocene–early Pliocene planktonic foraminiferal biostratigraphy and paleoceanography of low latitude marine sequences. *Marine Micropaleontology*, v. **6**, p. 71–90.

Turner, D.L., 1970: Potassium-argon dating of Pacific Coast Miocene foraminiferal stages. *In* Bandy, O.L. éd., *Paléontologic zontation and radiometric dating. Geol. Soc. Amer., Spec. Paper,* 124, p. 91–129.

Van Couvering, J.A. and Berggren, W.A., 1977: Biostratigraphical basis of the Neogene time scale. *In* Kauffman, E.G. and Hazel, J.E. eds., *Concepts and Methods in Biostratigraphy*, Dowden, Hutchinson and Ross, Inc., Stroudsburg, p. 283–306.

Vass, D., 1979: Review of activity-Working Group for radiometric age and paleomagnetism (1975–1978). *Ann. Geol. Pays. Hellen, Tome hors serie 1979,* fasc. III, p. 1427–1441.

Vass, D., 1975: Report of the working group on radiometric age and paleomagnetism. *6th Reg. Comm. Mediterranean Neogene Stratigraphy, Proc.*, p. 103–117.

Vass, D. and Bagdasarjan, G.P., 1977: A radiometric time scale for the Neogene of the Paratethys region. *Am. Assoc. Petrol. Geol., Studies in Geology*, no. 6, p. 179–203.

Vass, D. and Slavik, J., 1975: The radiometric calibration of Paratethys Neogene. *Geol. Prace. Zpr.*, v. **63**, p. 131–139.

Vass, D., Bagdasarjan, G.P., and Konecny, V., 1971: Determination of the absolute age of the West Carpathians Miocene. *Foeldt. Koezl*, v. **101**, no. 2–3, p. 321–327.

Vass, D., Slavik, J. and Bagdasarjan, G.P., 1975: Radiometric time scale for Neogene of Paratethys (toAugust 1, 1974): *6th Cong. Reg. Comm. Mediterranean Neogene Stratigraphy, Proc.*, p. 159–197.

Vincent, E., 1977: Indian Ocean Neogene planktonic foraminiferal biostratigraphy and its paleoceanographic implications. *In Indian Ocean Geology and Biostratigraphy* Heirtzler, J.R., Bolli, H.M., Davies, T.A., Saunders, J.B. and Sclater, J.G., eds., Amer. Geophysical Union, Washington, D.C., p. 469–584.

Postscript: Since the completion of this paper a new and revised Neogene geochronology has been developed by Berggren *et al.* (in press). It is an empirically based scale based upon the integration of, and first order correlation between, biostratigraphic and magnetostratigraphic data. Radiometric calibration points are based solely upon high temperature dates associated with magnetic polarity intervals which can be correlated to the standard geomagnetic polarity sequence derived from magnetic anomalies.

II - 2

Neogene Planktonic Foraminiferal Biostratigraphy and Biogeography: Atlantic, Mediterranean, and Indo - Pacific Regions

William A. BERGGREN

Abstract

A summary is presented of global Neogene planktonic foraminiferal distribution patterns primarily (but not exclusively) based on studies in the first 50 volumes of the Initial Reports of the Deep Sea Drilling Project. The major trends that can be determined include: 1) gradual, predominantly latitudinal provincialization of faunas, with the establishment of a clear-cut threefold faunal subdivision into tropical, temperate, and subarctic in the northern hemisphere and a truly Antarctic province in the southern hemisphere by late Miocene time; a truly Arctic province is not seen in the northern hemisphere until the initiation of polar glaciation in mid-Pliocene time (*ca.* 3 Ma); 2) establishment of disparate faunal distribution patterns in the Pliocene-Pleistocene owing to the elevation of the Isthmus of Panama in mid-Pliocene time; 3) relatively rapid faunal extinction during the late Pliocene, probably linked with dramatic climatic changes.

As a result of these trends in latitudinal biogeographic provincialism, a need for multiple biostratigraphic zonation schemes for the purpose of regional (inter- and intra-oceanic) correlation has arisen, and a suggested scheme is presented for the Atlantic and Indo-Pacific region.

Introduction

This is the second of a two-part contribution to the meeting of IGCP Project 114 (Biostratigraphic Datum Planes of the Pacific Neogene) held in Osaka and Kobe, Japan, between November 25–29, 1981. In the first paper I presented a review of current Neogene chronologies as a framework for regional correlation of Atlantic, Mediterranean and Indo-Pacific Neogene stratigraphies. This paper is devoted to a summary of global Noegene planktonic foraminiferal distribution patterns as recorded primarily, but not exclusively, in the Initial Reports of the Deep Sea Drilling Project (vols. 1–50).

The role of oceanic circulation patterns in paleobiogeographic provincialism and morphologic phenotypic variability among planktonic foraminifera has been discussed at length elsewhere (Berggren and Hollister, 1974, 1977; Kennett, 1976) and will be treated only briefly here (it is discussed at appropriate places in the text below). Several examples will be cited, here, however, by way of introduction to the main text which follows.

The influence of dynamic factors in oceanic circulation upon the paleobiogeographic distribution patterns of planktonic foraminifera are clearly shown by Saito (1977) in the successive reduction of the geographic (latitudinal) extent of the warm-water group of *Globorotalia fohsi* during the middle Miocene coincident with a major oxygen isotopic shift associated with the creation of a major Antarctic Ice Sheet (Savin *et al.*, 1975; Shackleton and Kennet, 1975) and the development of a modern-day deep-water circulation pattern (Schnitker, 1980).

In Middle Miocene time a major change also occurred in the NW Pacific paleocurrent system with the replacement by a temperate cold-water planktonic foraminiferal fauna of a nearly tropical early Miocene fauna north of 360 latitude (Saito, 1963). Similar faunal changes have been noted by Ingle (1973) in the California Neogene. This is also the time the *Globorotalia miozea* group appeared in the North and South Atlantic and became a predominant factor in subsequent Neogene paleobiogeography and biostratigraphy.

The presence of a southward flowing California current along the west coast of North America throughout the Neogene has resulted in an anomalously extensive southward penetration of subarctic to temperate microfaunas and microfloras which has precluded the application of low latitude zonation(s) to the California Neogene sections (Ingle, 1980). The use of

Department of Geology and Geophysics, Woods Hole Oceanographic Institution, Woods Hole, Massachusetts 02543 and Department of Geology, Brown University, Providence, Rhode Island 02912, U.S.A.

zonations based on siliceous microfossils (particularly diatoms) in low and high latitudes has made possible the correlation of the west coast (California) Neogene sections with the tropical and high latitude sequences astride the California Current province and thus with a Neogene chronologic scale (Berggren and van Couvering, 1974). Summaries of the correlation between Neogene zonations of California have been presented by a number of authors in the past decade; see in particular papers in Addicott, ed., 1978.

The pronounced, predominantly latitudinal, biogeographic provincialism of planktonic foraminifera during the Neogene emphasizes the need for multiple biostratigraphic zonation schemes for the purpose of regional (inter- and intra-oceanic) correlation. As a result a plethora of zonal schemes has developed over the past decade which has generally, but not necessarily exclusively, led to an improvement in resolution in regional biostratigraphic correlation (see, for instance, the lament expressed by Burckle *in* Burckle and Opdyke, 1977: 264, "of all these (diatom) zonal schemes which ones are most useful?"). I would agree with Burckle's expressed reliance on the criteria of good core-to-core correlation, use of conservative faunal elements, and paleomagnetic correlation.

In this paper I shall outline the general Neogene geographic distribution patterns of planktonic foraminiferal faunas based primarily on the Initial Reports of the Deep Sea Drilling Report (volumes 1–50). The Atlantic (and Gulf of Mexico), Pacific and Southern (Antarctic), Mediterranean Sea and Indian Ocean (and Red Sea) are treated in that order with the discussion proceeding generally from north to south. Comparison between various faunal zonation schemes are presented where appropriate in the text and at the conclusion a multiple, predominantly latitudinally dependent, zonation scheme is presented for the Atlantic and Indo-Pacific regions.

Temperate and Subarctic North Atlantic (> Lat. 45° N)

Temperate to Subarctic Neogene planktonic foraminiferal faunas (> Lat. 45° N) occur at sites drilled on DSDP Legs 12 (Berggren 1972b) and 38 and IPOD Legs 48 (Krasheninnikov, 1979) and 49 (Poore, 1979; Table 1).

A virtually complete Neogene stratigraphic sequence has been cored at Site 400 (Bay of Biscay, at 4400 m), and relatively complete sequences have been (discontinuously) cored at Sites 116 and 406 on Rockall Plateau. An essentially complete Middle Miocene to Recent sequence occurs at Site 408 (Lat. 63° N) and Upper Miocene-Recent sequences at Sites 407 (Lat. 64° M) and 410 (Lat. 45° N).

Representative Lower Miocene faunas occur at Sites 116 and 118, whereas Middle Miocene faunas occur at Sites 116, 119, 400, and 408. Particularly good Upper Miocene faunas occur at Sites 400 and 410 where the stratigraphic thickness is in excess of 150 m and 125 m, respectively. Representative Upper Miocene faunas also occur at Sites 116, 118, 119, 341 (the latter at Lat. 67° N, the northernmost Upper Miocene cored in the Atlantic), 407, and 408.

A complete Lower Pliocene occurs at Sites 111 (Labrador Sea, Lat. 50° N; subtropical fauna influenced by the Gulf Stream), 400 (Bay of Biscay, Lat. 47° N; temperate fauna), and 407 (Lat. 64° N; cold temperate fauna). Relatively complete Lower Pliocene sections also occur at Sites 408 (Lat. 63° N) and 410 (Lat. 45° N), in addition to Site 400 mentioned above.

The Upper Pliocene is well represented at Sites 111–116, 118, 119, 341, 344, 400, 407, 408, and 410.

Pleistocene occurs at most DSDP sites in the North Atlantic, but particularly complete recovery is noted at Sites 116, 341, and 407–410.

Lower Miocene faunas at Sites 116 and 118 are characterized by catapsydracids *(dissimilis, unicavus)* globoquadrinids (*baroemoenensis, dehiscens, praedehiscens*), *"Globorotalia" siakensis, Sphaeroidenellopsis seminulina,* and *Globigerinoides trilobus.* Praeorbulinids have been recorded at Sites 116, 118, and 406 indicating the presence of this stratigraphically important group at least as far north as Lat. 57° N.

Middle Miocene faunas are dominated by globigerinids (*druryi, nepenthes, praebulloides*), globoquadrinids (*dehiscens*), *Sphaeroidinellopsis* spp., orbulinids, and *"Globorotalia" mayeri* and *"G." siakensis.* Several lineages of keeled globorotaliids develop during the Middle Miocene. The *G. menardii* lineage *(G. archeomenardii* and *G. praemenardii*) occurs at several sites (116, 400, 407) between Lat. 57°–64° N. The *G. miozea-conoidea* group occurs at Sites 112 (Labrador sea) 118 (Bay of Biscay), and as far north as Lat. 63° at Site 408 (Fig. 1).

By Late Miocene time a distinct latitudinal differentiation is seen in North Atlantic planktonic foraminiferal faunas. Sites (341, 407, 408) north of Lat. 60° contain low diversity subarctic assemblages consisting of *Globigerina bulloides, Neogloboqadrina atlantica* (dextral), *N. acostaensis, Turborotalita quinqueloba,* orbulinids, and forms identified as *Neogloboquadrina pachyderma.* South of Lat. 55° N (Sites 400, 403, 404, 406, 410) these forms are joined by *Neogloboquadrina humerosa* and elements of the *Globorotalia tumida* (*merotumida-plesiotumida*), *G. menardii (G. pseudomiocenica),* and *G. conoidea-conomiozea* groups (Fig. 2). *Globoquadrina dehiscens* occurs in the Upper Miocene at Sites 111, 400, 403, 404, 406, and 410 at depths ranging from 1797–4399 m (*cf.* the erratic upper limit recorded for this taxon elsewhere). Upper Miocene faunas

Table 1. Temperate and Subarctic North Atlantic (> 45° N).

Leg	Area	Site No.	Latitude (N)	Longitude (W)*	Geographic Setting	Depth (m)	Tect. Hist.	Paleontol.	Remarks
12	North Atlantic	111	50°25.57′	46°22.05′	Orphan Knoll	1,797	S	W.A. Berggren	1. Predominantly temperate-subarctic faunas denominated by *N. atlantica.* 2. Subtropical faunas in S. Labrador Sea until mid-Pliocene (3 Ma). 3. *G. miozea-conoides* group common. 4. Appearance of subarctic faunas and glaciation at 3 Ma.
		112	54°01′	46°36′	S Labrador Sea	3,657	S		
		113	56°48′	48°20′	Axis, Labrador Sea	3,619	S		
		114	59°56′	26°48′	E Flank Mid-Atlantics Ridge	1,927	S		
		115	58°54.4′	21°07′	Basin E of Reykjanes Ridge	2,883	S		
		116	57°29.76′	15°55.46′	E side Hatton-Rockall Basin	1,151	S		
		118	45°02.65′	09°00.63′	Biscay Abyssal Plain	4,901	S		
		119	45°01.90′	07°58.49′	Cantabria Seamount, Bay of Biscay	4,447	S		
38	Norwegian-Greenland Sea	336	63°21.06′	07°47.27′	Iceland-Faroe Ridge	811	S	J.E. Van Hinte	1. Mid-Pliocene-Pleistocene subarctic faunas. 2. *N. atlantica-pachyderma* = dominant forms.
		337	64°52.30′	05°20.51′	Norway Basin	2,631	S		
		338	67°47.11′	05°23.26′E	Vøring Plateau	1,297	S		
		341	67°20.10′	06°06.64′E	Vøring Plateau	1,439	S		
		342	67°57.04′	04°56.02′E	Vøring Plateau	1,303	S		
		343	68°42.91′	05°45.73′E	Vøring Plateau	3,131	S		
		344	76°08.98′	07°52.52′E	Knipovich Ridge	2,156	S		
		345	69°50.23′	01°14.26′	Lofoten Basin	3,195	S		
		346	69°53.35′	08°41.14′	Jan Mayan Ridge	723	S		
		347	69°52.31′	08°41.80′	Jan Mayen Ridge	745	S		
		348	68°30.18′	12°27.72′	Iceland Plateau	1,763	S		
		349	69°12.41′	08°05.80′	Jan Mayen Ridge	915	S		
		350	67°03.34′	08°17.68′	E extension of Jan Mayen Ridge	1,275	S		
48	North Atlantic	399	47°23.4′	09°13.3′	Lower cont. slope Bay of Biscay	4,399	S–↓	D. Schnitker	1. Temperate faunas. 2. *N. atlantica* common in Pliocene.
		400	47°22.90′	09°11.90′	Meriadzek Escarp., Cont. Margin, Bay of Biscay	4,399	S–↓		
		401	47°25.65′	08°48.62′	Meriadzek Terr., Cont. margin, Bay of Biscay	2,495	S–↓		
		402	47°52.48′	08°50.44′	N. Cont. Margin, Day of Biscay	2,339	S-↓		
		403	56°08.31′	23°17.64′	SW Margin, Rockall Plateau	3,201	S		
		404	56°03.13′	23°14.95′	SW Margin, Rockall Plateau	2,306	S		
		405	55°20.18′	22°03.49′	SW Margin, Rockall Plateau	2,958	S		
		406	55°15.50′	22°05.41′	SW Margin, Rockall Plateau	2,907	S		
49	North Atlantic	407	63°56.32′	30°34.56′	W Flank Reykjanes Ridge (Anom. 13)	2,472	↓	R.Z. Poore	1. Temperate-subarctic faunas. 2. Subarctic faunas and glaciation appear at 3 Ma. 3. *G. conoidea-conomiozea* group common in late Miocene-Pliocene.
		408	63°22.63′	28°54.71′	W Flank Reyjanes Ridge (Anom. 6)	1,624	↓		
		409	62°36.98′	25°57.17′	W Crestal Area of Reykjanes Ridge (Anom. 2)	832	↓		
		410	45°30.51′	29°28.56′	Mid Atlantic Ridge (Anom. 5)	2,975	S–↓		

S = Stable; U = Uplift; ↓ = Subsidence during the Neogene.
* Unless otherwise noted.

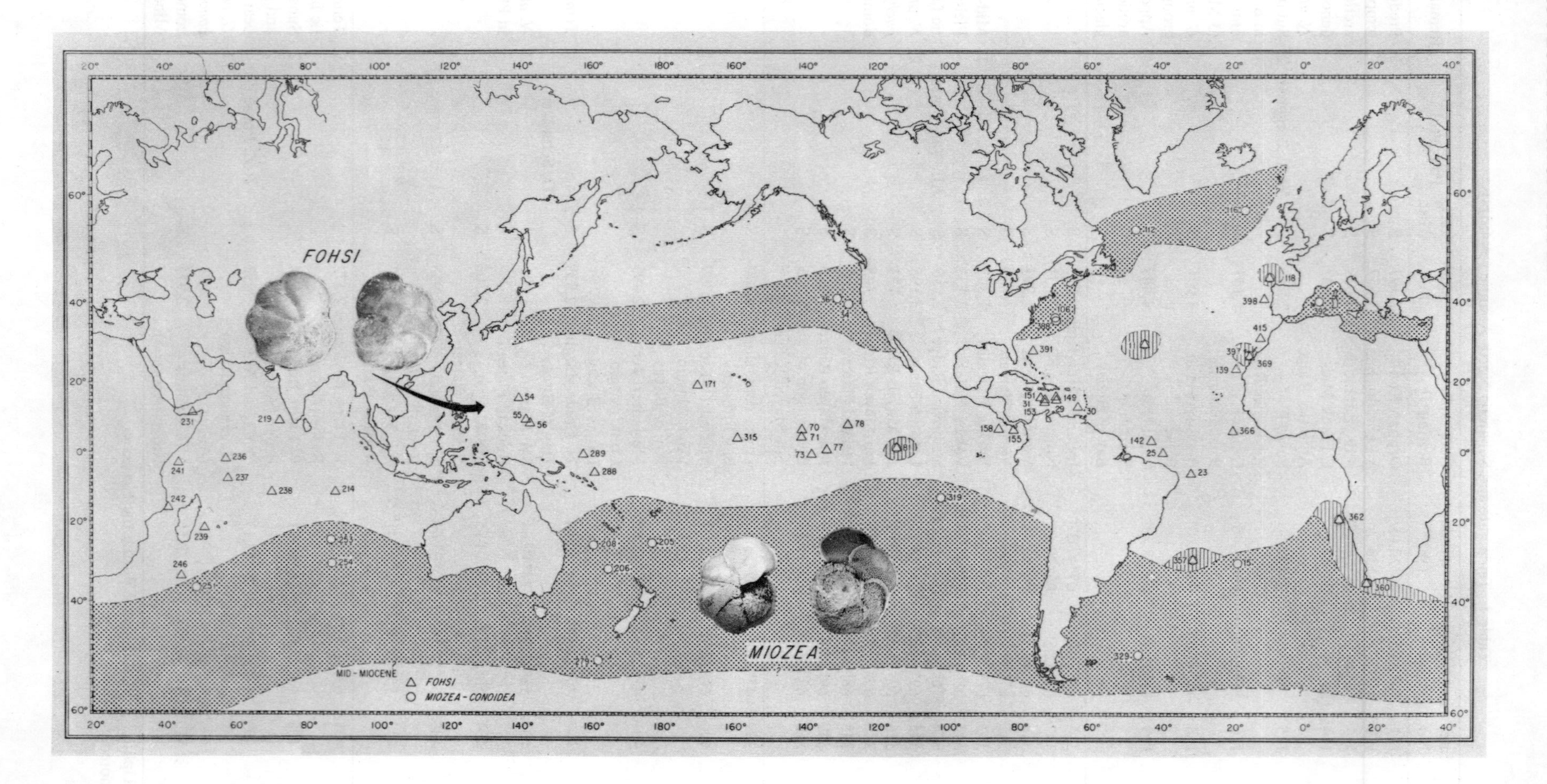

Fig. 1. Global oceanic distribution of *Globorotalia fohsi* (low to mid-latitude, blank) and *G. miozea-conoidea* (dotteed) groups in the Middle Miocene. Area of overlap of the two groups shown by vertical lines. Note southward penetration (to 30° S) of *G. fohsi* in western South Atlantic and northward penetration (to 20° S) of *G. miozea* in eastern South Atlantic reflecting the presence of warm and cool currents, respectively analogous to those of the present day. (above)

Fig. 2. Global oceanic distribution of *Globorotalia merotumida-G. plesiotumida,* and *G. menardiilimbata* groups (blank) and *G. miozea-conoidea-conomiozea* group (dotted) in the Late Miocene. Area of overlap shown by vertical lines. Note the penetration of the tropical taxa southwards and the temperate taxa northwards along the western and eastern margins of the South Atlantic, respectively, as in Fig. 1. Distributional patterns in Northern Hemisphere suggest sharper thermal gradients in the North Pacific than North Atlantic. (below)

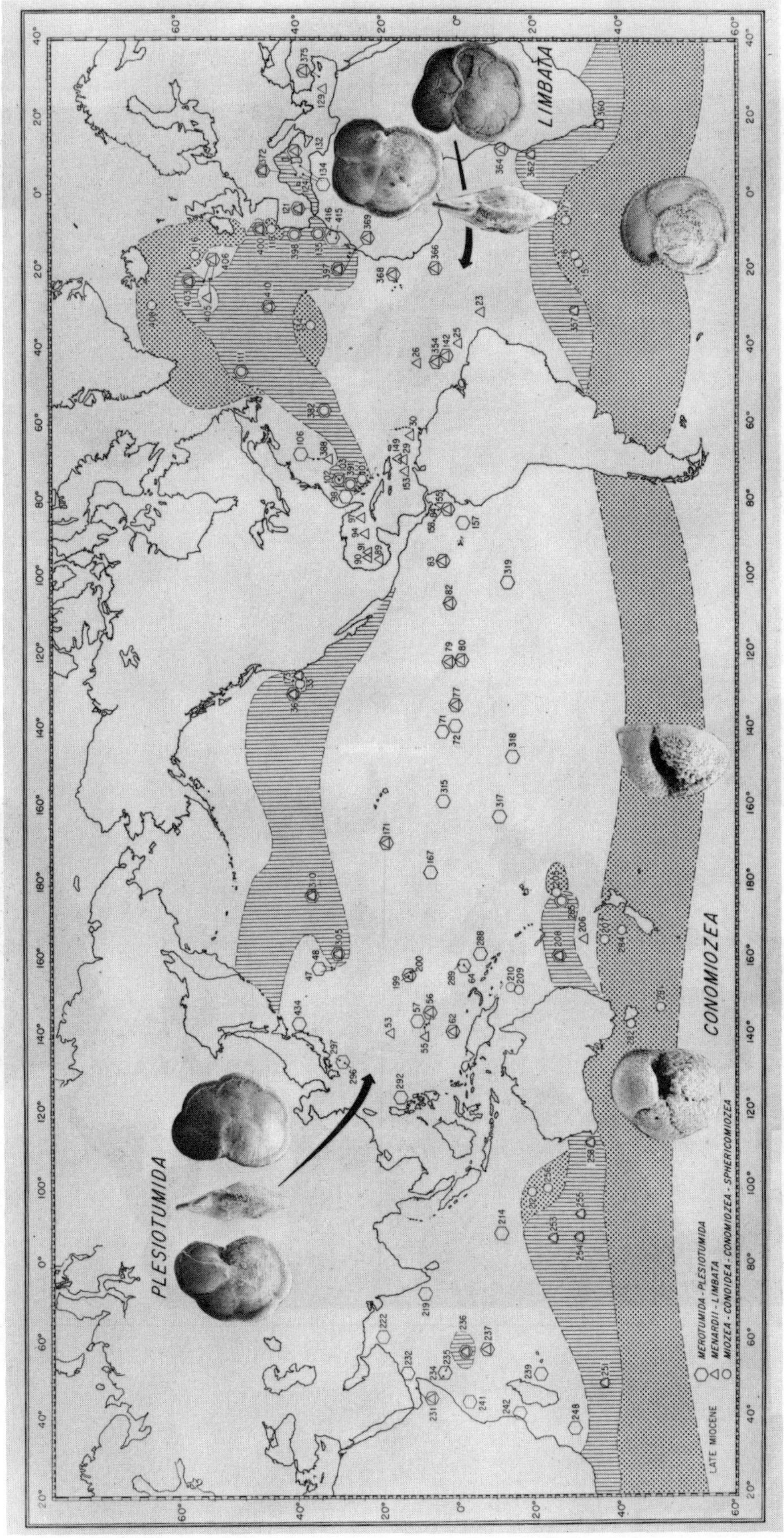
PLESIOTUMIDA
LIMBATA
CONOMIOZEA
LATE MIOCENE
MEROTUMIDA - PLESIOTUMIDA
MENARDII - LIMBATA
MIOZEA - CONOIDEA - CONOMIOZEA - SPHERICOMIOZEA

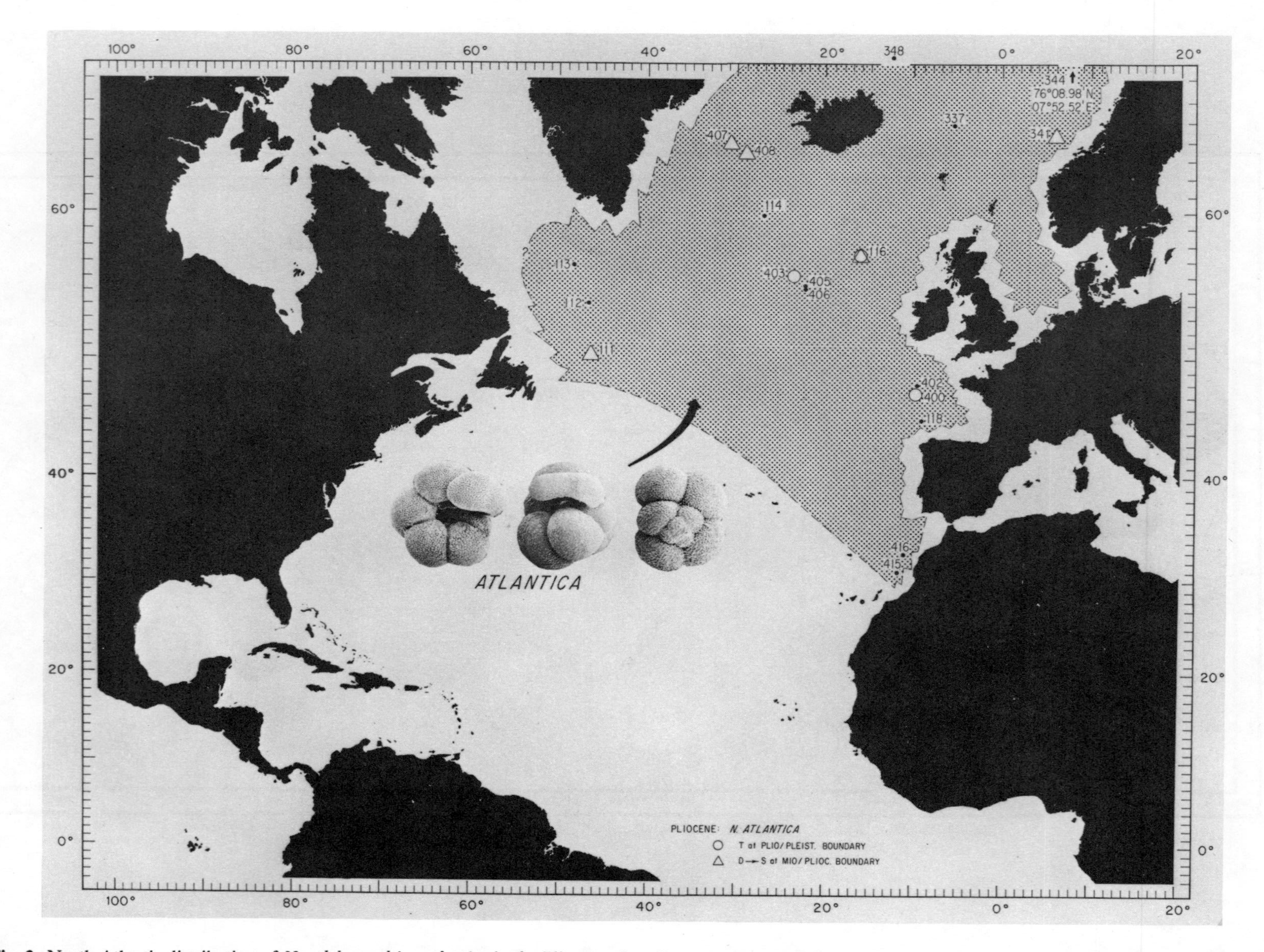

Fig. 3. North Atlantic distribution of *Neogloboquadrina atlantica* in the Pliocene. O = Presence; T = Extinction; D → S = Dextral-Sinistral coiling change.

intermediate between these temperate and subarctic faunas occur at Site 116 (Lat. 57° N): the *G. conoidea-conomiozea* group occurs with *G. bulloides, G. atlantica* (dextral), and *N. acostaensis,* but there are no elements of the *G. menardii* or *G. tumida* group.

The *Globorotalia conoidea-conomiozea* group occurs between Lat. 47°–57° N (Sites 111, 116, 400, 403, and 410) and is absent at Site 341 (Vøring Plateau, Norwegian-Greenland Sea) at Lat. 67° N, indicating that the northern limit of this group lies at about Lat. 60° N.

Early Pliocene faunas north of Lat. 60° (Sites 112, 341, 407, 408) are of low diversity and contain, *i.al., Globigerina bulloides, Neogloboquadrina pachyderma, N. acostaensis-humerosa* group, and *G. puncticulata.* Sites between Lat. 45°–57° N (116, 400, 403–406, 410) contain temperate assemblages including *Globorotalia margaritae,* the *Globorotalia conoidae-conomiozea* group, *Globigerina nepenthes,* and *Globigerinoides obliquus* in addition to the forms listed above. A typical and relatively abundant form (in some instances comprising 50–90% of the total fauna) common to Pliocene faunas north of Lat. 45° in the Atlantic Ocean is the apparently endemic sinistrally coiled *Neogloboquadrina atlantica* (Berggren, 1972b; Poore and Berggren, 1974, 1975) (Fig. 3). Site 111 (Lat. 50° N, Labrador Sea) contains a predominantly subtropical Early Pliocene faunal succession in which the sequential extinction of taxa mirrors that seen in paleomagnetically calibrated successions in the equatorial Indo-Pacific region (Hays *et al.*, 1969; Saito *et al.*, 1975). The evolutionary appearance of *Globorotalia tumida* at the base of the Pliocene at Site 111 is one of the rare instances in which the ancestral-descendant relationship of this taxon can be observed in the Atlantic Ocean (Berggren, 1972b; Poore and Berggren, 1974, 1975). Indeed the Early Pliocene faunal association at Site 111 is unique in that it is truly a transitional type of fauna with tropical (*Sphaeroidinellopsis* spp., *Globorotalia menardii-tumida* group, *Globigerinoides* spp.) and temperate (*Globorotalia conoides* group, *Neogloboquadrina atlantica-pachyderma)* elements occurring together. The northward extension of this subtropical-transitional fauna to Lat. 50° N in the Early Pliocene is attributed to the more northerly extent of the Gulf Stream prior to the inception of northern hemisphere glaciation (Berggren, 1972a).

The initiation of northern hemisphere glaciation, 3 Ma (Berggren, 1972a), had a pronounced effect on North Atlantic Late Pliocene planktonic foraminiferal faunas, primarily in the form of sharply reduced diversities. Faunas north of Lat. 45° are dominated by sinnistrally coiled *Neogloboquadrina atlantica.* North of Lat. 60° (Sites 337, 341, 344, 348, 407, 408) *N. atlantica* is accompanied by the morphologically similar *N. pachyderma* and less commonly *Globigerina bulloides.* Between Lat. 45°–60° N these elements are joined by *Globorotalia inflata, G. crassaformis, G. scitula,* and less commonly by *G. tosaensis* and *Globigerinita glutinata.* The initial appearance of *Globorotalia inflata* has been observed to coincide closely with the initiation of glaciation at Sites 111 (Labrador Sea Lat. 50° N) and 410 (Mid-Atlantic Ridge, Lat. 45° N). *Neogloboquadrina atlantica,* the dominant pliocene taxon in North Atlantic sites was not recored at Site 410 (Lat. 45° N), although it occurs at Sites 118 Bay of Biscay, Lat. 45° N) and 415 and 416 (off NW Africa, Lat. 32° N). North of this it is common, so this may be the southern limit of this species (Fig. 3).

During the Late Pliocene, *Neogloboquadrina atlantica* is gradually replaced by *N. pachyderma* as the dominant taxon in most North Atlantic assemblages and the extinction of *N. atlantica,* which has been shown to closely approximate the Pliocene/Pleistocene boundary (Berggren, 1972a) at Sites 111–116 (Lat. 50°–60° N), has been more recently recorded at sites ranging from Lat. 47°–76° N: 400 (47° N), 408 (63° N), 407 (64° N), 337 (65° N), 341 (67° N), 348 (68° N), and 334 (76° N). In addition, it disappears in discontinuously cored sections at Sites 402 (47° N), 403 (56 °N), and 406 (55° N) between the Pliocene and Pleistocene.

Although a detailed planktonic foraminiferal biostratigraphic subdivision of the Pleistocene of the North Atlantic is not possible at present, a broad pattern may be discerned which reflects the strong climatic control upon planktonic foraminiferal biogeographic distribution. Pleistocene faunas between Lat. 45°–50° N are generally of a temperate nature dominated by *Globorotalia inflata* and *Globigerina bulloides,* with incursions of *Neogloboquadrina pachyderma* (S + D); accessory species include *Globorotalia scitula, G. truncatulinoides, G. crassaformis* and *Turborotalita quinqueloba.*

Subarctic faunas with 3–5 species characterize latitudes 50°–69° N, and are dominated by *N. pachyderma; G. bulloides, G. scitula, G. inflata* and *T. quinqueloba* occur in generally subsidiary roles. A replacement of dextral populations of *N. pachyderma* by sinistral populations has been shown to occur within the mid-Pleistocene at these latitudes (Berggren, 1972b).

North of Lat 60° Pleistocene planktonic foraminiferal populations are essentially Arctic with sinistral *N. pachyderma* virtually the sole form present; less frequently *T. quinqueloba* and/or *G. bulloides* occurs with it. The presence of sinistral *N. pachyderma* throughout the Pleistocene at sites north of 60° and its appearance south of this in the mid-Pleistocene, suggests a southward displacement of the Arctic Polar Front in response to an intensification of northern hemisphere glaciation.

Eastern Equatorial to Temperate North Atlantic (~ Lat. 0°–40° N; ~ Long. 10°–30° W)

This region is represented by Sites 12 (Leg 2), 13 (Leg

Table 2. Eastern Equatorial to Temperate North Atlantic.

Leg	Area	Site No.	Latitude (N)	Longitude (W)	Geographic Setting	Depth (m)	Tect. Hist.	Paleontol.	Remarks
2	E. Subtr. N. Atl.	12	19°40′	26°01′	Cape Verde Basin	4,542	S	M.B. Cita	Plio-Pleist. subtrop. fauna; dissolution
3	Equat. E Atlantic	13	06°02.4′	18°13.71′	Sierra Leone Rise	4,588	S	T. Saito	Pliocene subtrop. fauna; dissolution
13	Temp. East. N Atlantic	120	36°41.39′	11°25.94′	Gorringe Bank	1,711	S	M.B. Cita	Early Miocene subtrop. fauna
		135	35°20.80′	10°25.46′	Horseshoe Abyssal Plain	4,152	S		
		136	34°10.13′	16°18.19′	Abyssal hill area N of Madeira	4,169	S		1. Subtropical faunas.
14	Subtrop. eastern N Atlantic	139	23°31.14′	18°42.26′	Mid-part cont. rise off NW Africa	3,047	S	J.P. Beckmann	2. Dissolution in Miocene-Early Pliocene.
		140	21°44.97′	21°47.52′	Base cont. rise off NW Africa	4,483	S		
		141	19°25.16′	23°59.91′	Diapiric struct. in Cape Verde Basin	4,148	S		3. Site 141 = strong dissolution in latest Miocene-Early Pliocene, decreasing upwards.
		366	05°40.7′	19°51.1′	Sierra Leone Rise	2,853	S		1. Tropical-subtropical Neogene faunas, sites 366 and 369 particularly complete.
		367	12°29.2′	20°02.8′	Base cont. rise, Cape Berde Basin	4,748	S		
41	Equatorial to subtrop. east. N Atlantic	368	17°30.4′	21°21.2′	Cape Verde Rise	3,366	S–? ↓	U. Pflaumann	
		369	26°35.5′	14°59.9′	Cont. slope off Cape Bojador, NW Africa	1,752	S		2. *G. fohsi, tumida, miocenica* groups well developed.
		370	32°50.2′	10°46.6′	Cont. margin off Morocco	4,214	S		
47	Subtrop.-temperate eastern N Atlantic	397	26°50.7′	15°10.8′	Uppermost cont. rise Cape Bajador, NW Africa	2,900	S	M.B. Cita	1. Subtropical-transitional faunas.
		398	40°57.6′	10°43.1′	Graben S of Vigo Seamount	3,910	S	S. Iaccarino	2. *G. fohsi* (397, 398). *conomiozea* (398).
50	Subtrop.-eastern N Atlantic	415	31°01.72′	11°39.11′	Moroccan Basin	2,794	S	E. Vincent	1. Subtrop.-temperate faunas.
		416	32°50.18′	10°48.06′	Cont. margin off Morocco	4,191	S		2. *N. atlantica* in Late Mioc.-Plioc. (31°–33°N)

3), 120 (Leg 13), 135, 136, 139–141 (Leg 14), 366–370 (Leg 41), 397, 398 (Leg 47, and 415, 416 (Leg 50) (Table 2).

Calcium carbonate dissolution has strongly affected the amount and quality of faunal recovery at deeper sites (~ below 3.5–4 km) in the Early Miocene at Sites 135, 136, 140, and 398 and to a lesser extent at about 3 km (Site 139). Late Miocene dissolution is also marked at Site 141 (~4 km) and dissolution cycles are noted within the Upper Miocene section at Site 397 (2,900 m).

An essentially complete Neogene sequence occurs at Sites 366 on the Sierra Leone Rise (~ Lat. 06° N) and 369 on the Continental Slope of Cape Bojador, West Africa (~ Lat. 26° N, 1752 m), and rich, tropical assemblages are present throughout the interval. At site 397, on the uppermost Continental Rise off Cape Bojador, West Africa (~ Lat. 27° N; 2,900 m), an essentially complete Middle Miocene-Pleistocene section has been cored, whereas a somewhat less complete Middle Miocene-Pleistocene section has been discontinuously cored at Site 398 in the axis of a half-graben, south of Vigo Seamount at ~ Lat. 40° N at 3,910 m water depth. A continuous Pliocene-Pleistocene section occurs at Site 141 on the Cape Verde Basin at Lat. 19° N in 4,148 m water depth.

Early Miocene faunas are generally strongly affected by calcium carbonate dissolution in this region. Typical faunal elements (remanents) include catapsydracids (*dissimilis, unicavus, stainforthi*), globoquadrinids (*dehiscens, praedehiscens, altispira, venezuelana*), globigerinids (*bollii, foliata*), *Globigerinoides* (*trilobus, subquadratus, primordius*), and, rarely, *Cassigerinella chipolensis.*

Middle Miocene faunas are best developed at Sites 368, 369 and less completely at Sites 366 and 398. Typical elements include globorotaliids (*siakensis, lenguaensis, praemenardii-archeomenardii* group), globigerinids (*drur-*

yi, nepenthes, decoraperta), and *Sphaeroidinellopsis* spp. *Clavigerinella bermudezi* is recorded at Site 366 and praeorbulinids are present at Sites 369, 398, and 415 (Lat. 27°–41° N). The *Globorotalia fohsi* group occurs from Lat. 06°–33° N at Sites 139, 366, 368, 370, and 415 and occurs together with *Globorotalia miozea* at Sites 397 (Lat. 27° N) and 398 (Lat. 41° N).

The Late Miocene is characterized by neogloboquadrinids (*acostaensis-humerosa* group), globigerinids (*nepenthes,*), *Globigerinoides* (*trilobus, obliquus* group), and *Sphaeroidinellopsis* spp. The *G. tumida* group (*merotumida-plesiotumida*) and *G. pseudomiocenica* occur at sites 366, 368, and 369. The *Globorotalia conoidea-conomiozea* group is present at Site 398 (Lat. 41° N).

The somewhat erratic occurrence of *Globoquadrina dehiscens,* noted elsewhere, is seen in this region as well. It occurs in Late Miocene assemblages at Sites 135 (4,152 m), 366 (2,853 m), 369 (1,752 m), 398 (3,910 m). It is (apparently) absent at comparable stratigraphic levels at Sites 268 (3,355 m), 370 (4,214 m), and 397 (2,900 m). *Globoquadrina dehiscens* occurs at Site 141 (4,148 m) in the Cape Verde Basin in the Late Miocene in a setting that represents the last stage of calcium carbonate dissolution in the deep-sea (Hayes *et al.*, 1972, p. 223; Beckmann, *in* Hayes *et al.*, 1972, p. 410) together with *G. venezuelana, G. nepenthes,* and *Sphaeroidinellopsis seminulinum.* This would suggest that *G. dehiscens* was one of the most resistant taxa to calcium carbonate dissolution during the Late Miocene.

An overlap of *Globoquadrina dehiscens* and *Globorotalia margaritae* is recorded in the terminal Miocene at Sites 369 (Lat. 27° N) and 415 (Lat. 31° N) and an overlap in these two taxa plus *Globigerinoides conglobatus* near the Miocene/Pliocene boundary is noted at Site 366 (Lat. 06° N)—all at depths less than 3 m (see in this connection a similar distribution pattern on the Rio Grande Rise, Berggren, 1977a,b).

Neogene planktonic foraminiferal faunas, which had remained relatively stable since the Early Miocene, exhibit a (climatically influenced) differentiation during the course of the Pliocene. Early Pliocene faunas from ~Lat. 06°–25° N (*i.al.* Sites 141, 366, 368, 369) contain tropical-subtropical assemblages characterized by the *Sphaeroidinellopsis* group, globorotaliids (*multicamerata, pseudomiocenica, margaritae*), globoquadrinids (*altispira*) and globigerinids (*nepenthes*). The brief occurrence and/or sinistral to dextral coiling change in the *Globorotalia cultrata-tumida* group and *Pulleniatina primalis,* observed elsewhere in the Early Pliocene, is recorded at Sites 139, 141, 366, 368, and 369. Between Lat. 20° and 30° N (Sites 369, 397) the keeled globorotaliids (exclusive of *G. margaritae)* diminish in frequency and are replaced by *Globorotalia puncticulata* and a greater frequency of the *Neogloboquadrina acostaensis-humerosa* group and *Globigerinoides obliquus* group.

Between Lat. 30° and 40° N (Sites 135, 139, 398, 415, and 416) *Globorotalia crassaformis* and *G. puncticulata* are common components of the Early Pliocene faunas. *Neogloboquadrina atlantica* (S) has been recorded from Sites 415 (Lat. 31° N) and 416 (Lat. 33° N) together with *N. pachyderma* (D) and *Globigerina bulloides.*

A pronounced latitudinal differentiation is seen in the Late Pliocene of this region. From ~ Lat. 6°–30° N (Sites 12, 13, 139–141, 366, 367, 368, 397), keeled globorotaliids of the *G. exilis-miocenica-multicamerata* complex occur together with *Sphaeroidinellospsis* and *Sphaeroidinella, Globigerinoides fistulosus,* and the *G. obliquus* group. The Late Pliocene reentry of *Pulleniatina obliquiloculata* has been documented in Site 141 to occur near the extinction level (*ca.* 2.2 Ma) of *Globorotalia exilis* (Beckmann *in* Hayes *et al.*, 1972) which agrees well with subsequently published data on this event (Saito *et al.*, 1975; Saito, 1976). The Late Pliocene reappearance of *P. obliquiloculata* is also seen at Site 366 on the Sierra Leone Rise.

Between ~ Lat. 30°–40° N (Sites 398, 416) Late Pliocene faunas are of a temperate-transitional nature with *Globorotalia inflata* and *G. crassaformis* the dominant elements. Orbulinids, *Globigerina bulloides, Neogloboquadrina pachyderma* (D), and *N. atlantica* (S) (at Site 415) occur commonly as well.

Tropical-subtropical Pleistocene faunas are found between Lat. 06°–20° N (Sites 366–368) with keeled globorotaliids (*menardii-tumida* group, *truncatulinoides*), *Globigerinoides* (*ruber, sacculifer*), *Pulleniatina obliquiloculata,* and *Neogloboquadrina dutertrei* as characteristic elements. Between Lat. 20°–30° N (Sites 141, 397) transitional elements such as *Globorotalia inflata, G. hirsuta,* and *G. truncatulinoides* assume a greater importance in the general faunal composition, whereas between Lat. 30° – 40° N (Sites 398, 415) *Globigerina bulloides Neogloboquadrina pachyderma* (D), and *Globorotalia inflata* are the predominant faunal components. The presence of pink colored *Globigerinoides ruber* and *Globigerina rubescens* at several sites (12, 368–370, 415, 416) confirms the presence of Late Pleistocene and/or Holocene.

Western Subtropical-Temperate North Atlantic (~ Lat. 20° – 40° N; ~ Long. 30° – 80° W)

This region is represented by Legs 1 (sites 4,7), 2 (9–11), 11 (98–102–104, 106), 37, 43 (382–397, exclusive of 384), 44 (388, 389, 391), and IPOD Leg 49 (411–413) (Table. 3).

Neogene recovery in this area has been very poor owing to a combination of factors: a) several sites located at depths below the CCD (in excess of 4–5 km); b)

Table 3. Western Subtropical-Temperate North Atlantic.

Leg	Area	Site No.	Latitude (N)	Longitude (W)	Geographic Setting	Depth (m)	Tect. Hist.	Paleontol.	Remarks
1	W Subtrop. N Atlantic	4	24°28.68′	73°47.52′	Between Hatteras	5,319		W.A. Berggren (W.H. Blow)	Late Pleistocene and/or Holocene transitional faunas
		5	24°43.59′	73°38.46′	Abyssal Plain and Bahama Platform	5,361	S		
		7	30°08.04′	68°17.80′	SW Flank Bermuda Rise	5,185	S		
2	Central Subtrop. N Atlantic	9	32°46.4′	59°11.7′	NE Flank Bermuda Rise	4,965	S	M.B. Cita	1. Subtropical faunas. 2. Strong mid-late Miocene dissolution
		10	32°51.73′	52°12.92′	W Flank Mid-Atlantic Ridge	4,697	↓		
		11	29°56.58′	44°44.80′	W Flank Mid-Atlantic Ridge	3,556	↓		
11	W Subtrop. N Atlantic	98	25°22.95′	77°18.68′	NE Providence Channel	2,769	S	C.W. Poag	1. Miocene-Pliocene subtropical faunas; Pleistocene cool temperate faunas. 2. *G. conoidea-conomiozea* group in Late Miocene-Pliocene.
		102	30°43.93′	74°27.14′	Blake-Bahama Outer Ridge	3,426	S		
		103	30°27.08′	74°34.99′	SW Flank Blake-Bahama Outer Ridge	3,964	S		
		104	30°49.65′	74°19.64′	NE Flank Blake-Bahama Outer Ridge	3,811	S		
		106	32°26.01′	69°27.69′	Lower cont. rise, E of Norfolk, Va.	4,500	S		
37	Central Subtrop. N Atlantic	332	36°52.72′	33°38.46′	Deep drill valley, 30 km W of Mid-Atlantic Ridge Axis	1,806	S-↓	G.A. Miles	1. Subtropical Mid-Pliocene faunas replaced by temperate Pleistocene faunas.
		333	36°50.45′	33°40.05′		1,666	S-↓		
		334	37°02.13′	34°24.87′	W Flank Mid-Atlantic Ridge (Anom. 5)	2,619	↓		
		335	37°17.74′	35°11.92′	W Flank Mid-Atlantic Ridge	3,188	↓		
43	W Subtrop. temperate N Atlantic	382	34°25.04′	56°32.25′	Nashville Seamount	5,526	S	A.G. Kaneps	1. Poor Neogene Recovery; solution common. 2. Subtropicl-temperate faunas.
		383	39°14.88′	53°21.18′	J. Anom. Ridge, Sohm Abyssal Plain	5,267	S		
		385	37°22.17′	60°09.45′	Vogel Seamount	4,936	S		
		386	31°11.21′	64°14.94′	Bermuda Rise	4,782	S		
		387	32°19.2′	67°40′	Hatteras Abyssal Plain	5,117	S		
44		388	35°31.33′	69°23.76′	Lower cont. rise off E Coast USA	4,919	S	F. Gradstein	1. Subtropical-temperate faunas. 2. *G. fohsi* (Site 391) and *G. miozea* (Site 388) present.
		389	30°08.54′	76°05.57′	Blake Nose, NE spur of Blake Plateau	2,724	S		
		391	28°13.7′	75°36.9′	Blake-Bahama Basin	4,963	S		
49	Central Temperate N Atlantic	411	36°45.97′	33°23.30′	Mid-Aglantic Ridge Crest (Jaramillo Anom.)	1,935	S	R. Z. Poore	Temperate (*G. inflata-N. pachyderma*) Pleistocene faunas.
		412	36°33.74′	33°09.96′	Mid-Atlantic Ridge, Fract. Zone B. FAMOUS area	2,609	S		
		413	36°32.59′	33°10.50′		2,598			

discontinuous drilling; c) poor core recovery.

Late Oligocene and/or Early Miocene faunas occur at Sites 10 (*Catapsydrax dissimilis, C. unicavus, Globigerina venezuelana,* and *Globorotalia nana* are the predominant constituents in an otherwise strongly dissolved assemblage), 98 (NE Providence Channel), 386 (Bermuda Rise), and J2 and J3 (Blake Plateau). Characteristic elements include *Globorotalia mendacis, G. kugleri, Globigerina praebulloides, G. angulisuturalis, Globigerinoides primordius* (Sites 98, 386), and catapsydracids (see above). The faunas are essentially of Chattian-Aquitanian age, and Burdigalian (~ Zone N6–8) faunas do not seem to have been recovered in the region.

Middle Miocene is found at sites 11, 103, 104, 106, 335, 388, 391, and J3 (Blake Plateau). Predominant faunal elements include globigerinids (*druryi, nepenthes, praebulloides*), globoquadrinids (*advena-dehiscens* group), *Sphaeroidinellopsis* group, neogloboquadrinids (*mayeri-*

siakensis group), and globorotaliids (*limibata-menardii* s.l. group, *fohsi* group—sites 11, 104, 106, 391, J3—and *miozea-conoidea* group—sites 11, 103, 106, 388). The co-occurrence of the *fohsi-miozea* groups at Sites 11 (Lat. 30° N) and 106 (Lat. 36° N) provides useful references points for trans-latitudinal interzonal correlation (see Fig. 1). The local disappearance of *Globoquadrina dehiscens* near the Middle/Upper Miocene boundary (*ca.* Zone N15) is noted at Sites 103, 104, 106 (all below 3,800 m) and, questionably, at Site 11, and is ascribed here to the rise in the CCD which occurred in the middle-late Miocene, although it has been recorded sporadically from the Upper Miocene at Sites 10 (~ 4,700 m), 386 (~ 4,800 m), and 388 (4919 m). At Site 391 (4,963 m; Blake-Bahama Basin, Lat. 28° N) *G. dehiscens* occurs in latest Miocene levels with *Globoquadrina altispira* and *Globigerina nepenthes* as it does on the Rio Grande Rise (Lat. 30° S; Berggren, 1977a,b). Preservation at these deep sites in the Late Miocene may be due to rapid burial.

Upper Miocene recovery is poor in this area, partial sequences being noted at Sites 98, 102, 103, 334, 335, 382, 386, 388, 391, and J1, J2, and J3. Faunas, in general, exhibit marked dissolution. Typical faunal elements include *Globigerina nepenthes, Sphaeroidinellopsis* spp., and *Neogloboquadrina acostaensis.* The *Globorotalia merotumida-plesiotumida* group has been recorded from Sites 98, 102, 103, 106, 382, 388, and 391 in the western North Atlantic, and the related *G. cultrata-limbata* group occurs at Sites 334 and 335 in the central North Atlantic attesting to the efficacy of the Gulf Stream in transporting tropical elements into mid-latitudes (>Lat. 37° N). *Globorotalia petaliformis* Boltovskoy, described originally from the Indian Ocean, is recorded from Site 334 (Miles, 1977, p. 943, pl. 6, Figs. 9–16) and interpreted here as an ecophenotypic (non-encrusted) form of the *Globorotalia miozea-conoidea-conomiozea* group (*cf.* Berggren, 1977a, pl. 3, Figs. 7, 18; pl. 4, Fig. 4; 1977b, pl. 2, Fig. 20; pl. 3, Figs. 11, 12) as seen in the similarly shallower (<3 km) Rio Grande Rise sediments. Similarity with the Rio Grande Rise is also seen in the presence at Site 334 of *Globorotalia lenguaensis* (recorded as *Globorotalia* sp. by Miles 1077, pl. 7, Figs. 11–15) and the common occurrence of *Globigerinoides mitrus* at Sites 48, 103, 334, and 335 (which is probably conspecific with, or closely related to, *Globigerinoides seigliei*). The latter form has an upper stratigraphic limit at, or only slightly above, the Miocene/Pliocene boundary in both the North and South Atlantic.

Early Pliocene faunas have been recovered at Sites 9–11, 98, 102, 103, and 335. Where not affected by dissolution the assemblages are virtually identical to those occurring at similar latitudes in the South Atlantic (see above). Typical elements include globigerinids (*nepenthes),* globorotaliids (*margaritae, miocenica,multicamerata,* and *puncticulata* [Site 10], globoquadrinids (*altispira*), *Globigerinoides* (*obliquus* group), and *Sphaeroidinellopsis.* Dextrally coiled *Pulleniatina primalis* is recorded at Sites 102 and 382 in association with *Globorotalia margaritae.* At Site 102 it occurs with, and ranges above, the termination of *Globigerina nepenthes. Pulleniatina* disappeared in the Atlantic Ocean during the interval between the LAD of *G. nepenthes* (3.7 Ma) and *G. margaritae* (3.3 Ma) and did not reappear until the Late Pliocene (*ca.* 2.3 Ma; Saito, 1976), whereas in the Pacific it continued its development. A relatively complete Upper Pliocene sequence occurs at Site 335, with partial recovery noted at Sites 332 and 333. Warm-water elements continue to occur in this area (*Globorotalia miocenica-exilis; Globigerinoides ruber-sacculifer*), but a decidedly temperate component became numerically significant during this interval (*Globigerina bulloides-falconensis* and *Globorotalia inflata*). The reentry of *Pulleniatina* in the North Atlantic is noted at Site 335 (Lat. 37° N) during the Late Pliocene (see Saito, 1976). Transitional Pliocene-Pleistocene has been recorded at Sites 7 (*Globorotalia tosaensis-G. truncatulinoides* transition) and 386 (*Globorotalia triangulata*).

Pleistocene faunas are present in the surficial veneer of sediments which is present at most sites, particularly those on Legs 11 and 43. Although no continuous Pleistocene sections were recovered in the course of these legs, the faunal fluctuations characteristic of the Pleistocene are apparent in the alternation between faunas dominated by cool-temperature forms (*Globigerina bulloides, Neogloboquadrina pachyderma,* and *Globorotalia inflata*) and warm-subtropical forms (*Globorotalia menardii* group, *Sphaeroidinella dehiscens, Pulleniatina obliquiloculata,* and, in Late Pleistocene levels, *Globigerinoides ruber* forma *rubrum*). *Globorotalia truncatulinoides* is present at most mid-latitude sites, and a coiling pattern in subbottom Holocene layers (~90% sinistral) similar to that on the surficial bottom (~86% sinistral) has been noted in the Sargasso Sea region (Poag, 1972).

Western Equatorial North Atlantic (Lat. 0–20° N, Long. 30–80° W)

The Western Equatorial North Atlantic, as delimited here, includes the area off the northeast coast of South America and extending into the Caribbean Sea. Neogene biostratigraphic coverage of this region is provided by parts of DSDP Legs 4 (26, 27, 29–21), 14 (142), and 39 (354) and all of Leg 15 (Table 4), and the discussion below is based on the reports of these data as well as a study by Berggren and Amdurer (1973). The classic planktonic foraminiferal zonations of the Neogene

Table 4. Equatorial Western North Atlantic (Lat. 0– 20°N; Long. 30–80°W).

Leg	Area	Site No.	Latitude (N)	Longitude (W)	Geographic Setting	Depth (m)	Tect. Hist.	Paleontol.	Remarks
39	W Equatorial N Atlantic	353	10°55′	44°02.25′	Vema Fracture Zone	5,165	S		1. Tropical faunas throughout Neogene.
		354	05°53.95′	44°11.78′	Ceara Rise	4,052	S-↓	A. Boersma	2. *G. menardii-tumida* well developed; *G. miozea* group absent
14		142	03°22.15′	42°23.49′	Ceara Abyssal Plain	4,372	S-↓	J.P. Beckmann	1. Tropical Neogene faunas. 2. *G. fohsi* group present. 3. *G. insueta* present in Lower Miocene
4	W Equatorial N Atlantic (incl. Caribbean Sea: Sites 29–31)	26	10°53.55′	44°02.57′	Vema Fracture Zone	5,169	S		1. Tropical faunas throughout Neogene.
		27	15°51.39′	56°52.76′	Demerara Abyssal Plain.	5,251	S		2. *Pulleniatina* and *Globorotalia tumida* restricted to Quaternary.
		29	14°47.11′	69°19.36′	Venezuela Basin	4,247	S	H.M. Bolli	
		30	12°52.92′	63°23.00′	Aves Ridge	1,218	S		
		31	14°56.60′	72°01.63′	Beata Ridge	3,369	?↓		
15	W Equatorial N Atlantic (incl. Caribbean	146	15°06.99′	69°22.67′		3,949			1. Tropical faunas throughout Neogene.
		149	15°06.25′	69°21.85′	Venezuela Basin	3,972	S-↓		2. *G. fohsi* group well represented.
		150	14°30.69′	69°21.35′		4,545			3. Diverse well preserved Pliocene faunas.
		147	10°42.48′	65°10.48′	Cariaco Trench	892	S		
		148	13°25.12′	63°43.25′	Aves Ridge	1,237	S-↓	H.M. Bolli and I. Premoli-Silva	4. Strong dissolution in Miocene.
		151	15°01.02′	73°25.58′	Beata Ridge	2,029	S	(F. Rögl)	5. *Pulleniatina* and *G. tumida* restricted to *latest Pliocene and Quaternary.*
		153	13°58.33′	72°26.08′	Aruba Gap between Beata Ridge and Venezuela	776	S		6. High resolution Quat. zonation at Site 147.

have been developed in the land-based sections of this area (Bolli, 1957, 1966; Blow, 1969) and subsequently modified based upon deep sea drilling data of Legs 4 (Bolli, 1970) and 15 (Bolli and Premoli-Silva, 1973; Rögl and Bolli, 1973).

The most prominent pattern that emerges from an analysis of the faunas in this region is the stability throughout the Neogene of tropical planktonic assemblages. It is only in the the Quaternary that distinct fluctuations in faunal assemblage patterns occur, a reflection of the periodic climatic changes occurring in the polar regions. The *Globorotalia fohsi* group (Middle Miocene) and the *Globorotalia menardii* group (Middle Miocene-Recent) are among the dominant and biostratigraphically useful elements in mid-late Neogene assemblages.

The Lower Miocene is substantially represented at Sites 149 and 151 with partial recovery at Sites 142, 150 (marked by strong dissolution), 152, and 354. Complete Middle Miocene sequences were recovered at Sites 30 and 149, with partial recovery achieved at Sites 142, 151, and 354.

The Late Miocene was characterized by intense dissolution, and as a result the planktonic foraminiferal faunas are virtually obliterated at sites below 3,000–3,500 m (29, 142, 149, 150, 354) and even at depths less than 1,000 m (Site 153 at 776 m). Yet these sites remain the only source of data on Late Miocene planktonic formainiferal faunas in the area, in addition to that which exists on land-based sections. There, however, a major Late Miocene regression resulted in the termination of planktonic sequences within the interval of Tortonian-Messinian time in many sections.

Dissolution characterizes the Early Pliocene as well, but representative Pliocene sections occur at Sites 29, 148, 149, and 154, with partial recovery at Sites 30, 31, 150, 151, 153, and 354.

Partial sections of the Pleistocene are present at most sites (except where surface coring was deliberately avoided), but notably complete recovery occurred at Sites 29, 148, and 154. At Site 147 a particularly thick (>300 m) section of Upper Pleistocene (<0.4 Ma) was recovered allowing a detailed biostratigraphic analysis (Rögl and Bolli, 1973).

Early Miocene faunas from the western equatorial North Atlantic are characterized by globigerinids (*venezuelana*), globorotaliids (*kugleri, peripheroronda*), *Neogloboquadrina siakensis*, catapsydracids (*dissimilis, unicavus, stainforthi*), *Globigerinoides* (*altiaperturus, primordius, trilobus*, and *bisphericus*). *Globigerinatella insueta*, described originally from the Caribbean region, and praeorbulinids, have been recorded at several sites (e.g., 29, 149, 150, 153, 354).

The Middle Miocene witnessed the development of the *Sphaeroidinellopsis* group, the *Globorotalia menardii* group, and, in particular, the *Globorotalia fohsi* group (Sites 29–31, 142, 149, 151, 153). Common elements include *Globigerina druryi, G. nepenthes*, globoquadrinids (*altispira, dehiscens*), *Globigerinoides subquadratus*, and neogloboquadrinids (*mayeri-siakensis*).

Strong dissolution has reduced the taxonomic diversity of Late Miocene faunas. Neogloboquadrinids (*acostaensis, humerosa* = "*dutertrei*"), *Globigerina nepenthes, Globigerinoides obliquus* s.l., *Sphaeroidinellopsis* spp., and some members of the *Globorotalia menardii* s.l. group appear to be the most resistant (i.e., preservable) forms.

Basal Pliocene assemblages exhibit marked dissolution also (e.g., Sites 29, 142, 354) and are comparable to the well-known *Sphaeroidinellopsis* faunas recorded from the circum-Mediterranean region (Cita *et al.*, 1973; see also Boersma, 1977, p. 508).

Early Pliocene assemblages contain a diverse assemblage of globorotaliids (*margaritae, menardii-pseudomiocenica, multicamerata*), globoquadrinids (*altispira*), globigerinids (*nepenthes*), *Sphaeroidinellopsis-Sphaeroidinella*, and *Globigerinoides* (*fistulosus, trilobus, ruber*). Of particular interest is the brief presence of *Pulleniatina primalis* in the Early Pliocen (*margaritae* Zone of Bolli, 1970) and the abrupt change in coiling direction from sinistral to dextral, followed abruptly by its (local) disappearance in this region (Sites 148, 154). This level corresponds closely to (but is older than) the LAD of *Globorotalia margaritae* and FAD of *G. miocenica*, which occurs over the interval of 3.3–3.5 Ma. Bolli and Premoli-Silva (1973) have pointed out that this level is correlative with Datum VI of Hays *et al.* (1969), just above the LAD of *Globigerina nepenthes* at 3.7 Ma in the Indo-Pacific region. In piston cores from the Indo-Pacific region as well as in the well Bodjenogoro 1 (Bolli, 1966b), the sinistral-dextral coiling change in *Pulleniatina primalis* is associated with a coiling change in the *Globorotalia menardii-tumida* complex (Hays *et al.*, 1969; Bolli, 1966b), whereas in the Caribbean region *Globorotalia tumida* is essentially restricted to the Pleistocene (Sites 29–31, 149, 151, 153). In fact the *Globorotalia tumida* lineage (*merotumida-plesiotumida* group) is poorly represented (i.e., sparse) in the western equatorial North Atlantic, supporting the suggestion (Berggren and Poore, 1974) that this group is essentially one of Indo-Pacific affinities. *Pulleniatina* (*obliquiloculata*) reappears in latest Pliocene-Pleistocene levels in the Caribbean following which its coiling pattern exhibits disjunct trends compared to the Indo-Pacific region (Saito, 1976).

The Late Pliocene is characterized by an essentially modern fauna in which the sequential extinction of several forms, *i.al.*, *Sphaeroidinellopsis* spp., *Globoquadrina altispira, Globigerinoides fistulosus, G. obliquus* s.l., *Globorotalia miocenica*, and *G. exilis* serve to denote important biostratigraphic datum levels useful in regional correlation. Discrepancies between the observed terminal limits of taxa recorded from the Caribbean (Bolli, 1970; Bolli and Premoli-Silva, 1973) and the Rio Grande Rise (Berggren, 1977a,b) may be ascribed to the effects of dissolution at depth in the former region.

Pleistocene faunas recorded from this region are essentially those of the present day. A five-fold subdivision of the Pleistocene-Holocene *Globorotalia truncatulinoides* Zone was proposed by Rögl and Bolli (1973; see Fig. 4, this paper). The incursion of cool-temperate elements (*Globorotalia inflata, Neogloboquadrina pachyderma*) at Site 147, Cariaco Trench, during the latter part of the Pleistocene (< 0.4 Ma) reflects alternating polar glacial-interglacial cycles and led Rögl and Bolli (1973) to propose a correlation of these zones with the climatic zones V-Z of Ericson and Wollin (1968) and the oxygen isotope chronology (Stages 1–7) of Emiliani (1966).

Gulf of Mexico

Neogene biostratigraphy of the Gulf of Mexico is represented by sites drilled during DSDP Legs 1 (1–3) and 10 (Table 5). Although Early Miocene faunas are only minimally represented, Middle Miocene to Pleistocene faunas are uniformly tropical-subtropical in character with temperate elements (*Globorotalia inflata, Globigerina bulloides*) appearing in the course of Pleistocene climatic oscillations.

The only Early Micoene faunas (Zones N4 and N8) recorded are those from Site 94. Representative Middle Miocene faunas occur in Sites 87, 80, 91, and 95, while Upper Miocene faunas occur in Sites 89, 90, 94, and 97.

Partial recovery of the Lower Pliocene was obtained at Sites 3, 88–91, and 94 and of the Upper Pliocene at Sites 3, 86, 89–92, and 94. Partial recovery of the Pleistocene-Holocene was obtained at Sites 1–3, 85, 86, and 88–97.

Gulf of Mexico Neogene faunas are essentially similar to those which occur in the Caribbean Sea and adjacent areas and will not be elucidated in detail here. It is sufficient to point to the presence of the *Globorotalia fahsi* group at Sites 87, 98–91, and 94, of the *G. mioce-*

ATLANTIC

EPOCH / SERIES	T in Ma	TEMPERATE: BERGGREN (1972), POORE AND BERGGREN (1975)	BLOW (1969)	TROPICAL: BOLLI AND PREMOLI SILVA (1973)	SUBTROPICAL TEMPERATE: BERGGREN (1973; 1977), BLOW (1969)
PLEISTOCENE: L, E	0–1.8	*Neogloboquadrina pachyderma*	N22	*Globorotalia truncatulinoides*: *G. fimbriata*, *G. bermudezi*, *G. calida*, *G. hessi*, *G. crassaformis viola*	N22 *Globorotalia truncatulinoides*
PLIOCENE: L	1.8–3	*Globorotalia inflata*	N21	*Globorotalia truncatulinoides* cf. *tosaensis*	PL6 *Globigerinoides obliquus extremus*
				Globorotalia miocenica: *G. exilis*	PL5 *Globorotalia miocenica*–*Globorotalia exilis*
				G. trilobus fistulosus	PL4 *Globoquadrina altispira*–*Globorotalia multicamerata*
PLIOCENE: E	3–5	*Globorotalia crassaformis*	? N20		PL3 *Sphaeroidinellopsis subdehiscens*–*Globoquadrina altispira*
			N19	*Globorotalia margaritae*: *evoluta*	PL2 *Globorotalia margaritae*, *Sphaeroidinellopsis subdehiscens*
		Globorotalia puncticulata		*Globorotalia margaritae*–*Globigerina nepenthes*; *margaritae*	PL1 c *Globorotalia crassaformis*; b *Globorotalia puncticulata* s.l.; a *Globorotalis cibaoensis*
			N18		*Globoquadrina dehiscens*–*Globorotalia margaritae*
MIOCENE: L	5–6+	*Globorotalia conomiozea*	N17	*Neogloboquadrina dutertrei*	*Globorotalia conomiozea*–*Globorotalia mediterranea*
		Neogloboquadrina acostaensis			*Globorotalia miozea*–*Globorotalia conoidea*

Fig. 4. Late Miocene-Pleistocene planktonic foraminiferal zones of the tropical to temperate Atlantic Ocean. Pre-Pleistocene zonation of tropical (Caribbean) region is by Bolli and Premoli-Silva (1973); 5-fold zonation of Caribbean region is by Rögl and Bolli (1973).

Table 5. Gulf of Mexico.

Leg	Area	Site No.	Latitude (N)	Longitude (W)	Geographic Setting	Depth (m)	Tect. Hist.	Paleontol.	Remarks
		1	25°51.5′	92°11.0′	Sigsbee Scarp	2,827	S		Tropical-subtropical
1		2	23°27.3′	92°35.2′	Challenger Knoll	3,572	S-? ↓	W.A. Berggren	Pliocene-Pleistocene
		3	23°01.0′	92°01.4′	Sigsbee Abyssal Plain	3,747	S	(W.H. Blow)	faunas.
		85	22°50.49′	91°25.37′	Campeche Scarp-Sigsbee Abyssal Plain	3,733	S		
		86	22°52.48′	90°57.75′	Campeche Scarp	1,462	S		
		87	23°00.90′	92°05.16′	Sigsbee Abyssal Plain	3,751	S		1. Tropical-subtropical
	Gulf of Mexico	88	21°22.93′	94°00.21′	Knoll in SW Sigsbee Abyssal Plain	2,532	S-? ↓		faunas throughout Neogene.
		89	20°53.41′	95°06.73′	E Mexican Cont. Rise	3,067	S		2. *G. fohsi* group and *G. miocenica-exilis*
10		90	23°47.80′	94°46.09′	Cont. Rise-Sigsbee Abyssal Plain	3,713	S		groups well represented in Middle
		91	23°46.40′	93°20.77′	Sigsbee Abyssal Plain	3,763	S	B.W. McNeely	Miocene and Late
		92	25°50.69′	91°49.29′	Salt ridge on Sigsbee Scarp	2,573	S	(J.H. Beard)	Pliocene, respectively. 3. Faunas essentially
		93	22°37.25′	91°28.78′	Valley in Campeche Scarp	3,090	S		identical to Caribbean region.
		94	24°31.64′	88°28.16′	NW Campeche Scarp	1,793	S		
		95	24°09.00′	86°23.85′	NW Campeche Scarp	1,633	S		
		96	23°44.56′	85°45.80′	Catoche Knoll	3,439	S		
		97	23°53.05′	84°26.74′	Florida Straits	2,930	S		

nica-exilis-multicamerata group at all Pliocene sites and of the anomalous rarity of such forms as *Globigerina nepenthes* and *Globigerinoides fistulosus* in Lower and Upper Pliocene, respectively. The presence of (sinistral) *Globorotalia tumida flexuosa* with *G. margaritae* in the Early Pliocene (between 3.3–3.7 Ma) at Site 2 (Challenger Knoll; Berggren, 1969, p. 595) mirrors the Early Pliocene occurrences of this form in the Labrador Sea (Site 111, Orphan Knoll, Laughton, Berggren, *et al.*, 1972, p. 113) and may be contrasted with its apparent absence from the Caribbean prior to the Late Pliocene-Pleistocene (Bolli and Premoli-Silva, 1973). Finally, the distribution of *Pulleniatina* is essentially identical in the Gulf of Mexico to that in the Caribbean: *Pulleniatina primalis* disappears within the Early Pliocene a short time after a dextral to sinistral change in coiling and reappears (as *Pulleniatina obliquiloculata*) with the Late Pliocene (Smith and Beard, 1973).

Tropical to Temperate South Atlantic (Lat. 0°–45° S)

Low and mid-latitude South Atlantic Neogene sites are represented by DSDP Legs 3 (excluding Site 13), 3 (Sites 23–25), 39 (excluding Sites 353 and 354) and 40 with the exception of several sites located north of the equator (Table 6). Mid-latitude sites (> Lat. 30° S), particularly those along the western margin of Africa (e.g., Site 360), contain mixed temperate-subtropical faunas throughout the Neogene (with a particularly noticeable contribution from the temperate SW Pacific-New Zealand region), indicating that present-day surface circulation patterns were already well established in the late Paleogene. Faunas from low-latitude sites (23–25) indicate that the equatorial region has remained a stable tropical regime throughout the Neogene.

The Lower Miocene is substantially represented at DSDP Sites 15, 17, 18, 357, 360, and 362 with partial recovery at Sites 22, 23, 24, 355, and 356. Virtually complete Middle Miocene sequences were recovered at Sites 360 and 362, whereas partial recovery was obtained at Sites 15, 25, and 357. An essentially complete Upper Miocene sequence occurs at Sites 360 and 362 with partial recovery at Sites 15, 16, 17, 357, and 359. Representative Pliocene sequences have been recovered at Sites 15, 16, 357, 359, 360, 362, and 364. Pleistocene (at least in part) has been recovered from most South Atlantic DSDP sites. Data from these sites, in addition to specific studies conducted in this region (Berggren, 1977a,b; Berggren and Amdurer, 1973; Berggren, work in progress), form the basis for the synthesis presented below.

Early Miocene faunas from mid-latitudes (> Lat. 30° S) are characterized by large, robust globoquadrinids of the *praedehiscens-dehiscens* group, *Globoquadrina baroemoenensis* and *G. globularis*, catapsydracids (*dissimilis, unicavus, stainforthi*), globorotaliids (*kugleri, peripheroronda*), *Globigerinoides primordius, G. trilobus, G. quadratus* and *G. altiapertурus*, and globigerinids (*occlusa, praebulloides, pseudobesa*). A notable contribution to these faunas from the temperate SW Pacific-New Zealand region occurred in the form of globiger-

Table 6. Tropical-Temperate South Atlantic (Lat. 10°–45° S).

Leg	Area	Site No.	Latitude (S)	Longitude (W)*	Geographic Setting	Depth (m)	Tect. Hist.	Paleontol.	Remarks
		14	28°19.89′	20°56.46′	W Flank Mid-Atlantic Ridge	4,343	S-↓		
		15	30°53.38′	17°58.99′	W Flank Mid-Atlantic Ridge	3,927	S-↓		
		16	30°20.15′	15°42.79′	W Flank Mid-Atlantic Ridge	3,527	S-↓		
	Subtropical to temperate	17	28°02.74′	06°36.15′	E Flank Mid-Atlantic Ridge	4,265	S-↓		1. Mixed subropical and temperate faunas.
3	S Atlantic (~ 30° S)	18	27°58.72′	08°00.07′	E Flank Mid-Atlantic Ridge	4,018	S-↓	T. Saito (W.H. Blow)	2. *G. miozea* group abundant.
		19	28°32.08′	23°40.63′	W Flank Mid-Atlantic Ridge	4,677	S-↓		
		20	28°31.47′	26°50.73′	W Flank Mid-Atlantic Ridge	4,518	S-↓		
		21	28°35.10′	30°35.85′	Rio Grande Rise	2,113	S-↓		
		22	30°00.31′	35°15.00′	Rio Grande Rise	2,134	S-↓		
		23	06°08.75′	31°02.60′	Outer edge of continental slope off Recife (Brazil)	5,079	S		1. Tropical faunas throughout Neogene.
4	W tropical S Atlantic	24	06°16.30′	30°53.53′	Outer edge of continental slope off Recife(Brazil)	5,148	S	H.M. Bolli (W.H. Blow)	2. *G. kugleri*, *G. fohsi* gp. and *N. acostaensis* characteristic of L. M. U Miocene, respectively.
		25	0°31.00′	39°14.40′	N Brazilian Ridge	1,916	↓		3. High diversity Pliocene and Q faunas. 4. *G. miozea* conspicuously absent.
		355	15°42.59′	30°36.03′	Brazil Basin	4,896	S		1. Mixed subtropical-temperate faunas.
	Subtropical	356	28°17.22′	41°05.28′	Sao Paulo Plateau	3,203	S-↓	A. Boersma	
	to temperate	357	30°00.25′	35°33.59′	Rio Grande Rise	2,109	S-↓	(W.A. Berggren)	2. *G. miozea* group abundant.
39	S Atlantic	359	34°59.10′	04°29.83′	Walvis Ridge-Seamount	1,548	S-↓		3. Co-occurrence of *G. miozea-G. fohsi* group at 30° S.
		360	35°50.75′	18°05.79′E	Continental rise off SW Africa	2,749	S		1. Mixed subtropical-temperate faunas.
	E subtropical	362	19°45.45′	10°31.95′E	Frio Ridge segment of Walvis Ridge	1,325	S-↓		2. *G. miozea* group abundant.
40	to temperate S Atlantic	363	17°38.75′	09°02.80′E	Frio Ridge segment of Walvis Ridge	2,248	S-↓	H.M. Bolli	3. Co-occurrence of *G. miozea-G. fohsi* group at 20° S.
		364	11°34.32′	11°58.30′E	Angola Basin	2,448	S		

* Unless otherwise noted.

inids (*brazieri, woodi, connecta*) and globorotaliids (*incognita, semivera, siakensis, zealandica, praescitula, miozea*; Saito, *in* Maxwell, Von Herzen *et al.*, 1972; Berggren and Amdurer, 1973; Bolli, Ryan *et al.*, 1978; Berggren, work in progress).

The Middle and (to a lesser extent) Late Miocene was a time of widespread hiatus, particularly in the South and Southwest Atlantic, due probably to the presence of active erosion and scour by bottom currents. As a result, recovery of Middle Miocene sequences in the South Atlantic has been rather poor, notable exceptions being the excellent recovery (noted above) at Sites 360 and 362 in the eastern South Atlantic. Middle Miocene faunas at mid-latitude sites are characterized by globoquadrinids (*altispira, dehiscens*), globorotaliids (*continuosa, mayeri, siakensis*), globigerinids (*druryi, nepenthes, woodi*), *Globigerinoides subquadratus, G. trilobus, Globigerinopsis* spp., (particularly at Site 357), *Sphaeroidinellopsis seminulinum, S. subdehiscens*), and globorotaliids of the *praescitula-scitula* group, *menardii-limbata* (= *pseudomiocenica*) group, the *fohsi* group (from Lat. 0° [Site 25] to 20° S [Site 362] on the eastern side and 30° S [Site 357] on the western side of the South Atlantic), and the *miozea-conoidea* group (essentially south of Lat. 30° S, but extending as far as Lat. ~20° N in the eastern South Atlantic [Site 362]). The co-occurrence of the *miozea-fohsi* group is noted at Site 357 (Lat. 30° S) with the *fohsi* group, exclusively, recorded, as expected, in the equatorial region (Site 25 at–Lat. 0°) and the *miozea* group, exclusively, at Sites 15, 16 (Mid-Atlantic Ridge), 359 (Walvis Ridge), and 360 (off Southwest Africa)–all between ~Lat. 30°–36° S, in addition to the northward penetration to Lat. 20° S at Site 362 noted above (see Fig. 1). The co-occurrence of these two groups at Site 357 in the Middle Miocene is important for it provides evidence of the climatically controlled low vs. high latitude biostratigraphic dichotomy that was taking place during this time among the keeled globorotaliids (as well as in other groups) and which was to continue—in descendant forms—during the Late Neogene.

The Late Miocene is characterized, particularly at sites located below 4 km, by marked dissolution resulting in a reduced faunal diversity. Upper Miocene sequences are likely to be best preserved on topographically shallow sites (e.g., Rio Grande Rise, Walvis Ridge), although the effects of erosional scour by currents, even at the relatively shallow depth of ~2km atop the Rio Grande Rise, tend to cancel out the effects of enhanced preservation which characterizes this area (see Berggren, 1977a,b).

Late Miocene faunas are characterized by the dominance of *Globoquadrina dehiscens* and *Sphaeroidinellopsis* spp. (although at sites near the CCD, *G. dehiscens* and *G. altispira* are frequently absent or sparse above the Middle Miocene), globigerinids (*nepenthes, pseudobesa*), neogloboquadrinids (*acostaensis, humerosa*), *Globigerinoides* (*ruber, seigliei, trilobus*), globorotaliids (*miozea-conoidea-conomiozea* group, and *menardii-limbata* (= *pseudomiocenica*) group with geographic distribution similar to the Middle Miocene discussed above, and *lenguaensis*). The co-occurrence of the *miozea* and *menardii* groups is noted at Sites 357 (Lat. 30° S) and 362 (Lat. 20° S), once again pointing to the respective further northward penetration of the temperate *miozea* group along the eastern margin, and the further southward penetration of the *menardii* group along the western margin of the South Atlantic (see Fig. 2).

Pliocene faunas are generically and specifically diverse and contain an admixture of subtropical and temperate elements which allows a rather accurate correlation from the equatorial to subantarctic regions. The Pliocene is characterized by the relatively rapid extinction of numerous taxa, some of which have formed the basis of a Pliocene subtropical-temperate zonation for the Atlantic (Berggren, 1973, 1977a,b). Typical forms include, *i.al.* globoquadrinids (*altispira*), globorotaliids (*menardii-limbata* group including *miocenica, multicamerata; margaritae;* the *conoidea-conomiozea* group (Fig. 5); the *cibaoensis-puncticulata-crassaformis-inflata* group), *Sphaeroidinellopsis* spp., and *Globigerinoides* (*conglobatus, obliquus, ruber, sacculifer, fistulosus*). The extension of the *conoidea-conomiozea* group into the mid-Pliocene at Sites 15, 16, 17, 357 (see also Berggren, 1977a,b, in this connection), 359, and 360 mirrors conditions recorded in the eastern temperate Pacific on DSDP Leg 5 (Olsson and Goll, 1970; Olsson, 1971) and temperate North Atlantic (Berggren, 1972a,b; Poore and Berggren, 1975). The restriction of *Pulleniatina* to the late Pliocene and Pleistocene is noted at low (Site 25) and mid (Site 357) latitude sites, mirroring its distribution in the North Atlantic (Saito, 1976) and attesting to the combined efficacy of the elevation of the Isthmus of Panama and the presence of the cool Benguela Current in excluding this form from the Atlantic Ocean during the early two-thirds of the Pliocene.

The Miocene/Pliocene boundary on the Rio Grande Rise has been shown (Berggren, 1977a,b) to be characterized approximately by the following multiple criteria: a) extinction of *Globorotalia mediterranea, Globoquadrina dehiscens, Globigerina pseudobesa, Globigerinoides seigliei;* b) initial appearance of *Globigerinoides conglobatus, Globorotalia cibaoensis, G. margaritae, Sphaeroidinella dehiscens immatura, G. scitula* (local) and the (local) reappearance of *Globoquadrina altispira.* The biochronology of the Late Miocene-Pliocene sequence of planktonic foraminiferal events is essentially identical to that delineated in the equatorial Pacific and elsewhere by Hays *et al.* (1969) and Ryan *et al.* (1974).

Pleistocene faunas are essentially comparable to those living in the region today, varying specific faunal composition being a reflection of changing climatic conditions during this time. Tropical elements (e.g., *Globorotalia menardii-tumida* group, *Pulleniatina, Sphaeroidinella, Globigerinoides ruber-sacculifer*) dominate in low latitudes (Leg 4), whereas temperate forms (*Globorotalia inflata, G. crassaformis, G. truncatulinoides, Globigerina bulloides*) are the dominant forms at mid-latitudes (= Lat. 30° S).

Mediterranean Sea

This region is represented by Sites 121–134 and 371–378 drilled on DSDP and IPOD Legs 13 and 42A, respectively (Table 7).

In general Pliocene-Pleistocene deep-sea faunal sequences are well known in the Mediterranean, whereas the Miocene assemblages are less well represented owing to the difficulties associated with penetrating the Late Miocene (Messinian) evaporite deposits.

A virtually complete Neogene sequence has been recovered at Site 372 (East Menorca Rise, 2,699 m). The upper part of the Lower Miocene, and Middle and Upper (pre-evaporitic) Miocene have also been recovered at Site 375 (Florence Rise, Levantine Basin, 1,900 m). A virtually complete Pliocene-Pleistocene

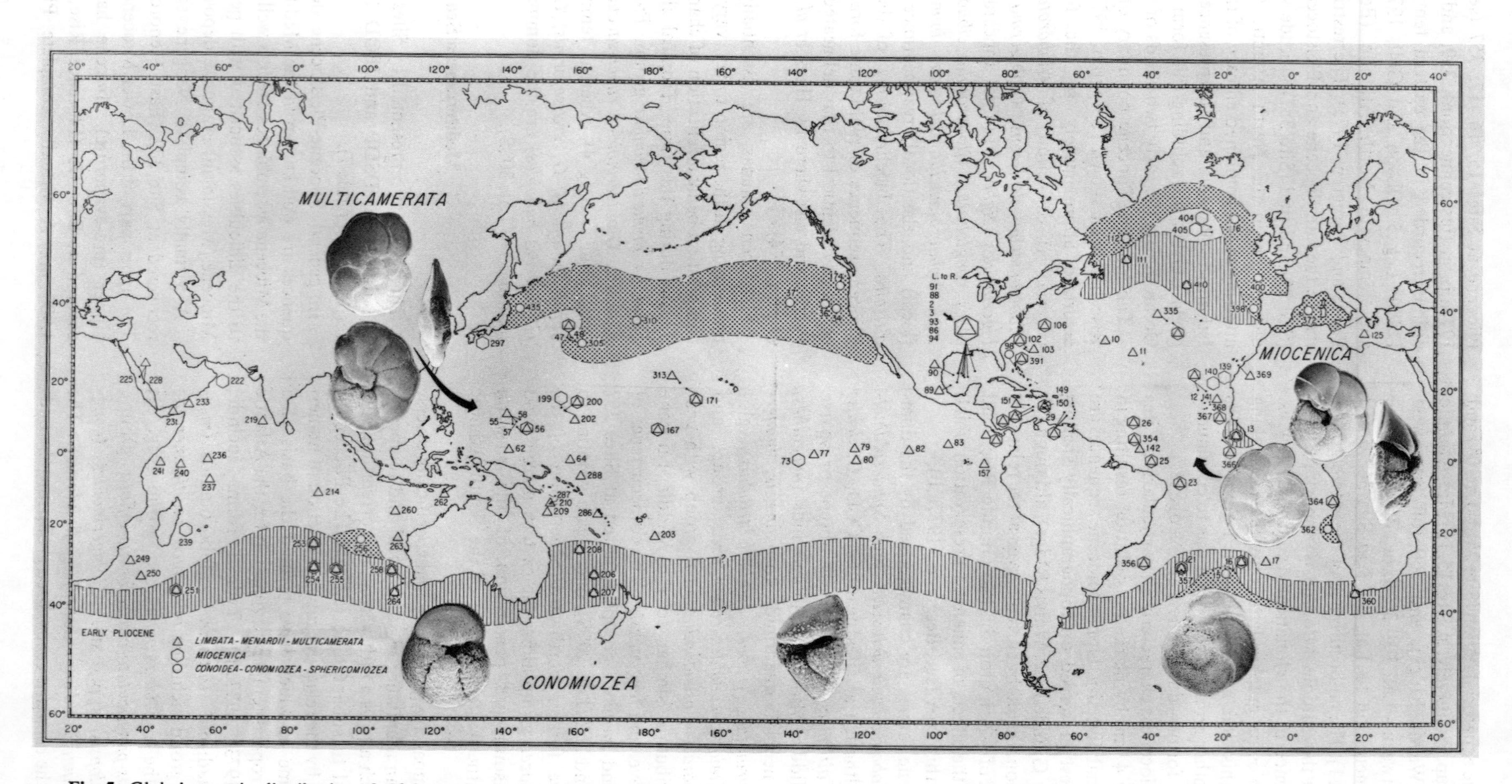

Fig. 5. Global oceanic distribution of *Globorotalia limbata-menardii-multicamerata* and *G. miocenica* groups (blank) and *G. conoidea-conomiozea-sphericomiozea* groups (dotted) in the Early Pliocene. Area of overlap shown by vertical lines. Records of *G. miocenica* in the Indo-Pacific region, although considered erroneous, are plotted as recorded.

Table 7. Mediterranean Sea.

Leg	Area	Site No.	Latitude (N)	Longitude (E)	Geographic Setting	Depth (m)	Tect. Hist.	Paleontol.	Remarks
		121	36°09.65′	04°22.43′	W Alboran Basin	1,163	S-↓		
		122	40°26.87′	02°37.46′	Valencia Trough	2.146	S		
		123	40°37.83′	02°50.21′	Valencia Basement Ridge	2,290	S		
		124	38°52.38′	04°59.69′	Balearic Rise	2,726	S		1. Predominantly temperate Pliocene-
		125	34°37.49′	20°25.76′	Mediterranean	2,782	↑		Pleistocene faunas
		126	35°09.72′	21°25.63′	Ridge	3,730			(particularly well 125 and 132).
		127	35°43.90′	22°29.81′		4,636			2. Mid-Miocene at
13		128	35°42.58′	22°28.10′	Hellenic Trench	4,634	S		sites 126, 129.
		129	35°20.96′	27°04.92′	Strabo Trench and Mountains	2,832–3,048	S	M.B. Cita	3. *G. margaritae*, *G. puncticulata-inflata* groups well developed.
		130	33°36.31′	27°51.99′	Mediterranean Ridge, Levantine Sea	2,979	S		
	Mediter-	131	33°06.39′	28°30.69′	Western Nile Cone	3,035	S		
	Ranean	132	40°15.70′	11°26.47′	Tyrrhenian Sea	2,835	S-↓		
	Sea	133	39°11.99′	07°20.13′	Boundary of	2,563			
		134	39°11.70′	07°18.25′	Sardinian Slope and Balearic Abyss. Plain	2,564	S		
		371	37°35.88′	05°15.55′	Balearic Abyssal Plain	2,792	S-↓		1. Representative
		372	40°01.86′	04°47.79′	East Menorca Rise	2,699	S		pre- and post-evaporite (Messinian)
		373	39°43.68′	12°59.56′	Seamount in Tyrr-henian Abyssal Plain Abyssal Plain	3,517	S		Neogene faunas. 2. Subtropical-tem-perate Early-Middle
42									Miocene faunas
		374	35°50.87′	18°11.78′	Ionian Abyssal Plain	4,078	S-↓	M.B. Cita G. Bizon	(372, 375).
		375	34°45.74′	30°45.58′	Florence Rise,	1,900	S		3. *G. miozea-conomiozea*
		376	34°52.32′	31°48.45′	Levantine Basing	2,101			in mid-late Mio-cene (372, 375).
		377	35°09.25′	21°25.86′	North Cretan	3,718 ↑			
		378	35°56.67′	25°06.97′	Basin	1,835 ↓			

sequence occurs at Site 132 (Tyrrhenian Sea, 2,835 m), mid-Pliocene–Pleistocene at Site 125 (Mediterranean Ridge, 2,782 m), and a relatively complete Pliocene–Lower Pleistocene sequence with numerous sapropels has been recovered at Site 374 (Ionian Abyssal Plain, 4,078 m).

Early Miocene faunas (Sites 372, 375) are characterized by subtropical assemblages including, *i.al.*, catapsydracids (*dissimilis*), globoquadrinids (*dehiscens*), globigerinids (*praebulloides, bollii*), *Globigerinoides* (*altiaperturus, trilobus*), and globorotaliids (*acrostoma, nana, peripheroronda, praescitula, siakensis*). Praeorbulinids occur at both sites, and *Globigerinoides bisphericus* and *Hastigerina praesiphonifera* characterize the lower-middle Miocene transition.

Middle Miocene faunas (Sites 372, 375) are characterized by globoquadrinids (*altispira, dehiscens*), *Globigerinoides* (*obliquus, trilobus*), globorotaliids (*menardii, peripheroronda, siakensis*, and the *miozea* group), the *Sphaeroidinellopsis* group, and globigerinids (*druryi, nepenthes*). Late Middle Miocene (Serravallian) faunas exhibit signs of the impending changes in paleogeography and paleo-oceanography in the Mediterranean Basin in the form of reduced diversity, local dissolution, and dwarfing of faunas.

The per-evaporitic Late Miocene (Tortonian-Early Messinian) faunas exhibit further evidence of the developing crisis, particularly dwarfing and relatively poor preservation. Characteristic elements include globoquadrinids (*altispira* and, rarely, *dehiscens*), *Globigerinoides* (*obliquus* group, *trilobus*), neogloboquadrinids (*acostaensis-humerosa* group), globigerinids (*nepenthes*), globorotaliids (*pseudomiocenica, merotumida-plesiotumida* group). The *Globorotalia miozea-conoidea-conomiozea* group, although rarely reported in deep sea

T in my	EPOCH	SERIES	ATLANTIC – TROPICAL: BERGGREN (1973) THIS PAPER	ATLANTIC – TROPICAL: PARKER (1973)	ATLANTIC – TEMPERATE: POORE and BERGGREN (1975)	MEDITERRANEAN: CITA (1973)	MEDITERRANEAN – CRETE: ZACHARIASSE (1975)
	PLEIST	E	(N 22)	V	Neogloboquadrina pachyderma	N 22	(NOT STUDIED)
2.0	PLIOCENE	L	PL6	IV	Globorotalia inflata	Globorotalia inflata	Globorotalia inflata
	PLIOCENE	L	PL5	IV	Globorotalia inflata	Globigerinoides obliquus extremus	Globorotalia inflata
3.0	PLIOCENE	L	PL4	IV	Globorotalia inflata	Globigerinoides obliquus extremus	Globorotalia bononiensis
	PLIOCENE	L	PL3	III	Globorotalia crassaformis	Sphaeroidinellopsis subdehiscens	UPPER
	PLIOCENE	E	PL2	II	Globorotalia crassaformis	Globorotalia margaritae evoluta	Globorotalia puncticulata – LOWER (margaritae)
4.0	PLIOCENE	E	PL1 c	II	Globorotalia puncticulata	Globorotalia margaritae evoluta	Globorotalia puncticulata – LOWER (margaritae)
	PLIOCENE	E	PL1 b	II	Globorotalia puncticulata	Globorotalia margaritae evoluta	Globorotalia puncticulata – LOWER (margaritae)
5.0	PLIOCENE	E	PL1 a	II	Globorotalia puncticulata	G. margaritae ss; Sphaeroidinellopsis ACME	Globorotalia margaritae; ACME Sphaeroidinellopsis
	MIOCENE	L	Globorotalia margaritae – Globoquadrina dehiscens	I	Globorotalia conomiozea	N17	Globorotalia conomiozea
6.0 – 8.0	MIOCENE	L	Globorotalia mediterranea – Globorotalia conomiozea	I	Globorotalia conomiozea	N17	Globorotalia conomiozea

Fig. 6. A comparison of some Late Miocene–Pliocene planktonic foraminiferal zones of the Mediterranean and Atlantic (from Berggren, 1977b, Fig. 16).

cores from the Mediterranean region (Site 372), is a common component of Late Miocene faunas in the circum-Mediterranean regions (Bizon and Bizon, 1972), from Greece (Bizon, 1967) to Crete (Zachariasse, 1975; 1979) to Italy (d'Onofrio *et al.*, 1975) and Sicily (Catalano and Sprovieri, 1969, 1971), Spain (Tjalsma, 1971) to the Atlantic coast of Morocco (Wernli, 1977). With the recognition that *Globorotalia miozea* has been recorded from Spain and Crete as *Globorotalia menardii* Form 4 by Tjalsma (1971; see Berggren, 1977a), an important link has been forged in mid-late Miocene planktonic foraminiferal biostratigraphic correlation between the Mediterranean and temperate regions of the Atlantic and Pacific Oceans.

The terminal Miocene isolation of the Mediterranean from the global ocean system (Hsu, Ryan *et al.*, 1973) resulted in the temporary elimination of marine planktonic faunas and floras for about 0.5 m.y. (Berggren and Haq, 1976; van Couvering *et al.*, 1976). Renewed connection with the Atlantic in the Early Pliocene (*ca.* 5.0 Ma) resulted in a repopulation of the Mediterranean but with a curiously depleted fauna. Among the planktonic foraminifera both tropical and temperate forms are noticeably absent or occur only rarely. To the latter category belongs *Globigerina nepenthes*, recorded in the Lower Pliocene (as a rare element) only at Sites 376 and 378 (Eastern Mediterranean) and in the stratotype Zanclean, in Sicily (Cita and Gartner, 1973). However, Zachariasse and Spaak (1979) recorded the presence of *G. nepenthes* in relatively high numbers near the base of the Pliocene, tapering off rapidly upwards, in sections examined on Crete, Sicily and mainland Greece. They suggested that this brief acme-interval may be a useful criterion in regional intra-Mediterranean correlation following the early Pliocene reconnection with the Atlantic. Tropical elements, notably absent in the Mediterranean Pliocene sequences include the *Globorotalia miocenica-multicamerata* group, *Globorotalia menardii-exilis* group, *Globorotalia tumida*, and *Globigerinoides fistulosus*. The temperate *Globorotalia conoidea-conomiozea* group did not return to the Mediterranean region in the Pliocene, although it continued to play an important biostratigraphic role in the Atlantic and Pacific Oceans.

A six-fold zonation of the Pliocene of the Mediterranean was formulated by Cita (1973; see Fig. 6, this paper) based upon DSDP Sites 125 and 132, the lower four of which were subsequently applied in a zonation of the stratotype Zanclean (Cita and Gartner, 1973). The sequential extinction of *Globorotalia margaritae* and the *Sphaeroidinellopsis* group is similar to that seen elsewhere in the world ocean and plays a basic role in the formulation of the Early Pliocene part of this zonation (*cf.* Berggren, 1973). The *Sphaeroidinellopsis* Acme-Zone, found at the base of the Pliocene transgression in the Mediterranean, has been recorded at Sites 132 and 376 in the Western and Eastern Mediterranean, respectively. Early Pliocene faunas in the Mediterranean are characterized by the occurrence, *i.al.*, of globorotaliids (*margaritae, puncticulata*), globoquadrinids (*altispira*), *Globigerinoides* (*obliquus* group, *trilobus-sacculifer* group), and the *Sphaeroidinellopsis* group. The concurrent range of *Globorotalia margaritae* and *G. puncticulata* (Sites 121–125, 132, 133, 371–376, 378) is a characteristic feature of Early Pliocene faunas in the Mediterranean, mirrors a similar concurrent range of these taxa over a large area within the subtropical-temperate belt of the world ocean, and serves as an important link in the development of biostratigraphic zones for typically tropical and temperate-subarctic/subantarctic regions where the two forms are mutually exclusive.

The Late Pliocene (Ciaranfi and Cita, 1973) and Pleistocene (Cita *et al.*, 1973) faunal record of the Mediterranean is strongly determined by the climatic fluctuations associated with northern hemisphere glaciations. Late Pliocene faunas are characterized by *Globorotalia crassaformis, G. inflata, G. scitula, Globigerina bulloides*, and *Neogloboquadrina pachyderma*. Cita (1973) indicates that *Globigerinoides obliquus extremus* became extinct within the Late Pliocene (Zone PL5) and utilizes the partial range of *Globorotalia inflata* between the (apparent) extinction of *G. obliquus extremus* and the initial appearance of *Globorotalia truncatulinoides* as the basic criterion of her latest Pliocene Zone (= PL6,). Other studies (Haq *et al.*, 1977; work in progress at Woods Hole on latest Pliocene sequences on Rio Grande Rise) suggest an overlap between the LAD and FAD of *G. obliquus extremus* and *G. truncatulinoides*, respectively. Because of the difficulty in distinguishing between Zones PL5 and 6 the suggestion (Cita, 1978) that a late Pleistocene hiatus may exist in several Mediterranean sites (e.g., 371, 376, 378) may be in error.

Pleistocene faunas exhibit marked fluctuations in assemblage components, but typical elements consist of globigerinids (*bulloides, rubescens*), globorotaliids (*inflata, crassaformis, scitula*), neogloboquadrinids (*pachyderma-dutertrei-eggeri* complex), *Globigerinoides* (*ruber, sacculifer, tenellus*), and *Turborotalita quinqueloba*. *Globorotalia truncatulinoides* is generally rare and exhibits an erratic distributional pattern, although its FAD has been shown at Site 125 (Mediterranean Ridge) to coincide closely with the LAD of discoasters and the Pliocene/Pleistocene boundary as generally recognized by most Mediterranean stratigraphers (see Haq *et al.*, 1977).

A comparison of some Miocene–early Pliocene planktonic foraminiferal zonation schemes of the Mediterranean with corresponding zones in the Atlantic is shown in Fig. 7.

EPOCH	SERIES	T in Ma	ATLANTIC: NORTH — BERGGREN (1972); POORE AND BERGGREN (1975)	ATLANTIC: TROPICAL — BLOW (1969)	ATLANTIC: SOUTH — SAITO (1970), BLOW (1970), BERGGREN (1977)	MEDITERRANEAN: BERTOLANI et al. (1968)	MEDITERRANEAN: BIZON (1967), BIZON AND BIZON (1972)	MEDITERRANEAN: d'ONOFRIO et al (1975)
	PLIOC.		Globorotalia puncticulata	N19	PL2	G. hirsuta — G. bononiensis	Globorotalia crassaformis	NOT STUDIED
					PL1 c	G. hirsuta — G. puncticulata	Globorotalia puncticulata	
		5	Globorotalia conomiozea	N18	PL1 b, a	G. hirsuta — Sph-opsis spp.	Globorotalia margaritae	
					G. dehiscens — G. margaritae	UNDEFINED		
MIOCENE	L			N17	G. conomiozea – G. mediterranea	Globorotalia menardii — G. psuedo-miocenica	Globorotalia dutertrei – Globorotalia humerosa	Globorotalia acostaensis — G. conomiozea; G. multiloba; G. mediterranea; G. suterae; G. obliquus extremus
		10	Neogloboquadrina acostaensis	N16	Globorotalia miozea – Globorotalia conoidea	Globorotalia menardii — G. scitula ventriosa	Globorotalia acostaensis	Globorotalia acostaensis — G. continuosa
	M		Orbulina suturalis	N15		Globorotalia menardii — G. praemenardii	Globorotalia menardii	
				N14				
				N13				
				N12		Orbulina universa — G. obliquus	Globorotalia mayeri	
		15		N11			Globorotalia peripheroronda Orbulina suturalis	
				N10		G. altispira		
				N9		Orbulina suturalis		
			Praeorbulina	N8		Praeorbulina	Praeorb. glomerosa	
	E		Globoquadrina dehiscens	N7		G. trilobus — G. dehiscens	Globigerinoides trilobus	
				N6				
		20	Globoquadrina praedehiscens	N5		Globigerinita dissimilis	Globigerinita dissimilis / Globigerinoides altiapertutus	
			Globoquadrina baroemoenensis	N4			Globigerinoides primordius	
	OLIGOC.	25		P22			Globorotalia kugleri	

Fig. 7. Comparison of some Miocene–early Pliocene planktonic foraminiferal zones of the Mediterranean with corresponding zones in the Atlantic.

Table 8. Temperate to Subarctic North Pacific Sites.

Leg	Area	Site No.	Latitude (N)	Longitude (W)*	Geogr. Setting	Depth (m)	Tect. Hist.	Paleontol.	Remarks
5	NE Pacific between 125°–140° W longitude, ~40° N Lat.	32	37°07.63′	127°33.38′	Distal Margin Delgada Fan	4,758	S	R.K. Olsson	Late Neogene (Plioc.–Pleist.) Assemblages; gradual change from subtropical-transit. assembl. in Early Pliocene to subarctic assembl. in Late Pleistocene
		33	39°28.48′	127°29.81′	West of base of Delgada Fan on upper slope of an abyssal hill	4,284	S		
		34	39°28.21′	127°16.54′	Distal portion of Delgada Fan on small abyssal plain	4,322	S		
		35	40°40.42′	127°28.48′	Escanaba Trough, median valley of Gorda Ridge	3,273	S		
		36	40°59.08′	130°06.58′	200 km W of ridge crest at Site 35	3,273	S		
		37	40°58.74′	140°43.11′	Abyssal hill N of Mendocino Ridge	4,682	S		
18	NE Pacific and Gulf of Alaska between 124°–148°W Long., 40°–58°N Lat.	173	39°57.71′	125°27.12′	Continental slope W of Cape Mendocino	2,927	S	J.C. Ingle	Relatively complete record of Neogene sub-arctic-cool-temperate faunas. Repeated migrations of surface water isotherms seen as well as correlation with Neogene climatic events in west coast outcrop sections.
		174	44°53.38′	126°20.80′	Distal portion Astoria Fan	2,185 2,799 (A)	S		
		175	44°50.2′	125°14.5′	Lower contiental slope off Oregon	1,999	S-↑ ~ 300 m)		
		176	45°56.0′	124°37.0′	Outer contiental shelf off Oregon	193	S		
		177	50°28.18′	130°12.30′	NW end of Paul Revere Ridge	2,006	↑		
18	NE Pacific and Gulf of Alaska between 124°–148°W Long., 40°–58°N Lat.	178	56°57.38′	147°07.86′	Western Alaskan Abyssal Plain	4,218	S	J.C. Ingle	
		179	56°24.54′	145°59.32′	Alaskan Abyssal Plain	3,781	S		
		180	57°21.76′	147°51.37′	Eastern Aleutian Trench	4,923	S		
		181	57°26.30′	148°27.88′	Lower contiental slope, ~200m above Aleutian Trench	3,086	↑		
		182	57°52.96′	148°43.39′	Upper continental slope off Kodiak Island	1,419	S		
		182A	57°52.88′	148°42.99′	Upper continental slope off Kodiak Island	1,434	S		
19	North Pacific and deep-water parts of Bering Sea between 170°W–170°E and ~55°N	184	53°42.64′	170°55.39′	SW Corner Umnak Plateau, SW Bering Sea	1,910	?↓	R.J. Echols	Sporadic Late Neogene subarctic faunas. Monospecific faunas in Middle Miocene. Calcareous sediments absent above Middle Miocene Bering Sea.
		185	54°25.73′	169°14.59′	Between Bristol and Bering Canyons, SE Bering Sea	2,110	?↓		
		188	53°45.21′	178°39.56′E	W Flank Bowers Ridge, Bering Sea	2,649	S-↓		
		189	54°02.14′	170°13.38′E	N Flank Aleutian Ridge, Bering Sea	3,437	↑		
		190	55°33.55′	171°38.42′E	SW Aleutian Basin, E of Shirshov Ridge, Sea	3,875	S		

* Unless otherwise noted.

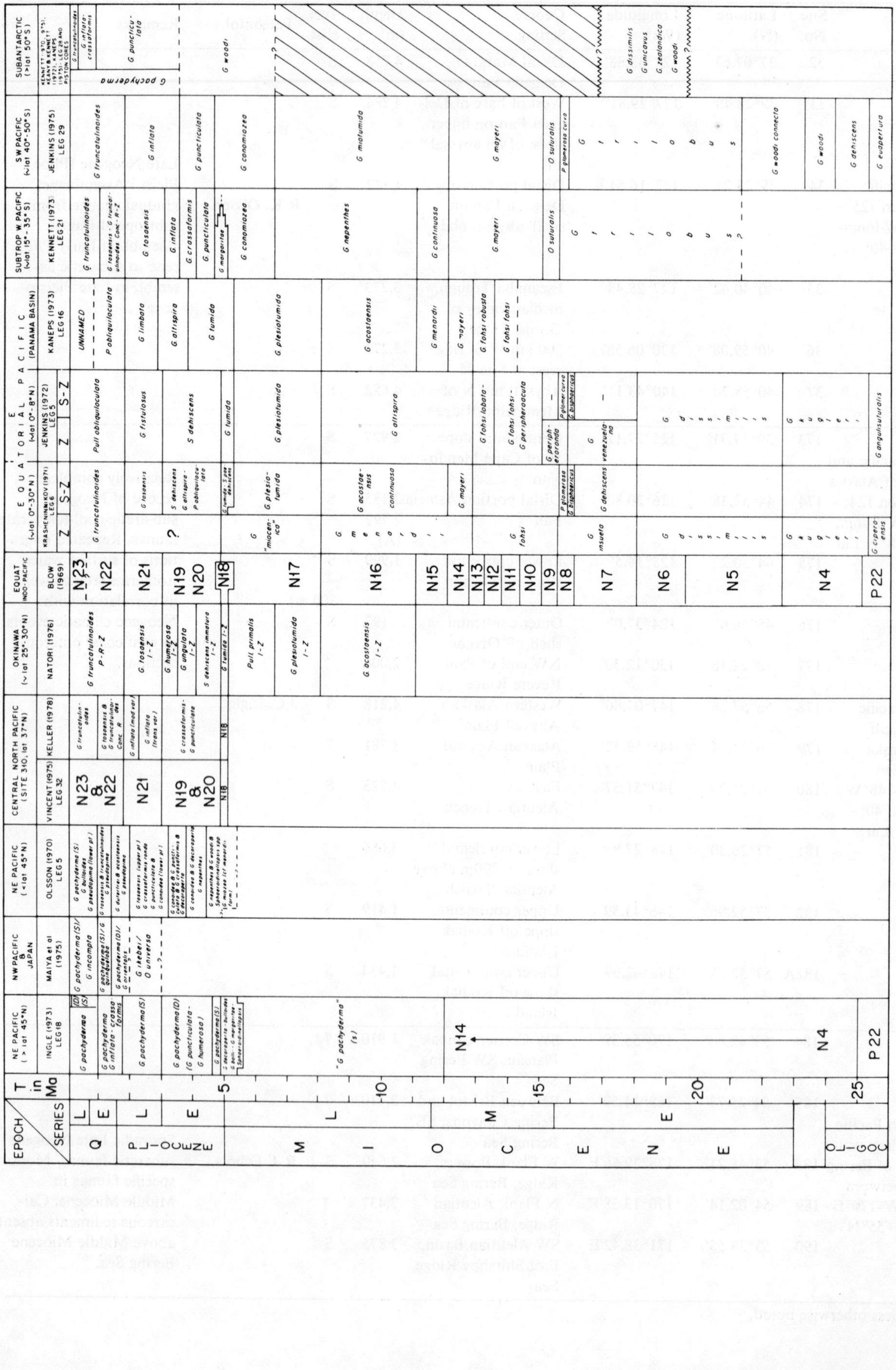

EPOCH / SERIES: PLIOCENE (L, E); Q (L, E); MIOCENE (L, M, E); OLIGOC

T in Ma: 5, 10, 15, 20, 25

NE PACIFIC (> lat 45°N) — INGLE (1973) LEG 18: G pachyderma (D)(S); G pachyderma G inflata-crassaformis; G pachyderma (S); G pachyderma (D) (G puncticulata - G humerosa); G pachyderma (S); G decoraperta - bulloides; G bolli - G margaritae - Sphaeroidinellopsis; "G pachyderma" (s); N14; N4; P22

NW PACIFIC & JAPAN — MAIYA et al (1975): G pachyderma (S)/ G incompta; G pachyderma (S)/ G quinqueloba; G pachyderma (D)/ G orientalis; G ikebei/ O universa

NE PACIFIC (< lat 45°N) — OLSSON (1970) LEG 5: G pachyderma (S) G bulloides G pseudopima (lower pt); G tosaensis & truncatulinoides G pseudopima; G dutertrei & acostaensis G pseudopima; G tosaensis (upper pt) G crassaformis ronda G puncticulata & G conoidea (lower pt); G conoidea & G puncticulata & G crassaformis & G decoraperta; G conoidea & G decoraperta G nepenthes; G nepenthes & G woodi & Sphaeroidinellopsis spp; G miozea (cf menardii form)

CENTRAL NORTH PACIFIC (SITE 310, lat 37°N) — VINCENT (1975) LEG 32: N23 & N22; N21; N19 & N20; N18

CENTRAL NORTH PACIFIC — KELLER (1978): G truncatulinoides; G tosaensis & G truncatulinoides Conc R; G inflata (mod var); G inflata (trans var); G crassaformis - G puncticulata; N18

OKINAWA (~ lat 25°-30°N) — NATORI (1976): G truncatulinoides P-R-Z; G tosaensis I-Z; G humerosa I-Z; G ungulata I-Z; S dehiscens immatura I-Z; G tumida I-Z; Pull primalis I-Z; G plesiotumida I-Z; G acostaensis I-Z

EQUAT INDO-PACIFIC — BLOW (1969): N23; N22; N21; N19; N20; N18; N17; N16; N15; N14; N13; N12; N11; N10; N9; N8; N7; N6; N5; N4; P22

EQUATORIAL PACIFIC (~lat 0°-30°N) — KRASHENINNIKOV (1971) LEG 6 (Z; S-Z): G truncatulinoides; ?; G tosaensis; S dehiscens; G altispira - P obliquiloculata; G tumida - S paedehiscens; G "miocenica"; G plesiotumida; G menardii; G acostaensis; continuosa; G mayeri; G fohsi; G insueta; P glomerosa; G bisphaericus; G dehiscens; G dissimilis; G kugleri; G ciperoensis

EQUATORIAL PACIFIC (~lat 0°-8°N) — JENKINS (1972) LEG 5 (Z; S-Z): Pull obliquiloculata; G fistulosus; S dehiscens; G tumida; G plesiotumida; G altispira; G fohsi lobata -; G fohsi fohsi - G peripheroacuta; G peripheroronda; –; P glomer curva; G bisphaericus; G venezuelana; –; G dissimilis; G kugleri; G angulisuturalis

EQUATORIAL PACIFIC (PANAMA BASIN) — KANEPS (1973) LEG 16: UNNAMED; P obliquiloculata; G limbata; G altispira; G tumida; G plesiotumida; G acostaensis; G menardii; G mayeri; G fohsi robusta; G fohsi fohsi

SUBTROP W PACIFIC (~lat 15°-35°S) — KENNETT (1973) LEG 21: G truncatulinoides; G tosaensis - G truncatulinoides Conc - R - Z; G tosaensis; G inflata; G crassaformis; G puncticulata; G margaritae; G conomiozea; G nepenthes; G continuosa; G mayeri; O suturalis; G trilobus ?

SW PACIFIC (~lat 40°-50°S) — JENKINS (1975) LEG 29: G truncatulinoides; G inflata; G puncticulata; G conomiozea; G miotumida; G mayeri; O suturalis; P glomerosa curva; G trilobus; G woodi connecta; G woodi; G dehiscens; G euapertura

SUBANTARCTIC (> lat 50°S) — KENNETT (1970, 1973); KEANY & KENNETT (1972); KANEPS (1975): LEG 28 AND PISTON CORES: G truncatulinoides; G inflata - crassaformis; G puncticulata; G pachyderma; G woodi; ?; G dissimilis; G unicavus; G zealandica; G woodi; ?

Fig. 8. Regional correlation of some Pacific Ocean Neogene planktonic foraminiferal zones.

Temperate and Subarctic North Pacific (>Lat. 40° N)

Neogene planktonic foraminiferal faunas of the North Pacific occur at sites drilled on DSDP Legs 5, 18, and 19 and indicate that temperate subarctic conditions were already established by Early Miocene time (Table 8). Siliceous sedimentation replaced the predominantly calcareous sedimentation of the Paleogene in the North Pacific.

A virtually complete Neogene sequence has been cored at Sites 173 (Delgada Fan, Lat. 40° N) and 192 (Meiji Seamount, Lat. 53° N); although sporadic and of low diversity, the planktonic faunal record at these sites is important for high-latitude Neogene biostratigraphy. A complete Pliocene-Pleistocene sequence was cored at Site 36 (Lat. 40° N), and a relatively thick and complete Pleistocene record has been cored at Site 174 (Lat. 45° N).

Low diversity faunas in the NE Pacific (Legs 18, 19) during Early Miocene time indicate that present-day circulation patterns (with a major southward flowing Eastern Boundary Current—the California Current) has already been established by Neogene time and perhaps even significantly earlier (Sliter, 1972). The overall tenor of the NE Pacific Neogene faunas is subantarctic-temperate with brief intervals during which the appearance of tropical-subtropical species allows direct correlation with "standard" tropical zonation(s).

Virtually monospecific assemblages of *Catapsydrax unicavus*, *C. dissimilis*, or *Globorotaloides suteri* characterize the Early Miocene of the Bering Sea (Site 192, Leg 19), although other species may have also been originally present. This is quite similar to Labrador Sea faunas (Leg 12, Site 112) and attests to the shallow level of the CCD during this time at high latitudes.

Early Miocene faunas of the mid-latitude NE Pacific (Leg 18) exhibit low diversity and are characterized by morphologically simple globigerinids and catapsydracids (*G. juvenilis*, *G. praebulloides*, *C. dissimilis*, *C. stainforthi*, *C. unicavus*, *i.al.*).

Late Neogene faunas reflect increasing climatic cooling and faunal provincialization. Mid-Late Miocene faunas are characterized by temperate to subarctic faunas dominated by globigerinids (Legs 18 and 19). The *Globorotalia miozea* group (*G. conoidea*, *G. miozea* s.l.), *Globorotalia puncticulata*, and the *Sphaeroidinellopsis* group characterize the mid-latitude (40° N; Site 36, Leg 5) Pliocene, whereas representatives of this group are absent further to the north (Legs 18 and 19). At higher latitudes (>45° N, Legs 18, 19) Pliocene faunas are characterized by subarctic taxa: *i.al.*, *Neogloboquadrina "pachyderma," Globigerina bulloides*, *Globorotalia scitula*, *Turborotalita quinqueloba*, *Globigerinita uvula*, and *G. glutinata*.

Similar faunas are seen in the Late Pliocene of Legs 5 and 19 with *N. pachyderma* (D), *N. humerosa*, *Globorotalia puncticulata*, *G. crassaformis*, and *Globigerina decoraperta* as characteristic elements.

Fluctuating faunal patterns have been observed in the late Neogene of the North Pacific (Legs 18 and 19). Ice-rafted detritus is found beginning in the late Pliocene (~3 Ma) and younger in this region. Sinistral populations of *N. pachyderma* replace Early Pleistocene dextral assemblages at about the Brunhes/Matuyama boundary (Olsson, 1971) or Jaramillo Event (Kent *et al.*, 1971) (Table 8). This faunal boundary—the limit of the subarctic fauna—has not been found north of its present postition (~45° N) since its inception, although it has shifted far to the south several times during cold cycles in the Late Pleistocene (Bandy, 1960).

During the Early Pleistocene, faunas south of 45° N (Leg 5) are characterized by *Globigerina bulloides*, *Globorotalia crassaformis*, *G. inflata*, *G. tosaensis*, and *G. truncatulinoides*, whereas to the north Subarctic faunas predominate (*Globigerina bulloides*, *Neogloboquadrina pachyderma* (D), *Globigerinita uvula*, *G. glutinata*, *Turborotalita quinqueloba*, *Globorotalia scitula*, *i.al.*). Further to the north (> 50°–55° N) Pleistocene fauna are rare and dominated by virtually monospecific faunas of *N. pachyderma* (S).

Relatively warm intervals with dextral *N. pachyderma* occur near the Pliocene/Pleistocene boundary in the North Pacific (Leg 19). Similarity of Pleistocene faunas under the California Current System (Sites 173–176, Leg 19) and the Subarctic (Alaskan) Gyre System (Site 177–181) suggests synchrony of faunal and paleotemperature events under the two current systems (Ingle, 1973).

The southward flowing California Current system is a major eastern boundary current that carries cool water as far south as latitude 24° N in the marginal northeastern Pacific. This has resulted in a relatively low diversity fauna in this area during the Neogene and in the inability to establish the Neogene planktonic foraminiferal stratigraphy of this area as well as in the California Neogene in terms of the "standard" tropical-subtropical zonation applicable in equatorial regions (Ingle, 1980).

Since Early Miocene time calcareous plankton have contributed only minor amounts to the sedimentary records, their place having been taken by iliceous organisms concomitant with the gradual reduction in surface water temperatures.

While the present-day biogeographic distribution of planktonic foraminiferal species in the North Pacific (Bradshaw, 1959) may have been established only recently (*i.e.*, during the last 1 m.y.), it is clear that temperate-subarctic conditions were established in this area during the Early–Middle Miocene and that the gradual cooling of the northern hemisphere has merely enhanced the biogeographic provincialization already clearly established in the early Neogene (t =20–25 Ma).

Tropical zonation systems are applicable (with difficulty) up to Middle Miocene time in the southern area (south of Lat. 45° N), but since this time (~last 10–12 m.y.) the application of tropical zonations is quite impossible in the North Pacific (Fig. 8). A regional biostratigraphic zonation based on climatic control of planktonic foraminiferal assemblages is the only way possible for the biostratigraphic subdivision of the Late Neogene in this region.

Western Equatorial and Northwest Mid-Latitude Pacific (Lat. 0° - 40° N; Long. 177° W-133° E)

This area is represented by Legs 6, 7, 20, 31, and 32 and spans an area from approximately Long. 177° W to 113° E and from Lat. 0–40° N (Table 9). Most of the sites contain tropical–subtropical Neogene faunas reflecting the stable influence of tropical–subtropical water masses during the past 20–25 m.y. However, in the NW area (Japan Sea and adjacent land sections) subarctic faunas are recorded as early as late Middle Miocene (*ca.* 12–15 Ma; Saito, 1963; Takayanagi and Oda, 1966; Shinbo and Maiya, 1971; Takayanagi *et al.*, 1975; Maiya *et al.*, 1975; Ingle, 1975), reflecting in turn the establishment at that time of essentially modern circulation patterns in the area.

Tropical-subtropical assemblages are present throughout the Neogene in the W Equatorial region (Legs 7, 20) and Phillipine Sea (Leg 31). Early Miocene assemblages are dominated by globoquadrinids (*altispira, baroemoenensis, dehiscens, langhiana, praedehiscens*), *Globigerinoides, Globorotalia kugleri*, and elements of the *Orbulina* bioseries (Krasheninnikov, 1973; Bronnimann and Resig, 1970). Less frequent are catapsydracids (*dissimilis, unicavus, stainforthi*) and globigerinids (*bradyi, woodi, juvenilis*). Middle and Late Miocene faunas contain typical representatives of keeled globorotaliids (*fohsi* bioseries, *menardii* group), *Sphaeroidinellopsis, Globigerinoides*, and *Globoquadrina*. Conspicuously absent are representatives of the *Globorotalia miozea* group.

Similar faunas occur in the NW mid-latitude sites (Legs 6, 32) with the difference that the influence of temperate faunas is seen beginning in the latest Miocene and Pliocene with the appearance of elements of the *G. miozea* group (*G. conoidea, G. conomiozea*) and *G. puncticulata-crassaformis-inflata* group, as well as temperate species of *Globigerina* (*bulloides, parabulloides*) and *Globigerinita* (*glutinata*).

The abrupt, retarded appearance of *Globorotalia conoidea* near the Miocene/Pliocene boundary in the NW mid-latitude Pacific (Site 305: Shatsky Rise and Site 310: Hess Rise) has been ascribed to dissolution (Vincent, 1975, p. 786) and may be associated with the relatively shallow CCD during mid-late Miocene time which appears to be connected with expansion of Antarctic glaciation and concomitant changes in deep ocean circulation patterns (Ryan *et al.*, 1974; van Couvering *et al.*, 1976; van Andel, 1974; Berger, 1975). In general biostratigraphic distribution and extinction patterns in this area are similar to those delineated earlier in tems of paleomagnetic stratigraphy (Hays *et al.*, 1969; Saito *et al.*, 1975).

The locations of Sites 305 and 310 make them ideal for late Neogene paleoclimatic investigation of the North Pacific because they lie close to the present-day boundary between the subtropical central water mass and the transitional (temperate) water mass. Shifts of this boundary over the sites should be recorded in the Hess and Shatsky rise sections.

A recent study (Keller, 1978) of Site 310 (Hess Rise) reveals, indeed, that temperate faunal assemblages had been established in the central North Pacific at least by latest Miocene time (Fig. 7). The *Globorotalia conoidea-conomiozea* and *G. puncticulata-inflata* groups were found to be abundantly represented and of greatest use in biostratigraphic zonation in the Pliocene, mirroring their distribution patterns in the middle latitude Southwest Pacific (Kennett, 1973), North Atlantic (Berggren, 1972b; Poore and Berggren, 1974, 1975), and South Atlantic (Berggren, 1977a,b).

Late Pliocene and Pleistocene climatic fluctuations at Site 310 were delineated by frequent oscillations of cool, intermediate, and warmer water assemblages of *Neogloboquadrina pachyderma, Globigerina*, and *Globigerinita*. Cold events were denoted at *ca.* 4.7 Ma, 3.0 Ma, 2.6–1.8 Ma, and 1.2 Ma (Keller, 1978). The use of the Blow (1969) zonation adopted by Vincent (1975) was modified by Keller (1978) to reflect the temperate nature of the late Neogene faunas at Site 310. The extinction patterns of the *G. conoidea-conomiozea* group at Site 310 (Hess Rise) mirror those seen at other temperate-mid-latitude areas such as the NE Pacific (Leg 5), N Atlantic (Leg 12), and S Atlantic (Leg 39).

Tropical planktonic faunas occur throughout the Neogene in the Phillipine Sea (Leg 31) as well as on the Ryuku Islands and Southern Japan (Kyushu) to the north (between Lat. ~25°-32° N) (Natori, 1976). However, the gradual replacement of tropical faunal elements by temperate elements (*G. miozea-conoidea* group and various globigerinids) during the late Miocene and Pliocene reflects the gradual cooling during the Late Neogene seen elsewhere.

To the north, tropical faunas extended as far north as Hokkaido (Lat. 41° N) during the Early and Middle Miocene to be abruptly replaced, following the extinc-

Table 9. Western Equatorial and NW Mid-Latitude Pacific Sites.

Leg	Area	Site No.	Latitude (N)*	Longitude (E)*	Geogr. Setting	Depth (m)	Tect. Hist.	Paleontol.	Remarks
		47	32°26.9′	157°42.7′	Shatsky Rise	2,689	S-↓		
	W Equatorial-NW Mid-Latitude Pacific	48	32°24.5′	158°01.3′	Shatsky Plateau	2,619	S-↓		Subtropical temperate assemblages; small differences between Caroline Ridge (9°N) and Shatsky Rise (32°N)
6		53	18°02.0′	141°11.5′	Flank of Iwo Jima Ridge	4,629	S-↓	V.A. Krasheninnikov	
		54	15°36.6′	140°18.1′	West of Iwo Jima Ridge	4,990	S-↓		
		55	09°18.1′	142°32.9′	Caroline Ridge	2,850	S-↓		
		305	32°00.13′	157°51.00′	Shatsky Rise	2,903	S-↓		Subtropical-temperate Late Neogene assemblages; Late Miocene cooling event and intra-Quaternary faunal fluctuations
32	NW Mid-Latitude Pacific	310	36°52.11′	176°54.09′	Hess Rise	3,516	S-↓	E. Vincent	
		313	20°10.52′	170°57.15′W	Basin in Mid-Pacific Mountains	3,484	S-↓		
		199	13°30.78′	156°10.34′	East Margin, Caroline Abyssal Plain, north of Caroline Islands	6,100	S		
		200	12°50.20′	156°46.96′	Ita Maitai Guyot	1,479			Tropical assemblages (Early Miocene-Q; N6–N22)
20	W Equatorial Pacific	200A	12°50.20′	156°46.96′	Ita Maitai Guyot	1,479		V.A. Krasheninnikov	
		201	12°49.9	156°44.6′	Ita Maitai Guyot	1,554			
		202	12°48.90′	156°57.15′	Ita Maitai Guyot	1,515			
		290	17°44.85′	133°28.08′	West flank of Palau-Kyushu Ridge, West Phillipine Basin	6,062.5	S-↓		
		291	12°48.43′	127°49.85′	Outer Arch, Phillipine Trench	5,217	S		
		292	15°49.11′	124°39.05′	Benham Rise, West Phillipine Basin	2,943	S-↓		
		293	20°21.25′	124°05.65′	NW Corner of W Phillipine Basin	5,599	S		1. Phillipine Sea: subtropical temperate Neogene assemblages
31	NW Pacific (Phillipine Sea; Sea of Japan) (15° N–40°N)	296	29°20.41′	133°31.52′	N end of Palau-Kyushu Ridge adjacent to Nankai Through	2,920	S-↓	H. Ujiié and J.C. Ingle, Jr.	2. Sea of Japan: Subarctic late Pleistocene-Holocene assemblages dominated by *N. pachyderma* (S) and *G. bulloides*
		297	30°52.36′	134°09.89′	Shikoku Basin	4,458	S-↓		
		298	31°42.93′	113°36.22′	Nankai Trough, off Shikoku Island, Japan	4,628	S		
		299	39°29.69′	137°39.72,	NE Yamato Basin, Sea of Japan	2,599	S-↓		
		300	41°02.96′	136°06.30′	E. Central Japan Abyssal Plain,	3,427	S		
		301	41°03.75′	134°02.86′	adjacent to Yamato Rise, Sea of Japan	3,520	S		
		302	40°20.13′	136°54.01′	N end Yamato Rise, Sea of Japan	2,399	S		
		62.0	1°52.2′	141°56.3′	Eurapik	2,602	S-↓		
		62.1	1°52.2′	141°56.3′	Ridge	2,607			
		63.0	0°50.13′	147°53.39′	East Caroline Ridge	4,486	S		
7	W Equatorial Pacific (1°S–4°N; 140°–177°E)	63.1	0°50.13′	147°53.39′		4,486			
		63.2	0°50.13′	147°53.39′		4,486		P. Brönnimann & J. Resig	Tropical Neogene assemblages
		64.0	1°44.53′S	158°36.58′		2,060			
		64.1	1°44.53′S	158°36.58′	Ontong-Java	2,060	S-		
		65.0	4°21.21′	176°59.14′	Plateau	6,142			
		65.1	4°21.21′	176°59.16′		6,142			

* Unless otherwise noted.

tion of *Globorotalia fohsi, ca.* 13 Ma, by temperate to cool-water faunas in the northern half of Japan (Saito, 1963). The Sea of Japan, extending from approximately Lat. 35–50° N, straddles mid- to high latitude biogeographic zones and is a marginal basin formed primarily during mid- to late Cenozoic periods of increased subduction and back are extension in the western Pacific. Yet it is unique in that, although over a quarter of the area is presently deeper than 3,000 m, it is bathymetrically and topographically isolated from the adjacent Pacific Ocean by narrow straits and shallow sills (120–130 m). Subarctic water moves south along the western margin of the Sea of Japan, whereas some subtropical water, termed the Tsushima, enters the sea from the south as a branch of the Kuroshio, and contains a warm temperate to subtropical fauna (Ujiie, 1973). A North Atlantic analogue of this situation would seem to be the NE flowing Gulf Stream and its subtropical fauna (Cifelli and Smith, 1970), and the Labrador Current which flows southwards to the south of Newfoundland and Nova Scotia along the NE coast of North America.

The Sea of Japan faunas recovered on Leg 31 represent essentially Late Pleistocene time (a brief interval of mid-Pliocene is represented in Site 299) and consists of subarctic elements dominated by sinistral and dextral *Neogloboquadrina pachyderma* and *Globigerina bulloides* s.l. Early Pleistocene and Pliocene time are represented by siliceous sediments, the result, probably, of a shallow CCD (Ingle, 1975). The climatic record of the past 1 m.y. in the Sea of Japan essentially mirrors that recorded in the faunal record (Leg 18) as well as in the ice-rafted detritus (Kent *et al.*, 1971) of the North Pacific.

In a recent study (Maiya *et al.*, 1975), the Late Pliocene-Pleistocene planktonic foraminiferal biostratigraphy of temperate-subarctic faunas of the NW Pacific and northern Japan have been compared and a zonation reflecting local distribution patterns erected. Elements of the *Globorotalia inflata* complex, the *Globoquadrina asanoi–G. kagaensis* group (which closely resemble *Neogloboquadrina atlantic* Berggren of the North Atlantic), and coiling changes in *Neogloboquadrina pachyderma* were found useful in regional correlation as well as biozonation.

Among the interesting conclusions reached by the authors is the observation that sinistral populations of *N. pachyderma* have persisted in the area of Lat. 46° N since at least 1.2 Ma and were preceded by dextrally coiled populations. This is taken to indicate a severe temperature reduction ($< 7°$ C) at the time and agrees well with the ice rafting data of Kent *et al.* (1971) which shows a major increase in ice rafting at this time.

A complete Neogene succession occurs at Site 77, Lower-Middle Miocene at Site 78, and Upper Miocene-Pleistocene at Sites 82 and 84. Representative Middle Miocene has been recovered at Site 319 (Bauer Deep), uppermost Miocene-lower Pliocene, and lower Pliocene-Pleistocene at Sites 157 and 158, respectively, in the Panama Basin. An abundant and diverse tropical fauna characterizes the Neogene throughout this area. In general the standard zonation of Blow (1969) appears to be applicable, at least to most of the pre-Pliocene part of the Neogene sequence in this region (Fig. 8). However, alternative zonations for the Neogene have been proposed in the Eastern Equatorial Pacific based on solution-resistant taxa (Jenkins and Orr, 1973; Orr and Jenkins, 1977; 1980) and for the Pliocene-Pleistocene in the Panama Basin and the Eastern Central Equatorial Pacific based on the absence of several of Blow's (1969) index species (Kaneps, 1973, 1976) (see Fig. 8).

The stratigraphic distributions of major faunal elements appear to agree well within this area, and in the Pliocene-Pleistocene the sequential extinction of various taxa and coiling change with *Pulleniatina* have been found to agree well with those delineated by Hays *et al.* (1969) and Saito *et al.* (1974) in paleomagnetically dated piston cores.

Central and Eastern Equatorial Pacific (~15° N–15° S; Lat. 80° W–180° E)

This area encompasses the equatorial belt (~15° N–15° S) extending from roughly longitude 80° W (Panama Basin) to 180° (central Pacific Basin) and includes sites drilled on DSDP Legs 8, 9, 16, 17, 33, and 34 (Table 10). A large part of this region lies within the equatorial high productivity belt. The axis of this belt (characterized by high rates of calcareous sedimentation) has shifted some 3–5° S since early-middle Miocene time. The equatorial current system contains two divergences: at the equator and at 10° N. The equatorial divergence and a shallow thermocline at the north edge of the equatorial counter current combine to bring nutrient rich deeper waters into the surface region which results in high standing crops of zooplankton. This is reflected in the high sedimentation rates and concentrations of calcareous microfossils. A concomitant problem is the high degree of dissolution on the ocean floor which tends to destroy solution-susceptible taxa, and Neogene recovery of planktonic foraminifera in this region has been relatively poor as a result. The general stability of this region during the Neogene is attested to by the fact that tropical planktonic faunas are recorded at nearly all sites during this interval.

Particularly useful faunal elements in this region include species of *Globoquadrina*, *Sphaeroidinellopsis*, *Pulleniatina*, and keeled *Globorotalia* (*kugleri*, *fohsi*, and *menardii* groups). The general rarity of the *Neogloboquadrina acostaensis-humerosa-dutertrei* group in

Table 10. Central and Eastern Equatorial Pacific Sites.

Leg	Area	Site No.	Latitude	Longitude (W)*	Geogr. Setting	Depth (m)	Tect. Hist.	Paleontol.	Remarks
17	Central Equatorial Pacific	167	07°04.01′N	176°49.05′	Magellan Rise	3,176	S	R.G. Douglas	Tropical Neogene Assemblages; moderate to strong dissolution; representative cosmopolitan benthonic foram assemblages; incomplete and irregular record
		168	10°42.02′N	173°35.09′E	West edge Central Basin, ~300 km E Mejit Island, in Marshall Island Chain	5,420	S-?↓		
		171	19°07.09′N	169°27.06′	Horizon Guyot	2,290	S-↓		
8	East Central Equatorial (16°N–12°S; 140°W)	69	6°00.00′N	152°51.93′	~600 km NE Christmas Island, between Clarion & Clipperton Fr. Zones	4,978	S-↓	J.P. Beckmann	Tropical Neogene Assemblages; notable dissolution
		70	6°20.08′N	140°21.72′	~ 30 km N Clipperton Fr. Zone	5,059	S-↓		
		71	4°28.28′N	140°18.91′	~ 85 km S Clipperton Fr. Zone	4,419	S-↓		
		72	0°26.49′N	138°52.02′	~ 520 km S Clipperton Fr. Zone	4,326	S-↓		
		73	1°54.58′S	137°28.12′	~ 800 km S Clipperton Fr. Zone	4,387	S-↓		
		74	6°14.20′S	136°05.80′	~ 415 km NE Marquesas Islands	4,431	S-↓		
		75	12°31.00′S	134°16.00′	~ 500 km SE Marquesas Islands	4,181	S-↓		
9	East Equatorial Pacific	76	14°05.90′S	145°39.64′	N of Tuamotu Ridge	4,598	S-↓	D.G. Jenkins and W. Orr	Tropical Neogene Assemblages; notable dissolution
		77	00°28.90′N	133°13.70′	S flank sediment pile, tropical Pacific	4,291	S-↓		
		78	07°57.00′N	127°21.35′	N of Clipperton Fr. Zone	4,378	S-↓		
		79	02°33.02′N	121°34.00′	Crest of equatorial Pacific sediment belt	4,574	S-↓		
		80	00°57.72′S	121°33.22′	S of Site 79	4,411	S-↓		
		81	01°26.49′N	113°48.54′	W flank of E Pacific Rise	3,865	S-↓		
		82	02°35.48′N	106°56.52′	W flank (near crest) of E Pacific Rise	3,707	S-↓		
		83	04°02.8′N	95°44.25′	E flank of E Pacific Rise; N of Galapagos Rift Zone	3,646	S-↓		
		84	05°44.92′N	82°53.29′	NW of Panama Basin, 400 km S of Panama	3,096	S-↓		
16	East Equatorial Pacific (01°S–13°N; 81–122°W	155	06°07.38′N	81°02.62′	Coiba Ridge, N part of Panama Basin	2,752	S-?↑	A.G. Kaneps	1. Tropical Late Neogene (mid-Miocene-Quat.) faunas in Panama Basin. 2. Late Oligoene-early Miocene faunas, western flank of East Pacific Rise; below CCD during Neogene.
		156	01°40.80′S	85°24.06′	S Flank of Carnegie Ridge	2,369	S-?↑		
		157	01°45.70′S	85°54.17′	S Flank of Carnegie Ridge	1,591	S-?↑		
		158	06°37.36′N	85°14.16′	Cocos Ridge, W edge Panama Basin	1,953	S-?↑		
		159	12°19.22′N	122°17.27′	W flank E Pacific Rise, between Clarion and Clipperton Fracture Zone	4,484	S-↓		
		315	4°10.26′N	158°31.54′	Fanning Fan East, south central part Line Island Chain	4,152	S		1. Tropical Neogene Assemblages (moderate dissolution

Leg	Area	Site No.	Latitude	Longitude (W)*	Geogr. Setting	Depth (m)	Tect. Hist.	Paleontol.	Remarks
33	East Central Equatorial Pacific (4°–16°S; 146°–162° W)	316	0°05.44′N	157°07.71′	S end Line Island Chain	4,451	S	A.G. Kaneps (Takayanagi and Oda)	throughout) 2. Late Miocene zonation difficult owing to scarcity of *N. acostaensis* and *G. plesiotumida*
		317	11°00.09′S	162°15.78′	Manihiki Plateau	2,598	S		
		318	15°59.63′S	146°51.51′	NW end of Tuamotu Chain	2,641	S		
34	East Equatorial Pacific (~ 9°–13° S; 80–100° W)	319	13°01 04′S	101°31.46′	W Flank Galapagos Rise (Bauer Deep)	4,296	S	P.G. Quilty J.M. Resig	1. Tropical Neogene Assemblages 2. Temperate Q Assemblages
		320	9°00.40′S	83°31.80′	E edge Nazca Plate (Peru Basin)	4,487			

* Unless otherwise noted.

the equatorial Pacific and its common occurrence in the Southern Panama Basin (Site 157; near the junction of the equatorial and west Peru currents) suggests that this group may prefer highly productive areas associated with eastern boundary currents and associated upwelling or other types of convergences (Kaneps, 1973, 1976). A similar pattern has been observed in the Upper Miocene (Andalusian) of western Spain (Berggren and Haq, 1976; Berggren *et al.*, 1977) associated with near coastal upwelling (see under discussion of Atlantic Ocean). The *Globorotalia miozea-conoidea-conomiozea* group is conspicuously absent in this area, save for the occurrence of *G. miozea* at Site 319 (Bauer Deep, Lat. 13° S) which may reflect the (peripheral) influence of the northward flowing Humboldt Current.

Common to the Pliocene sequences in this area, as well as elsewhere in the tropical-subtropical regions of the Pacific, is the absence of the *Globorotalia exilis-pertenuis* group and *G. miocenica* (rare records of these forms are incorrectly identified) and the occurrence of simultaneous changes in coiling patterns in *Pulleniatina* which serve as useful datum levels in regional correlation (Hays *et al.*, 1969; Saito, 1976).

Southwest Tropical-Temperate Pacific (~Lat. 0°-45° S)

This area is represented by DSDP Legs 21 and 30, and 29 (Sites 283, 284), which traversed the SW Pacific between latitudes 0°–45° S (Table 11). This area has proved to be of great value in delineating a latitudinal gradation between tropical to subtropical-temperate Neogene faunas.

An essentially complete Neogene sequence has been recovered in tropical (289: Ontong-Java Plateau), warm subtropical (208: N Lord Howe Rise), and temperate-subtropical (206: New Caledonia Basin) regions, while a virtually complete Middle Miocene-Pleistocene temperate-subtropical record at Site 207 (S Lord Howe Rise; 37° S), Upper Miocene-Pleistocene at Site 284 (Challenger Plateau; 40° S) and subtropical Upper Pliocene-Pleistocene at Site 209 (Queensland Plateau, 16° S) provides complementary coverage of the Neogene.

Early Miocene faunas are essentially similar in this region and consist of predominantly tropical elements (Blow, 1969). Of considerable interest is the absence (in Site 289) of praeorbulinids and the Neogene faunal dominance of globoquadrinids and the rarity of *Sphaeroidinellopsis* and *Globigerina nepenthes* (Saito *in* Andrews, Packham *et al.*, 1975, p. 248). The Middle Miocene of the more southerly region (Sites 206–208) is characterized by a relatively homogeneous fauna with a preponderance of globigerinids (*decoraperta, druryi, nepenthes, woodi, falconensis*), neogloboquadrinids (*continuosa*), non-keeled (*mayeri-siakensis* group) and keeled (*miozea-conoidea* group) globorotaliids. Part of the *Globorotalia fohsi* group has been recorded at Sites 206 and 208. In the equatorial region (Sites 209, 289), equivalent levels are characterized by the common occurrence of elements of *Sphaeroidinellopsis*, *Globigerinoides*, keeled globorotaliids of the *fohsi* and/or *menardii* groups for which the Blow (1969) zonation appears adequate (Kennett, 1973; Saito, 1975).

The relative uniformity of Middle Miocene faunas in areas of warm (Sites 206, 208) and cool (Site 207) subtropical regions appears to have broken down in the Late Miocene and Pliocene with increased provincialization of planktonic faunas (*i.e.*, reduction in diversity). Kennett (1973) has pointed out that this change either reflects a narrowing of faunal belts due to increasing global climatic deterioration or an increase in tropical influence after the Middle Miocene due to northward shift of the Tasman Sea region associated with the Cenozoic northward drift of Australia.

While typical tropical faunal assemblages continued in the equatorial region (Site 289), Late Miocene assemblages at mid-latitudes (Sites 206–208) contain mixed assemblage with temperate globigerinids (*nepenthes, bulloides, woodi, decoraperta*) more common to the south (Sites 284; 40° S and 207: 37° S) and tropical keeled

Table 11. S.W. Temperate–Equatorial Pacific Sites.

Leg	Area	Site No.	Latitude (S)	Longitude (E)*	Geogr. Setting	Depth (m)	Tect. Hist.	Paleontol.	Remarks
		203	22°09.22′	177°37.77′W	Lau Basin	2,720	S		
		205	25°30.99′	177°53.95′	S. Fiji Basin	4,320	S-?↓		
		206	32°00.75′	165°27.15′	New Caledonia Basin	3,196	S-↓		
21	S.W. temperate to Equatorial Pacific	207	36°57.75′	165°26.06′	S-Lord Howe Rise	1,389	S	J.P. Kennett	Warm to cool-subtropical Neogene faunas
		208	26°06.61′	161°13.27′	N-Lord Howe Rise	1,545	S		
		209	15°56.19′	152°11.27′	Queensland Plateau	1,428	S-↓		
		210	13°45.99′	152°53.78′	Coral Sea	4,543	S-?↓		
		283	43°54.60′	154°16.96′	Central Tasman Sea	4,729	S	D.G. Jenkins, J.P. Kennett, and P. Vella	
29	S.W. temperate Pacific	284	40°30.48′	167°40.81′	Challenger Plateau	1,066	S		Temperate Late Neogene faunas
		285	26°49.16′	175°48.24′	S. Fiji Basin	4,548	S-↓		
30	S.W. temperate to Equatorial Pacific	286	16°31.92′	166°22.18′	Gap between N&S New Hebrides trenches	4,465	S-↓	T. Saito	Tropical Neogene assemblages, particularly on Ontong-Java Plateau (Sites 288 and 289)
		287	15°54.67′	153°15.93′	Coral Sea	4,632	S		
		288	5°58.35′	161°49.53′	Eastern salient of Ontong-Java Plateau	3,000	S-↓		
		289	00°29.92′	158°30.69′	Ontong-Java Plateau	2,206	S-↑		

* Unless otherwise indicated.

globorotaliids of the *menardii* group, *Sphaeroidinellopsis, Neogloboquadrina acostaensis,* and *Globigerinoides* more common towards the equatorial region (Site 208: 26° S). A distinctly cool interval was noted at Site 207 (Lord Howe Rise) in the Late Miocene which indicates the northward migration of subantarctic waters over this area at that time (Kennett, 1973).

The *Globorotalia miozea-conoidea-conomiozea* group has been shown to be of use in Middle Miocene–Early Pliocene cool-subtropical and temperate biostratigraphy in this area (Fig. 7), and elements of this group have been recorded from the Lord Howe Rise (Sites 207, 208), New Caledonia Basin (Site 206), the Challenger Plateau (Site 284), and the South Fiji Basin (Site 285)—all of which are south of the 25° parallel. These forms were originally described from the New Zealand Miocene sequence (see Walters, 1965; Kennett, 1966; Jenkins, 1971) at latitudes in excess of 40° S.

In the Pliocene a distinct differentiation can be seen across the 40° latitudinal span represented by sites in this area. At one extreme (Sites 209, 210, 289), tropical assemblages dominate (*Globigerinoides, Pulleniatina, Sphaeroidinella-Sphaeroidinellopsis,* keeled globorotaliids (*menardii-tumida* groups), whereas at the other (Sites 207 and 284, at 37° and 40°, respectively), temperate assemblages (globigerinids: *bradyi, bulloides, decoraperta, falconensis;* globorotaliids: *puncticulata-inflata-crassaformis* group, *G. scitula; Neogloboquadrina pachyderma, Turborotalita quinqueloba, Globigerinita glutinata*) predominate. Sites located at intermediate latitudes (206, 208, 264) contain mixed assemblages with tropical elements predominating at 208, temperate forms at 206 aod 264. The association of *Globorotalia puncticulata* and *G. margaritae* in temperate-subtropcal latitudes (Sites 206–208, 264), and the occurrence of *G. margaritae* alone in equatorial (Sites 210, 289) and of *G. puncticulata* alone in temperate (Site 284) latitudes mirrors conditions seen in the North Pacific and in the North and South Atlantic and points to the importance of the latitudinal overlap (~Lat. 25°–35° N and S) of these two taxa as a link between low (< 25°) and high (> 35°–40°) zonations.

Pleistocene faunas are essentially identical to those at the present, and only minor changes in frequencies attest to changes associated with glacial–interglacial climatic fluctuations.

Southern Ocean: Circum-Antarctic (> 45° S)

Sites which are situated within the present-day limits of the Antarctic and Subantarctic water masses are included here in the Circum-Antarctic region. They are represented by some of the sites drilled on DSDP Legs 28, 29, 35, and 36 (see Table 12). The presence of

Table 12. Southern Ocean: Circum-Antarctic (> 45°S) Sites.

Leg	Area	Site No.	Latitude (S)	Longitude	Georgr. Setting	Depth (m)	Tect. Hist.	Paleontol.	Remarks
28	Antarctic	265	53°32.45′	109°56.74′E	S. Flank of SE Indian Ridge	3,582	S	A.G. Kaneps, J.P. Kennett	1. Polar Neogene faunas, low diversity 2. Late Neogene dominated by *N. pachyderma* (S) 3. Early Neogene dominated by *C. dissimilis* and *C. unicavus* 4. *G. miozea* group important in late Miocene–early Pliocene 5. Diverse Early Neogene calcareous and agglutinated benthonic assemblages
		266	56°24.13′	110°06.70′E	S Flank of SE Indian Ridge	4,173	S		
		267	59°15.74′	104°06.70′E	S Flank of SE Indian Ridge	4,564	S		
		268	63°56.99′	105°09.34′E	Lower Cont. Rise, N of Knox Coast of Antarctica	3,544	S-?↓		
		269	61°40.57′	140°04.21′E	SE edge of S Indian Abyssal Plain	4,285	S		
		270	77°26.48′	178°30.19′E	SE Ross Sea	634	S-↓		
		272	77°07.62′	176°45.61′E	SE Ross Sea	629	S-↓		
		273	74°32.29′	174°37.57′E	W Central part Ross Sea	495	S		
29	Subantarctic-Antarctic	275	50°26.34′	176°18.99′E	E edge Campbell Plateau	2,800	S	D.G. Jenkins, J.P. Kennett, and P. Vella	1. Temperate Neogene assemblages 2. *G. miozea* group biostratigraphically important in Late Miocene-Pliocene
		276	50°48.11′	176°48.40′E	SW Pacific Basin, SE of Campbell Plateau	4,671	S		
		277	52°13.43′	166°11.48′E	S Campbell Plateau	1,214	S		
		278	56°33.42′	160°04.29′E	S Emerald Basin	3,675	S		
		279	51°20.14′	162°38.10′E	N Macquarie Ridge	3,341	S		
		280	48°57.44′	147°14.08′E	S Tasman Rise	4,176	S		
		281	47°59.84′	147°45.85′E	S Tasman Rise	1,591	S-↓		
		282	42°14.76′	143°29.18′E	S Tasman Rise	4,202	S		
35	Subantarctic (S. Pacific)	322	60°01.45′	79°25.49′W	E end of Bellingshausen Abyssal Plain	5,026	S	F. Rögl	Low diversity, polar Neogene assemblages
		323	63°40.84′	97°59.69′E	Bellingshausen Abyssal Plain	4,993	S		
		324	63°03.21′	98°47.20′E	Lower Cont. Rise of Antarctica	4,449	S		
		325	65°02.79′	73°40.40′W	Antarctic Continental Rise	3,745	S		
36	Subantarctic (S. Atlantic)	326	56°35.00′	65°18.20′W	Drake Passage	3,812	S	R.C. Tjalsma	1. Low diversity, polar faunas since Early Miocene 2. *G. miozea-conoidea* group absent in Late Miocene
		327	50°52.28′	46°47.02′W	Falkland plateau	2,401	S		
		329	50°39.31′	46°05.73′W	E end of Falkland Plateau	1,519	S		

two major oceanographic boundaries, the Antarctic Convergence with a marked temperature gradient varying roughly befween 50°–60° S and the Subtropical Convergence which separates subantarctic and temperate water masses at about 45° S, sets the Circum-Antarctic region off as a biogeographic province distinct from areas to the north (Belyaeva, 1975).

The stratigraphic record of the Neogene is rather poor in this region. In the Subantarctic region (~45–60° S) a representative Lower-Middle Miocene section is present at Site 279 (North Macquarie Ridge).

The Neogene record in the Antarctic region (> 60° S) is poor, and in that which has been recovered, planktonic foraminifera form but a minor part of planktonic assemblages. Distributional data are supplemented by piston core studies of the Late Neogene (Kennett, 1970; Keany and Kennett, 1972) in this area (Fig. 8). Neogene planktonic foraminifera biogeography of the Circum-Antarctic region has been treated as part of detailed syntheses on the Cenozoic evolution of this area by Kennett (1977, 1978).

Early Neogene faunas (Sites 265, 266, 268, 269 of Leg 28; Sites 279, 280, 281 of Leg 29) are of low diversity (~3–5 species), owing in part to the effects of dissolution, and characterized in the Early Miocene by catapsydracids (*dissimilis, unicavus*), globigerinids (*woodi*, cf. *bulloides*), globorotaliids (*nana, pseudocontinuosa, praescitula, zealandica*), and *Globorotaloides suteri*. At Site 329, forms transitional between *G. praescitula*. and *G. miozea* have been recorded (Tjalsma, 1976).

Dominant elements in Middle to Late Miocene and Pliocene high latitude assemblages are the *Globorotalia miozea-conoidea* and *G. puncticulata-crassaformis-inflata* group, respectively. Middle Miocene Subantarctic faunas (Site 279) are characterized by globigerinids (*woodi*, cf. *bulloides*), *Globorotalia mayeri, G. praescitula*, and *Neogloboquadrina continuosa*. Praeorbulinids have been recorded at Site 279. In the Antarctic region (Site 266), *G. zealandica, G.* cf. *siakensis*, and *G. woodi* have been recorded.

Late Miocene faunas occur at Sites 281 and 329 and include the *Neogloboquadrina continuosa-acostaensis* group, and forms similar to *N. pachyderma*. Dominant forms include *Globigerina bulloides* and *Globorotalia scitula*. The *Globorotalia conoidea-conomiozea* group is conspicuously absent at Site 329 (Lat. 50° S), whereas it is present at Site 281 (48° S) which suggests that the southernmost limit of this group may be related to the position of the Subtropical Convergence. *Neogloboquadrina pachyderma* has not been recorded below the Pliocene from mid–high latitude Indian and Southern Ocean sites.

Subantarctic Pliocene faunas (Legs 28 and 29; above 45° S) are characterized by low diversity faunas: *Neogloboquadrina pachyderma* and *Globorotalia puncticulata, G. crassaformis, G. inflata, G. scitula*, and *Globigerina bulloides*. Keeled globorotaliids are absent here as well as in the Pleistocene, with the exception of sporadic occurrences of *G. truncatulinoides*. At higher latitudes (> Lat. 65° S), *N. pachyderma* and *Turborotalita quinqueloba* occur rarely in the Pliocene (Site 324, Leg 35).

In the Pleistocene, *G. crassaformis* and *G. inflata* migrated into the Subantarctic region at about 0.7 Ma (Kennett, 1970; Keany and Kennett, 1972).

Late Pleistocene faunas were recovered at Sites 326, 327, and 329. Very young assemblages were recovered at all three sites and belong to the *Globorotalia truncatulinoides* Zone (*sensu* Kennett, 1970) based on the occurrence of the nominate taxon (which appeared in the subantarctic region *ca.* 0.2 Ma), in association with *Globorotalia inflata, G. scitula, Globigerina bulloides, Globigerinita glutinata, G. uvula*, and *Neogloboquadrina pachyderma* (sinistral). The presence of *Globorotalia puncticulata* (*sensu* Kennett), with essentially the same faunal elements at somewhat lower levels at Site 329, indicates the presence of the *Globorotalia inflata* Zone (*sensu* Kennett, 1970).

Polar-subpolar Pleistocene assemblages are dominated by *Neogloboquadrina pachyderma* and *Globigerina bulloides*, whereas temperate assemblages are characterized by *Globorotalia trunatulinoides* and *G. inflata*, in the same manner as present day distribution patterns (Bé and Hutson, 1977).

Major changes occurred in planktonic foraminiferal ecology of the Southern Ocean near the Eocene/Oligocene boundary. Relatively diverse Eocene planktonic faunas are replaced by low diversity Oligocene faunas (e.g., Site 267; 59° S, Leg 28). This trend continues throughout the Neogene. Antarctic planktonic foraminiferal assemblages have had a polar aspect since Oligocene time (contrast this with the North Atlantic where this change occurred in mid-Pliocene time some 20 m.y. later).

The poor recovery of Neogene sediments and low faunal diversity on the Falkland Plateau (Leg 36) may be attributed to the role played by the Circumpolar Current which was initiated following the opening of the Drake Passage, 25–30 Ma (Kennett *et al.*, 1975; Kennett, 1977, 1978).

A diachronous calcareous/siliceous sediment boundary occurs over the interval of Late Oligocene–Early Miocene, becoming younger to the north. This reflects the northward movement of the Antarctic Convergence and the growth of glaciation on the Antarctic continent. This boundary was relatively stable during the Miocene (t = ~24–5 Ma) and shifted northward with respect to the SE Indian Ocean Ridge again in the Pliocene. Above 60° S siliceous organisms became the dominant faunal and floral elements by Middle Miocene time and

Table 13. Indian Ocean and Red Sea.

Leg	Area	Site No.	Latitude (N)*	Longitude (E)	Geographic Setting	Depth (m)	Tect. Hist.	Paleontol.	Remarks
		211	9°46.53′S	102°41.95′					1. Essentially
					Wharton				tropical
		212	19°11.34′S	99°17.84′	Basin	6,243			Neogene
									assemblages
22	Central	214	11°20.21′S	88°43.08′	Ninetyeast Ridge	1,655		B. McGowran	2. Sites 214, 216
	Indian							(W.A. Berggren	good for late
	Ocean	216	1°27.73′	90°12.48′	Ninetyeast	2,247		R.Z. Poore	Miocene–
					Ridge			G.P. Lohmann)	Pliocene bio-
		217	8°55.57′	90°32.33′		3,020			stratigraphy.
									3. *Globorotalia*
									miozea-conoi-
		218	5°00.42′	86°16.97′	Bengal Fan	3,759			*dea* group
									important at
									Site 212
		219	9°01.75′	72°52.67′		1,764	S-↓		Arabian Sea:
		220	6°30.97′	70°59.02′		4,036	S		1. Stable, tropi-
									cal faunas
									throughout
		221	7°58.18′	68°24.37′		4,650	S		Neogene.
					Arabian				2. Blow (1969)
					Sea				zonation
		222	20°05.49′	61°30.56′		3,546	S		difficult to
									apply in post-
									middle Miocene
		223	18°44.98′	60°07.78′		3,633	S		Red Sea
23	Arabian (219–	224	16°32.51′	59°42.10′		2,500	S-↑	R.L. Fleisher	1. Restricted
	224) and Red	225	21°18.58′	38°15.11′		1,228	S	(W.H. Akers)	(surface
	(225–230) Seas	227	21°19.86′	38°07.97′		1,795	S		dwelling)
									Pliocene-
									Pleistocene,
									globigerinid,
									Globigerin-
									oides faunas.
		228	19°05.16′	39°00.20′	Red	1,038	S		2. Gradually
		229	14°46.09′	42°11.47′	Sea	852	S		increasing
									diversity
									related to
									decreasing
									salinity
		231	11°53.41′	48°14.71′	Gulf of Aden	2,161	S		1. Tropical
		232	14°28.93′	51°54.87′	NW Flank Alula-	1,758	?S		Neogene
					Fartak Trench				faunas
		233	14°19.68′	52°08.11′	E Flank Alula-	1,860	?S		2. Early Neogene
	Western				Fartak Trench				=marked
	Equatorial								dissolution.
24	Indian Ocean	234	4°28.96′	51°13.48′	W Margin	4,738	?S	E. Vincent	3. Blow (1969)
	(11°S–15°N)				Somali Basin			(W.E. Frerichs	zonation dif-
								M.E. Heiman)	ficult to apply
									in late Neo-
		235	3°14.06′	52°41.64′	Somali Basin	5,146	S		gene (post
									middle Mio-
									cene)
		236	1°40.62′S	57°38.85′	Foothills SW of	4,504	S		
					Carlsberg Ridge				
		237	7°04.99′S	58°07.48′	Mascarene	1,640	S-↓		
					Plateau				
		238	11°09.21′S	70°31.56′	NE end Argo	2,844	S		
					Fracture Zone				
					Central Indian				
					Ridge				

* Unless otherwise noted.

Table 13. Indian Ocean and Red Sea (cont.)

Leg	Area	Site No.	Latitude (S)	Longitude (E)	Geographic Setting	Depth (m)	Tect. Hist.	Paleontol.	Remarks
		239	21°17.67′	51°40.73′	Mascarene Basin	4,971	S		
		240	3°29.24′	50°03.22′	Somali Basin	5,082	S		
		241	2°22.24′	44°40.77′	East Africa Cont. Rise, W Somali Basin	4,054	S-?↓		
		242	15°50.65′	41°49.23′	Mozambique Channel (Davie Ridge)	2,275	S		
25	Western Equatorial Indian Ocean (0°–35°S)	243	22°54.49′	41°23.99′	Mozambique Channel (Zambezi Canyon floor)	3,879	D	B. Zobel	Tropical Neogene faunas
		244	22°55.87′	41°25.98′	Mozambique Channel (Zambezi Canyon flank)	3,768	S		
		245	31°32.02′	52°18.11′	S Madagascar Basin	4,857	S-↓		
		246	33°37.21′	45°09.60′	Madagascar Ridge	1,030	↑		
		247	33°37.53′	45°00.68′	Madagascar Ridge	944	↑		
		248	29°31.78′	37°28.48′	Mozambique Basin	4,794	S		
		249	29°56.99′	36°04.62′	Mozambique Ridge	2,088	S		
		250	33°27.74′	39°22.15′	Mozambique Basin	5,119	S		
		251	36°30.26′	49°29.08′	SW Indian Ridge	3,489	S		
		252	37°02.44′	59°14.33′	Crozet Basin	5,032	S		
	Temperate to	253	24°52.65′	97°21.97′		1,962			1. Temperate
26	subtropical Indian Ocean (lat. 24°–37°S)	254	30°58.15′	87°53.72′	Ninety East Ridge	1,253	S-↓	E. Boltovskoy	faunas, except Site 253 (=subtropical)
		255	31°07.87′	93°43.72′	Broken Ridge	1,144	S		2. Blow (1969)
		256	23°27.35′	100°46.46′	Wharton Basin	5,361	S		zonation difficult to use
		257	30°59.16′	108°20.99′		5,278	S		in postmid
		258	33°47.69′	112°28.42′	Naturaliste Plateau	2,793	S		Miocene
		259	29°37.05′	112°41.78′		4,712	S		
	Eastern	260	16°08.67′	110°17.92′	E Edge Wharton Basin	5,709	S	H.M. Bolli (F. Rögl	Tropical-subtropical faunas
27	Indian	261	12°56.83′	117°53.56′		5,687	S	R.K. Olsson)	except Site 259
	Ocean	262	10°52.19′	123°50.78′	Timor Trough	2,315	↓		(=temperate
		263	23°19.43′	110°58.81′	E Edge Wharton Basin	5,065	S		Q fauna)
28	Temperate-subtrop. Indian Ocean	264	34°58.13′	112°02.68′	Naturaliste Plateau	2,873	S	A.G. Kaneps	

planktonic foraminifera are virtually absent due to dissolution.

The earliest evidence of ice-rafting in the southern hemisphere occurs in the Late Oligocene or Early Miocene (Legs 28 and 35). North of 63° S, ice rafting occurs only in post-Miocene sediments at Leg 28 and 35 sites. However, the earliest evidence south of this is in the Early Miocene, *ca.* 20 Ma, although evidence of earlier ice rafting is seen at Sites 268 and 274 (Leg 28).

Leg 28 data suggest a basal age of 25–26 Ma for earliest ice rafting in the Ross Sea, *i.e.*, inception of major glaciation in Ross Sea area in mid-late Oligocene time.

The Indian Ocean and Red Sea

For the purpose of this discussion the demarcation line between the Indian Ocean and Pacific Ocean south of Australia is made along the 150° E meridian following Vincent (1977) and between the Indian Ocean and Southern Ocean along the Subtropical Convergence which separates subantarctic and temperate water masses at about 45° S. As a result, parts of Legs 28 and

29 are included under the Indian Ocean and parts under the South Pacific, resulting in a slight, but unavoidable, duplication (Table 13).

Over 12,000 meters of Neogene sediments were drilled in the Indian Ocean at nearly 50 sites on DSDP Legs 22–28. In addition, some 1,200 meters were drilled (673 m recovered) at 4 sites in the Red Sea. The data from these sites (20° N to ~ 50° S) form the basis for the summary discussion presented below. The discussion, in turn, is divided into two parts: a) tropical-subtropical Indian Ocean (~ Lat. 20° N–45° S: Legs 22–27); b) Red Sea (~ Lat. 15°–20° N: Leg 23).

Tropical-Subtropical Indian Ocean (Legs 22–28)

An essentially complete Neogene succession has been recovered at Site 214 (although dissolution becomes pronounced in Lower Miocene levels) and at Site 253 (strongly condensed). Middle Miocene to Pleistocene is represented at Site 231, Upper Miocene to Pleistocene at Sites 219 and 236, upper Upper Miocene (Zone N17) to Pleistocene at Site 237 and 253, Lower Pliocene to Pleistocene at Site 232, and Upper Pliocene-Pleistocene at Sites 233 and 262; upper Lower Miocene (Zone N7) to lower Upper Miocene (Zone N17) occurs at Site 216 and complements that part of the section at nearby Site 214.

The Indian Ocean has remained relatively stable during the last 25 million years and tropical faunas, essentially similar to those found in the equatorial Pacific, are present throughout the Neogene.

Early Neogene faunas exhibit a tendency towards low diversity caused by carbonate dissolution, whereas Late Neogene faunas are diverse, numerically common to abundant, and generally well preserved. In the early Miocene, *Catapsydrax dissimiliis* and *C. unicavus, Globorotalia peripheroronda, G. siakensis* (= *G. mayeri auct.*), and various globigerinids and globoquadrinids dominate the faunas.

It has been possible, although rather difficult, to apply Blow's (1969) tropical zonation scheme to low latitude Indian Ocean sites. This appears to be due in part to the apparent rarity, or absence, of some zonal taxa, such as *Catapsydrax stainforthi,* the strong dissolution noted above, and the fact that lower Miocene sections, where encountered, are strongly condensed and appear to have been deposited at rates of sedimentation of less than 0.5 cm/1000 yrs (Vincent, 1977). *Globorotalia kugleri* (Zone N4) has been found to be relatively common and widespread in the tropical Indian Ocean (e.g., Sites 214, 216, 238), and the nominate taxa of early Miocene Zones N5–N8 have been found at Sites 214, 216, 238, among others. The extension of *Globigerina binaiensis* beyond the upper limit of *G. sellii* and/or *Globorotalia kugleri* (e.g. Sites 214, 216; Leg 22, Sites 246, Leg 25, and 252, Leg 26) is a useful criterion in recognizing the presence of Zone N5 and attests to the utility of this taxon in Early Miocene biostratigraphy of the Indian Ocean. Distinction between Zones N7 and N8 has been difficult in the Indian Ocean because of the virtual absence of *Catapsydrax stainforthi,* and because of the anomalous stratigraphic record of *Globigerinoides sicanus* at Sites 253 and 254 (Vincent, 1977).

Accurate determination of the Lower/Middle Miocene boundary is precluded at both low and mid-latitude sites by the absence of the *Orbulina* bioseries owing to dissolution.

Middle Miocene sediments of low latitude Indian Ocean cores (particularly Legs 22–24) are characterized by elements of the *Globorotalia fohsi* and *menardii* (*archeomenardii, praemenardii*) lineages, *Globigerinoides* (*bollii, obliquus, subquadratus*), the nascent *Sphaeroidinellopsis* group, *Neogloboquadrina* (*mayeri, siakensis*), *Globoquadrina,* and *Globigerina* (*decoraperta, druryi* and *nepenthes*). The *G. miozea-conoidea* group is particularly well developed in the Middle Miocene at Sites 212 (where it occurs with *Neogloboquadrina mayeri, N. siakensis,* and elements of the *G. menardii* group; Vincent, 1974), 251, 254 (where it occurs with elements of the *G. fohsi* lineage), 255, and 256 (Boltovskoy, 1975). The co-occurrence of the *G. miozea-conoidea* group with elements of the *G. fohsi* and/or *G. menardii* groups forms an important link in mid–late Neogene stratigraphic zonation between low and high latitudes. Associated Middle Miocene elements include globigerinids of the *woodi* and *bulloides* groups. *Neogloboquadrina acostaensis* and *Globoquadrina altispira* (Sites 251, 254, 255) are commonly associated with the *G. miozea-conoidea* group.

An interesting similarity with the Middle Miocene of the South Atlantic (Berggren, 1977a, b) is seen in the occurrence in tropical Site 253 of *Globigerinopsis aguasayensis, Globigerinoides amplus,* and *G. bulloides* in Zones N13–14 (Boltovskoy, 1975).

Late Miocene assemblages are characterized by keeled globorotaliids of the *Globorotalia tumida (merotumida-plesiotumida*), and *G. menardii* lineages, *Neogloboquadrina acostaensis, Sphaeroidinellopsis, Globigerina nepenthes,* and *G. decoraperta, Candeina,* and in the latest Miocene by the initial appearance of *Pulleniatina* (e.g., at Sites 214, 219, 236, 237, and 238 apparently at a level coincident with its demonstrated appearance in the early part of paleomagnetic Epoch 5, *ca.* 6 Ma; Saito *et al.,* 1975). Taxonomic problems associated with recognizing morphologic boundaries within the *G. tumida* lineage, a lament shared by the shipboard micropaleontologists of Legs 22–26, rendered it less than useful in late Miocene zonation of the tropical Indian Ocean, although Berggren and Poore (1974), in making a more detailed investigation of this group at Site 214, indicated that, at the very best, the initial appearance of *Globorotalia tumida* occurs rather abruptly and is a relatively distinct, unequivocal event associated with the Miocene/Pliocene boundary (see also Saito *et al.*, 1975).

On the other hand, the apparent latitudinal and/or dissolution control on the last occurrence of *Globoquadrina dehiscens* (ranging from late Middle Miocene to questionably occurrences within the Pliocene; cf. Vincent, 1977) would appear to exclude this taxon from the biostratigraphic utility it has in the Pacific (Saito *et al.*, 1975) and Atlantic (Berggren, 1977a, b) oceans.

Pliocene assemblages are numerically rich, taxonomically diverse, and essentially the same as those found in the tropical Pacific with one notable exception: the apparent absence of the early Pliocene taxon *Pulleniatina spectabilis.* The sequential extinctions and/or initial appearances of various taxa such as *Globigerina nepenthes, Globorotalia margaritae, Sphaeroidinellopsis* spp., *Globigerinoides fistulosus, Globoquadrina altispira,* mirror those seen in equatorial Pacific and Atlantic cores and are calibrated there as well as in Indian Ocean core V20–163 to the paleomagnetic stratigraphy (Hays *et al.*, 1969; Saito *et al.*, 1975). Coiling change patterns in *Pulleniatina* during the Pliocene-Pleistocene mirror those seen in the Pacific (Saito, 1976), indicating the uninterrupted continuity of the Indo-Pacific biogeographic province during the late Neogne.

The initial appearance of *Globigerinoides fistulosus* occurs consistently between the extinction of *Globorotalia margaritae* (3.4 Ma) and that of *Globoquadrina altispira* (2.8 Ma) at Sites 214, 219, 237, 238, and 253, as well as in paleomagnetically calibrated core V20–153 (Saito *et al.*, 1975). This level is essentially coincident with the initial appearance of *Globorotalia tosaensis* (*ca.* 3.1–3.2 Ma) and the use of *Globigerinoides fistulosus* in concert with, or as a substitute for, *G. tosaensis* (which is relatively rare in low latitudes) may be suggested as a suitable alternative to the demonstrable shortcoming of Zone N21 (Blow, 1969) in Late Pliocene low latitude biostratigraphy of the Pacific and Indian Ocean and of Zone PL 5 (Berggren, 1973) since the nominate taxa, *G. miocenica-exilis,* do not occur outside of the Atlantic.

Of general interest is the observation that the initial appearance of *Globigerinoides conglobatus* coincides closely with the Miocene/Pliocene boundary at various sites in the Indian Ocean (e.g. 253) as it does in the South Atlantic (Rio Grande Rise; Berggren, 1977a, b).

Temperate Pliocene faunas of the Indian Ocean are characterized by globorotaliids of the essentially unkeeled *puncticulata-crassaformis-inflata* group (here treated as phylogenetically distinct from the keeled *Globorotalia conoidea-conomiozea* group; Berggren, 1977b) and various globigerinids and *N. pachyderma. G. crassaformis* and *G. inflata* occur rarely in low latitude sites and form an important link in low-high latitude zonation in the Pliocene, as does *G. puncticulata,* which has not been recorded from low latitude sites. *Globorotalia puncticulata, G. crassaformis,* and *inflata* have been shown to evolve at approximately 4.6, 4.0 and 3.0 Ma, respectively (Hays *et al.*, 1969; Berggren, 1972, 1977b; Berggren and Poore, 1974; Saito *et al.*, 1975) in the Atlantic and Pacific Oceans, and the sequential appearance of these taxa in the Indian Ocean (Vincent, 1977) would appear to indicate their interoceanic synchrony.

Pleistocene assemblages are essentially the same as those occurring today in this area and are dominated by *Globorotalia menardii, G. tumida, Globigerinoides ruber, G. sacculifer, Sphaeroidinella, Pulleniatina, i.al. Globorotalia inflata* and *G. crassaformis* occur as accessory species, and the ratio of the former to *G. menardii* has been used to derive a generalized climatic curve for the Pleistocene of the tropical Indian Ocean (Boltovskoy, 1975; cf. Vincent, 1977).

The Red Sea

Neogene planktonic foraminiferal biostratigraphy of the Red Sea is restricted to the last 5 m.y. (post late Miocene evaporitic phase). The sill at the Straits of Bab el Mandab control, and essentially restrict, the inflow of water from the Indian Ocean–Arabian Sea to surface and near surface (~100 m) waters which prohibits a free circulation of deep water between the Red Sea and the Gulf of Aden; this in turn has restricted planktonic faunas to inhabitants of the upper 100 m of the sea surface (Herman, 1965; Berggren and Boersma, 1969; Fleisher, 1975).

Late Neogene (Pliocene-Pleistocene) faunas consist of species of *Globigerinoides (obliquus, quadrilobatus, ruber, sacculifer), Globigerina (decoraperta), Globigerinita glutinata, Turborotalita quinqueloba, Hastigerina siphonifera,* and, rarely, *Globorotalia menardii* and related keeled faunas (Fleisher, 1975). Since the Early Pliocene, there has been a general increase in faunal diversity with short-term fluctuations which Fleisher (1975) has utilized to make a generalized 4-fold assemblage zonation of the Pliocene-Pleistocene and has interpreted as being related to a gradual decrease in salinity.

Late Pleistocene assemblages have been shown to be strongly controlled by fluctuations in salinity (evaporation) resulting from northern hemisphere polar glacial-interglacial cycles (Berggren and Boersma, 1969;

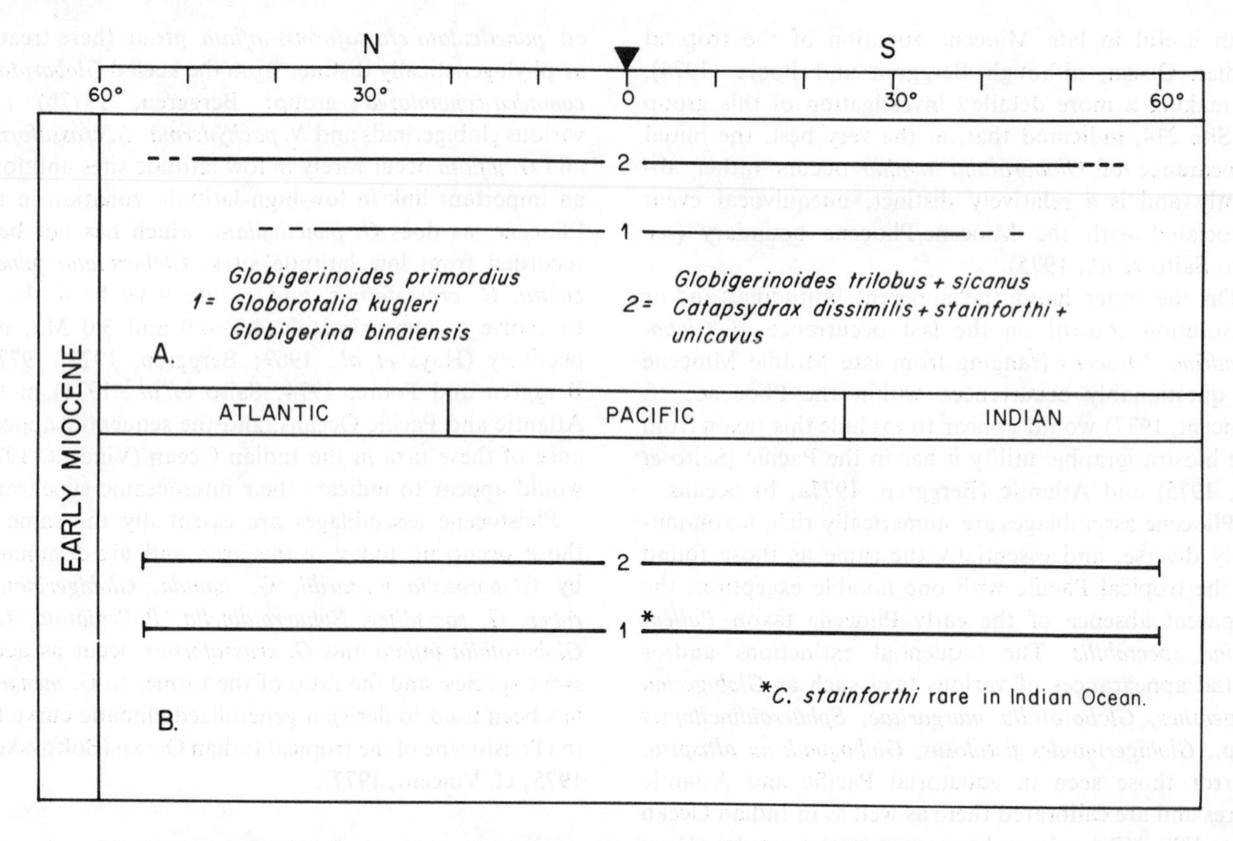

Fig. 9. Geographic distribution of stratigraphically important planktonic foraminiferal taxa: Early Miocene. A–latitudinal distribution; B–interoceanic distribution.

Berggren, 1969).

Summary of Planktonic Foraminiferal Biogeography

Early Miocene

Earliest Miocene planktonic foraminiferal faunas exhibit a marked uniformity over a large latitudinal extent (Fig. 9). Between latitudes 45° North and South faunas in the Atlantic, Pacific, Indian Oceans, and Mediterranean Sea were characterized by globoquadrinids *(baroemoenensis, dehiscens, praedehiscens)*, catapsydracids *(dissimilis, unicavus* and—with the exception of fhe Indian Ocean—*stainforthi*), and *Globigerinoides* (*primordius, trilobus, altiaperturus* group). Less common were the globigerinids (*juvenilis, praebulloides*) and globorotaliid (*siakensis*). Locally, and particularly at low latitude sites, globorotaliids of the *kugleri* and *peripheroronda* groups were common. Praeorbulinids are common in various sites, but noticeably absent in the Indian Ocean due to dissolution. In the South Atlantic a noticeable contribution to the fauna from the temperate SW Pacific and New Zealand region is seen in the fauna of various globigerinids (*brazieri, woodi, connecta*) and globorotaliids (*pseudocontinuosa, semivera, zealandica*).

In the North Pacific (Bering Sea) and North Atlantic (Labrador Sea), Early Miocene faunas exhibit low diversity (3–5 species) and are characterized by catapsydracids (*dissimilis, unicavus*) and/or *Globorotaloides suteri*, which attest to the shallow level of the CCD at this time. Similar faunas occur at Southern Ocean sites, with the addition of various globorotaliids (*nana, pseudocontinuosa, praescitula, zealandica*).

Middle Miocene

Whereas interoceanic (longitudinal) distribution was still cosmopolitan, Middle Miocene planktonic foraminiferal faunas exhibit a continuing trend towards latitudinal differentiation (Fig. 10). Common elements at low and middle latitudes in all oceans included the globoquadrinids (*altispira, dehiscens*), *Globigerinoides* (*obliquus, subquadratus, trilobus*). globigerinids (*druryi, nepenthes*), *Sphaeroidinellopsis* group, orbulinids, and various non-keeled (*mayeri-siakensis*) and keeled (*menardii, fohsi*) globorotaliids. The latitudinal differentiation taking place at this time is perhaps best exemplified by the essentially low vs. mid-high latitude distribution patterns of *Globorotalia fohsi* and *G. miozea* groups, respectively. The co-occurrence of these two groups at several sites in the Atlantic, Pacific, and Indian Oceans serves as a useful link in high vs. low latitude

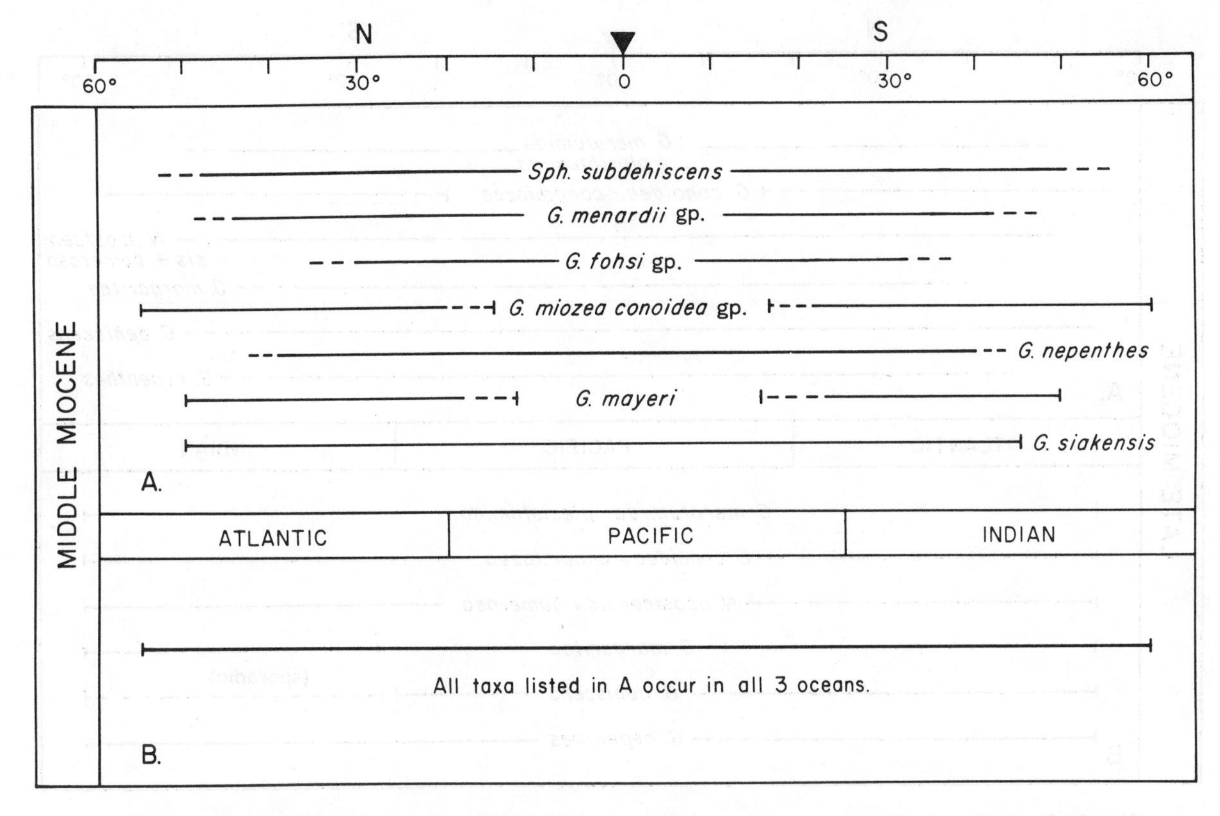

Fig. 10. Geographic distribution of stratigraphically important planktonic foraminiferal taxa: Middle Miocene. A–latitudinal distribution; B–interoceanic distribution.

biostratigraphic correlation.

Globigerinids (*woodi*, cf. *bulloides*) and globorotaliids (*continuosa*, *praescitula*, *zealandica*) characterize the Southern Ocean region, while globigerinids (N of 50°) and globorotaliids (*miozea-conoidea* group; 40° N) characterize the eastern North Pacific. In its present position is seen in the distribution of subtropical faunas almost to Lat. 60° N (Rockall area).

Late Miocene

Continued latitudinal provincialization in plaoktonic foraminiferal faunas occurred in the Late Miocene (Fig. 11). While the tropical-subtropical belt remained relatively stable and was characterized by keeled globorotaliids (*plesiotumida* and *menardii* lineages), neogloboquadrinids (*acostaensis-humerosa* group), *Globigerinoides* (*obliquus, trilobus*), globigerinids (*nepenthes*), and *Sphaeroidinellopsis*, mid- and high latitudes exhibited a more pronounced differentiation than that seen in the Middle Miocene.

In the North Atlantic mid-latitude faunas are characterized by keeled globorotaliids (*menardii* and *conoidea-conomiozea* groups), *Globigerinoides*, and *Globoquadrina dehiscens*. North of Lat. 55°, faunas are characterized by the temperate-subarctic faunas with globigerinids (*bulloides*), neogloboquadrinids (*acostaensis, atlantica*), and turborotalitids (*quinqueloba*). Similar subtropical fuanas occur in the South Atlantic; *Globoqudrina dehiscens* and *Sphaeroidinellopsis* spp. are particularly abundant in the Late Miocene on the Rio Grande Rise (at Lat. 30° S).

Late Miocene Indo-Pacific equatorial planktonic foraminiferal faunas are characterized by keeled globorotaliids (*plesiotumida, menardii* groups), globigerinids (*nepenthes*), *Sphaeroidinellopsis*, and—particularly in the Indian Ocean—by *Candeina* and the initial appearance of *Pulleniatina*. The *Globorotalia plesiotumida-tumida* group is predominantly of Indo-Pacific origin and distribution.

An interesting phenomena is the erratic termination of *Globoqudrina dehiscens*. Synthesis of a large amount of distributional data suggests that its (local) termination in various sites from latest Middle Miocene time (late Serravallian) to the Miocene/Pliocene boundary (Messinian/Zanclean boundary) may be a function of depth-related changes in the CCD. In general, at shallower sites (< 3 km) it became extinct at the Miocene/Pliocene boundary.

Planktonic foraminiferal faunas are rare in the high latitude North Pacific and Southern Ocean. In the former area globigerinids are the dominant elements

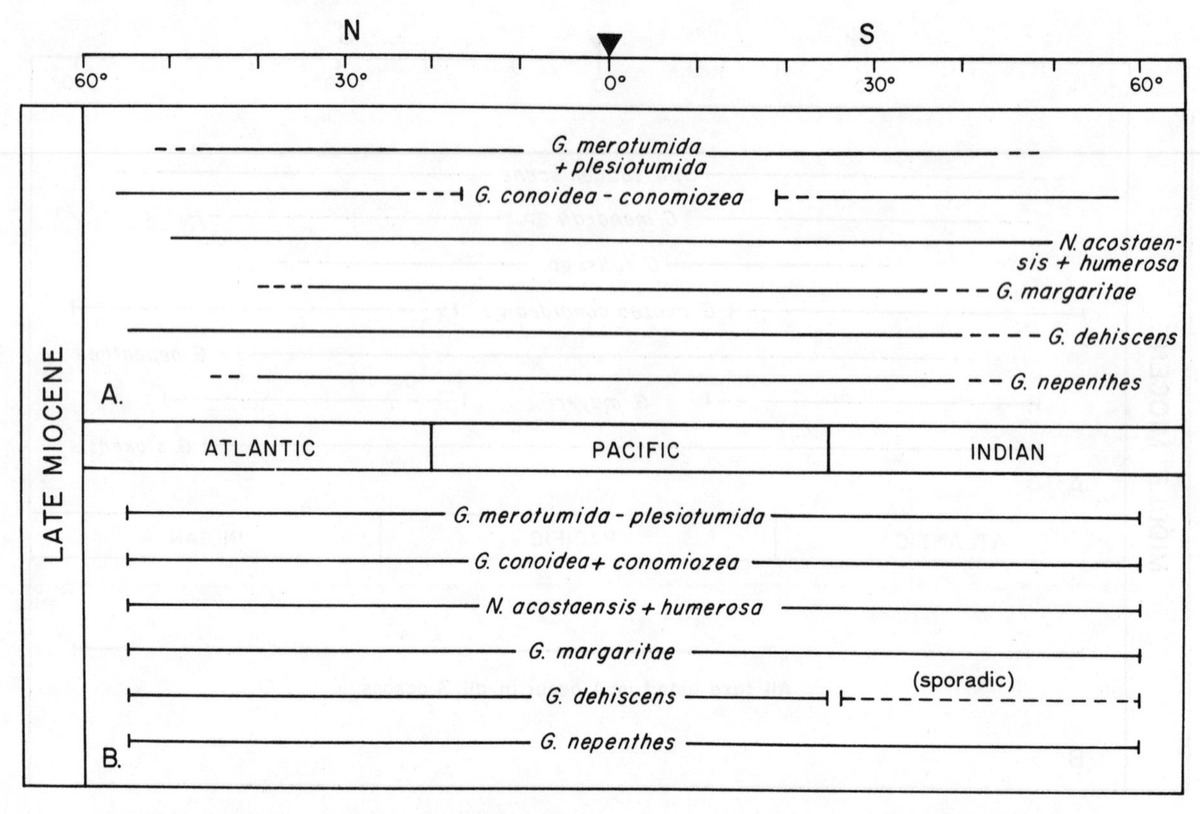

Fig. 11. Geographic distribution of stratigraphically important planktonic foraminiferal taxa: Late Miocene. A–latitudinal distribution; B–interoceanic distribution.

north of Lat. 40°; keeled globorotaliids of the *conoidea-conomiozea* group occur across the North Pacific at about Lat. 40° N and in the Indo-Pacific between Lats. 20°–50° S. The Southern Ocean is characterized by neogloboquadrinids (*acostaensis, continuosa*), globigerinids (*bulloides*) and globorotaliids (*scitula*). Distributional data suggest that the southern limit of the *G. conoidea-conomiozea* group may be linked with the position of the Subtropical Convergence.

Thus a three-fold subdivision of planktonic foraminiferal faunas into tropical, temperate, and subarctic may be made in the northern hemisphere, while a fourth, truly Antarctic province can be differentiated in the southern hemisphere. A truly Arctic province is not seen in the northern hemisphere until the initiation of polar glaciation in mid-Pliocene time (3 Ma). In actual fact we are unable to determine unequivocally whether a true Arctic faunal existed earlier because we do not know what was happening in the Arctic proper. Well preserved, continuous sequences of Late Miocene faunas are rather rare owing to the shallow CCD during this time. An understanding of the historical development of faunas at this time is best obtained through study of sites located on shallow rises (Rio Grande Rise, Walvis Ridge) and banks (Rockall Bank).

Pliocene

Planktonic foraminiferal faunal provincialization continued in the Pliocene (Figs. 12, 13). Whereas climatic deterioration continued to compress essentially latitudinally distributed faunal belts, geographic barriers, such as the elevation of the Isthmus of Panama, led to distinct, disjunct patterns of global biogeography.

Tropical faunas continued to evolve unabated in low-latitude regions. Keeled globorotaliids and species of *Globigerinoides, Sphaeroidinellopsis,* and *Pulleniatina* were the dominant elements. The elevation of the Isthmus of Panama in Early Pliocene time resulted in the development of endemic Atlantic faunal elements (*Globorotalia miocenica-exilis* group) and to the temporary (*ca.* 1 m.y.) elimination of *Pulleniatina* from the Atlantic during the mid-late Pliocene. That the Indo-Pacific region remained a homogenous biogeographic province is seen in the stratigraphic continuity and temporal correlation of coiling changes in *Pulleniatina* and globorotaliids of the *G. tumida* group.

In the Mediterranean Sea, the reconnection with the Atlantic Ocean *ca.* 5 Ma resulted in a repopulation of the region, but with a fauna curiously depleted in both tropical (globorotaliids of the *menardii-tumida*; *exilis* and *miocenica-multicamerata* groups; *Pullenia-*

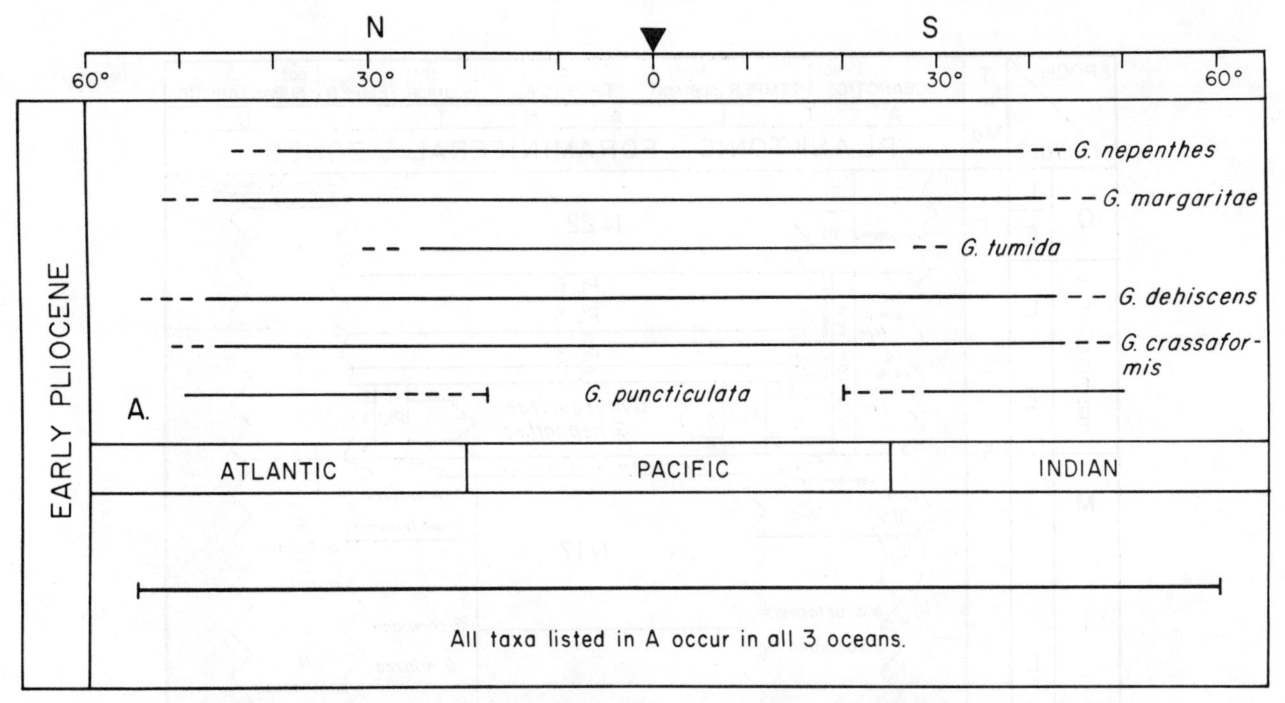

Fig. 12. Geographic distribution of stratigraphically important planktonic foraminiferal taxa: Early Pliocene. A–latitudinal distribution; B–interoceanic distribution.

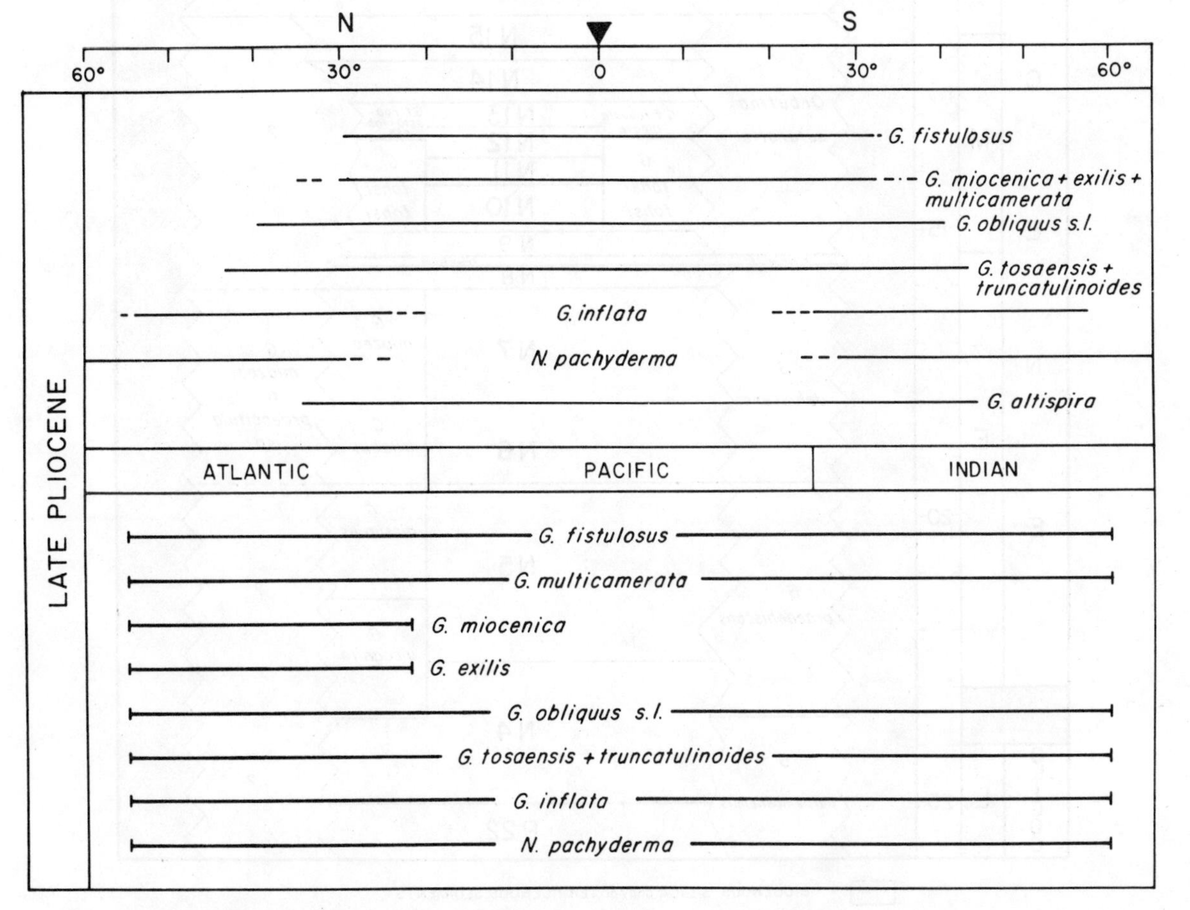

Fig. 13. Geographic distribution of stratigraphically important planktonic foraminiferal taxa: Late Pliocene. A–latitudinal distribution; B–interoceanic distribution.

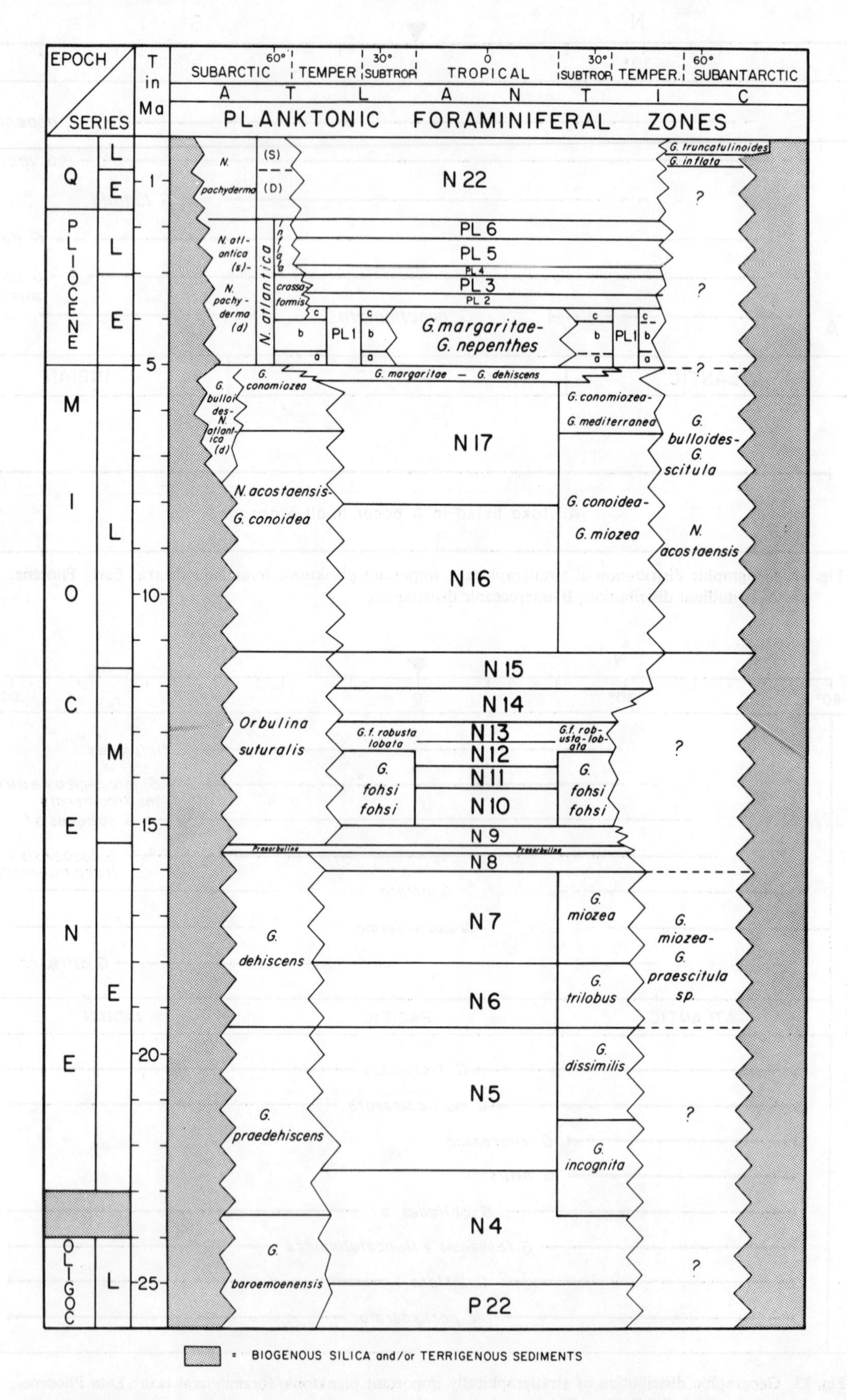

Fig. 14. Multiple, predominantly latitudinal, Neogene planktonic foraminiferal zonation: Atlantic Ocean.

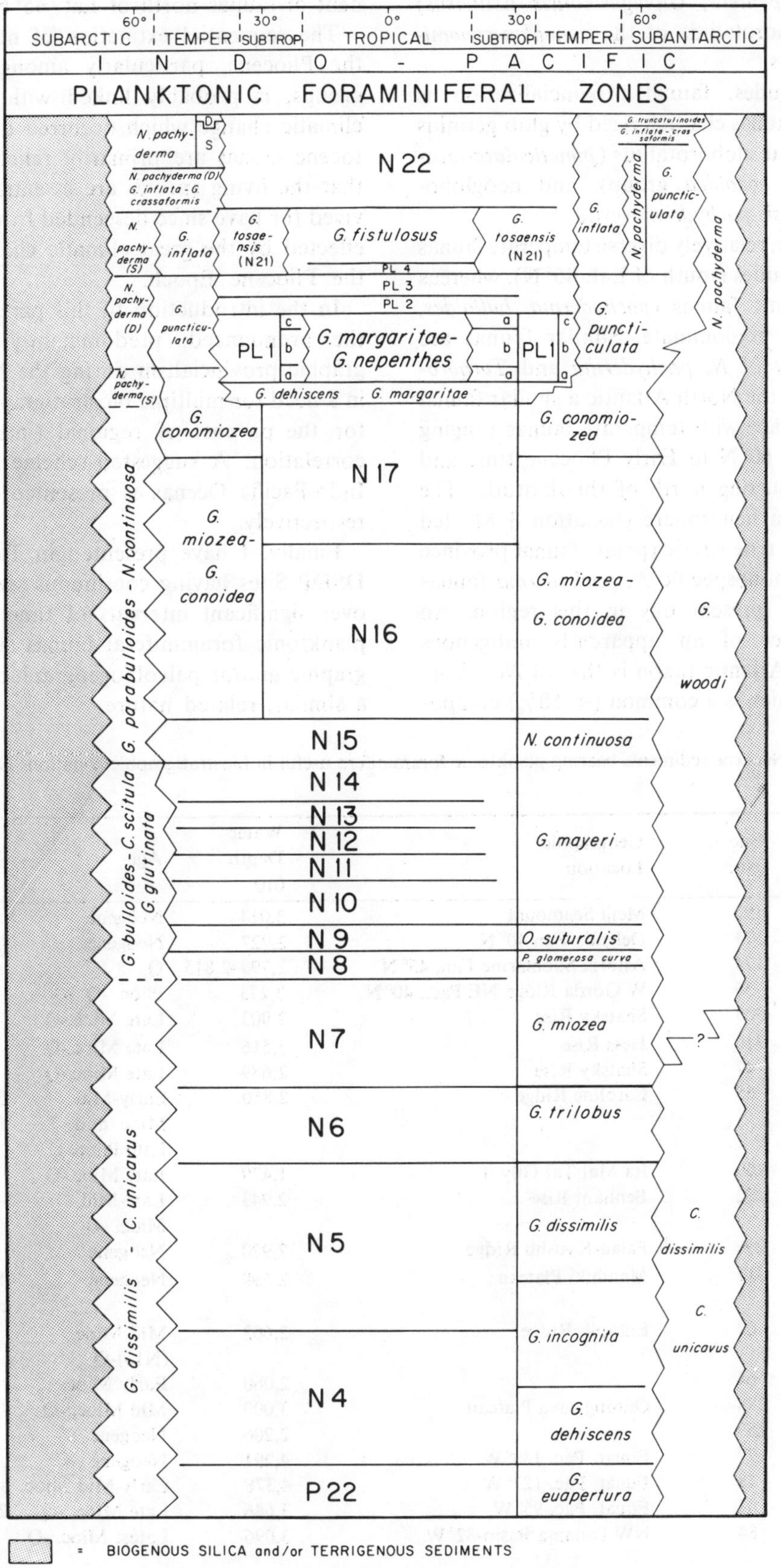
60°
30°
0°
30°
60°
SUBARCTIC
TEMPER
SUBTROP
TROPICAL
SUBTROP
TEMPER
SUBANTARCTIC
I N D O - P A C I F I C
PLANKTONIC FORAMINIFERAL ZONES
N. pachyderma
D
S
N. pachyderma (D)
G. inflata - G. crassaformis
N. pachyderma (S)
G. inflata
G. tosaensis (N21)
G. fistulosus
G. tosaensis (N21)
N22
G. truncatulinoides
G. inflata - crassaformis
G. inflata
G. puncticulata
N. pachyderma
PL 4
PL 3
PL 2
N. pachyderma (D)
G. puncticulata
PL1
G. margaritae - G. nepenthes
G. puncticulata
N. pachyderma (S)
G. dehiscens
G. margaritae
G. conomiozea
G. conomiozea
N17
G. miozea - G. conoidea
G. miozea - G. conoidea
N16
G. woodi
N15
N. continuosa
N14
N13
N12
N11
N10
G. mayeri
N9
O. suturalis
N8
P. glomerosa curva
N7
G. miozea
?
N6
G. trilobus
N5
G. dissimilis
C. dissimilis
G. incognita
C. unicavus
N4
G. dehiscens
P22
G. euapertura
G. bulloides - C. scitula G. glutinata
G. parabulloides - N. continuosa
C. unicavus
G. dissimilis
= BIOGENOUS SILICA and/or TERRIGENOUS SEDIMENTS

Fig. 15. Multiple, predominantly latitudinal, Neogene planktonic foraminiferal zonation: Indo-Pacific Ocean.

tina, Globigerina nepenthes, Globigerinoides fistulosus) and temperate (globorotaliids of the *conoidea-conomiozea* group) elements.

At mid-high latitudes, faunal provincialization is seen in the form of faunas characterized by globigerinids (*bulloides*), non-keeled globorotaliids (*puncticulata-crassaformis-inflata* and *conoidea* group), and neogloboquadrinids (*acostaensis-pachyderma* group).

In the North Pacific, relatively diverse temperate faunas characterize mid-latitudes (south of Lat. 45° N), whereas north of this subarctic faunas (*pachyderma, bulloides, scitula, quinqueloba*) predominate. Similar faunas are composed essentially of *N. pachyderma* and *Turborotalita quinqueloba*. In the North Atlantic a similar faunal development took place with temperate faunas ranging as far north as Lat. 60° N in Early Pliocene time and subarctic faunas occurring north of this latitude. The inception of northern hemisphere glaciation 3 Ma led to the formation of a true Arctic (polar) faunal province and the essentially monospecific *N. pachyderma* faunas characteristic of the present day in this region. An interesting occurrence of an apparently indigenous high-latitude North Atlantic taxon is that of *Neogloboquadrina atlantica* which is a common (< 50%) component of faunas north of Lat. 45° N in the Pliocene.

The sequential extinction of numerous taxa during the Pliocene, particularly among tropical-subtropical groups, is probably linked with the relatively rapid climatic change which occurred during this time. Pleistocene faunas are primarily relict faunas in the sense that the living species are essentially those which survived (or have since descended from) the disequilibrium effected by the great climatic changes wrought during the Pliocene Epoch.

In the introduction to this paper I pointed out that the pronounced, predominantly latitudinal, biogeographic provincialism during the Neogene has resulted in a need for multiple biostratigraphic zonation schemes for the purpose of regional (inter- and intraoceanic) correlation. A suggested scheme for the Atlantic and Indo-Pacific Ocenas is presented in Figs. 14 and 15, respectively.

Finally, I have presented in Tables 14–20 a list of DSDP Sites having continuous sediment accumulation over significant intervals of time with well preserved planktonic foraminiferal faunas suitable for biostratigraphic and/or paleobiogeographic studies, or those of a similar, related nature.

Table 14. Age range of Neogene sediments bearing planktonic foraminifera useful in biostratigraphic zonation: North Pacific (DSDP drill sites).

Region	Leg	Site No.	Geographic Location	Water Depth (m)	Age	Remarks
	19	192	Meiji Seamount	3,014	Neogene	
	18	173	Delgada Fan 40° N	2,927	Neogene	Temperate-subarctic faunas
		174	Astoria Submarine Fan, 45° N	2,799–2,815	Q	
	5	36	W Gorda Ridge NE Pac., 40° N	3,273	Plioc. -Q	
	32	305	Shatsky Rise	2,903	Late Mioc.-Q	
		310	Hess Rise	3,516	Late Mioc.-Q	
	6	47	Shatsky Rise	2,689	Late Mioc.-Q	
North Pacific		55	Caroline Ridge	2,850	Early-Mid Mioc. and Late Plioc.-Q	Subtropical-temperate faunas
	20	200	Ita Mai Tai Guyot	1,479	Late Mioc.-Q	
	31	292	Benham Rise	2,943	Late-Mid Mioc. -Q	
		296	Palau-Kyushu Ridge	2,920	Neogene	
	33	317	Manihiki Plateau	2,598	Neogene	Mid Mioc. = poor recovery
	7	62	Eurapik Ridge	2,602	Mid Mioc. (N13)-Q	
		64		2,060	Early Mioc.	
	30	288	Ontong Java Plateau	3,000	Mid Mioc. -Q	
		289		2,206	Neogene	
		77	Equat. Pac. 133° W	4,291	Neogene	
	9	78	Equat. Pac. 127° W	4,378	Early-Mid Mioc.	1. Tropical faunas
		83	Equat. Pac. 95° W	3,646	Late Mioc. -Q	2. Leg 9 sites = strong dissolution
		84	NW Panama Basin-82° W	3,096	Latest Mioc. -Q	
	16	157	Panama Basin, Carnegie Ridge	1,591	Late-Mid Mioc. -Early Q	
		158	Panama Basin, Cocos Ridge	1,953	Plioc. -Q	

Table 15. Age range of Neogene sediments bearing planktonic foraminifera useful in biostratigraphic zonation: South Atlantic (DSDP-IPOD drill sites).

Region	Leg	Site No.	Geographic Location	Water Depth (m)	Age	Remarks
		15		3,927	Early Miocene	
					and Pliocene	
	3	17	Mid-Atlantic Ridge	4,265 4,360	Early Miocene	
		18		4,018	Early Miocene	
		357	Rio Grande Rise	2,109	Neogene	
South	39	359	Walvis "Seamount"	1,658	Pliocene	
Atlantic		360	Cont. Rise off	2,949	Mid Mioc.	Subtropical-
			S. Africa		-Plioc.	temperate faunas
		362	Frio Ridge	1,325	Mid Miocene-	
	40		segment of Walvis Ridge		Plioc.	
		364	Angola Basin	2,448	Pliocene	

Table 16. Age range of Neogene sediments bearing planktonic foraminifera useful in biostratigraphic zonation: Mediterranean Sea (DSDP-IPOD drill sites).

Region	Leg	Site No.	Geographic Location	Water Depth (m)	Age	Remarks
	13	125	Mediterranean Ridge	2,782	Plioc. -Q	
		132	Tyrrhenian Sea	2,835	Plioc. -Q	
Mediter-		372	East Menorca Rise	2,699	Neogene	Subtrop.-
ranean					(excluding	temperate
					Messinian)	faunas
	42	374	Ionian Abyssal Plain	4,078	Pliocene	
		378	North Cretan Basin	1,835	Plioc. -Q	

Table 17. Age range of Neogene sediments bearing planktonic foraminifera useful in biostratigraphic zonation: North Atlantic (DSDP-IPOD drill sites).

Region	Leg	Site No.	Geographic Location	Water Depth (m)	Age	Remarks
	38	341	Vøring Plateau	1,439	Mid-Plioc. -Q	
		407		2,472	Late Mioc. -Q	Subarctic faunas
	49	408	Mid-Atlantic Ridge	1,624	Mid Mioc. -Q	(65–67° N)
		410		2,975	Late Mioc. -Q	
		116	Hatton-Rockall Basin	1,151	Neogene	1. Temperate-
						subtropical
	12					2. Discont. coring
		111	Orphan Knoll	1,797	Pliocene	Subtrop. Early
						Plioc. to subarctic
						Late Plioc.
	11	102	Blake-Bahama Outer	3,426	Late Mioc. -Q	Subtropical
			Ridge			
	48	400	Bay of Biscay	4,400	Neogene	Subtrop. -temperate
North		147		892	Late Q	
Atlantic		148		1,232	Pliocene-Q	1. Tropical-
		149	Caribbean Sea	3,972	Early Miocene	subtropical
	15				and E Pliocene	faunas
		151		2,029	Early-Mid Mioc.	2. Dissolution in
					and Early Plioc.	Early Plioc.
					and Q	
		154		3,338	Pliocene-Q	
	4	29		4,247	Pliocene-Q	
		366	Sierra Leone Rise	2,853	Neogene (exc.	Trop. -subtrop.
					early Mid-Mioc.)	
	41	369	Off Cape Bojador	1,752	Mid-Miocene-	Subtropical faunas
					Mid-Pliocene	
		397	Off Cape Bojador,	2,900	Mid-Mioc. -Q	1. Subtrop. fauna
	47		W Africa			2. Discont. coring
		398	S of Vigo Seamount	3,910	Mid-Mioc. -Q	Less complete than 397
	14	141	Cape Verde Basin	4,148	Plioc. -Q	Strong dissolution in
						Late Mioc.- earliest
						Plioc.

Table 18. Age range of Neogene sediments bearing planktonic foraminifera useful in biostratigraphic zonation: South Pacific (DSDP drill sltes).

Region	Leg	Site No.	Geographic Location	Water Depth (m)	Age	Remarks
South Pacific	21	206	Lord Howe Rise	3,196	Mid Mioc. -Q	Subtropical-temperate faunas
		207		1,389	Mid Mioc. -Q	
		208		1,545	Mid Mioc. -Q	
		209	Queensland Plateau	1,428	Late Plioc. -Q	
	34	319	Bauer Deep Galapagos Ridge	4,296	Middle Miocene	Temperate faunas
	29	284	Challenger Plateau	1,066	Late Mioc. -Q	

Table 19. Age range of Neogene sediments bearing planktonic foraminifera useful in biostratigraphic zonation: Southern Ocean (DSDP drill sites).

Region	Leg	Site No.	Geographic Location	Water Depth (m)	Age	Remarks
Southern Ocean (circum-Antarctic-subant-arctic)	29	279	N. Macquarie Ridge	3,341	Early-Middle Miocene	
		281	S. Tasman Rise	1,591	Neogene	Low diversity, polar dissembl.
	36	329	Falkland Plateau	1,599	Early Mioc. (pt) and Late Mioc. (pt)	

Table 20. Age range of Neogene sediments bearing planktonic foraminifera useful in biostratigraphic zonation: Indian Ocean (DSDP drill sites).

Region	Leg	Site No.	Geographic Location	Water Depth (m)	Age	Remarks
Indian Ocean	22	214	Ninetyeast Ridge	1,655	Neogene	Tropical-subtropical faunas
		216		2,247	Late Early Mioc. (N7) to Early Late Mioc. (N15)	
	24	231	Gulf of Aden	2,161	Mid Mioc. -Q	
		232	NW Flank Alula-Fartak Trench	1,758	Early Plioc. -Q	
		233	E flank Alula-Fartak Trench	1,860	Late Plioc. -Q	
		236	SE Carlsberg Ridge	4,504		
		237	Mascarene Plateau	1,640	Late Miocene -Q	
		238	Central Indian Ridge	2,844		
	26	253	Ninetyeast Ridge	1,962	Late Mioc. -Q	
	27	262	Timov Trough	2,315	Late Plioc. -Q	

Acknowledgements

This is the second of a two-part contribution to the meeting held in Osaka, Japan, of the IGCP Project 114 (Pacific Neogene Datum Planes) Working Group. I wish to thank Dr. N. Ikebe and Dr. M. Chiji, President and Secretary, respectively, of IGCP Project 114 for their invitation to attend and generous hospitality during the meetings held in Osaka.

I should like to thank my colleagues Drs. E. Vincent (Scripps Institution of Oceanography, La Jolla), J.P. Kennett (University of Rhode Island, Kingston), J. C. Ingle (Stanford University, Stanford), and R.Z. Poore (U.S. Geological Survey, Reston) for comments and constructive critique of a draft manuscript of this paper.

This research has been supported by grants No. OCE 78–19769 and OCE 80–23728 from the Submarine Geology and Geophysics Branch, Oceanography Division of the National Science Foundation. This is Woods Hole Oceanographic Institution Contribution No. 5130.

References

Addicott, W.A., ed., 1978: Neogene biostratigraphy of selected areas in the California Coast Ranges. IGCP Project 114 International Field Conference (21–24 June, 1978), *U.S. Geol. Survey Open-File Report*, 78–446, 109 p.

Andrews, J.E., Packham, G., *et al.*, 1975: *DSDP, Initial Reports*, Washington U.S. Government Printing Office., v. **30**, p. 3–753.

Bandy, O., 1960: The geologic significance of coiling ratios in the foraminifera *Globigerina pachyderma* (Ehrenberg). *J. Paleontol.*, v. **34**, no.4, p. 671–681.

Bé, A. and Hutson, W.H., 1977: Ecology of planktonic foraminifera and Indian Ocean. *Micropaleontology*, v. **23**, no. 4, p. 369–414.

Beckmann, J.P., 1972: The Foraminifera and some associated microfossils of Sites 135 to 144. *In* Hays, D., *et al.*, *DSDP, Initial Reports*, Washington, U.S. Government Printing Office, v. **14**, p. 389–420.

Berggren, W.A., 1969: Micropaleontologic investigations of Red Sea cores–summation and synthesis of results. *In* Degens E.T. and Ross D.A., eds., *Hot Brines and Recent Heavy Mineral Deposits in the Red Sea*, Springer-Verlag, New York, p. 329–335.

Berggren, W.A., 1972a: Late Pliocene-Pelistocene glaciation. *In* Laughton, A.S., Berggren, W.A., *et al.*, eds., *DSDP, Initial Reports*, Washington U.S. Government Printing Office, v. **12**, p. 953–965.

Berggren, W.A., 1972b: Cenozoic biostratigraphy and paleobiogeography of the North Atlantic. *In* Laughton, A.S., Berggren, W.A., *et al.*, eds., *DSDP, Initial Reports*, Washington, U.S. Government Printing Office, v. **12**, p. 965–1001.

Berggren, W.A., 1973: The Pliocene time scale: calibration zones. *Nature*, v. **243**, no. 5407, p. 391–397.

Berggren, W.A., 1977a: Late Neogene planktonic foraminiferal biostratigraphy of DSDP Site 357 (Rio Grande Rise). *In* Supko, P.R., Perch-Nielsen, K., *et al.*, eds., *DSDP, Initial Reports*, Washington, U.S. Government Printing Office, v. **39**, p. 591–614.

Berggren, W.A., 1977b: Late Neogene planktonic foraminiferal biostratigraphy of the Rio Grande Rise (South Atlantic). *Marine Micropaleontology*, v. **2**, p. 265–313.

Berggren, W.A., 1978: Recent advances in Cenozoic planktonic foraminiferal biostratigraphy, biochronology and biogenography: Atlantic Ocean. *Micropal.*, v. **24**, p. 327–330.

Berggren, W.A., 1981: Correlation of Atlantic, Mediterranean and Indo-Pacific stratigraphies: Geochronology and chronostratigraphy. *In* Ikebe, N., Chiji, M., Tsuchi, R., Morozumi, Y., eds., *IGCP 114 International Workshop on Pacific Neogene Biostratigraphy*, 1981, Osaka, *Proc.*, p. 29–60.

Berggren, W.A. and Amdurer, M., 1973: Late Paleogene (Oligocene) and Neogene planktonic foraminiferal biostratigraphy of the Atlantic Ocean (Lat. 30°N to Lat. 30°S). *Ital. Paleont. Riv.*, v. **79**, no. 3, p. 337–392.

Berggren, W.A. and Boersma, A., 1969: Late Pleistocene and Holocene planktonic Foraminifera from the Red Sea. *In* Degens, E.T., and Ross, D.A., eds., *Hot Brines and Recent Heavy Mineral Deposits in the Red Sea*, Springer-Verlag, New York, p. 282–298.

Berggren, W.A. and Haq, B., 1976: The Andalusian Stage (Late Miocene): biostratigraphy, biochronology and paleoecology. *Palaeogeogr., Palaeoclimatol., Palaeoecol.*, v. **20**, p. 67–129.

Berggren, W.A. and Hollister, C.D., 1974: Paleogeography, paleobiogeography and the history of circulation of the Atlantic Ocean. *In* Hay, W. W. ed., *Sympos. of Geologic History of the Oceans, Soc. Econ. Pal. Min., Spec. Mem.*, v. **20**. p, 126–186.

Berggren, W.A. and Hollister, C.D., 1977: Plate tectonics and paleocirculation: commotion in the ocean. *Tectonophysics*, **38**, 11–48.

Berggren, W.A. and Poore, R.Z., 1971: Late Miocene-Early Pliocene planktonic foraminiferal biochronology: *Globorotalia tumida* and *Sphaeroidinella dehiscens* lineages. *Ital. Paleont. Riv.*, v. **80**, no. 4, p. 689–698.

Berggren, W.A. and van Couvering, J.A., 1974: The Late Neogene: biostratigraphy, geochronology, and paleoclimatology of the last 15 million years in marine and continental sequences. *Palaeogeogr., Palaeoclimatol., Palaeoecol.*, v. **16**, p. 1–216.

Bertolini, V., Borsetti, A.M., Cati, F., Cinelli, D., Colalongo, M.L., Crescenti, U., Dallan, L., de Francisco, A., Dondi, L., d'Onofrio, S., Giannelli, L., Papetti, I., Pomesano-Cherchi, A., Salvatorini, G., Sampo, M., Sartoni, S., and Tedeschi, D., 1968: Proposal for a biostratigraphy of the Neogene in Italy based on planktonic foraminifera. *Gior. Geol.*, v. **35**, no. 2, p. 23–30.

Bizon, G., 1967: Contribution à la connaissance des foraminifères planctoniques d'Epire et des îles Ioniennes (Grèce occidentale) depuis le Paléogène supérieur jusqu' au Pliocène. *Inst. Franc. du Pétrole, Edit. Technip.*, 114 p., 29 pls.

Bizon, G. and Bizon, J.J., 1972: Atlas des principaux Foraminifères planctoniques du bassin méditerranean: Oligocène à Quaternaire. *Edit. Technip.*, Paris, 316 p. numerous (unnumbered) plates.

Blow, W.H., 1969: Late Middle Eocene to Recent planktonic foraminiferal biostratigraphy. *In* Bronnimann, R. and Renz, H.H. eds., *First International Conference on Planktonic Microfossils, Geneva, 1967, Proc.*, p. 199–421.

Boersma, A., 1977: Cenozoic planktonic foraminifera-DSDP Leg 39 (South Atlantic). *In* Supko, P.R., Perch-Nielsen, K., *et al.*, eds., *DSDP, Initial Reports*, Washington U.S. Government Printing Office, v. **39**, p. 567–590.

Bolli, H.M., 1957: Planktonic foraminifera from the Oligocene-Miocene Cipero and Lengua Formations of Trinidad, B.W.I. *U.S. Nat. Mus., Bull.*, v. **215**, p. 97–123, pls. 22–29.

Bolli, H.M., 1966a: Zonation of Cretaceous to Pliocene marine sediments based on planktonic foraminifera. *Bol. Inf. Asoc. Venez. Geol. Min. Pet.*, v. **8**, p. 119–149.

Bolli, H.M., 1966b: The planktonic foraminifera in well Bodjonegoro-1 of Java. *Ecolg. Geol. Helv.*, v. **59**, no. 1, p. 449–465.

Bolli, H.M., 1970: The foraminifera of Sites 21–31, Leg 4. *In* Bader, R.G., *et al.*, eds., *DSDP, Initial Reports*, Washington, U.S. Government Printing Office, v. **4**, p. 577–643.

Bolli, H.M. and Bermudez, P.J., 1965: Zonation based on planktonic foraminifera of middle Miocene to Pliocene warm-water sediments. *Bol. Inf. Asoc. Vanez. Geol. Min. Pet.*, v. **8**, p. 119–149.

Bolli, H.M. and Premoli-Silva, I., 1973: Oligocene and Recent planktonic foraminifera and stratigraphy of the Leg 15 sites in the Caribbean Sea. *In* Edgar, W.T., Saunders, J.B., *et al.*, eds., *DSDP, Initial Reports*, Washington, U.S. Government Printing Office, v. **15**, p. 475–497.

Bolli, H.M., Ryan, B.F., *et al.*, 1978: *DSDP Initial Reports*, Washington, U.S. Government Printing Office, v. **40**, p. 5–1079.

Boltovskoy, E., 1974: Neogene planktonic foraminifera of the Indian Ocean (DSDP Leg 26). *In* Barker, P.F., Dalziel, I.W.D., *et al.*, eds., *DSDP, Initial Reports*, Washington U.S. Government Printing Office, v. **26**, p. 675–741.

Bradshaw, J., 1959: Ecology of living planktonic foraminifera in the north and equatorial Pacific Ocean. *Cushman Found. Foram. Res., Contr.*, v. **10**, no. 2, p. 25–64.

Bronnimann, P. and Resig, J., 1971: A Neogene globigerinacean biochronologic time scale of the southwestern Pacific. *DSDP, Initial Reports*, Washington, U.S. Government Printing Office, v. **7**, p. 235–1469.

Burckle, L.H. and Opdyke, N.D., 1977: Late Neogene diatom correlations in the Circum-Pacific. *In* Saito, T. and Ujiie, H., eds., *First Intern. Congress Pacific Neogene Stratigraphy, Proc., Tokyo, 1976*, p. 255–284.

Catalano, R. and Sprovieri, R., 1969: Stratigrafia e micropaleontologia dell' intervallo tripolaceo di tozzente Rossi (Enna). *Atti Accad. Gioenia, Sci. Nat., Catania*, v. **7**, no. 1, p. 513–517.

Catalano, R. and Sprovieri, R., 1971: Biostratigrafia di alcune Serie Saheliane (Messiniano in feriore) in Sicilia. *In* Farinacci, A., ed., *2nd Planktonic Conf. Roma, 1970, Proc.*, Edizioni Tecno-Scienza, Rome, v. **1**, p. 211–249.

Ciaranfi, N. and Cita, M.B., 1973: Paleontological evidence of changes in the Pliocene climates. *In* Ryan, W.J.F., Hsu, K.J., *et al.*, eds., *DSDP, Initial Reports*, Washington, U.S. Government Printing Office, v. **13**, p. 1387–1399.

Cicha, I. *et al.*, 1975: Biozonal division of the upper Tertiary basins of the eastern Alps and West Carpathians. IUGS Comm. Stratigraphy, Subcomm. Neogene Stratigr., *Regional Comm. Mediterranean Neogene Stratigr., Proc.*, 6th Congress, Bratislava, 1975, 147 p., Geol. Survey, Prague.

Cicha, I., Marinescu, F. and Senes, J., 1975: Correlations du Néogène de la Paratéthys centrale. Stratigraphic Correlation Tethys-Paratethys Neogene. *IGCP Proj. 25*, **33** p., Geol. Survey, Prague.

Cifelli, R. and Smith, R., 1970: Distribution of planktonic foraminifera in the vicinity of the North Atlantic Current. *Smithsonian Contrib. Paleobiology*, no. 4: 52 p.

Cita, M.B., 1973: Pliocene biostratigraphy and chronostratigraphy. *In* Ryan, W.J.F., Hsu, K.J., *et al.*, eds., *DSDP, Initial Reports*, Washington, U.S. Government Printing Office, v. **13**, p. 1343-1379.

Cita, M.B., 1975: Planktonic foraminiferal biozonation of the Mediterranean Pliocene deep sea record: A version. *Ital. Paleont. Riv.*, v. **81**, no. 4, p. 527–544.

Cita, M.B., 1976: Planktonic foraminiferal biostratigraphy of the Mediterranean Neogene. *In* Takayanagi, Y. and Saito, T., eds., *Progress in Micropaleontology*, Micropaleo. Press Spec. Publ., p. 47–68.

Cita, M.B., Chierici, M.A., Ciampo, G., Moncharmont Zei, M., d'Onofrio, S., Ryan, W.B.F., and Scorziello, R., 1973: The Quaternary record in the Tyrrhenian and Ionian Basins of the Mediterranean Sea. *In* Ryan, W.J.F., Hsu, K.J., *et al.*, eds., *DSDP, Initial Reports*, Washington, U.S. Government Printing Office, v. **13**, p. 1263–1339

Emiliani, C., 1966: Paleotemerature analysis of the Caribbean cores P6304–8 and P6304–9 and a generalized temperature curve for the last 425,000 years. *Jour. Geol.*, v. **74**, no. 6, p. 109–126.

Ericson, D. and Wollin, G., 1968: Pleistocene climates and chronology in deep-sea sediments. *Science,* **162**, p. 1227–1234.

Fleisher, R., 1974a: Cenozoic planktonic Foraminifera and biostratigraphy, Arabian Sea, Deep Sea Drilling Project, Leg 23A. *In* Whitmarsh, R.B., Weser, O.E., Roso, D.A., *et al.*, eds., *DSDP, Initial Reports*, Washington, U.S. Government Printing Office, v. **23**, p. 1001–1072.

Fleisher, R., 1974b: Preliminary report on Late Neogene Red Sea foraminifera, Deep Sea Drilling Project, Leg 23B. *In* Davies, T.A., Luyendyk, B.P., *et al.*, eds., *DSDP, Initial Reports*, Washington, U.S. Government Printing Office, v. **26**, p. 985–1011.

Hayes, D., Pimm, A.C., *et al.*, 1972: *DSDP, Initial Reports*, v. **14**, Washington, U.S. Government Printing Office, p. 3–975.

Hays, J.D., Saito, T., Opdyke, N.D., and Burckle, L.H., 1969: Pliocene-Pleistocene sediments of the equatorial Pacific: Their paleomagnetic, biostratigraphic, and climatic record. *Geol. Soc. Amer., Bull.*, v. **80**, p. 1481–1514.

Hays, J.D. *et al.*, 1972: *Initial Reports of the Deep Sea Drilling Project*, v. **9**, Washington, U.S. Government Printing Office, p. 3–1205.

Haq, B., Berggren, W.A. and van Couvering, J.A., 1977: Corrected age of the Pliocene/Pleistocene boundary. *Nature*, v. **169**, no. 5628, p. 483–488.

Herman, Y., 1965: Etude des sédiments quaternaires de la Mer Rouge. Doctoral thesis. Univ. Paris, Ser. A., 1123. *Ann. Inst. Oceanogr., Monaco*, Masson and Co., Paris, 2, 341 p.

Ingle, J.C., Jr., 1973: Neogene foraminifera from the northeastern Pacific Ocean, Leg 18, Deep Sea Drilling Project. *In* Kulm, L.D., von Huene, R., *et al.*, eds., *DSDP, Initial Reports*, Washington, U.S. Government Printing Office, v. **18**, p. 517–567.

Ingle, J.C., Jr., 1975: Pleistocene and Pliocene Foraminifera from the Sea of Japan, Leg 31, Deep Sea Drilling Project. *DSDP, Initial Reports*, Washington, U.S. Government Printing Office, v. **31**, p. 622–701.

Ingle, J.C., Jr., 1980: Cenozoic paleobathymetry and depositional history of selected sequences within the southern California continental borderland. *Cushman Found. Spec. Publ.*, no. 19, Memorial to Orville L. Bandy, p. 163–195.

Jenkins, D.G., 1971: New Zealand Cenozoic planktonic framinifera. *New Zealand Geol. Survey, Pal. Bull.*, no. 41, p. 1–278.

Jenkins, D.G. and Orr, W.N., 1972: Planktonic foraminiferal biostratigraphy of the eastern equatorial Pacific, Leg 9. *In* Hays, J.D., *et al.*, eds, *DSDP, Initial Reports*, Washington, U.S. Government Printing Office, v. **9**, p. 1059–1196.

Kaneps, A., 1973: Cenozoic planktonic foraminifera from the eastern equatorial Pacific Ocean. *DSDP, Initial Reports*, Washington, U.S. Government Printing Office, v. **16**, p. 713–745.

Kaneps, A., 1976: Cenozoic planktonic foraminifera, equatorial Pacific Ocean, Leg 33, DSDP. *DSDP, Initial Reports*, Washington, U.S. Government Printing Office, v. **33**, p. 361–367.

Kaneps, A., 1979: Gulf Stream: velocity fluctuations during the Late Cenozoic. *Science*, 204, p. 297–301.

Keany, J. and Kennett, J.P., 1972: Pliocene-early Pleistocene paleoclimatic history recorded in Antarctic-Subantarctic deep-sea cores. *Deep-Sea Res.*, v. **19**, p. 529–548.

Keller, G., 1978: Late Neogene biostratigraphy and paleoceanography of DSDP Site 310 Central North Pacific and correlations with the Southwest Pacific. *Marine Micropaleontology*, v. **3**, p. 97–119.

Keller, G., 1980: Middle to Late Miocene planktonic foraminiferal datum levels and paleoceanography of the North and Southeastern Pacific Ocean. *Marine Micropaleontology*, v. **5**, p. 249–281.

Keller, G., 1981: Early to Middle Miocene planktonic foraminiferal datum levels of the equatorial and subtropical Pacific. *Micropaleontology*, v. **26**, no. 4, p. 372–391.

Kennett, J.P., 1966: The *Globorotalia cressaformis* bioseries in North Westland amd Marlborough, New Zealand. *Micropaleontology*, v. **12**, no. 2, p. 235–245.

Kennett, J.P., 1970: Pleistocene paleoclimates and foraminiferal biostratigraphy in subantarctic deep-sea cores. *Deep Sea Res.*, v. **17**, p. 125–140.

Kennett, J.P., 1973: Middle and Late Cenozoic planktonic foraminiferal biostratigraphy of the southwest Pacific-DSDP Leg 21. *DSDP, Initial Reports*, Washington, U.S. Government Printing Office, v. **21**, p. 575–639.

Kennett, J.P., 1976: Phenotypic variation in some Recent and Late Cenozoic planktonic foraminifera. *In* R.H. Healey and C.G. Adams, eds., *Foraminifera,* Academic Press, v. **2**, p. 60.

Kennett, J.P., 1972: Cenozoic evaluation of Antarctic glaciation, the Circum-Antarctic Ocean, and their impact on global paleoceanography. *Jour. Geophys. Res.*, v. **82**, no. 7, p. 3843–3860.

Kennett, J.P., 1978: The development of planktonic biogeography in the Southern Ocean during the Cenozoic. *Marine Micropaleontology*, v. **3**, p. 301–345.

Kennett, J.P., Houtz, R.E., *et al.*, 1975: Cenozoic paleoceanography in the southern Pacific Ocean, Antarctic glaciation, and the development of the Circum-Antarctic Current. *In* Kennett, J.P., Houtz, M.E., *et al.*, eds., *DSDP, Initial Reports*, Washington, U.S. Government Printing Office, v. **29**, p. 1155–1169.

Kent, D.U., Opdyke, N.D. and Ewing, M., 1971: Climatic change in the North Pacific using ice-rafted detritus as a climatic indicator. *Geol. Soc. Amer., Bull.*, v. **82**, p. 2741–2754.

Krasheninnikov, V.A., 1979: Stratigraphy and planktonic foraminifers of Cenozoic deposits of the Bay of Biscay and Rockall Plateau, DSDP Leg 48. *In* Montadert, L., Roberts, D.G., *et al.*, eds., *DSDP, Initial Reports*, Washington, U.S. Government Printing Office, v. **48**, p. 431–450.

Krasheninnilkov, V.A. and Hoskins, R.H., 1973: Late Cretaceous, Paleogene and Neogene planktonic foraminifera. *In* Heezen, B.C., MacGregor, I.d., *et al.*, eds., *DSDP, Initial Reports*, Washington, U.S. Government Printing Office, v. **20**, p. 105–203.

Laughton, A.S., Berggren, W.A., *et al.*, 1972: *DSDP, Initial Reports*, Washington, U.S. Government Printing Office, v. **12**, p. 3–1243.

Maxwell, A.E., von Herzen, R.P., *et al.*, 1970: *DSDP., Initial Reports*, Washington, U.S. Government Printing Office, v. **3**, p. 3–806.

Maiya, S., Saito, T. and Sato, T., 1975: Late Cenozoic planktonic foraminiferal biostratigraphy of northwest Pacific sedimentary sequences. *In* Saito, T. and Burckle, L.H., eds., *Late Neogene Epoch Boundaries*, Micropaleontology Press, Spec. Publ., no. 1, p. 395–422.

Miles, G.A., 1977: Planktonic foraminifera from Leg 37 of the Deep Sea Drilling Project. *DSDP, Initial Reports*, Washington, U.S. Government Printing Office, v. **37**, p. 929–961.

Natori, H., 1976: Planktonic foraminiferal biostratigraphy and datum planes in the late Cenozoic sedimentary sequence in Okinawa-jima, Japan. *In* Takayanagi, Y. and Saito, T. eds., *Progress in Micropaleontology*, Micropaleontology Press, Spec. Publ., p. 214–243.

Olsson, R.K., 1971: Pliocene-Pleistocene planktonic foraminiferal biostratigraphy of the northeastern Pacific. *In* Farinacci, A., ed., *2nd International Planktonic Conference, Roma (1970), Proc.*, p. 921–928.

Olsson, R.K. and Goll, R., 1970: Biostratigraphy. *In* McManus, D.A., *et al.*, eds., *DSDP, Initial Reports*, Washington, U.S. Government Printing Office, v. **5**, p. 557–567.

Orr, W. N. and Jenkins, D.G., 1977: Cainozoic planktonic foraminifera zonation and selective test solution. *In* Ramsey, A.T.S., ed., *Oceanic Micropaleontology*, Academic Press, v. **1**, p. 163–203.

Orr, W.N. and Jenkins, D.G., 1980: Eastern equatorial Pacific Pliocene and Pleistocene biostratigraphy. *Cush-*

man Found. Spec. Publ. no. 19, Memorial to Orville L. Bandy, p. 278–286.

Poag, C.W., 1972: Neogene planktonic foraminiferal biostratigraphy of the western North Atlantic. *In* Ewing, J., Hollister, C.D., *et al.*, eds., *DSDP, Initial Reports*, Washington, U.S. Government Printing Office, v. **11**, p. 483–543.

Poore, R.Z., 1979: Oligocene through Quaternary planktonic foraminiferal biostratigraphy of the North Atlantic: DSDP Leg 49. *In* Luyendyk, B.P., Cann, J.R., *et al.*, eds., *DSDP, Initial Reports*, Washington, U.S. Government Printing Office, v. **49**, p. 447–517.

Poore, R.Z. and Berggren, W.A., 1974: Pliocene biostratigraphy of the Labrador Sea: Calcareous plankton. *Jour. Foram. Res.*, v. **4**, no.3, p. 91–108.

Poore, R.Z. and Berggren, W.A., 1975: Late Cenozoic planktonic foraminiferal biostratigraphy and paleoclimatology of Hatton-Rockall Bank: DSDP Site 116. *Jour. Foram. Res.*, v. **5**, no.4, p. 270–293, pls. 1–5.

Riedel, W.R. and Sanfilippo, A., 1978: Stratigraphy and evolution of tropical Cenozoic radiolarians. *Micropaleontology*, v. **24**, no. 1, p. 61–96.

Rögl, F. and Bolli, H.M., 1973: Holocene to Pleistocene planktonic foraminifera of Leg 15, Site 147 (Cariaco Basin [Trench], Caribbean Sea). *In* Edgar, N.J., Saunders, J.B., *et al.*, eds., *DSDP, Initial Reports*, Washington U.S. Government Printing Office, v. **15**, p. 553–615.

Ryan, W.B.F., Hsu, K.J., *et al.*, 1973: *DSDP, Initial Reports*, Washington U.S. Government Printing Office, v. **13**, p. 3–1447.

Ryan, W.B.F., Cita, M.B., Dreyfus, R.M., Burckle, L.H., and Saito, T., 1974: A paleomagnetic assignment of Neogene stage boundaries and the development of isochronous datum planes between the Mediterranean, the Pacific and Indian Oceans in order to investigate the response of the world ocean to the Mediterranean "salinity crisis". *Ital. Paleont., Riv.*, v. **80**, no. 4, p. 631–688.

Saito, T., 1963: Miocene planktonic foraminifera from Honshu, Japan. *Tohoku Univ., Sci. Rept., 2nd Ser. (Geol.)*, v. **35**, no. 2, p. 123–209.

Saito, T., 1976: Geologic significance of coiling direction in the planktonic foraminifera *Pulleniatina*. *Geology*, v. **4**, no. 11, p. 305–309.

Saito, T., 1977: Late Cenozoic planktonic foraminiferal datum levels: the present stage of knowledge toward accomplishing pan-Pacific correlation. *In* Saito, T. and Ujiie, H., eds., *First International Congress of Pacific Neogene Stratigraphy, Proc.*, Tokyo, 1976, p. 61–80.

Saito, T., Burckle, L.H. and Hays, J.D.,1975: Late Miocene to Pleistocene biostratigraphy of equatorial Pacific sediments. *In* Saito, T. and Burckle, L.H., eds., *Late Neogene Epoch Boundaries,* Micropaleontology, Spec. Publ., No. 1, p. 226–244.

Savin, S., Douglas, R.G. and Stehli, F.G., 1975: Tertiary marine paleotemperatures. *Geol. Soc. Amer., Bull.*, v. **86**, p. 149–1510.

Shinbo, K. and Maiya, S., 1971: Neogene Tertiary palnktonic foraminiferal zonation in the oil-producing provinces in Japan. *In* Stratigraphic correlation between sedimentary basins of the ECAFE region (vol. 2). *United Nations, Mineral Resources Development Series*, no. 36, pt. B, p. 135–142.

Schnitker, D., 1980: North Atlantic oceanography as a possible cause of Antarctic glaciation and eutrophication. *Nature*, v. **284**, p. 615–616.

Sliter, W.V., 1972: Upper Cretaceous planktonic foraminiferal zoogeography and ecology-eastern Pacific margin. *Palaeogeog. Palaeoclimatol. Palaeoecol.*, v. **12**, p. 15–31.

Smith, L.A. and Beard, J.H., 1973: The Late Neogene of the Gulf of Mexico. *In* Worzel, J.L., Bryant, W., *et al.*, eds., *DSDP, Initial Reports*, Washington, U.S. Government Printing Office, v. **10**, p. 663–677.

Takayanagi, Y., Takayama, T., Sakai, T., Oda, M., and Kitazato, H., 1975: Microbiostratigraphy of some Middle Miocene sequences in northern Japan. *In* Takayanagi, Y. and Saito, T., eds., *Progress in Micropaleontology*, Micropaleontology, Spec. Publ., Micropaleontology Press, New York, p. 356–381.

Takayanagi, Y. and Oda, M., 1976: Shore laboratory report on Cenozoic planktonic foraminifera: Leg 33. *In* Schlanger, S.O., Jackson, E.D., *et al.*, eds., *DSDP, Initial Reports,* Washington, U.S. Government Printing Office, v. **33**, p. 451–465.

Tjalsma, R.C., 1971: Stratigraphy and foraminifera of the Neogene of the eastern Guadalquivir Basin (southern Spain). *Utrecht Micropal. Bull.*, v. **4**, 161 p.

Tjalsma, R.C., 1976: Cenozoic foraminifera from the South Atlantic, DSDP Leg 36. *In* Barker, P.F., Dalziel, I.W.D., *et al.* eds., *DSDP, Initial Reports*, Washington, U.S. Government Printing Office, v. **36**, p. 493–517.

Ujiie, H., 1973: Sedimentation of planktonic foraminifera shells in the Tsushima and Korean Straits between Japan and Korea. *Micropaleontology*, v. **19**, no. 4, p. 444–460.

Ujiie, H., 1975: Planktonic foraminiferal biostratigraphy in the western Phillipine Sea, Leg 31 of DSDP. *In* Karig, D.E., Ingle, J.C., Jr., *et al.*, eds., *DSDP, Initial Reports*, Washington, U.S. Government Printing Office, v. **31**, p. 677–691.

Van Andel, T., 1974: Cenozoic migration of the Pacific plate, northward shift of the axis of deposition, and paleobathymetry of the central equatorial Pacific. *Geology*, v. **2**, p. 507–510.

Van Andel, T., 1975: Mesozoic/Cenozoic calcite compensation depth and global distribution of calcareous sediments. *Earth and Planet. Sci. Letters*, v. **26**, p. 187–194.

Van Couvering, J.A., Berggren, W.A., Drake, R.E., Aguirre, E., and Curtis, G.H., 1976: The terminal Miocene event. *Marine Micropaleontology*, v. **1**, p. 263–286.

Vincent, E., 1974: Cenozoic planktonic biostratigraphy and paleooceanography of the tropical western Indian Ocean. *In* Fisher, R.L., Bunce, E.T., *et al.* eds., *DSDP, Initial Reports*, Washington, U.S. Government Printing Office, v. **24**, p. 1111–1150.

Vincent, E., 1975: Neogene planktonic foraminifera from the Central North Pacific, Leg 32, Deep Sea Drilling

Project. *In* Larson, R.L., Moberly, R., *et al.* eds., *DSDP, Initial Reports*, Washington, U.S. Government Printing Office, v. **32**, p. 765–801.

Vincent, E., 1977: Indian Ocean Neogene planktonic foraminiferal biostratigraphy and its paleoceanographic implications. *In* Heirtzler, J.R., Bolli, H.M., Davies, T.A., Saunders, J.B., and Sclater, J.G., eds., *Indian Ocean Geology and Biostratigraphy*. Amer. Geophysical Union, Washington, D.C., p. 469–584.

Walters, R., 1965: The *Globorotalia zealandica* and *G. miozea* lineages. *New Zealand Jour. Geol. Geophys.*, v. **8**, p. 109–127.

Wernli, R., 1977: Les foraminifères planctoniques de al limite mio-pliocéne dans les environs de Rabat (Maroc.). *Eclog. Geol. Helv.*, v. **70**, no. 2, p. 143–191.

Zachariasse, W.J., 1975: Planktonic foraminiferal biostratigraphy of the late Neogene of Crete (Greece). *Utrecht Micropal. Bull.*, **11**, 171 p.

Zachariasse, W.J., 1979: The origin of *Globorotalia conomiozea* in the Mediterranean and the value of its entry level in biostratigraphic correlations. *Ann. Geol. Pays Hellen.*, Tome Hors Serie, fasc., v. **3**, p. 1281–1292.

Zachariasse, W.J. and Spaak, P., 1979: The frequency distribution of *Globigerina nepenthes* in the Mediterranean lowermost Pliocene. *Ann. Geol. Pays Hellen.*, Tome Hors Serie, fasc., v. **3**, p. 1293–1301.

III

Neogene Biostratigraphy and Chronology of Selected Areas in the Pacific Region

III - 1

Biochronology of the Northern Pacific Miocene*

Richard Z. POORE, John A. BARRON and Warren O. ADDICOTT

Introduction

One result of the Deep Sea Drilling Project (DSDP) has been the recovery and study of a large number of Neogene pelagic and hemipelagic sequences throughout the world's oceans. Paleontologic analyses of these sections are done routinely and the occurrence of stratigraphically diagnostic taxa are usually recorded for each major planktonic microfossil group, as appropriate, in separate chapters of each DSDP Initial Reports Volume. Besides advancing knowledge of individual microfossil groups, the biostratigraphic zonations of different microfossil groups. Moreover, the study of a large number of sections and calibration of integrated biostratigraphies to an absolute time scale by paleomagnetic stratigraphy and radiometric dates allow construction of planktonic biochronology models or standards that can be used to reliably date and correlate geologic processes and products over a wide geographic area. In this report we present an integrated planktonic biochronology standard for the Miocene and show how this model can be applied to the eastern and western margins of the North Pacific Basin.

Planktonic Biochronology Standard

In the standard shown on Fig. 1, estbblished planktonic foraminifer (Blow, 1969), coccolith (Bukry, 1975), and diatom (Barron, in press) zonations are correlated to one another using experience gained in a number of DSDP and land sections. Construction of the standard incorporates the work of a number of researcheɪs over the last few years; however, the details of correlations of the zonations shown on Fig. 1 were developed during DSDP Leg 63 by Barron and others (in press). The integrated zonations of Fig. 1 represent a series of integrated paleontologic events or datum events in the different fossil groups. Zone and subzone boundaries are defined and recognized by one or two discrete events or datum planes. This model incorporates planktonic foraminifer and coccolith zones that have general use in low to middle latitudes with diatom zones that are applicable to middle to high latitudes. Correlation of the integrated microfossil zonations to European Stages essentially follows the correlations of Ryan and others (1975) and relies on coccolith, planktonic foraminifer, and, in some cases, paleomagnetic data from European sections. Calibration to an absolute time scale is done through paleomagnetic stratigraphy and a few direct radiometric dates on zone boundaries or datum events. The model shown on Fig. 1 is continuously being tested by various workers, and we expect that minor changes will occur as more data are accumulated. We are confident, however, that the basic patterns and correlations shown on Fig. 1 are correct, and we accept this as a standard.

U.S. Pacific Coast Miocene Provincial Stages

Provincial stages are commonly used to date and correlate Miocene marine units along the U.S. Pacific coast. The Saucesian, Relizian, Luisian, Mohnian, and Delmontian Stages were defined by Kleinpell (1938), and although type sections were designated for each stage, they are based in large part on association of benthic foraminifers. There has been controversy for some time as to how these stages relate to internationally recognized subdivisions of the Miocene and whether or not these stage are useful for chronostratigraphic correlations. Planktonic microfossils associated with the type sections of these benthic foraminifer stages can be used to correlate to European Stages via the planktonic biochronology standard of Fig. 1. Data for the calibrations shown on Fig. 1 are summarized in Table 1.

Coccoliths from and immediatedly below the type

U.S. Geological Survey, 345, Middlefield Rd., MS15, Menlo Park, California 94025, U.S.A.

* *Reprinted from Proceedings of International Workshop on Pacific Neogene Biostratigraphy, Osaka Japan (1981) by the courtesy of the authors.* [*Editor's note*]

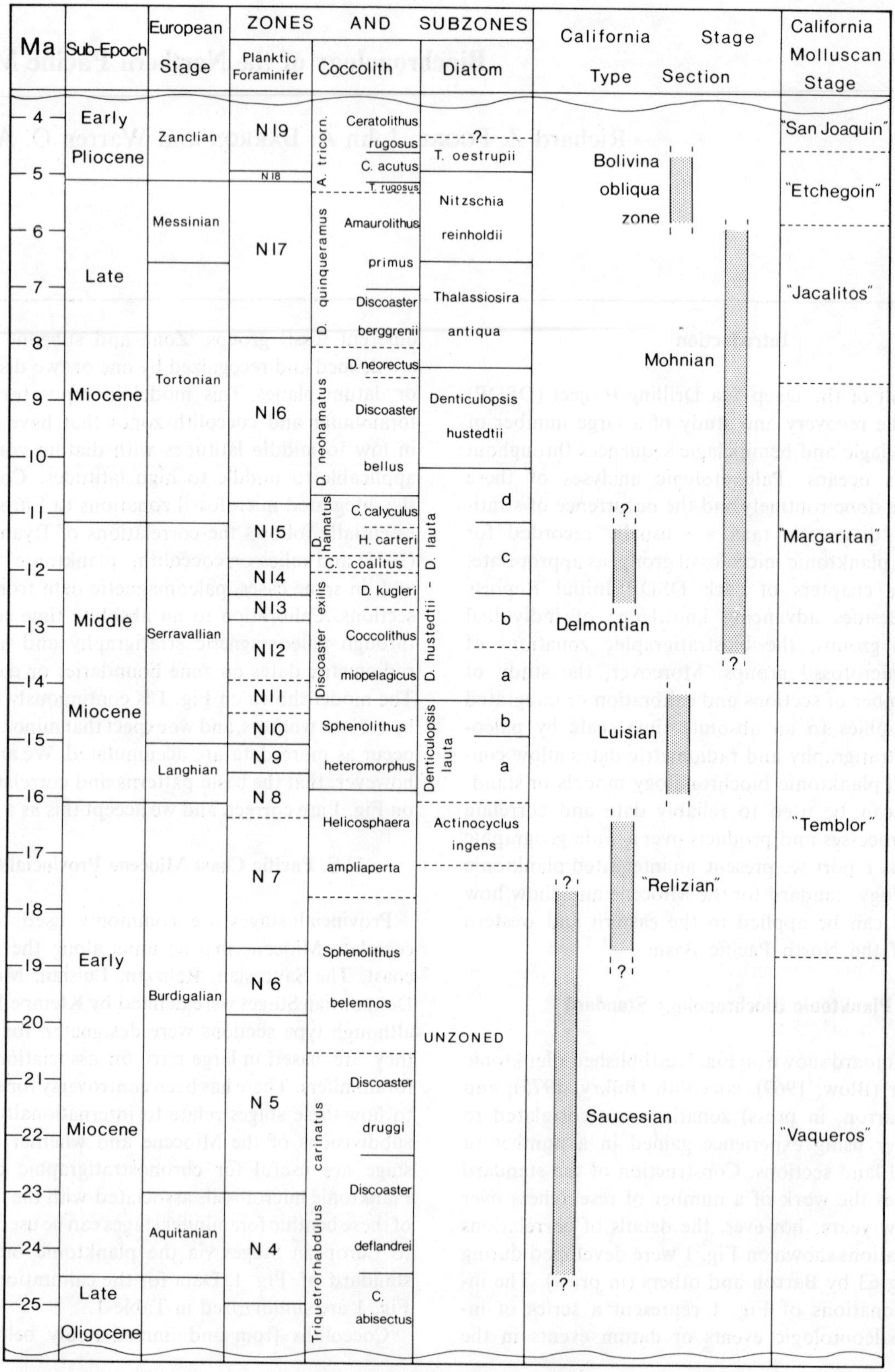

Ma | Sub-Epoch | European Stage | ZONES AND SUBZONES: Planktic Foraminifer, Coccolith, Diatom | California Stage: Type Section | California Molluscan Stage

Ma: 4, 5, 6, 7, 8, 9, 10, 11, 12, 13, 14, 15, 16, 17, 18, 19, 20, 21, 22, 23, 24, 25

Sub-Epoch: Early Pliocene; Late Miocene; Middle Miocene; Early Miocene; Late Oligocene

European Stage: Zanclian; Messinian; Tortonian; Serravallian; Langhian; Burdigalian; Aquitanian

Planktic Foraminifer: N19; N18; N17; N16; N15; N14; N13; N12; N11; N10; N9; N8; N7; N6; N5; N4

Coccolith: A. tricorn.; D. quinqueramus; D. neohamatus; C. hamatus; C. exilis; Discoaster; Triquetrorhabdulus carinatus; Ceratolithus rugosus; C. acutus; T. rugosus; Amaurolithus primus; Discoaster berggrenii; D. neorectus; Discoaster bellus; C. calyculus; H. carteri; C. coalitus; D. kugleri; Coccolithus miopelagicus; Sphenolithus heteromorphus; Helicosphaera ampliaperta; Sphenolithus belemnos; Discoaster druggi; Discoaster deflandrei; C. abisectus

Diatom: ? T. oestrupii; Nitzschia reinholdii; Thalassiosira antiqua; Denticulopsis hustedtii; D. hustedtii - D. lauta (d, c, b, a); Denticulopsis lauta (b, a); Actinocyclus ingens; UNZONED

California Stage, Type Section: Bolivina obliqua zone; Mohnian; ?; Delmontian; ?; Luisian; "Relizian"; ?; ?; Saucesian; ?

California Molluscan Stage: "San Joaquin"; "Etchegoin"; "Jacalitos"; "Margaritan"; "Temblor"; "Vaqueros"

Fig. 1. Miocene planktonic biochronology standard for Northern Pacific. See text for discussion of calibration of California benthic foraminifer and molluscan stages to the standard.

Table 1. Summary of planktonic microfossil data from type sections of California Provincial Stages.

Stage/Data	Source
SAUCESIAN STAGE	
Upper *Triquetrorhabdulus carinatus* Zone to *Sphenolithus belemnos* Zone	Lipps and Kalisky, 1972
Discoaster druggi Subzone of *T. carinatus* Zone	Warren and Newell, 1980
Discoaster druggi Subzone to *Helicosphaera ampliaperta* Zone	Warren, 1981,
underlying strata-*Sphenolithus ciperoensis* Zone to ?lower *T. carinatus* Zone	Lipps and Kalisky, 1972 Warren and Newell, 1980
RELIZIAN STAGE	
Sphenolithus belemnos Zone to *Helicosphaera ampliaperta* Zone	Poore and others 1981
LUISIAN STAGE	
Sphenolithus heteromorphus Zone to *Coccolithus miopelagicus* Subzone of *Discoaster exilis* Zone	Poore and others, 1981
Denticulopsis lauta Zone to lower Subzone a of *D. hustedtii–D. lauta* Zone	
MOHNIAN STAGE	
Coccolithus miopelagicus Subzone of *D. exilis* Zone (lower part)	Poore, unpublished
overlying strata–*Nitzschia reinholdii* Zone to *Thalassiosira oestrupii* Zone (= type *Bolivina obliqua* Zone)	Barron 1976
DELMONTIAN STAGE	
Subzone c and ? Subzone d of *Denticulopsis hustedtii–D. lauta* Zone	Barron, 1976

section of the Saucesian Stage indicate that the type Saucesian is early Miocene and correlates with the Aquitanian Stage and most of the Burdigalian Stage (ca. 24–17.5 Ma). Coccoliths from the vicinity of the type section of the Relizian Stage show that the Relizian Stage is early Miocene and correlates with the upper part of the Burdigalian Stage of Europe (*ca.* 16.5–18.5 Ma). Because the calibrations were not done on the actual type section of the Relizian, it is placed in quotes on Fig. 1. Note that the "type" Relizian correlates with part of the type Saucesian.

Stratigraphically diagnostic diatoms and coccoliths occur in the type section of the Luisian Stage. Both microfossil groups indicate that the Luisian Stage is middle Miocene and correlates with the Langhian Stage and lower part of the Serravallian Stage (*ca.* 16–14 Ma). The observed correlation of diatom and coccolith zones in the type section of the Luisian Stage corresponds to the correlation predicted by our biochronology standard and thus constitutes a successful test of the model.

Coccoliths from the type section of the Mohnian Stage and diatoms from strata directly and conformably overlying the type Mohnian indicate that the Mohnian Stage is middle and late Miocene and correlates with the upper Serravallian Stage, the Tortonian Stage, and part of the Messinian Stage (*ca.* 14–6 Ma). Diatoms in the type section of the Delmontian Stage show that it is middle Miocene and correlates with the upper part of the Serravallian Stage (*ca.* 12.5–11 Ma). The type Delmontian Stage is thus equivalent to the lower part of the type Mohnian Stage.

The correlations shown on Fig. 1 demonstrate that the planktonic biochronology standard can be applied directly to California onshore sections and show that most of the Miocene is represented by Kleinpell's Stages. The correlations also reveal several problem with the California Stages. A new stage should be designated in post-Mohnian rocks to replace the Delmontian Stage and additional work is necessary to define a base for the Relizian Stage so that it doesn't overlap the Saucesian Stage.

The last column of Fig. 1 shows calibration of California Miocene molluscan stages of Addicott (1972) to the planktonic standard. Calibration of the "Vaqueros" through "San Joaquin" Stages is done indirectly via benthic foraminifers. Benthic foraminifers associated with molluscan assemblages are used to first assign the molluscan stages to the benthic foraminifer stages of Keinpell (1938), and then these assignments are used to correlate to the planktonic standard. Note that recent studies (e.g., Crouch and Bukry, 1979) show that the stratigraphic distributions of benthic foraminifers diagnostic of Kleinpell's Stages are facies controlled, and stage assignments based on benthic foraminifers can vary widely in age. Therefore the indirect calibrations of molluscan stages to the biochronology standard are only approximations and must be used with caution. More work is needed to define the age assignments and test the reliability of the California molluscan stages.

Northwest Pacific Margin

Figure 2 illustrates how the planktonic biochronology standard can be applied directly along the western Pacific margin. This figure incorporates the coccolith and diatom zonations from Fig. 1 with a late Miocene to early Pliocene zonation developed especially for Japan and the northwest Pacific by Barron (1980) from Koizumi's (1975) zonation. The main difference between the two diatom zonations is due to the geographic distribution of *Denticulopsis kamtschatica,* which do-

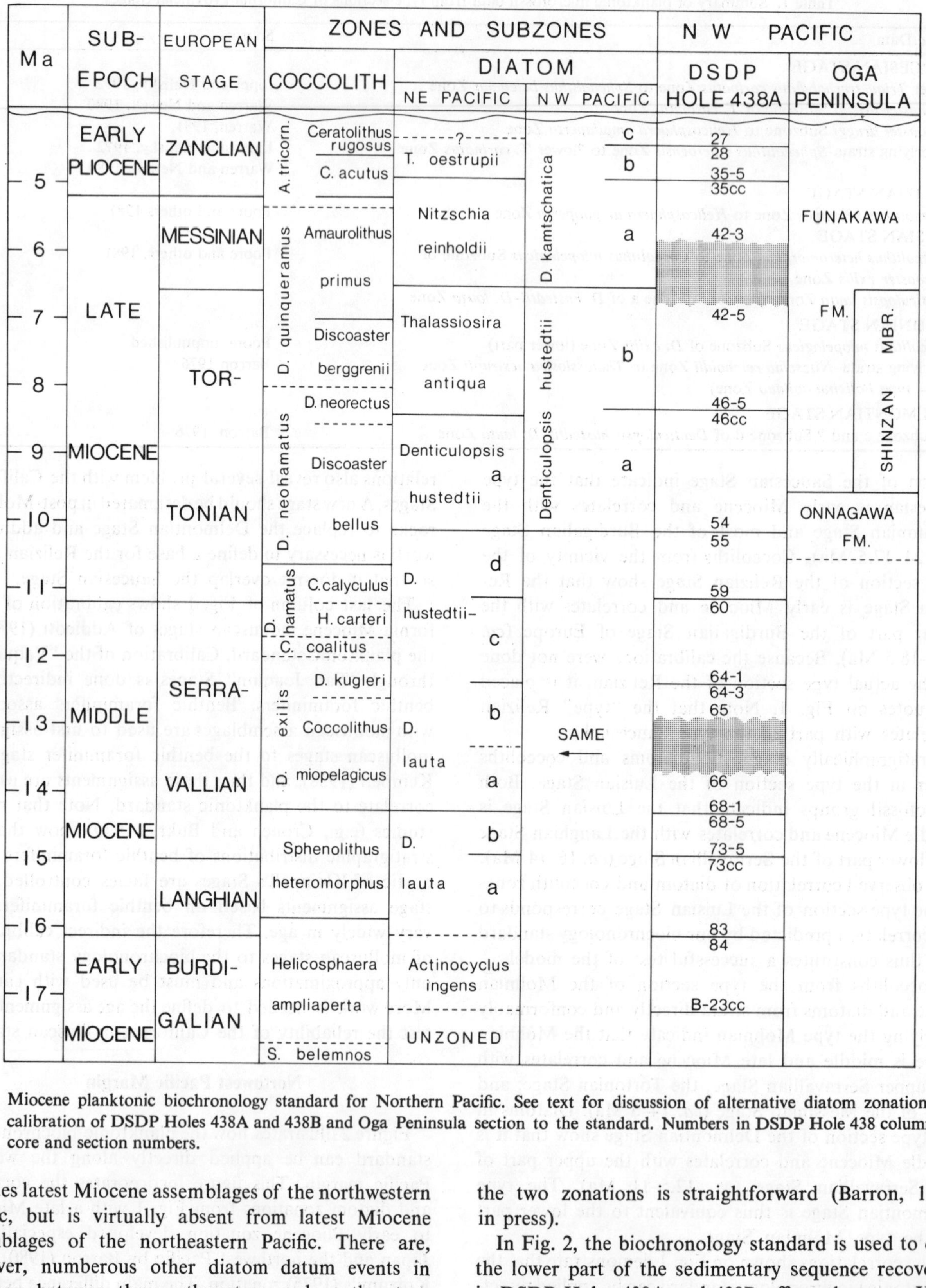

Fig. 2. Miocene planktonic biochronology standard for Northern Pacific. See text for discussion of alternative diatom zonation and calibration of DSDP Holes 438A and 438B and Oga Peninsula section to the standard. Numbers in DSDP Hole 438 column are core and section numbers.

minates latest Miocene assemblages of the northwestern Pacific, but is virtually absent from latest Miocene assemblages of the northeastern Pacific. There are, however, numberous other diatom datum events in common between the northwestern and northeastern Pacific over the interval of the *Thalassiosira antiqua* through *T. oestrupii* Zones, and correlation between the two zonations is straightforward (Barron, 1980, in press).

In Fig. 2, the biochronology standard is used to date the lower part of the sedimentary sequence recovered in DSDP Holes 438A and 438B off northeastern Honshu. The correlations shown are done with diatoms (Barron, 1980) and, with a few minor exceptions, are compatible with independent age assignments based on coccoliths (Shaffer, 1980). Note that the correlations allow recognition of a short hiatus in Core 42 (*ca.* 7–6 Ma) and a second short hiatus between Cores 65 and 66 (*ca.* 13 Ma). In addition, diatom data of Koizumi (1977) and Akiba (written commun., 1978) are used to correlate the Oga Peninsula section of Japan to the biochronology standard. The correlations on Fig. 2 show that the Onnagawa Formation is early late Miocene, whereas the Funakawa Formation extends from the late Miocene into the early Pliocene. The boundary between the two formations is in Subzone a of the *Denticulopsis hustedtii* Zone at a level estimated at 9.5 Ma.

In summary coccoliths, diatoms, and the biochronology standard shown on Fig. 1 have proved to be powerful tools for dating and correlating Miocene marine sections along the North Pacific margins. The standard can be used to make reliable correlations over a wide geographic area and in a variety of depositional environments.

References

Addicott, W.O., 1972: Provincial middle and late Tertiary molluscan stages, Temblor Range, California. *In Symposium on Miocene biostratigraphy of California*: Pacific Section Society of Economic Paleontologists and Mineralogists, p. 1–26.

Barron, J.A., 1976: Marine diatom and silicoflagellate biostratigraphy of the type Delmontian Stage and the type *Bolivina obliqua* Zone, California. *U.S. Geological Survey Journal of Research*, v. **4**, p. 339–351.

Barron, J.A., 1980: Lower Miocene to Quaternary diatom biostratigraphy of Leg 57, off northeastern Japan, Deep Sea Drilling Project. *In* Scientific Party, eds., *DSDP, Initial Reports*, Washington, U.S. Government Printing Office, v. **56**, **57**, p. 641–685.

Barron, J.A., (in press): Late Cenozoicdiatom biostratigraphy and paleoceanography of the middle-latitude eastern North Pacific, Deep Sea Drilling Project Leg 63. *In* Scientific Party, eds., *DSDP, Initial Reports*, Washington, U.S. Government Printing Office, v. **63**.

Barron, J.A., Poore, R. Z. and Reinhard, W. (in press): Biostratigraphic summary, Deep Sea Drilling Project Leg 63. *In* Scientific Party, eds., *DSDP, Initial Reports*, Washington, U.S. Government Printing Office, v. **63**.

Blow, W.H., 1969: Late middle Eocene to Recent planktonic forminiferal biostratigaphy. *In* Bronnimann, P. and Renz, H. H., eds., *First Planktonic Conference, Proceedings*, Leiden, E. J. Brill., p. 199–422.

Bukry, D., 1975: Coccolith and silicoflagellate stratigraphy, northwestern Pacific Ocean, Deep Sea Drilling Project Leg 32. *In* Larson, R.L., and others, eds., *DSDP, Initial Reports,* Washington, U.S. Government Printing Office, v. **32**, p. 677–701.

Crouch, J.K. and Bukry, D., 1979: Comparison of Miocene provincial foraminiferal stages to coccolith zones in the California Continental Borderland. *Geology,* v. **7**, p. 211–215.

Kleinpell, R. M., 1938: Miocene stratigraphy of California. American Association of Petroleum Geologists, Tulsa, Okla., 450 p.

Koizumi, I., 1975: Late Cenozoic diatom biostratigraphy in the circum-North Pacific. *Geological Society of Japan, Jour.*, v. **81**, p. 611–627.

Koizumi, I., 1977: Diatom biostratigraphy in the North Pacific region. *In* Saito, T. and Ujiie, H., eds., *First International Congress on Pacific Neogene Stratigraphy, Proceedings*, Tokyo, Kaiyo Shuppan, p. 235–253.

Lipps, J. H. and Kalisky, M., 1972: California Oligo-Miocene calcareous nannoplankton biostratigraphy and paleoecology. *In Symposium on Miocene biostratigraphy of California*, Pacific Section, Society of Economic Paleontologists and Mineralogists, p. 239–254.

Poore, R. Z., McDougall, K., Barron, J. A., Brabb, E. E., and Kling, S. A., 1981: Microfossil biostratigraphy and biochronology of the type Relizian and Luisian Stages of California. *In The Monterey Formation and related siliceous rocks of California*, Pacific Section, Society of Economic Paleontologists and Mineralogists, p. 15–41.

Ryan, W. B. F., Cita, M. B., Rawson, M. D., Burckle, L. H., and Saito, T., 1975: A paleomagnetic assignment of Neogene stage boundaries and the development of isochronous datum planes between the Mediterranean, the Pacific and Indian Oceans in order to investigate the response of the world ocean to the Mediterranean "Salinity Crisis": *Italiana di Paleontologia, Rivista*, v. **80**, p. 631–688.

Shaffer, F. L., 1980: Calcareous-nannofossil biostratigraphy of Japan Trench transect, Deep Sea Drilling Project Leg 57. *In* Scientific Party, eds., *DSDP, Initial Reports*, Washington, U.S. Government Printing Office, v. **56**, **57**, p. 875–886.

Warren, A. D., 1981: Calcareous nannoplankton biostratigraphy of Cenozoic marine stages in California. *Studies in Geology,* American Association of Petroleum Geologists, no. 11, p. 60–69.

Warren, A. D. and Newell, J.H., 1980: plankton biostratigraphy of the Refugian and adjoining stages of the Pacific Coast Tertiary. *Cushman Foundation Special Publication*, no. 19, p. 233–251.

III-2

Neogene Biostratigraphic-Chronostratigraphic Scale for the Northeastern Pacific Margin

John M. ARMENTROUT*, Ronald J. ECHOLS* and James C. INGLE, Jr.**

Abstract

In recent years numerous students of northeastern Pacific stratigraphy have contributed new paleontologic and radiometric data and have improved correlations of the provincial biostratigraphic standards both one to another and to the radiometric time scale. Critical review and synthesis of this work is the basis of a new regional biostratigraphic-chronostratigraphic scale for the northeastern Pacific margin. The new scale revises part of the 1944 chart on "Correlation of the Marine Cenozoic Formations of Western North America (Weaver *et al.*, 1944) and provides a basis for chronostratigraphic correlations within the context of the current worldwide geologic time scale. The new northeastern Pacific Neogene biostratigraphic-chronostratigraphic scale is applicable to the stratigraphy of Baja California, California, Oregon, Washington, British Columbia, and Alaska.

Introduction

The northeastern Pacific margin is tectonically dynamic and has been throughout the Cenozoic Era. Driven by subduction and transformational fault tectonics, the continental margin has been divided into numerous basins having distinct sedimentological regimes with associated faunas often reflecting distinct climatologic regimes (Fig. 2). Rock-stratigraphic correlation from one basin to another, and even within one basin, has been virtually impossible. As a consequence, biostratigraphy has been the principle tool for establishing the age relationship of rock units and stratigraphic correlation between separate geographic areas (Fig. 1) (see Armentrout and Echols, 1981, Fig. 7–12).

Northeastern Pacific margin Neogene age assignments are traditionally based upon megafossil and benthic foraminiferal biostratigraphy (Weaver *et al.*, 1944). The early workers developed a time-stratigraphic framework consisting of ten megafossil "formational" units and ten benthic foraminiferal "stages" (Weaver *et al.*, 1944; Natland, 1952). These were crucial for initial resolution of the Neogene history of the region and remain important tools for correlation where used within relatively homogenous paleoenvironments. However, further research has revealed some problems with traditional "stage" and "formational" units and the correlations based upon them. Type sections were sometimes poorly defined or lacked super-positional faunal control (e.g., Addicott, 1972; Poore, McDougall, Barron, Brabb, and Kling, 1981). Age diagnostic megafaunal and microfaunal assemblages seldom were collected within the same stratigraphic units hampering correlation of the separate biochronologies (Addicott, 1972, 1976; Poore, Barron, and Addicott, 1981). Original definitions of faunal zones erected in this region included large numbers of species (e.g., Kleinpell, 1938; Durham, 1944). The ranges of some of the species have subsequently been modified as new areas have been studied. In addition, various workers have placed emphasis on different species for zonal identification. These two practices in the defin tion and application of zones have sometimes led to inconsistent correlations due to presence or absence of individual taxa or inconsistent taxonomy (see McDougall, 1980). Finally, local Neogene faunas commonly exhibit marked endemism which hinders correlation between different geographic areas (e.g., Addicott, 1977; Allison, 1978).

Current Work

Resolution of the above problems has been pursued in four phases: 1) collection of fossils from nearly continuous stratigraphic intervals representing age equivalent but depositionally different environments; 2) definition or redefinition of zones and stages based upon detailed biostratigraphy encompassing a broad ecologic

* *Mobil Exploration and Producing Services, Inc., P.O.Box 900 Dallas, Texas 75221, U.S.A.*
** *Department of Geology, School of Earth Sciences, Stanford University, Stanford, California 94305, U.S.A.*

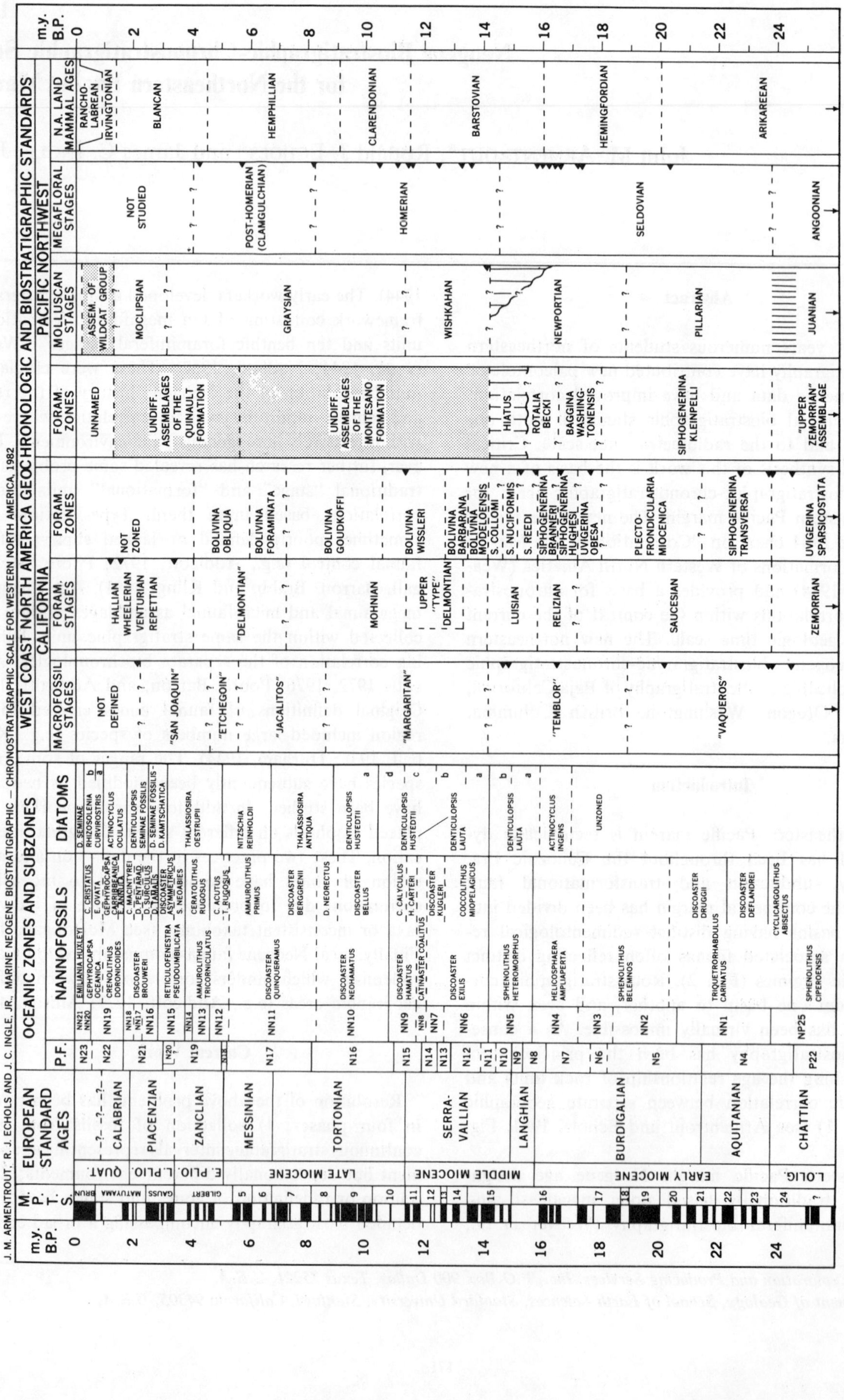

J. M. ARMENTROUT, R. J. ECHOLS AND J. C. INGLE, JR., MARINE NEOGENE BIOSTRATIGRAPHIC — CHRONOSTRATIGRAPHIC SCALE FOR WESTERN NORTH AMERICA, 1982

m.y. B.P.: 0, 2, 4, 6, 8, 10, 12, 14, 16, 18, 20, 22, 24

M. P. T. S.: BRUN, MATUYAMA, GAUSS, GILBERT, 5, 6, 7, 8, 9, 10, 11, 12, 13, 14, 15, 16, 17, 18, 19, 20, 21, 22, 23, 24, ?

EUROPEAN STANDARD AGES: QUAT. — ?— ?— ?—; CALABRIAN; L. PLIO. PIACENZIAN; E. PLIO. ZANCLIAN; LATE MIOCENE: MESSINIAN, TORTONIAN; MIDDLE MIOCENE: SERRA-VALLIAN, LANGHIAN; EARLY MIOCENE: BURDIGALIAN, AQUITANIAN; L. OLIG. CHATTIAN

OCEANIC ZONES AND SUBZONES

P.F.: N23, N22, N21, N20, N19, N18, N17, N16, N15, N14, N13, N12, N11, N10, N9, N8, N7, N6, N5, N4, P22

NANNOFOSSILS: NN21 EMILIANIA HUXLEYI; NN20 GEPHYROCAPSA OCEANICA; C. CRISTATUS; E. OVATA; NN19 CRENOLITHUS DORONICOIDES; GEPHYROCAPSA CARRIBBEANICA; E. ANNULA; NN18; NN17; NN16 DISCOASTER BROUWERI; C. MACINTYREI; D. PENTARAD.; D. SURCULUS; D. TAMALIS; NN15 RETICULOFENESTRA PSEUDOUMBILICATA; DISCOASTER ASYMMETRICUS; S. NEOABIES; NN14; NN13 AMAUROLITHUS TRICORNICULATUS; CERATOLITHUS RUGOSUS; NN12; C. ACUTUS; T. RUGOSUS; NN11 DISCOASTER QUINQUERAMUS; AMAUROLITHUS PRIMUS; DISCOASTER BERGGRENII; D. NEORECTUS; NN10 DISCOASTER NEOHAMATUS; DISCOASTER BELLUS; NN9 DISCOASTER HAMATUS; C. CALYCULUS; H. CARTERI; NN8 CATINASTER COALITUS; NN7 DISCOASTER KUGLERI; NN6 DISCOASTER EXILIS; COCCOLITHUS MIOPELAGICUS; NN5 SPHENOLITHUS HETEROMORPHUS; NN4 HELICOSPHAERA AMPLIAPERTA; NN3; NN2 SPHENOLITHUS BELEMNOS; NN1 TRIQUETRORHABDULUS CARINATUS; DISCOASTER DRUGGII; DISCOASTER DEFLANDREI; CYCLICARGOLITHUS ABISECTUS; NP25 SPHENOLITHUS CIPEROENSIS

DIATOMS: D. SEMINAE; RHIZOSOLENIA CURVIROSTRIS b, a; ACTINOCYCLUS OCULATUS; DENTICULOPSIS SEMINAE FOSSILIS; D. SEMINAE FOSSILIS - D. KAMTSCHATICA; THALASSIOSIRA OESTRUPII; NITZSCHIA REINHOLDII; THALASSIOSIRA ANTIQUA; DENTICULOPSIS HUSTEDTII a; d; DENTICULOPSIS HUSTEDTII c; b; DENTICULOPSIS LAUTA a; DENTICULOPSIS LAUTA b, a; ACTINOCYCLUS INGENS; UNZONED

WEST COAST NORTH AMERICA GEOCHRONOLOGIC AND BIOSTRATIGRAPHIC STANDARDS

CALIFORNIA

MACROFOSSIL STAGES: NOT DEFINED; — ? — ? —; "SAN JOAQUIN"; — ? — ? —; "ETCHEGOIN"; — ? — ? —; "JACALITOS"; — ? — ? —; "MARGARITAN"; — ? — ? —; "TEMBLOR"; — ? — ? —; "VAQUEROS"

FORAM. STAGES: HALLIAN; WHEELERIAN; VENTURIAN; REPETTIAN; "DELMONTIAN"; MOHNIAN; UPPER "TYPE" DELMONTIAN; LUISIAN; RELIZIAN; — ? — ? —; SAUCESIAN; ZEMORRIAN

FORAM. ZONES: NOT ZONED; BOLIVINA OBLIQUA; BOLIVINA FORAMINATA; — ? — ? —; BOLIVINA GOUDKOFFI; BOLIVINA WISSLERI; BOLIVINA BARBARANA; BOLIVINA MODELOENSIS; S. COLLOMI; S. NUCIFORMIS; S. REEDI; SIPHOGENERINA BRANNERI; SIPHOGENERINA HUGHESI; UVIGERINA OBESA; PLECTO-FRONDICULARIA MIOCENICA; — ? — ? —; SIPHOGENERINA TRANSVERSA; UVIGERINA SPARSICOSTATA

PACIFIC NORTHWEST

FORAM. ZONES: UNNAMED; UNDIFF. ASSEMBLAGES OF THE QUINAULT FORMATION; UNDIFF. ASSEMBLAGES OF THE MONTESANO FORMATION; HIATUS; ROTALIA BECKI; BAGGINA WASHING-TONENSIS; SIPHOGENERINA KLEINPELLI; "UPPER ZEMORRIAN" ASSEMBLAGE

MOLLUSCAN STAGES: ASSEM. OF WILDCAT GROUP; MOCLIPSIAN; GRAYSIAN; WISHKAHAN; NEWPORTIAN; PILLARIAN; JUANIAN

MEGAFLORAL STAGES: NOT STUDIED; POST-HOMERIAN (CLAMGULCHIAN); HOMERIAN; SELDOVIAN; ANGOONIAN

N.A. LAND MAMMAL AGES: RANCHO-LABREAN; IRVINGTONIAN; BLANCAN; HEMPHILLIAN; CLARENDONIAN; BARSTOVIAN; HEMINGFORDIAN; ARIKAREEAN

m.y. B.P.: 0, 2, 4, 6, 8, 10, 12, 14, 16, 18, 20, 22, 24

Fig. 1. Marine Neogene biostratigraphic-chronostratigraphic scale for western North America. References for each biostratigraphic and chronostratigraphic component of the chart and the method of chart construction are presented in the text. The dot-pattern denotes informal units. Solid triangles represent radiometric dates from rocks within fossiliferous sequences. P.F. = Planktonic Foraminifera.

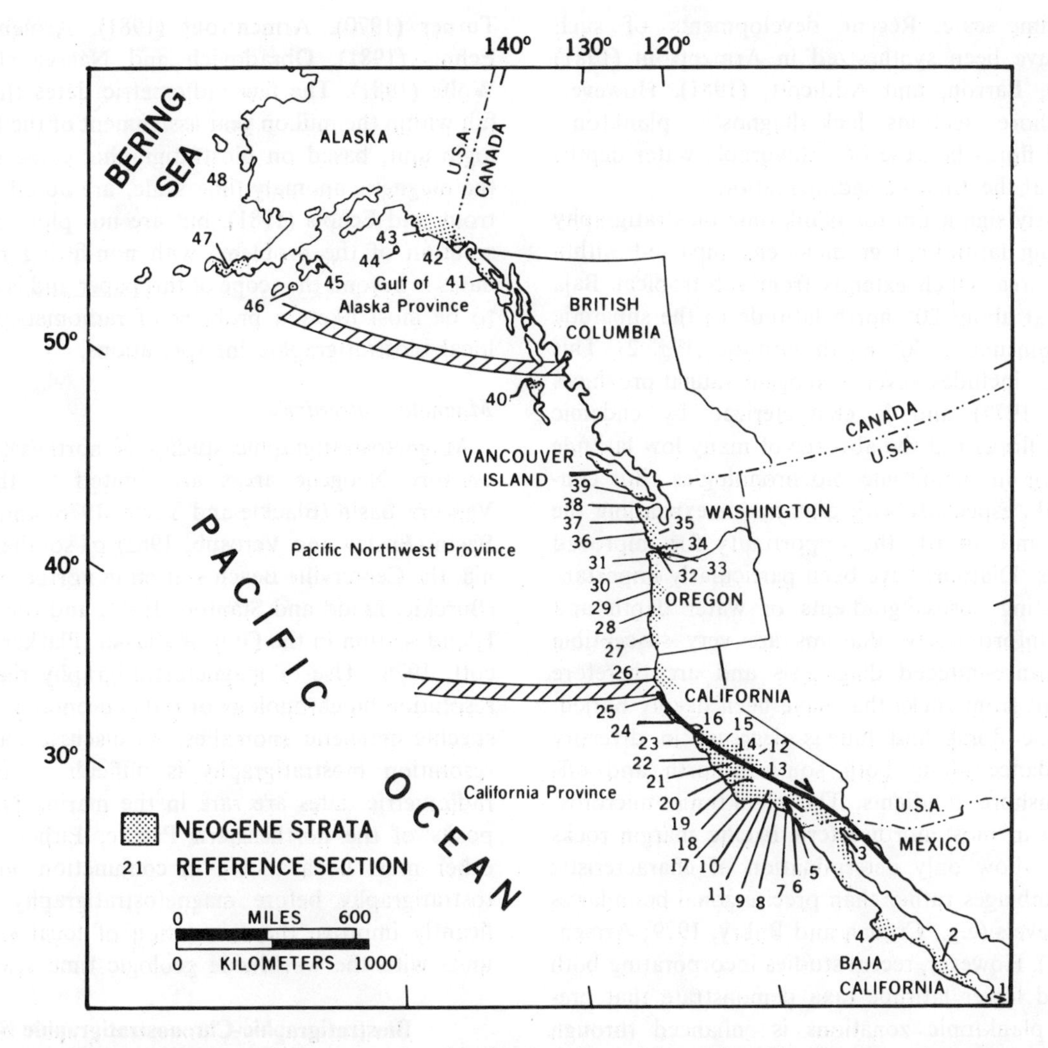

Fig. 2. Biostratigraphic, radiometric, and magnetostratigraphic data from forty-eight stratigraphic reference sections were used to construct the biostratigraphic-chronostratigraphic scale presented as Fig. 1. The reference sections occur in Neogene basins (dot-pattern) in Baja California, California, Oregon, Washington, British Columbia, and southern Alaska. The fossil assemblages from these basins have been used to define three Neogene faunal provinces: California Province, Pacific Northwest Province, Gulf of Alaska Province (Addicott, 1977). Stratigraphic information and correlation of the forty-eight reference sections are presented in Armentrout and Echols (1981).

spectrum; 3) correlation of provincial biochronologies one with another within the same stratigraphic sequence; and 4) correlation of the provincial biochronologies with one another and with the worldwide geologic timescale using planktonic biostratigraphy, magnetostratigraphy, and radiometric dating.

The detailed studies of provincial faunas used in defining the biochronologies correlated on Fig. 1 include: Addicott (1972, 1976), Allison (1978), and Armentrout (1981) for molluscs; Kleinpell, Hornaday, Warren, and Tipton (1980), Haller (1980), Crouch and Bukry (1979), Rau (1981) and Poore, Barron and Addicott (1981) for Foraminifera; Wolfe (1981) for megafloras; and Woodburne and Robinson (1977), Berggren, McKenna, Hardenbol, and Obradovich (1977), and Tedford, Galusha, Skinner, Taylor, Fields, Macdonald, Patton, Rensberger, and Whistler (in press) for land mammals. These references should be consulted for details of definition and correlation.

Significant progress in correlation of the provincial biochronologies with the evolving worldwide geologic time-scale has been made in recent years using planktonic biostratigraphy, magnetostratigraphy, and radiometric dating. However, all of these disciplines have inherent limitations with the result that confident resolution of some correlation problems is not yet available.

Planktonic biostratigraphy

Planktonic biostratigraphy offers the most potential for correlation of the northeastern Pacific margin marine Neogene biochronologies with the worldwide geologic time scale. Recent developments of such research have been synthesized in Armentrout (1981) and Poore, Barron, and Addicott, (1981). However, many onshore sections lack diagnostic planktonic faunas and floras because of unfavorable water depths or climate at the time of sedimentation.

Particularly significant for planktonic biostratigraphy is the strong latitudinal gradient encompassed within the study area which extends from sub-tropical Baja California at about 20° north latitude to the subarctic Alaska Peninsula at 60° north latitude (Fig. 2). This broad area includes several Neogene faunal provinces (Addicott, 1977) and is characterized by endemic faunas and floras and the absence of many low latitude species used in planktonic biochronologies. But continued work, especially with diatoms, is expanding the data base and affords the opportunity for improved correlations. Diatoms have been particularly important for correlating across gradients of water depth and latitude. Unfortunately, diatoms are very susceptible to temperature-induced diagenesis and are therefore often absent from rocks that have been deeply buried.

Planktonic floras and faunas decrease in diversity and abundance along both south-to-north and offshore-to-onshore gradients. The planktonic microfossils present in most northeastern Pacific margin rocks commonly allow only determination of characteristic zonal assemblages rather than precise zonal boundaries or datum levels (e.g., Crouch and Bukry, 1979; Armentrout, 1981). However, recent studies incorporating both middle and lower latitude taxa demonstrate that precision of planktonic zonations is enhanced through combined use of several planktonic groups and correlation of deep sea and onshore sequences (e.g., Keller and Barron, 1981).

Radiometric dates

The inherent limitation in age resolution by radiochronologic methods is well known (Berggren and Van Couvering, 1978). The assignment of precise ages to most biostratigraphic datum levels is impossible for datums older than about four million years because the experimental error in the dating methods exceeds the discrimination available from biologic events (Berggren, 1981). For this reason, radiometric dates from rocks interbedded within strata containing age diagnostic fossils are plotted in Fig. 1 as triangles only where the appropriate million year position falls within the appropriate biochronolgic unit. All potassium-argon data used in preparing this paper were calibrated and recalibrated to the decay and abundance constants recommended by Steiger and Jäger (1977), using the conversion factors recommended by Dalrymple (1979). Data sources for the plotted dates are summarized in Turner (1970), Armentrout (1981), Armentrout and Echols (1981), Obradovich and Naeser (1981), and Wolfe (1981). The few radiometric dates that did not fall within the million year assignment of the biochronologic unit, based on biostratigraphic correlations with the magnetic anomaly time scale, are noted in Armentrout and Echols (1981), but are not plotted here. Resolution of the problems with non-fitting radiometric dates is beyond the scope of this paper and is considered to be most likely a problem of radiometric dating or local biostratigraphic interpretations.

Magnetostratigraphy

Magnetostratigraphic studies of northeastern Pacific onshore Neogene areas are limited to the western Ventura Basin (Blackie and Yeats, 1976) and the Ridge Basin (Ensley and Verosub, 1982) of southern California, the Centerville Beach section in northern California (Burckle, Dodd and Stanton, 1980), and the Middleton Island section in the Gulf of Alaska (Plafker and Addicott, 1976). Use of magnetostratigraphy requires high resolution biochronology or radiochronology to identify specific magnetic anomalies. As discussed above, high resolution biostratigraphy is difficult to obtain and radiometric dates are rare in the marine Neogene deposits of the northeastern Pacific. Either one or the other must be available in conjunction with magnetostratigraphy before magnetostratigraphy can significantly improve the correlation of local stratigraphic units with the worldwide geologic time scale.

Biostratigraphic-Chronostratigraphic Scale

The correlation of northeastern Pacific Ocean margin Neogene biostratigraphic and chronostratigraphic units with the worldwide geologic time scale and oceanic zones and subzones is shown in Fig. 1. The correlations shown have been updated since the original figure was published (Armentrout and Echols, 1981, Fig. 6) by the incorporation of newly published data and discussions at the 1981 IGCP-114 Osaka meetings. The following discussion focuses on the construction of Fig. 1.

Worldwide scale

The Magnetic Anomaly Timescale of Ness, Levi, and Crouch (1980) has been adopted for the age of magnetic anomalies. The Ness *et al.* (1980) time scale is based upon the synthesis of published magnetic anomaly data from numerous sources which have been carefully correlated, calibrated with the most reliable radiometric dates, and assumed constant rates of seafloor spreading between major events of rate change. The result is a Magnetic Anomaly Timescale which allows for calibrations of approximately $\pm$ 0.1 m.y.

for the Neogene. These factors led to the adoption of the Ness *et al.* (1980) time scale by the IGCP-114 working committee on magnetostratigraphy during the 1981 Osaka meeting (W. Berggren, H. Nakagawa, T. Saito, N. Shackleton, and K. Shibata, 1981, written communication).

Correlation for the oceanic zones and subzones is taken from the chart of Barron, Poore, and Wolfart (1981) developed on Leg 63 of the Deep Sea Drilling Program adjacent to Baja California. The Barron *et al.* (1981) correlation framework incorporates planktonic foraminifer and calcareous nannoplankton zones and subzones that have general use in low to middle latitudes with diatom zones and subzones that are applicable to middle and high latitudes. Construction of the Barron *et al.* (1981) planktonic biochronologic framework proceeded by 1) correlating the magnetic anomaly time scale of Mankinen and Dalrymple (1979) to the north Pacific diatom zonation of Barron (1981); 2) correlating Barron's (1981) north Pacific diatom zonation to the calcareous nannoplankton zones and subzones of Bukry (1973, 1975) and the planktonic foraminifer zonation of Blow (1969). We have added the calcareous nannofossil zonation of Martini (1971) and Martini and Worsley (1970) as correlated to Bukry's (1973, 1975) calcareous nannofossil zonation.

Correlation of the integrated oceanic zone and subzone framework to the European stages follows Ryan, Cita, Rawson, Burckle, and Saito (1975). Correlation of the biochronologic framework to the Magnetic Anomaly Timescale of Ness *et al.* (1980) follows Berggren (1981) with some slight adjustments recommended by Berggren at the 1981 IGCP-114 Osaka meetings. Berggren (1981) used the magnetostratigraphic and magnetobiostratigraphic framework established by Ryan *et al.* (1975) with minor adjustments as suggested by Dalrymple (1979) and Harrison, McDougall, and Watson (1979) to reflect the recently proposed ^{40}K decay constants (Steiger and Jäger, 1977). Use of the decay constants causes the slight downward expansion of the Neogene time scale with the result that the ages of the various biostratigraphic datum levels and chronostratigraphic boundaries have been re-estimated. The re-estimated ages average about 0.1 m.y. older in the Pliocene and 0.3 m.y. older in the early Miocene (Berggren, 1981, p.45).

The resultant integrated biostratigraphic and chronostratigraphic scale provides a geologic time scale which serves as a worldwide correlation standard for the Neogene.

Northeastern Pacific margin geochronologic and biostratigraphic standards

The right half of Fig. 1 consists of six provincial biostratigraphic-chronostratigraphic standards for the northeastern Pacific margin. The provincial standards are correlated with the worldwide geologic time scale discussed above. Each biostratigraphic standard is separated from the others by a narrow empty column–this is to graphically represent the fact that the stratotypes of each are correlated one to another through geographically intermediate reference sections and not within the same stratigraphic section. Stage, zone, and age names follow the style of the data sources cited above under current work. The dot-patterned areas denote informal units. Horizontal boundaries between stages, zones, and ages are qualified by the relative confidence in correlating provincial boundaries with units of the worldwide geologic time scale. Thus, an unbroken boundary is well controlled by planktonic biostratigraphy, radiometric ages, or magnetostratigraphy: a dashed and question-marked boundary is less well controlled. The molluscan Juanian Stage/Pillarian Stage boundary is denoted by vertical lines and question marks to alert users to a difference of boundary placement. Addicott (1976) and Addicott and Poore (1979) place the Juanian/Pillarian boundary within the lowermost part of the foraminiferal Saucesian Stage. Armentrout (1973, 1975, 1981) places the Juanian Stage/Pillarian Stage boundary coincident with the Zemorrian Stage/Saucesian Stage boundary.

Summary

Biostratigraphy will continue to be the most useful tool for correlation of Neogene stratigraphic units along the northeastern Pacific margin. Continued study is resolving problems of stratotype recognition, ecologically diverse faunal and floral assemblages, and calibration of provincial biostratigraphic frameworks to the worldwide geologic time scale.

The northeastern Pacific margin Neogene biostratigraphic-chronostratigraphic scale presented here provides a basis for time-stratigraphic correlations within the context of the current worldwide geologic time scale. The precision of correlations using the biostratigraphic-chronostratigraphic scale of Fig. 1 depends on the degree of resolution of the faunal and floral biostratigraphy used. Employment of an interdisciplinary approach and careful attention to the superpositional relationship of samples will result in improved resolution.

Although the worldwide geologic time scale and the provincial biochronolgies are certain to undergo further refinement, the time scale presented here reflects the current collective knowledge of most northeastern Pacific margin biostratigraphers.

Acknowledgements

The authors thank Professor N. Ikebe, chairman of IGCP-114, for the opportunity to participate in the 1983 Osaka workshop on Pacific Neogene biostratigraphy. The manuscript was reviewed and improved thanks to comments from W. O. Addicott, R. C. Allison, J. A. Barron, M. B. Lagoe, K. McDougall, J. D. Obradovich, and W. W. Rau.

References

Addicott, W. O., 1972: Provincial middle and late Tertiary molluscan stages, Temblor Range, California. *In* Stinemeyer, E. H., ed., *Symposium on Miocene biostratigraphy of California,* SEPM, Pacific Section, Bakersfield, California, p. 1–26, pls. 1–4.

Addicott, W.O., 1976: Neogene molluscan stages of Oregon and Washington. *In* Fritsche, A. E. and others, eds., *Neogene Symposium,* SEPM, Pacific Section, San Francisco, California, p. 95–115.

Addicott, W. O., 1977: Neogene chronostratigraphy of nearshore marine basins of the eastern North Pacific: northwestern Mexico to Canada. *In* Saito, T. and Ujiie, H., eds, *I-CPNS, Tokyo 1976, Proc.*, p. 151–175.

Addicott, W. O. and Poore, R.Z., 1979: The Paleogene/Neogene boundary in eastern North Pacific marine sequences. *XIV Pac. Sci. Cong., Khabarovsk, USSR, Abst.*, v. **2**, p. 5–6.

Allison, R. C., 1978: Late Oligocene through Pleistocene molluscan faunas in the Gulf of Alaska region. *The Veliger*, v. **21**, no. 2, p. 171–188.

Armentrout, J. M., 1973: Molluscan paleontology and biostratigraphy of the Lincoln Creek Formation, Late Eocene-Oligocene, southwestern Washington (Ph.D. dissertation). Seattle, University of Washington, 478 p.

Armentrout, J. M., 1975: Molluscan biostratigraphy of the Lincoln Creek Formation, southwest Washington. *In* Weaver, D. E., and others, eds., *Future energy horizons of the Pacific Coast*, Paleogene symposium and selected technical papers, AAPG, SEPM, SEG, Pacific Sections, Annual Meeting, Long Beach, California, p. 14–28.

Armentrout, J. M., 1981: Correlation and ages of Cenozoic chronostratigraphic units in Oregon and Washington. *In* Armentrout, J. M., ed., *Pacific Northwest Cenozoic Biostratigraphy*, Geol. Soc. Amer. Special Paper 184, p. 137–148.

Armentrout, J. M. and Echols, R. J., 1981: Biostratigraphic-chronostratigraphic scale of the northeastern Pacific Neogene. *In* Ikebe, N., and others, eds., *IGCP-114 Internatl. Workshop on Pacific Neogene Biostrat. Osaka, Japan, 1981, Proc.*, p. 7–27.

Barron, J. A., 1981: Late Cenozoic diatom biostratigraphy and paleoceanography of the middle-latitude eastern North Pacific, DSDP Leg 63. *In* Yeats, R. S., Haq, B. U., and others, eds., *DSDP, Initial Reports*, Washington, D. C., U. S. Government Printing Office, v. **63**, p. 507–538.

Barron, J.A., Poore, R. Z. and Wolfart, R., 1981: Biostratigraphic Summary, DSDP Leg 63. *In* Yeats, R. S., Haq, B. U., and others, eds., DSDP, Initial Reports Washington, D. C., U. S. Government Printing Office, v. **63**, p. 927–941.

Berggren, W. A., 1981: Correlation of Atlantic, Mediterranean and Indo-Pacific Neogene stratigraphies: Geochronology and chronostratigraphy. *In* Ikebe, N., and others, eds., *IGCP-114 Internatl. Workshop on Pacific Neogene Biost. Osaka, Japan, 1981, Proc.*, p. 29–59.

Berggren. W. A., McKenna, M. C., Hardenbol, J., and Obradovich, J. D., 1977: Revised Paleogene polarity time scale. *Journal of Geology*, v. **86**, p. 67–81.

Berggren, W. A. and van Couvering, J.A., 1978: Biochronology. *A A P G, Studies in Geology*, no. 6, p. 39–55.

Blackie, G. W. and Yeats, R. S., 1976: Magnetic reversal stratigraphy of Pliocene-Pleistocene producing section of the Saticoy Oil Field, Ventura Basin, California. *AAPG, Bull.*, v. **60**, p. 1985–1992.

Blow, W.H., 1969: Late Middle Eocene to Recent planktonic forminiferal biostratigraphy. *First International Conference Planktonic Microfossils, Geneva, 1967, Proc.* p. 199–421, 43 figs., 54 pls.

Bukry, D., 1973: Low-latitude coccolith biostratigraphic zonation. *In* Edgar, N. T., Saunders, J.B., and others, eds., *DSDP, Initial Reports*, Washington D. C., U. S. Government Printing Office, v. **15**, p. 685–703.

Bukry, D., 1975: Coccolith and silicoflagellete stratigraphy, northwestern Pacific Ocean. Deep Sea Drilling Project Leg 32, *In* Larson, R. L. Moberly, R., and others, eds., *DSDP*, *Initial Reports*, Washington D. C., U. S. Government Printing Office, v. **32**: p. 677–701.

Burckle, L. H., Dodd, J. R. and Stanton, R. J., Jr., 1980: Diatom biostratigraphy and its relationship to paleomagnetic stratigraphy and molluscan distribution in the Neogene Centerville Beach section, California. *Jour. Pal.*, v. **54**, no. 4, p. 664–674.

Crouch, J. K. and Bukry, D., 1979: Comparison of Miocene provincial stages to coccolith zones in the California continental borderland. *Geology*, v. **7**, p. 211–215.

Dalrymple, G. B., 1979: Critical tables for conversion of K-Ar ages from old to new constants. *Geology*, v. **7**, p. 588–560.

Durham, J. W., 1944: Megafaunal zones of the Oligocene of northwestern Washington. *California University Publications, Dept. Geol. Sci., Bull.*, v. **27**, no. 5, p. 84–104.

Ensley, R. A. and Verosub, K. L., 1982: A magnetostratigraphic study of the sediments of the Ridge Basin, southern California and its tectonic and sedimentologic implications. *Earth and Planetary Science Letters*, v. **59**, p. 192–207.

Haller, C. R., 1980: Pliocene biostratigraphy of California. *AAPG, Studies in Geology*, no. 11, p. 183–341.

Harrison, C. G. A., McDougall, I. and Watkins, N. D., 1979: A geomagnetic field reversal time-scale back to 13.0 million years before present. *Earth and Planetary Science Letters*, v. **42**, p. 143–152.

Keller, G. and Barron, J. A., 1981: Integrated planktic foraminiferal and diatom biochronology for the northeast Pacific and Monterey Formation. *In* Garrison, R. E., and Douglas, R. G., eds., *The Monterey Formation and related siliceous rocks in California*. SEPM Pacific Section, Los Angeles, California, p. 43–54.

Kleinpell, R.M., 1938: *Miocene Stratigraphy of California*. AAPG, Tulsa, Oklahoma, p. 1–450.

Kleinpell, R. M., Hornaday, G., Warren, A. D., and Tipton, A., 1980: The Miocene Stratigraphy of California Revisited. *AAPG, Studies in Geology*, no. 11, p. 1–182.

Mankinen, E. A. and Dalrymple, G. B., 1979: Revised geomagnetic polarity time scale for the interval 0–5 m.y.B.P. *Jour. Geophys. Res.*, v. **84**, p. 615–626.

Martini, E., 1971: Standard Tertiary and Quaternary calcareous nannoplankton zonation. *Second International Conference Planktonic Microfossils, Roma, 1970, Proc.*, p. 739–785, 4 pls.

Martini, E. and Worsley, T., 1970: Standard Neogene calcareous nannoplankton zonation. *Nature*, v. **225**, p. 289–290.

McDougall, K., 1980: Paleoecological evaluation of late Eocene biostratigraphic zonation of the Pacific coast of North America. *Jour. Pal.*, v. **54**, no. 4, p. 1–75, 29 pls.

Natland, M.L., 1952: Pleistocene and Pliocene stratigraphy of southern California (Ph.D. dissertation). Los Angeles, University of California, 165 p., 20 pls.

Ness, G., Levi, S. and Couch, R., 1980: Marine magnetic anomaly timescales for the Cenozoic and Late Cretaceous. A precis, critique, and synthesis. *Geophysics and Space Physics, Rev.*, v. **18**, no. 4, p. 753–770.

Obradovich, J.D. and Naeser, C. W., 1981: Geochronology bearing on the age of the Monterey Formation and siliceous rocks in California. *In* Garrison, R. E., and Douglas, R. G., eds., *The Monterey Formation and related siliceous rocks in California*, SEPM, Pacific Section, Los Angeles, California, p. 87–95.

Plafker, G. and Addicott, W. O., 1976: Glaciomarine deposits of Miocene through Holocene age in the Yakataga Formation along the Gulf of Alaska margin, Alaska. *U. S. Geological Survey Open-File Report 76-84*, 36 p.; reprinted *in* Proceedings of Symposium on Recent and Ancient sedimentary environments in Alaska: Alaska Geol. Soc. Anchorage, Alaska, 1976 p. Q1–Q23.

Poore, R. Z., Barron, J. A. and Addicott, W. O., 1981: Biochronology of the northern Pacific Miocene. *In* Ikebe, N., and others, eds., *IGCP-114 Internatl. Workshop on Pacific Neogene Biostrat. Osaka, Japan, 1981, Proc*, p. 91–97.

Poore, R. Z., McDougall, K., Barron, J. A., Brabb, E. E., and Kling, S. A., 1981: Microfossil biostratigraphy and biochronology of the type Relizian and Luisian Stages of California. *In* Garrison, R. E., and Douglas, R. G., eds., *The Monterey Formation and related siliceous rocks of California*. SEPM, Pacific Section, Los Angeles, California, p. 15–41.

Rau, W. W., 1981: Pacific Northwest Tertiary benthic foraminiferal biostratigraphic framework–An overview. *In* Armentrout, J. M., ed., *Pacific Northwest Cenozoic Biostratigraphy*, Geol. Soc. Amer. Spec. Paper, 184, p. 67–84.

Ryan, W. B. F., Cita, M. B., Rawson, M. D., Burckle, L.H., and Saito, T., 1974: A paleomagnetic assignment of Neogene stage boundaries and the development of isochronous datum planes between the Mediterranean, the Pacific and Indian Oceans in order to investigate the response of the world ocean to the Mediterranean "Salinity Crisis". *Italiana di Paleontologia e Stratigrafia, Riv.*, v. **80**, p. 631–688.

Steiger, R. H. and Jäger, E., 1977: Subcommission on geochronology: Convention to the use of decay constants in geo- and cosmochronology. *Earth and Planetary Science Letters*, v. **36**, p. 359–363.

Tedford, R. H., Galusha, T., Skinner, M. F., Taylor, B., Field, R. W., MacDonald, J. R., Patton, T. H., Rensberger, J. M., and Whistler, D. P.: *In* Woodburne, M. O., ed., *Vertebrate Paleontology as a Discipline in Geochronology*: University of California Press. (in press).

Turner, D. L., 1970: Potassium-argon dating of Pacific coast Miocene foraminiferal stages. *In* Bandy, O. L., ed., *Radiometric Dating and Paleontologic Zonation. Geol. Soc. Amer., Spec. Paper*, 124, p. 91–129.

Weaver, C. E. and others, 1944: Correlation of the marine Cenozoic formations of western North America. *Geol. Soc. Amer., Bull.*, v. **55**, p. 569–598.

Wolfe, J. A., 1981: A chronologic framework for Cenozoic megafossil floras of northwestern North America and its relation to marine geochronology. *In* Armentrout, J.M., ed., *Pacific Northwest Cenozoic Biostratigraphy*: Geol. Soc. Amer. Spec. Paper 184, p. 39–47.

Woodburne, M. O. and Robinson, P. T., 1977: A new late Hemingfordian mammal fauna from the John Day Formation, Oregon, and its stratigraphic implications. *Jour. Pal.*, v. **51**, p. 750–757.

III-3

Neogene Stratigraphy and Planktonic Foraminiferal Datum Planes in New Zealand

Norcott de B. HORNIBROOK

Outline of Stratigraphy

The Neogene in New Zealand is mainly distributed throughout seven basins of marine sedimentation (Katz, 1968; Suggate *et al.*, 1979) (Fig. 1). Some of these basins have formed in the tectonically mobile setting of the Indian-Pacific plate junction (Walcott *et al.*, 1981). Consequently, they vary greatly in stratigraphic and sedimentary character and especially in the thickness of the sedimentary formations which are considerable in places (e.g., 7,000 m of Lower Miocene-Lower Pleistocene sandstone and mudstone, northern part of East Coast Basin). Shelf limestones predominate in the upper Oligocene and basal Miocene but thereafter sandy and muddy facies are more usual. Coquinas are prominent in the Pliocene of the East coast basin having formed as extensive barnacle banks along the margins of a narrow seaway during repeated movements of the plate boundary zone. Correlations are usually expressed in New Zealand marine stages (Fig. 2) based principally on foraminifera in the Miocene and lower Pliocene and on mollusca in the upper Pliocene and Pleistocene in stratotypes and reference sections (Suggate *et al.*, 1979).

Northland Basin

This basin rapidly filled up during the early Miocene. The sediment type (Waitemata Group) varies from subtidal conglomerate and limestone lenses to sandstone and bathyal mudstone, lying unconformably on a basement of Permian and Mesozioc. Vigorous volcanic activity contributed much flow and tuffaceous material. Sedimentary thickness varies greatly reaching a maximum of over 1,000 m. Extensive areas of mainly Cretaceous rock overlie Miocene, apparently as olistostromes.

Waikato Basin

Shelf limestone of the Te Kuiti Group is overlain by up to 800 m of lower Miocene sandstone and mudstone of the Waitemata Group. Shallow water Pliocene and Pleistocene sands occur in patches.

Taranaki and Wanganui basins: This large basin-complex extends for a considerable distance offshore to reach the northwest corner of the South Island. Much of the northwestern part is covered by young volcanics. The area is one of the most prospective for petroleum exploration, and drilling has provided much information on subsurface stratigraphy.

Because of its wide extent, lateral variation in facies and thickness, and alternative stratigraphic nomenclatures, it is only possible to summarize the stratigraphy in this brief account.

The lower Miocene formations (Mahoenui Mudstone and Limestone; Mokau Sandstone and Mudstone) are comformable on the Oligocene Te Kuiti Group, being exposed on land and also present in the subsurface where they attain a thickness of 1,000 m and 350 m, respectively.

The middle and upper Miocene formations (Mohakatino Tuffaceous Sandstone and Mudstone; Waikieke Tuffaceous Mudstone and Sandstone; Tongaporutu Tuffaceous Mudstone and Sandstone; Urenui Mudstone; Matemateonga Sandstone (in part)) consist of rapidly deposited, mainly shelf sediments which attain an aggregate subsurface thickness of about 3,000 m. The Tongaporutu and Urenui Formations, exposed on the coast of the North Taranaki Bight, comprise the stratotype of the Tongaporutuan Stage.

The lower to mid Pliocene formations (Matemateonga Sandstone (in part), and Tangahoe Mudstone) attain an outcrop thickness of about 3,500 m in the Wanganui and Tangahoe Rivers.

A classic sequence of upper Pliocene and mid Pleistocene, divided into 9 groups and 57 formations with an aggregate thickness of 800 m is well exposed along the Wanganui coast. The predominantly sandy lithology is rich in mollusca characteristic of inner to mid-shelf environments. This sequence is the standard for most of the Wanganui Series including the Waito-

c/o New Zealand Geological Survey, P.O.Box 30368, Lower Hutt, New Zealand

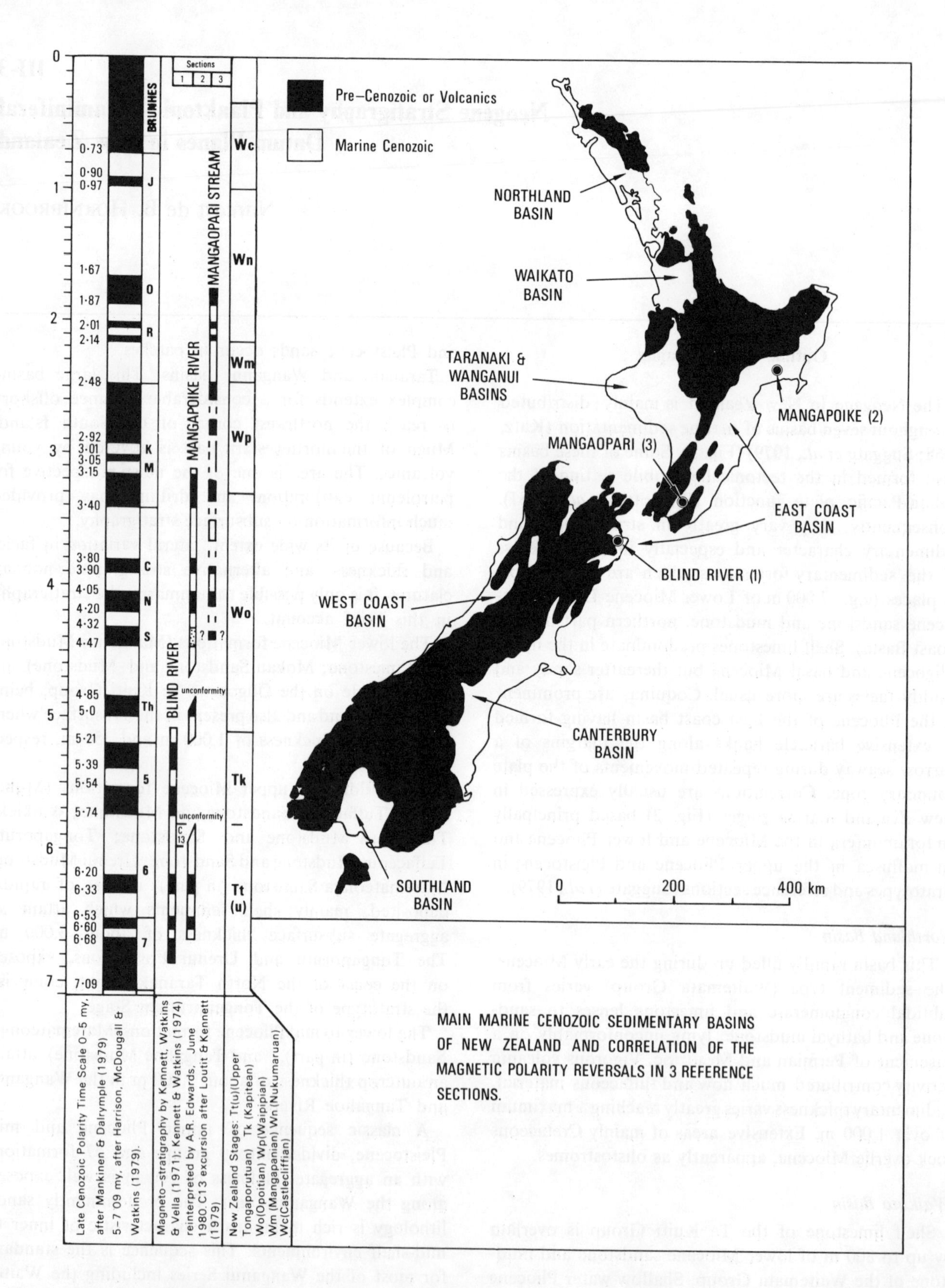

Fig. 1. Main marine Cenozoic sedimentary basins of New Zealand and correlation of magnetic polarity reversals in the Blind River, Mangapoike River, and Mangaopari Stream sections after Edwards and Hornibrook (1980).

SERIES	STAGES		MAP SYMBOL		INTERNATIONAL	
Hawera						
	Putikian	Castlecliffian	Wu	Wc	Pleistocene	L
	Okehuan		Wk			
	Marahauan		Wa			
Wanganui	Hautawan	Nukumaruan	Wh	Wn		E
	Mangapanian	Waitotaran	Wm	Ww	Pliocene	
	Waipipian		Wp			
	Opoitian			Wo		
Taranaki	Kapitean			Tk		
	Tongaporutuan			Tt		L
	Waiauan			Sw		
Southland	Lillburnian			Sl	Miocene	
	Clifdenian			Sc		M
	Altonian			Pl		
Pareora	Otaian			Po		E
	Waitakian			Lw		
Landon	Duntroonian			Ld	Oligocene	
	Whaingaroan			Lwh		
	Runangan			Ar		
Arnold	Kaiatan			Ak		L
	Bortonian			Ab		
	Porangan			Dp	Eocene	M
	Heretaungan			Dh		
Dannevirke	Mangaorapan			Dm		E
	Waipawan			Dw		
	Teurian			Dt	Paleocene	

Fig. 2. Series and stages of the New Zealand Cenozoic.

taran (Waipipian and Mangapanian), Nukumaruan, and Castlecliffian Stages which are based principally on mollusca.

East Coast Basin

Exceedingly thick monotonous sandstone and mudstone in the northern part of this long complex basin have made lithostratigraphic classifications difficult to apply and New Zealand stages, recognized principally by foraminifera in the Miocene and lower Pliocene, and by mollusca and foraminifera in the upper Pliocene and lower Pleistocene, have been used as mapping units. Generalized thicknesses for the northern part of the basin are: lower Miocene, 2,500 m; middle Miocene, 1,500 m; upper Miocene, 2,000 m; lower to middle Pliocene, 1,000 m.

Nearer to the middle of the basin, in Hawkes Bay, the Mangapoike River section which has been paleomagnetically studied (Epoch 7 to basal Gauss, Kennett and Watkins, 1974) is the stratotype for the Opoitian Stage (lower Pliocene). Two unconformities introduce uncertainties in correlating the polarity sequence. Three cycles of inner shelf barnacle coquina formation (Te Aute Limestone facies) alternating with deeper water mudstones dominate Pliocene stratigraphy (Beu *et al.*, 1980).

The uppermost Pliocene–lower Pleistocene is characterized by cyclic sedimentation with intertidal shelly deposits alternating with mudstones containing distinctly deeper water molluscan assemblages. The aggregate thickness is 300 m. In the southern part of the North Island, cycles of similar age have been described (Vella, 1963). The Mangaopari Stream section, which exposes a good lower Pliocene to lower Pleistocene sequence, is a reference section for magnetic polarity reversals (Kennett *et al.*, 1971), although correlation of the lower part (Gauss or Gilbert?) is uncertain. In the southernmost part of the basin, in the northeastern corner of the South Island, Blind River cuts a long sequence through upper Miocene to lower Pliocene which is also a reference section for magnetic polarity reversals from Epoch 7 to basal Gilbert (Kennett and Watkins, 1974; Loutit and Kennett, 1979).

West Coast Basin

Marine sedimentation in this long complex basin commenced in the middle Eocene, and a fairly complete Eocene to Pliocene sequence is preserved in several small rapidly subsiding sub-basins. Widespread limestone deposition (Cobden Limestone) in the late Oligocene was followed by a change to predominantly sandy and muddy sediments in the early Miocene (Blue Bottom Group) with lenses of limestone in the middle Miocene. The uppermost Miocene typically consists of sandstone, glauconitic at the base, and thin coal measures. A short marine sedimentary hiatus within the upper Miocene is followed by sandstone, glauconitic at the base, and muddy micaceous fine sandstone in the uppermost Miocene and lower Pliocene. Recorded thicknesses in the Greymouth sub-basin are lower Miocene (limestone and marls) 100 m, middle Miocene (mudstone) 400 m, upper Miocene (mudstone and sandstone) 400 m, and Pliocene (muddy sandstone) 900m. The stratotype of the Kapitean Stage (Eight Mile Formation; uppermost Miocene to lowermost Pliocene) is in the Greymouth sub-basin.

Canterbury Basin

This long narrow basin developed in a comparatively stable tectonic setting. The Neogene, characterized by shelf sediments, is thickest and most complete in the north and generally overlies Paleogene sequences, some of which are fairly complete. Almost throughout the whole basin the upper Oligocene consists of shelf limestones, and limestone facies persist into the Waitakian Stage, which is basal Miocene in part (Weka Pass Limestone, Craigmore Limestone, Otekaike Limestone). Sediment becomes glauconitic then predominantly sandy and muddy in the lower Miocene (Grey Marls, Bluecliffs Silt, Southburn Sand, Gee Greensand, Rifle Butts Mudstone, Caversham Sandstone) with limestone in places (Mt Brown Limestones, Goodwood Limestone). Sedimentation was patchy and shallow-water or non-marine in type from middle Miocene onwards. In the northern part of the basin, shelly limestone (Glenmark Limestone) formed in the upper Miocene followed by conglomerates (Kowai Gravels) and shelly sandstone in the lower Pliocene and lower Pleistocene. In the Weka Greek section, the aggregate thickness of the Miocene and Pliocene is 800 m. The stratotypes of the Waitakian Stage (Otekaike Limestone) and Otaian Stage (Bluecliff Silt and Southburn Sands in part) are near the southern end of the basin which was more tectonically stable with more discontinuous and thinner sedimentation than in the north.

Southland Basin

This isolated basin developed as a wedge in a basement of Paleozoic igneous complex in early Oligocene time. Neogene sequences accumulated in three contrasting sedimentary environments:

1) A northeastern platform containing extensive upper Oligocene to lower Miocene estuarine and freshwater lignitic deposits.

2) An intermediate area of shelf sedimentation with lower Miocene bryozoan limestones, conglomerate, and sands.

3) A western geosyncline. At Clifden, the Waiau

River exposes a continuous section of approximately 350 m of lower, middle, and upper Miocene (Clifden Limestone, Otuhu Siltstone, Merton Sandstone, Nga Pari Mudstone, Park Bluff Sandstone), which is the stratotype of the Altonian, Clifdenian, Lillburnian, and Waiauan Stages. The aggregate thickness of Miocene on the margins of the geosyncline is about 3,000 m.

Offshore

DSDP Leg 29, Site 284, drilled on the Challenger Plateau, 600 km west of Wellington, penetrated 208 m of calcareous biogenic sediment of late Miocene (Tongaporutuan) to Pleistocene age (Castlecliffian). Planktonic foraminifera in this deep-sea sequence, which is a complete sedimentary record of the late Neogene in the area, have been studied by Kennett and Vella (1975) and Hornibrook (1982). Thick deposits of Neogene sediments have been drilled offshore for petroleum exploration in most of the main sedimentary basins and also in the Campbell Plateau to the south of New Zealand. Reports on the biostratigraphy are held by the New Zealand Geological Survey, Lower Hutt.

Sedimentary hiatuses

Hiatuses of various duration are present in most Neogene sequences in New Zealand, particularly those deposited at shelf depths. It is unlikely, however, that any hiatus extends throughout the whole area. Loutit and Kennett (1981) attribute many hiatuses to eustatic sea level changes and try to demonstrate a correlation between New Zealand stages and marine sedimentary cycles.

Eustatic sea level control seems a likely cause of cyclic sedimentation in the Pleistocene of the Wanganui and East Coast Basins. However, because of the generally mobile tectonic setting of many of the sedimentary basins, much detailed stratigraphic analysis remains to be done in order to confidently distinguish eustatically from tectonically caused hiatuses.

Planktonic Foraminiferal Datum Levels

A number of datum levels in 3 paleomagnetically studied sections (Kennett *et al.*, 1971; Kennett and Watkins, 1974) are tied to parts of the geomagnetic polarity scale with fair probability. Others are interposed using sedimentation rates, principally in DSDP Site 284 (Hornibrook, 1981 a, b, 1982).

The LADs of *Globigerina nepenthes* and *Globorotalia margaritae* are synchronous in the S.W. Pacific and equatorial Pacific. Other datum levels are not synchronous, e.g., the LAD of *Globoquadrina dehiscens* is earlier and the FAD of *Globorotalia tosaensis* is later in the New Zealand area. As no ties to the polarity scale are known in New Zealand below epoch 7, the datum levels are located against the adopted correlation with Blow's N-zones on the largely untested assumption that they are all synchronous in the S.W. Pacific and the equatorial Pacific.

A taxonomic question still not satisfactorily resolved is the identity of *Globorotalia mayeri*. The species identified as *Globorotalia mayeri mayeri*, which evolves from *Globorotalia peripheroronda* in the upper Miocene in New Zealand (Jenkins, 1971), with its FAD at the base of the Waiauan Stage (Hoskins, 1978), is apparently not the same taxon that is recorded as *Globorotalia mayeri* with a much earlier FAD elsewhere.

The FAD of *Globoquadrina dehiscens* at the base of the Waitakian Stage is below the LAD of the coccolith *Reticulofenestra scissura-bisecta* (Chattian ?) in the Waitakian stratotype (Hornibrook and Edwards, 1971). Thus the Paleogene-Neogene boundary probably falls within the Waitakian.

References

Beu, A.G., Grant-Taylor, T.L. and Hornibrook, N.de B., 1980: The Te Aute Limestone Facies, Poverty Bay to Northern Wairarapa. 1:250,000. New Zealand Geological Survey Miscellaneous Series Map 13(2 sheets) and notes (36 p.). New Zealand Department of Scientific and Industrial Research, Wellington.

Edwards, A.R. and Hornibrook, N. de B., 1980: An integrated bio- and magneto-stratigraphy of the late Miocene to Pleistocene of the New Zealand region. *Geological Society of New Zealand, Christchurch Conference 1980, Abstracts*, p. 35.

Harrison, C.G.A., McDougall, I. and Watkins, N.D., 1979: A geomagnetic field reversed time scale back to 13.0 million years before present. *Earth and Planetary Science Letters*, v. **42**, no. 2, p. 143–152.

Hornibrook, N. de B., 1981a: Evaluation of Pacific southern mid latitude and equatorial latitude planktonic foraminiferal datum levels. *IGCP-114 International Workshop on Pacific Neogene Biostratigraphy, Osaka, Nov. 25-29, 1981, Proc.*, p. 61–71.

Hornibrook, N. de B., 1981b: *Globorotalia* (planktic foraminiferida) in the late Miocene and early Pleistocene of New Zealand. *New Zealand Journal of Geology and Geophysics*, v. **24**, no. 2, p. 263–292.

Hornibrook, N. de B., (1982): Late Miocene to Pleistocene *Globorotalia* (Foraminiferida) from DSDP Leg 29, Site 284, S.W.Pacific. *New Zealand Journal of Geology and Geophysics*, v. **25**, p. 83-99.

Hornibrook, N. de B. and Edwards, A.R., 1971: Integrated planktonic foraminiferal and calcareous nannoplankton datum levels in the New Zealand Cenozoic. *In* Farinacci, A., ed., *Second International Conference Planktonic Microfossils, Roma, 1970, Proc.*, v. **1**, p. 645–657.

Hoskins, R.H., 1978: "New Zealand Middle Miocene Foraminifera: The Waiauan Stage". Ph.D. Thesis, Univer-

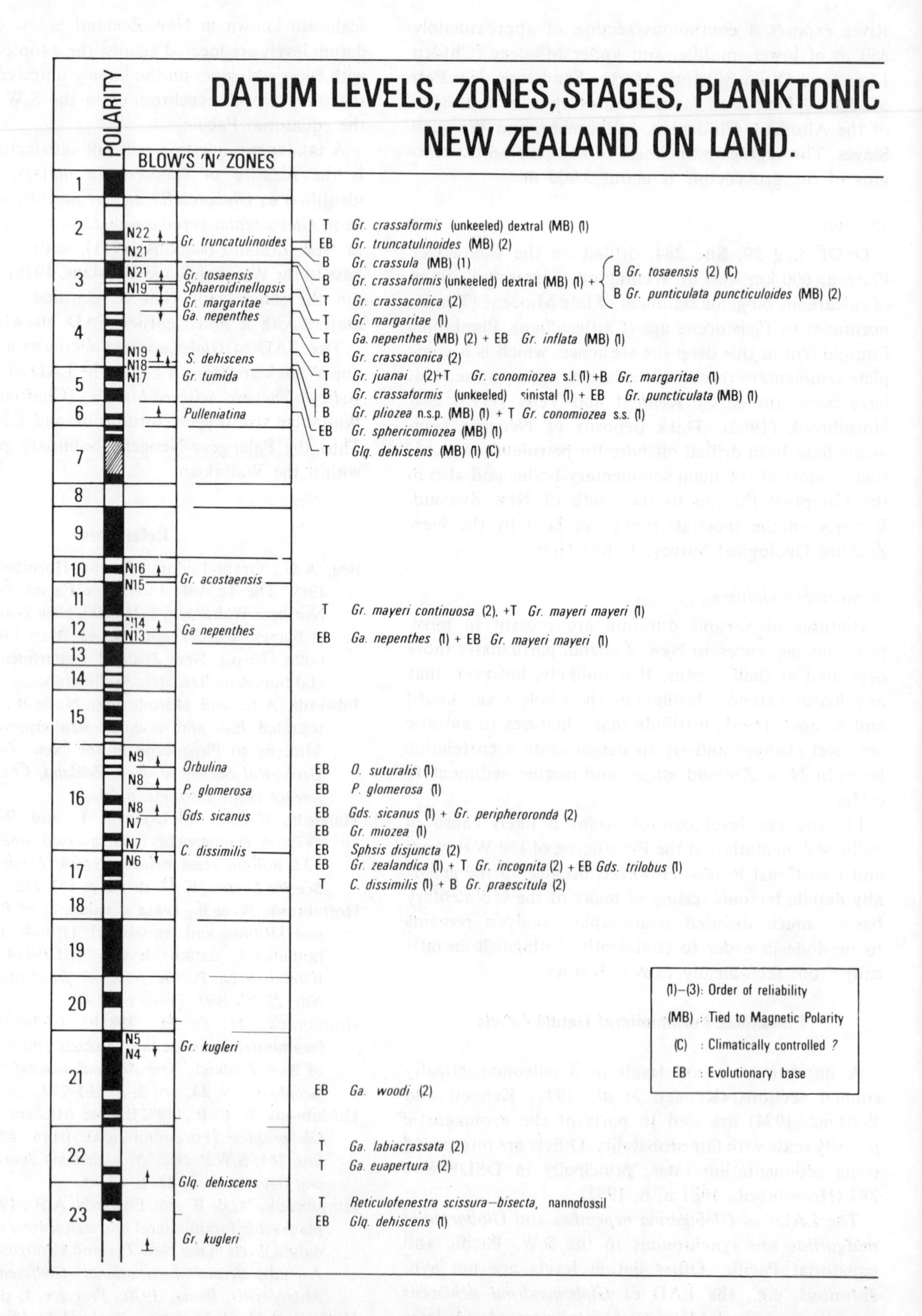

DATUM LEVELS, ZONES, STAGES, PLANKTONIC
NEW ZEALAND ON LAND.
POLARITY
BLOW'S 'N' ZONES
1
2
3
4
5
6
7
8
9
10
11
12
13
14
15
16
17
18
19
20
21
22
23
N22
N21
N21
N19
N19
N18
N17
N16
N15
N14
N13
N9
N8
N8
N7
N7
N6
N5
N4
Gr. truncatulinoides
Gr. tosaensis
Sphaeroidinellopsis
Gr. margaritae
Ga. nepenthes
S. dehiscens
Gr. tumida
Pulleniatina
Gr. acostaensis
Ga nepenthes
Orbulina
P. glomerosa
Gds. sicanus
C. dissimilis
Gr. kugleri
Glq. dehiscens
Gr. kugleri
T Gr. crassaformis (unkeeled) dextral (MB) (1)
EB Gr. truncatulinoides (MB) (2)
B Gr. crassula (MB) (1)
B Gr. crassaformis (unkeeled) dextral (MB) (1) + B Gr. tosaensis (2) (C)
B Gr. punticulata puncticuloides (MB) (2)
T Gr. crassaconica (2)
T Gr. margaritae (1)
T Ga. nepenthes (MB) (2) + EB Gr. inflata (MB) (1)
B Gr. crassaconica (2)
T Gr. juanai (2)+T Gr. conomiozea s.l. (1) +B Gr. margaritae (1)
B Gr. crassaformis (unkeeled) sinistal (1) + EB Gr. puncticulata (MB) (1)
B Gr. pliozea n.s.p. (MB) (1) + T Gr. conomiozea s.s. (1)
EB Gr. sphericomiozea (MB) (1)
B Glq. dehiscens (MB) (1) (C)
T Gr. mayeri continuosa (2), +T Gr. mayeri mayeri (1)
EB Ga. nepenthes (1) + EB Gr. mayeri mayeri (1)
EB O. suturalis (1)
EB P. glomerosa (1)
EB Gds. sicanus (1) + Gr. peripheroronda (2)
EB Gr. miozea (1)
EB Sphsis disjuncta (2)
EB Gr. zealandica (1) + T Gr. incognita (2) + EB Gds. trilobus (1)
T C. dissimilis (1) + B Gr. praescitula (2)
EB Ga. woodi (2)
T Ga. labiacrassata (2)
T Ga. euapertura (2)
T Reticulofenestra scissura–bisecta, nannofossil
EB Glq. dehiscens (1)
(1)–(3): Order of reliability
(MB) : Tied to Magnetic Polarity
(C) : Climatically controlled ?
EB : Evolutionary base

Fig. 3. Correlation of planktonic foraminiferal datum levels, zones, and stages with

FORAMINIFERA, N. de B. Hornibrook (author)

ZONES		N.Z. STAGES.
Globorotalia crassula	Modified HORNIBROOK (1981)	CASTLECLIFFIAN (Wc)
		NUKUMARUAN (Wn)
Gr. crassaformis – crassula overlap		
Gr. crassaformis dextral		MANGAPANIAN (Wm)
Globorotalia crassaformis sinistral		WAIPIPIAN (Wp)
		OPOITIAN (Wo)
Gr. sphericomiozea		KAPITEAN (Tk)
Gr. conomiozea		
Globorotalia miotumida (conoidea)	JENKINS (1971)	TONGAPORUTUAN (Tt)
Gr. mayeri mayeri		WAIAUAN (Sw)
Orbulina suturalis		LILLBURNIAN (Sl)
P. glomerosa curva		CLIFDENIAN (Sc)
Globigerinoides trilobus		ALTONIAN (Pl)
Globigerina woodi woodi & *Globigerina woodi connecta*		OTAIAN (Po)
		WAITAKIAN (Lw)
Globoquadrina dehiscens		
Globigerina euapertura (inpart)		DUNTROONIAN (Ld)

the geomagnetic polarity scale. New Zealand, on land.

sity of Exeter, U.K., 2 vols. 388 p.

Jenkins, D.G., 1971: New Zealand Cenozoic planktonic foraminifera, *N.Z. Geological Survey Paleontological Bulletin*, 42, 278 p.

Katz, H.R., 1968: Potential oil formations in New Zealand, and their stratigraphic position as related to basin evolution. *New Zealand Journal of Geology and Geophysics*, v. **11**, p. 1077–133.

Kennett, J.P., Watkins, N.D. and Vella, P., 1971: Paleomagnetic chronology of Pliocene-early Pleistocene climates and the Plio-Pleistocene boundary in New Zealand. *Science*, v. **171**, p. 276–279.

Kennett, J.P. and Watkins, N.D., 1974: Late Miocene–early Pliocene paleomagnetic stratigraphy, paleoclimatology, and biostratigraphy in New Zealand. *Geological Society of America, Bulletin*, v. **85**, p. 1385–1398.

Loutit, T.S. and Kennett, J.P., 1979: Application of Carbon Isotope stratigraphy to late Miocene shallow marine sediments, New Zealand. *Science,* **204**: p. 1196–1199.

Loutit, T.S. and Kennett, J.P., 1981: Australasian sedimentary cycles, global sea level changes and the deep sea sedimentary record. *AAPG. Bulletin*, v. **65**, p. 1586–1601.

Mankinen, E.A. and Dalrymple, G.B., 1979: Revised geomagnetic polarity time scale for the interval 0–5 m.y. B.P. *Journal of Geophysical Research*, v. **84**, p. 615–626.

Suggate, R.P., Stevens, G.R. and Te Punga, M.T., eds., 1978: *The Geology of New Zealand.* Government Printer, Wellington, 2 vols., 820 p.

Vella, P. 1963: Plio-Pleistocene cyclothems, Wairarapa, New Zealand. *Royal Society of New Zealand, Transactions, Geology*, v. **2**, p. 15–50, 12 figs.

Walcott, R.I., Christoffel, D.A. and Mumme, T.C., 1981: Bending within the axial tectonic belt of New Zealand in the last 9 My from paleomagnetic data. *Earth and Planetary Science Letters*, v. **52**, p. 427–34.

III-4

Neogene Datum Planes: Foraminiferal Successions in Australia with Reference Sections from the Ninetyeast Ridge and the Ontong-Java Plateau

Robert S. HEATH* and Brian McGOWRAN**

Introduction

The Australian Marine Neogene is treated for present purposes as a two-part record:

(i) Trpopical sections on the western and northeastern continental margins encountered during offshore petroleum exploration. The northern margin of the continent is, of course, the tectonically active region of Papua New Guinea and West Irian. That region is not included here.

(ii) Extratropical sections present along the southern margin. Again, the succession is found to be most complete in offshore petroleum exploration, although the foundations of foraminiferal biostratigraphy were developed in a skillful integration of onshore sections.

This paper cannot summarize the numerous, scattered outcrops and subsurface sections around the long and tectonically stable coastline of Australia. Instead, and in full awareness that a composite presentation prevents the interested reader from analyzing data in ways that he or she may deem important, we prepared the enclosed charts (Tables 1–4) around three objectives and used the biostratigraphic/geochronologic scale of Berggren (1981):

(i) The comparison of the northwest margin with the southern margin is a comparison of two composite successions. The data for the former come from Heath and Apthorpe (1981) with reference to Wright (1973) and Chaproniere (1981b). Earlier studies are summarized by Quilty (1974).

The biostratigraphy of the southern margin has been discussed and summarized on several occasions (among others: Wade, 1964; Carter, 1964; Ludbrook and Lindsay, 1969; McGowran *et al.*, 1971). More recent studies drawn upon here include Lindsay (1973, 1976), Heath (1979), and Mallett (1978).

(ii) Studies by one of us (RSH) on the Northwest Shelf have been a test of his earlier investigations on oceanic sections on the Ninetyeast Ridge and the Ontong-Java Plateau (Heath, 1979). Thus, the succession of biostratigraphic events at Site 214 and the very similar succession at Site 289 (developed independently of Srinivasan and Kennett, 1981, and of Kennett and Srinivasan, 1981) are presented here as prime references for the Australian context.

(iii) The distribution of the "larger" foraminifera is considered to be meaningful in terms of "climatic" fluctuations, which by mutual correlations and by correlations with oxygen isotope curves can be shown to be bipolar (McGowran, 1979 b, c). The important recent studies are by Chaproniere (1980, 1981 a). Earlier records from the northeast margin are by Palmieri (1975), whose main topic was the planktonics (not compiled here).

For completeness, the following unpublished theses are cited: Chaproniere (1977), Mallett (1977), Heath (1979), Lindsay (1981).

McGowran (1979b) discussed aspects of stratigraphic sequence, of biogeographic and isotopic signals of "climate," and of tectonism with fairly rigorous attention to correlation and age determination. However, he did not outline the basis for correlation, i.e., the succession of foraminiferal events in the tropical and extratropical parts of the continent.

Oceanic Sections

Most events listed at DSDP Site 214 can be matched at DSDP Site 289. Particularly interesting in that regard are coiling changes in *Globorotalia, Neogloboquadrina,* and *Pulleniatina,* and abundance changes in *Globigerinoides, Sphaeroidinella,* and *Globoquadrina.* The sharp decrease in abundance of *Globoquadrina dehiscens* at top Zone N15 is followed by a top occurrence of *G.* cf. *dehiscens* near top Zone N17 (Tables, 1,2).

Perhaps the only noteworthy offset between the two sections in separate oceans is in the horizon of *Globigerina nepenthes,* which appears—on the constraints provided by other markers—somewhat later at Site 289.

By no means all the defining events of the N zones are observed or recorded. That is in keeping with numerous recent observations, but it is no reason for modifying

* *Woodside Offshore Petroleum Pty. Ltd., Box D188, Perth 6001, Australia*
** *Department of Geology, The University of Adelaide, Adelaide 5001, Australia*

Table 1. Selected early Neogene planktonic foraminiferal events, Ninetyeast Ridge and Ontong Java Plateau, demonstrating close parallelism between Indian and Pacific Oceans.

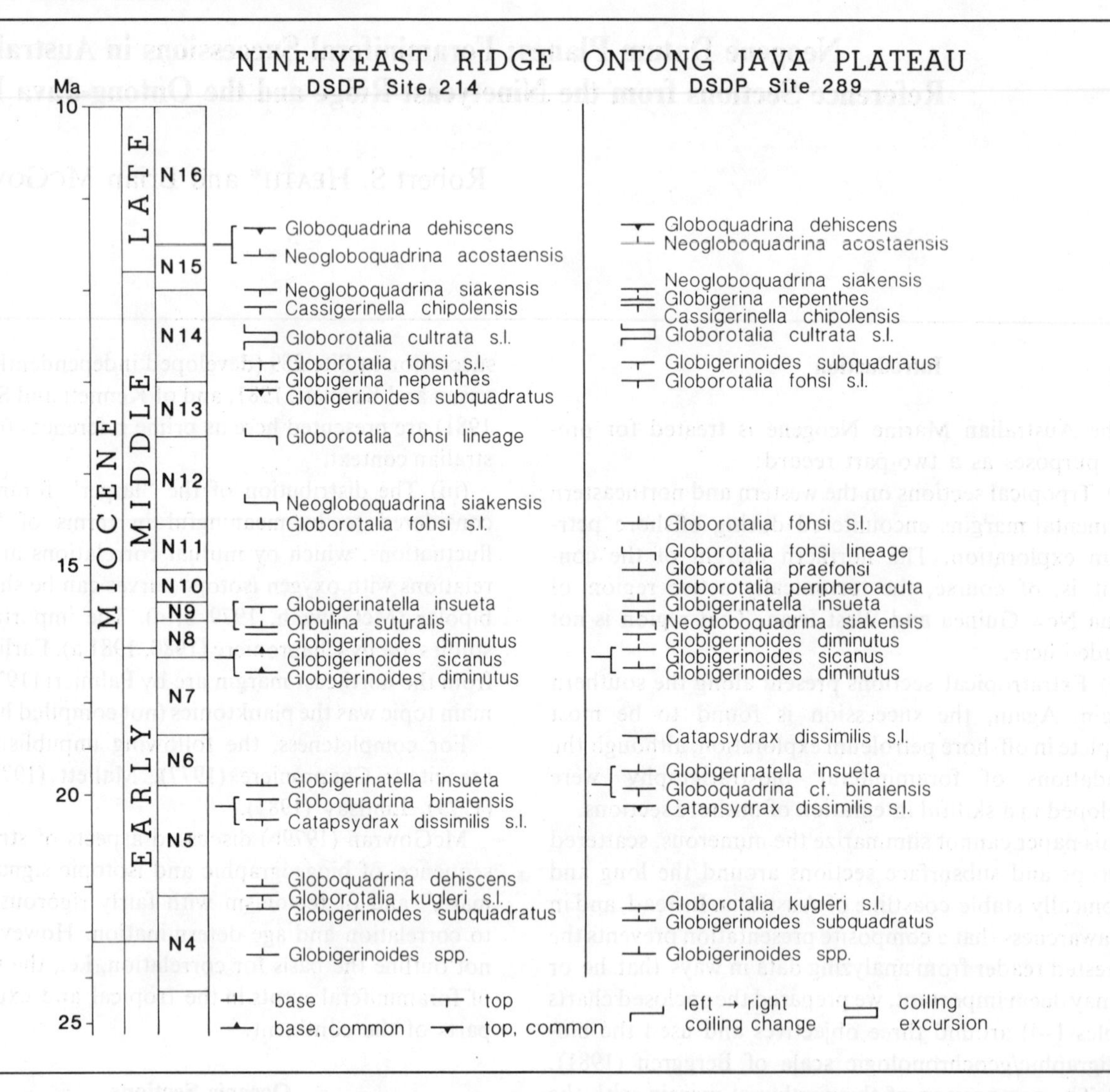

drastically the scheme or even erecting an alternative. On the other hand, Zones N17, N19, and N22 can be subdivided in a way that may be useful, since the events concerned are in the same place in the succession in the two oceans.

Continental Margins

Predictably enough, the succession compiled for northwest Australia is very similar to the oceanic sections at low latitudes (Table 3).

We draw attention to biostratigraphic events in the vicinity of the Oligocene/Miocene boundary. The first appearance of *Globigerinoides subquadratus* consistently precedes the last appearance of *Globorotalia kugleri. Globoquadrina dehiscens,* however, appears below the disappearance of *G. kugleri,* as in southern Australia, but comes in just above that event at Site 214.

In southern Australia the sequence is, of course, somewhat different for climatic/biogeographic reasons. However, we see no evidence of "diachronism," although there can be problems from restricted ranges. Thus, Mallett (1978) takes the first appearance of *Globorotalia plesiotumida* to identify base Zone N17, forcing the (earlier) first appearance of *G. conomiozea* back into Zone N16. We suggest that *G. conomiozea* marks a level in Zone N17, as elsewhere (e.g., Berggren, 1981) and that *G. plesiotumida* records a warming just before the latest Miocene glaciation.

"Larger" Foraminifera

The correlation of the Indo-Pacific letter "Stages" (i.e. zones) with the N-Zones (Tables 3, 4) is from Adams (1983

Table 2. Selected late Neogene planktonic foraminiferal events, Ninety East Ridge and Ontong Java Plateau, demonstrating close parallelism between Indian and Pacific Oceans.

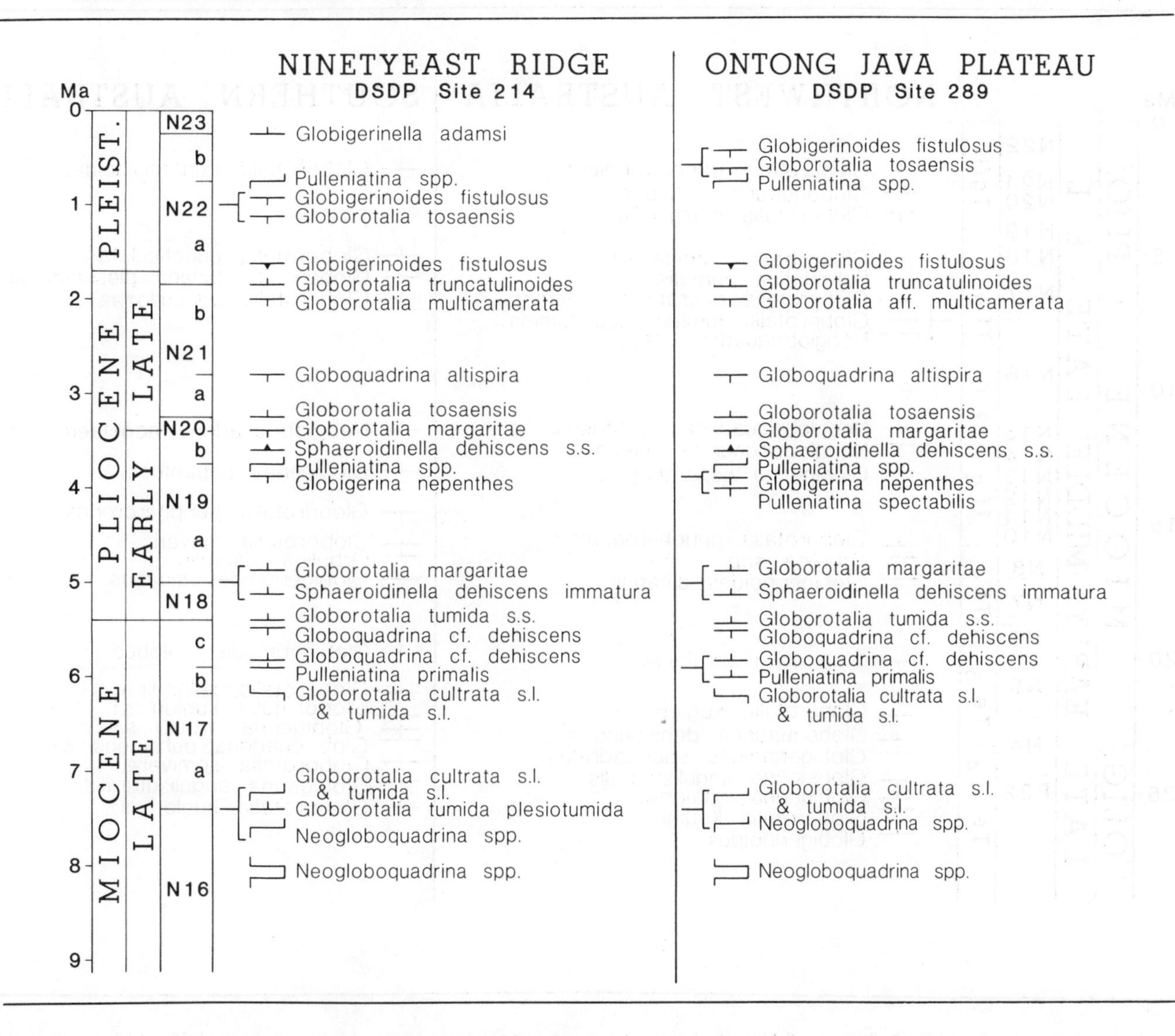

and pers. comm.). Larger foraminiferal assemblages are included here as evidence of climatically stimulated extratropical excursions (McGowran, 1979 a, 1979 b, 1979 c) which have biostratigraphic significance additional to their more traditional meaning based on ranges and assemblages.

LF1 to LF8 in Table 4 are larger foraminiferal associations recognized with nominate species by Chaproniere (1981) whose Table 1 lists the complete assemblages. In the other columns we list characteristic species.

It is noteworthy that *Eulepidina* occurred in northwest Australia, to at least 30° S palaeolatitude, until the time the genus went extinct. Again, *Austrotrillina howchini* occurred in LF7 and LF8 and also extended around the margin to southern Australia as a final flourish before going extinct. There is no evidence of cross-latitudinal "diachronous" extinction in these tropical-type genera.

References

Adams, C.G., 1983: Speciation, phylogenesis, tectonism, climate and eustasy: factors in the evolustion of Cenozoic larger foraminiferal bioprovinces. *In* Sims, R.W., Price, J.H. and Whalley, P.E.S., eds., *Evolution, Time and Space: The Emergence of the Biosphere,* Systematics Association Special Volume, no. 23, p. 255–289.

Berggren, W.A., 1981: Correlation of Atlantic, Mediterranean and Indo-Pacific Neogene stratigraphies: geochronology and chronostratigraphy. *IGCP-114 Internat. Workshop Pac. Neog. Stratigr., Osaka, Proc.*, p. 29–60.

Carter, A.N., 1964: Tertiary foraminifera from Gippsland, Victoria and their stratigraphical significance. *Geol. Surv. Vic., Mem.*, v. **23**, p. 1–154.

Chaproniere, G.H.C., 1977: Studies on Foraminiferida from Oligo-Miocene sediments, north-west Western Australia. Unpubl. Ph.D. thesis, Univ. Western Australia, 486 p.

Table 3. Selected Neogene planktonic foraminiferal events on the Australian continental margins: Northwest (more or less tropical) and southern (extratropical); both successions are composite.

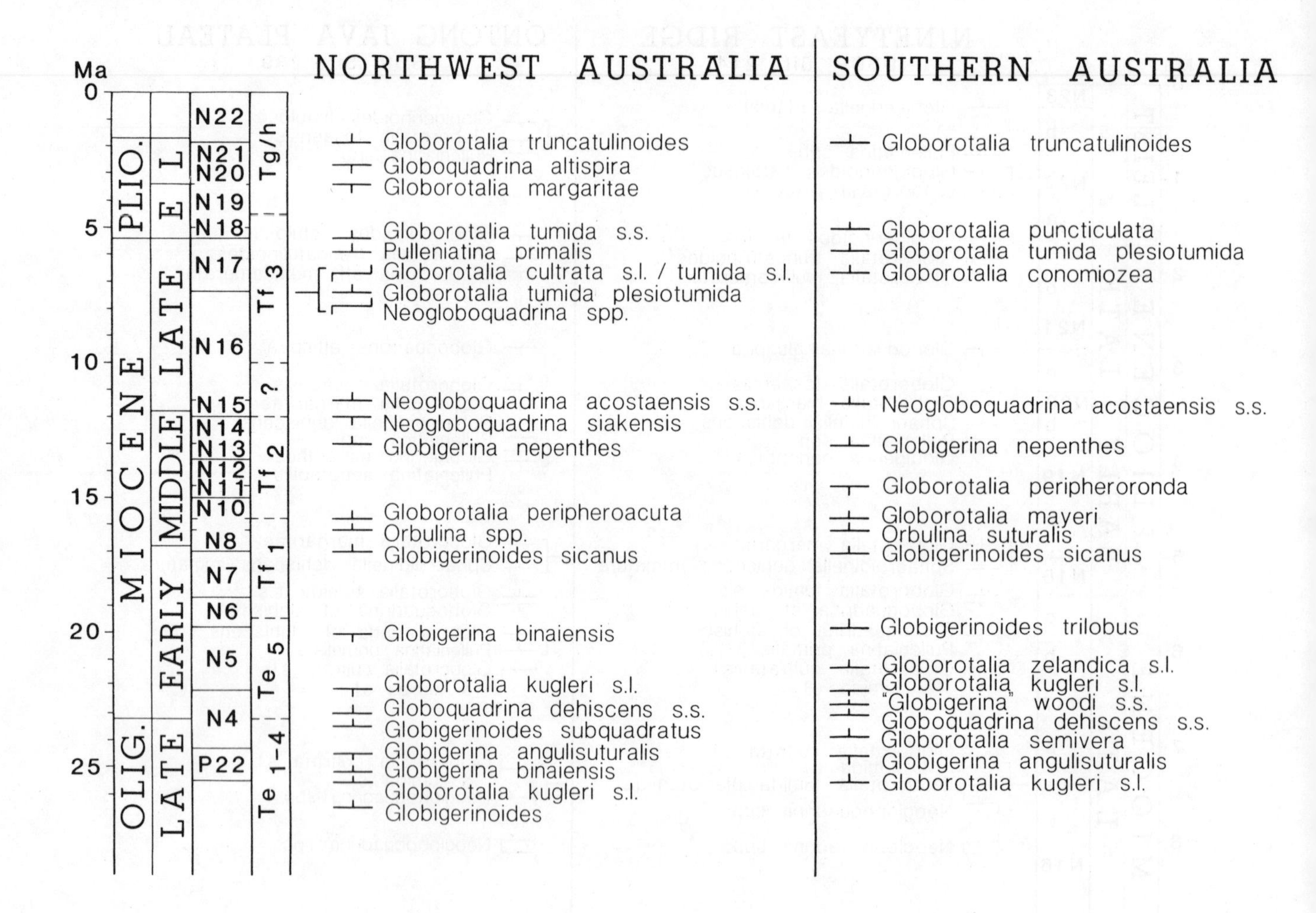

Chaproniere, G.H.C., 1980: Influence of plate tectonics on the distribution of late Palaeogene to early Neogene larger foraminiferids in the Australasian region. *Palaeogeogr., Palaeoclimatol., Palaeoecol.*, v. **31**, p. 299–317.

Chaproniere, G.H.C., 1981a: Australasian mid-Tertiary larger foraminiferal associations and their bearing on the East Indian Letter Classification. *B.M.R. Jour. Aust. Geol. Geophys.*, 6, p. 145–151.

Chaproniere, G.H.C., 1981b: Late Oligocene to early Miocene planktic Foraminiferida from Ashmore Reef No. 1 Well, northwest Australia. *Alcheringa*, 5, p. 103–131.

Heath, R.S., 1979: Neogene planktonic foraminifera: studies on Indo-Pacific oceanic sections. Unpubl. Ph.D. thesis, University of Adelaide, 185 p.

Heath, R.S. and Apthorpe, M., 1981: Tertiary foraminiferal biostratigraphy of the North West Shelf, Western Australia. *5th Aust. Geol. Conv., Abstr.*, p. 68.

Kennett, J.P. and Srinivasan, M.S., 1981: Neogene equatorial to subantarctic planktonic foraminiferal biostratigraphy and datum levels: South Pacific. *IGCP-114 Internatl. Workshop Pac. Neog. Biostrat., Osaka, Proc.*, p. 73–90.

Lindsay, J.M., 1973: Oligocene in South Australia. *45th ANZAAS Congress, Perth, Sect. 3, Abstracts.* p. 105–106.

Lindsay, J.M., 1976: Tertiary history of South Australia–the foraminiferal record. *25th Int. Geol. Congr., Sydney, Abstracts.* p. 329–330.

Lindsay, J.M., 1981: Tertiary stratigraphy and foraminifera of the Adelaide City area, St. Vincent Basin, South Australia. Unpubl. M.Sc. thesis, University of Adelaide, 684 p.

Ludbrook, N.H. and Lindsay, J.M., 1969: Tertiary foraminiferal zones in South Australia. *In*, Bronnimann, P. and Renz, H.H. eds., *1st Internat. Conf. Planktonic Microfossils*, Geneva 1967, *Proc.*, E.J. Brill, Leiden, 2, p. 366–374.

Mallett, C.W., 1977: Studies in Victorian Tertiary forami-

Table 4. Neogene larger foraminiferal assemblages on the Australian continental margins correlated with Indo-Pacific letter "Stages" and with planktonic foraminiferal N zones. Note scale changes.

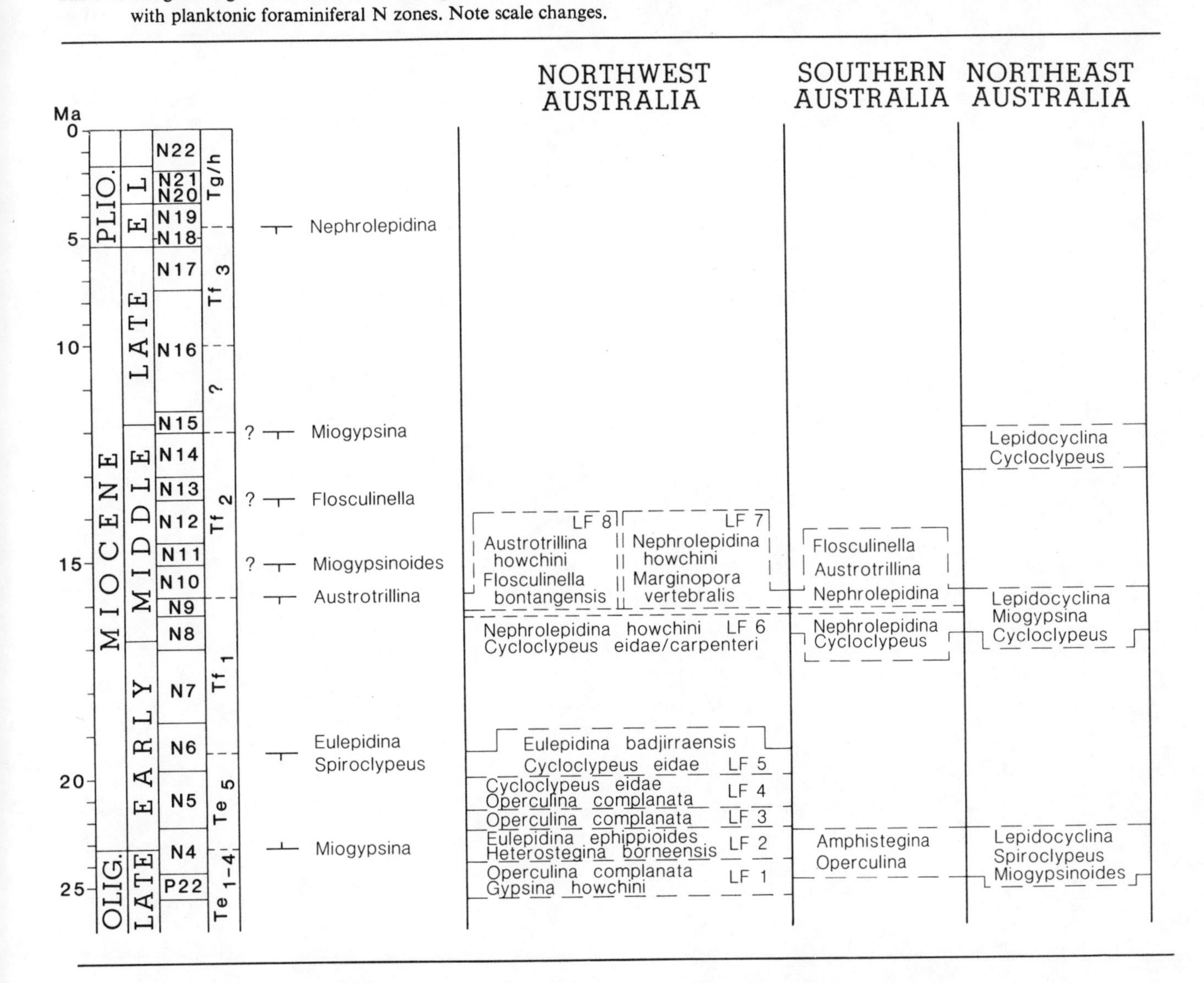

nifera, Neogene planktonic faunas. Unpubl. Ph.D. thesis, University of Melbourne.

Mallett, C.W., 1978: Sea level changes in the Neogene of southern Victoria. *The APEA Journal*, p. 64–69.

McGowran, B., 1979a: Australian Neogene sequences and events. *IGCP-114 2nd Working Group Meeting, Bandung, Proc.*, p. 165–167.

McGowran, B., 1979b: The Tertiary of Australia: foraminiferal overview. *Mar. Micropal.*, v. **4**, p. 235–264.

McGowran, B., 1979c: Some Miocene configurations, from an Australian standpoint. *Ann. Geol. Pays Hellen., Tome hors series*, fasc. II, p. 767–779.

McGowran, B., Lindsay, J.M. and Harris, W.K., 1971: Attempted reconciliation of Tertiary biostratigraphic systems, Otway Basin. *In* Wopfner and Douglas, J.G., eds., *The Otway Basin in Southeastern Australia. Geol. Surv. S. Aust. Vict., Spec. Bull.*, p. 273–281.

Palmieri, V., 1975: Planktonic Foraminiferida from the Capricorn Basin, Queensland. *Geol. Surv. Queensland, Publ.* 362, p. 1–47.

Quilty, P.G., 1974: Tertiary stratigraphy of Western Australia. *Geol. Soc. Aust., Jour.*, p. 301–318.

Srinivasan, M.S. and Kennett, J.P., 1981: A review of Neogene planktonic foraminiferal biostratigraphy: applications in the equatorial and South Pacific. *Soc. Econ. Paleont. Mineral., Spec. Publ.*, 32, p. 395–432.

Wade, M., 1964: Application of the lineage concept to biostratigraphic zoning based on planktonic foraminifera. *Micropaleontology*, v. **10**, p. 273–290.

Wright, C.A., 1977: Distribution of Cainozoic foraminifera in the Scott Reef No. 1 Well, Western Australia. *Geol. Soc. Aust., Jour.*, v. **24**, p. 24, p. 269–277.

III-5

Biostratigraphy of Selected Neogene Sequences in Indonesia

Darwin KADAR* and Soemoenar SOEKA**

Northwest Java Basin

The subsurface Neogene sequence in the Northwest Java Basin comprises the Cibulakan, Parigi, and Cisubah Formations (Fig. 3). The Cibulakan Formation unconformably overlies the volcanic Jatibarang Formation which ranges in age from late Eocene to early Oligocene (Arpandi and Padmosukismo, 1975). The Cibulakan Formation consists of a lower part with shale, siltstone and intercalations of coal, a middle limestone with interbedded shale and marl, and an upper part of claystone and shale interbedded with sandstone. The overlaying Parigi Formation consists of fossiliferous limestone with occasional dolomitic limestone, rich in larger foraminifera. The Cisubah Formation, which conformably overlies the Parigi Formation, consists mainly of clay with sandstone intercalations. Planktonic foraminifera are common in the lower part and decrease upwards. The planktonic foraminiferal biostratigraphy of this basin was established by Harsono *et al.* (1978), and correlation with the subsurface units is based on Kadar (1979).

The sedimentary sequence of the Northwest Java Basin ranges from N4 to N19 in terms of Blow's (1969) zonal scheme. Nine planktonic foraminiferal datum planes have been recognized in the time interval represented by the Cibulakan to Cisubah Formations (Fig. 3). These are the first appearance datums of 1. *Globigerinoides primordius*, 2. *Praeorbulina sicana*, 3. *Orbulina suturalis*, 4. *Neogloboquadrina acostaensis*, 5. *Globorotalia plesiotumida*, 6. *Globorotalia margaritae* and the last occurrence datums of 7. *Globigerinoides subquadratus*, 8. *Globorotalia siakensis*, and 9. *Globoquadrina altispira*. In addition, Harsono *et al.* have recognized the top datums of *Spiroclypeus* and *Miogypsina* in the lower and upper part of the Cibulakan Formation below the *P. sicana* and the *G. subquadratus* datums. The top datum of *Lepidocyclina* occurs in the Parigi Formation below the *Globorotalia plesiotumida* datum.

West Progo Mountains, South Central Java

The Neogene sequences of the West Progo Mountains comprise the Jonggrangan and Sentolo Formations (Fig. 4). These units transgressively overlie the Oligocene volcanic rocks of the Old Andesite Formation of van Bemelen (1949). This volcanic unit is considered to form the basement of the Miocene South Central Java Basin. The Jonggrangan Formation consists predominantly of reef limestone and marly tuff breccia as well as tuff sandstone with lignite seams in its lower part. It yields molluscs and larger foraminifers but lacks planktonic foraminifera. The Sentolo Formation crops out to the east and southeast of the West Progo Mountains and exhibits a homoclinal structure which gently dips southeastwards. In places, the sediments are folded and form small anticlines trending in an ENE–WSW direction. The Sentolo Formation, about 1,000m thick, consists in the lower part of the Karanganyar, Genung, and Tanjunggunung Members in stratigraphical ascending order. The Karanganyar Member consists of tuff, tuffaceous sandstone, and marl with an intercalation of slump block. The Genung Member is composed of vitreous tuff, marl, and tuff breccia, whereas the Tanjunggunung Member is predominantly marl with thin intercalations of tuff and limestone. The upper part of the Sentlo Formation consists of interfingering marl and limestone. This formation was assigned to the Burdigalian-Pliocene by Harsono (1968) on planktonic foraminifera. The planktonic foraminiferal biostratigraphy of this unit was also studied by Kadar (1981) who confirmed its age as Early Miocene to Late Pliocene.

The sedimentary sequence of the West Progo Mountains yields an almost continuous section covering planktonic foraminiferal zones from the upper part of N5 to N21. However, as a hiatus representing Zones

* *Paleontology Laboratory, Geological Research and Development Centre, Diponegoro no. 57, Bandung, Indonesia*
** *Stratigraphy Laboratory, Indonesian Petroleum Institute, Lemigas, Jakarta, Indonesia*

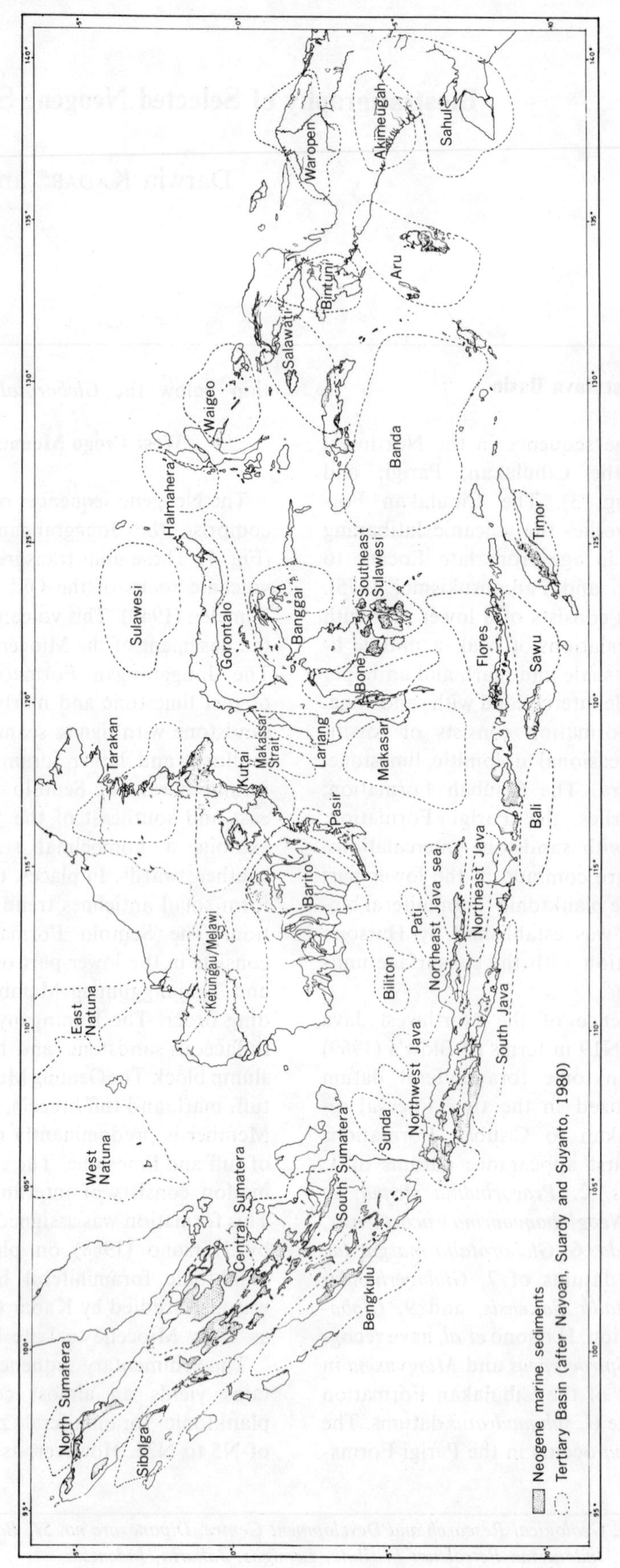

Fig. 1. Exposures of Neogene marine sediments in Indonesia and locations of the Tertiary sedimentary basins.

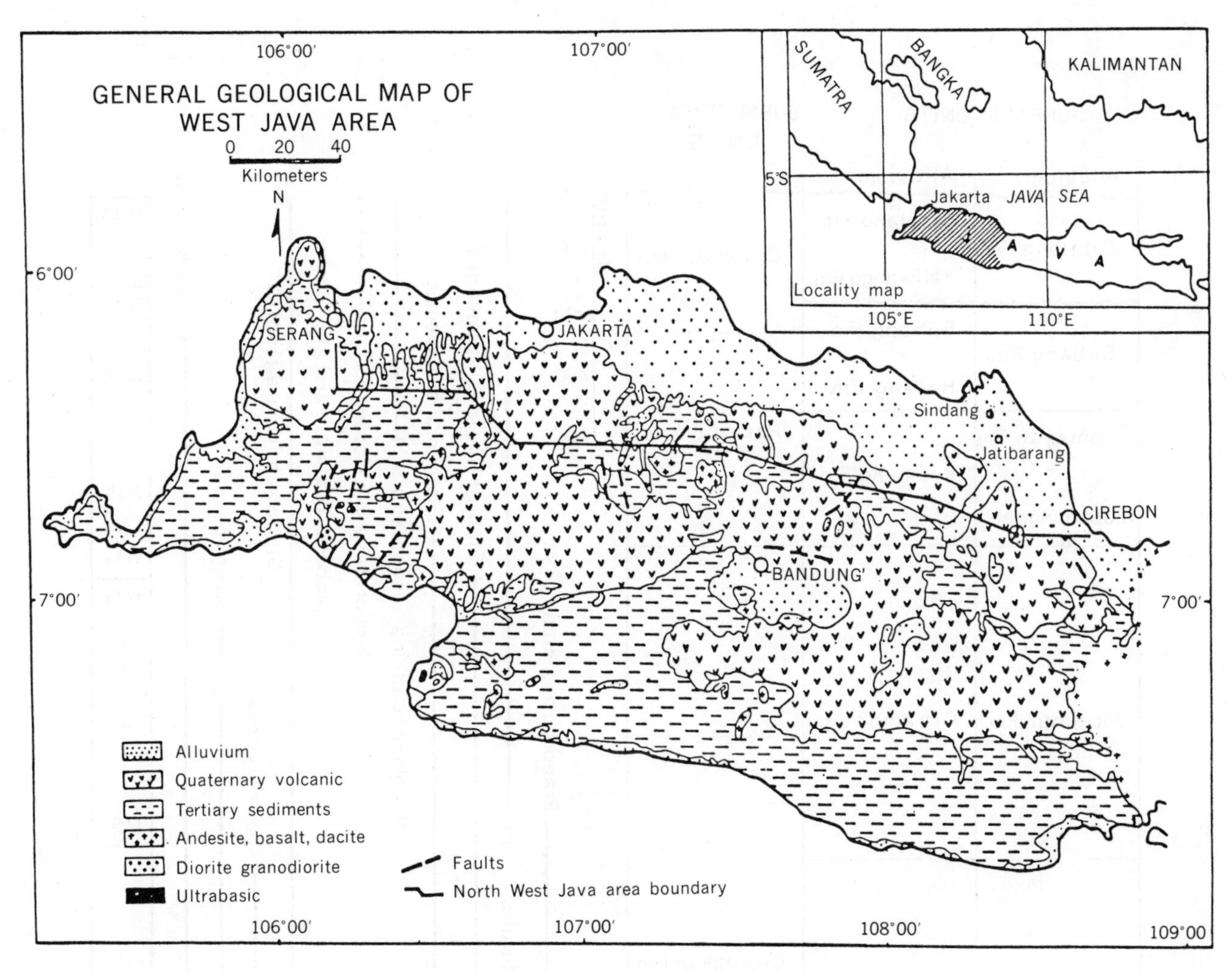

Fig. 2. General geological map of West Java area.

N14–N16 occurs between the Tanjunggunung Member and the upper part of the Sentolo Formation, twelve datum planes have been recongnized in the Sentolo Formation (Fig. 5), namely, the first appearance datums of 1. *Globigerinatella insueta,* 2. *Praeorbulina sicana,* 3. *Orbulina suturalis,* 4. *Globorotalia peripheroacuta,* 5. *Globorotalia praefohsi,* 6. *Globorotalia fohsi,* 7. *Sphaeroidinellopsis subdehiscens subdehiscens* 8. *Globorotalia plesiotumida,* 9. *Pulleniatina primalis,* 10. *Globorotalia tumida,* 11. *Sphaeroidinella dehiscens,* and 12. *Globorotalia tosaensis.*

Cepu Area, East Java

The Neogene sequence of Cepu area comprises the OK (orbitoidal limestone), Ngrayong, Wonocolo, Ledok, Mundu, and Tambakromo Formations (Fig. 6). These units crop out in an anticlinoria and synclinoria, the axes of which generally trend in NW–SE and E–W directions. The OK Formation, more than 1,000 meters thick, unconformably underlies the Ngrayong Formation. It is composed of clay, shale, marl, and orbitoidal limestone and sandstone intercalations. The Ngrayong Formation, predominantly sandstone, is conformably overlain by the Wonocolo Formation. Its thickness is about 200 meters. The Wonocolo Formation, more than 500 meters thick, consists of *Globigerina* marl with occassional glauconitic sandstone interbeds. The Ledok Formation, conformably underlain by the Wonocolo Formation, consists of glauconitic sandstone with alternations of platy limestone and sandstone. Its thickness is about 200 meters. The Mundu Formation, about 200 meters thick, consists of *Globigerina* marl. It is conformably underlain by the Ledok Formation and is in turn overlain by the Tambakromo Formation, which consists predominantly of shallow-marine clay. The OK, Ngrayong, Wonocolo, and Ledok Formations are assigned to the Miocene, whereas the Mundu and Tambakromo are respectively regarded as Pliocene and Pleistocene (Udin, 1972).

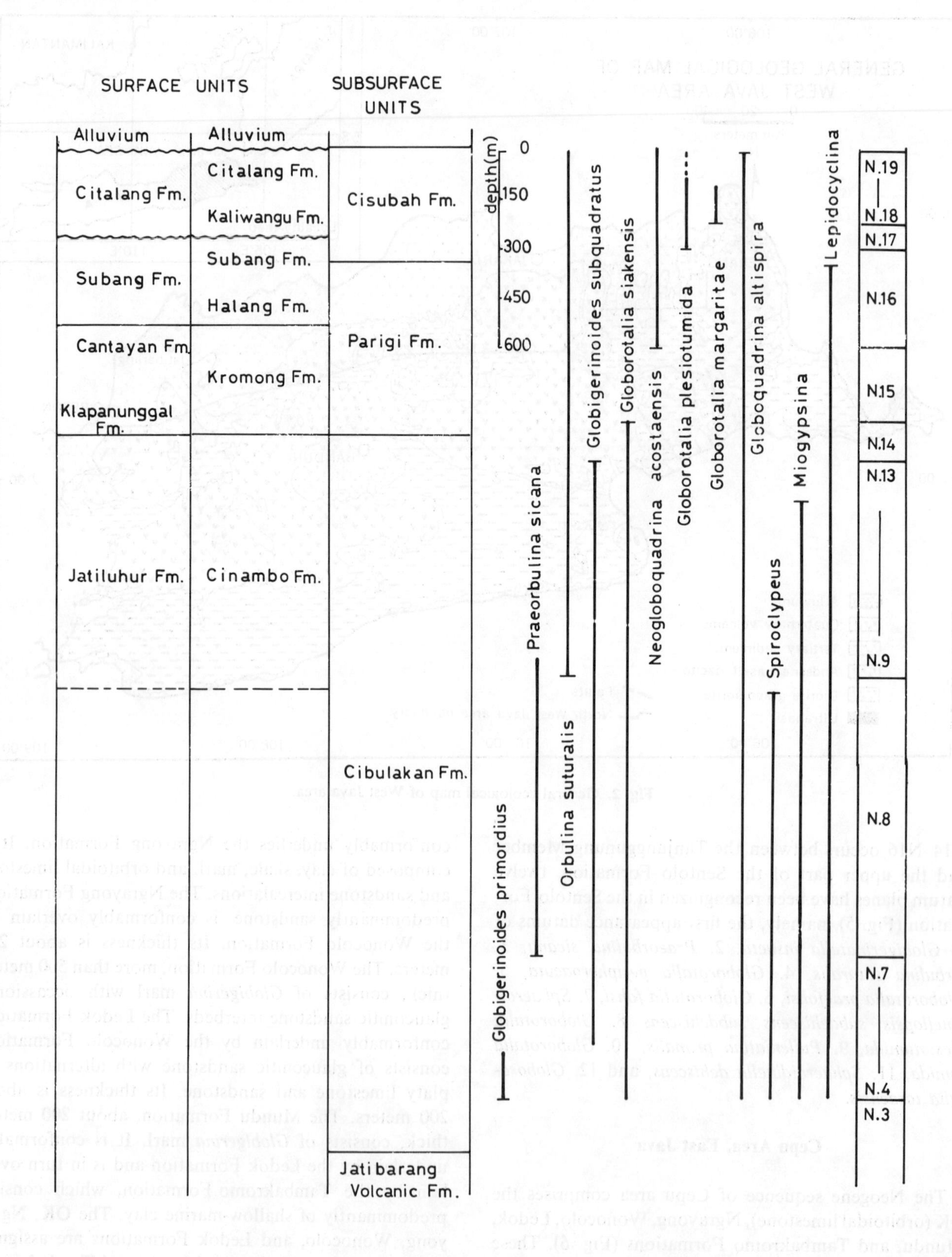

Fig. 3. Litho- and biostratigraphy of the Northwest Java Basin.

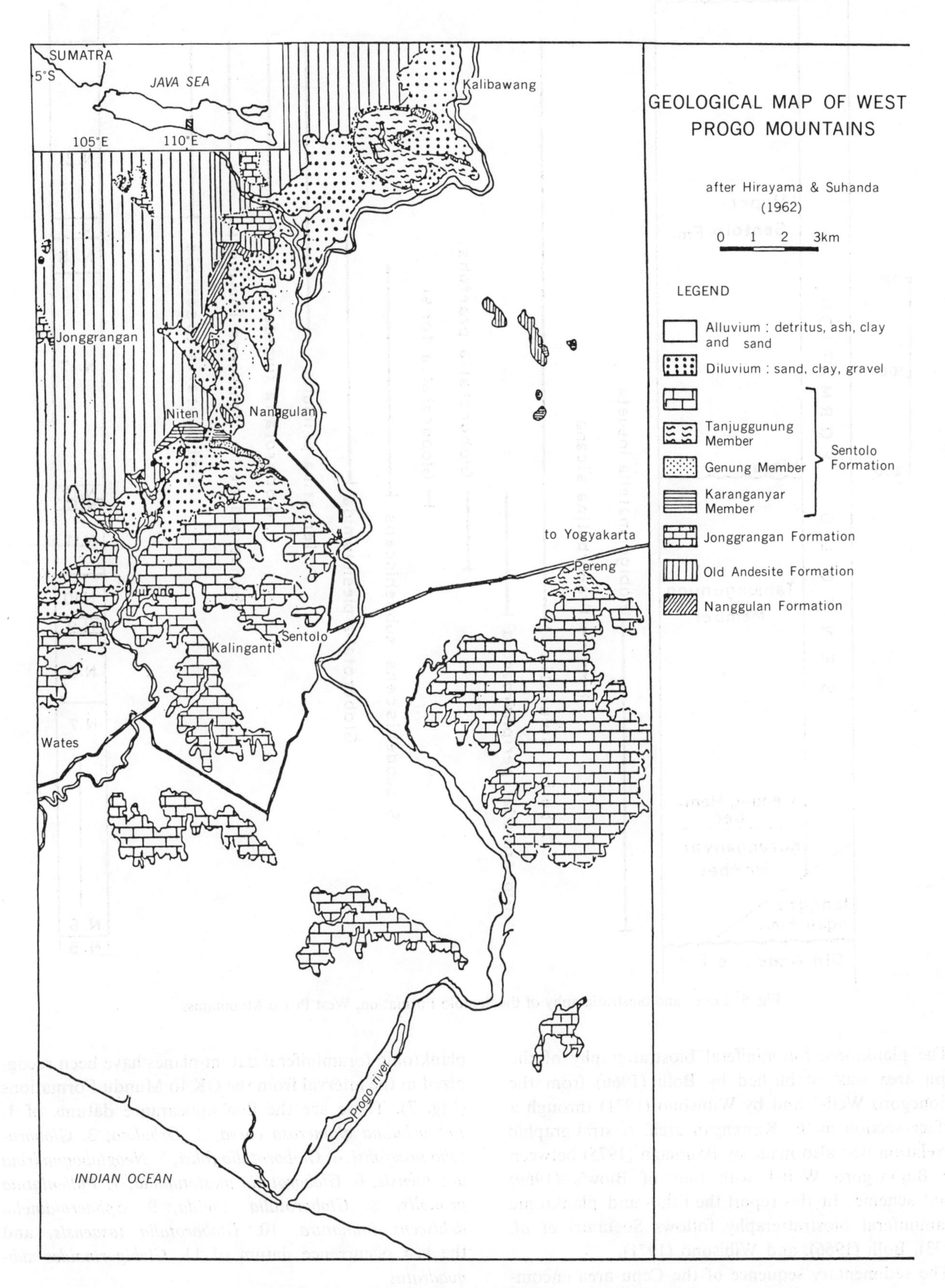

Fig. 4. Geological map of the West Progo Mountains, after Hirayama and Suhanda (1962).

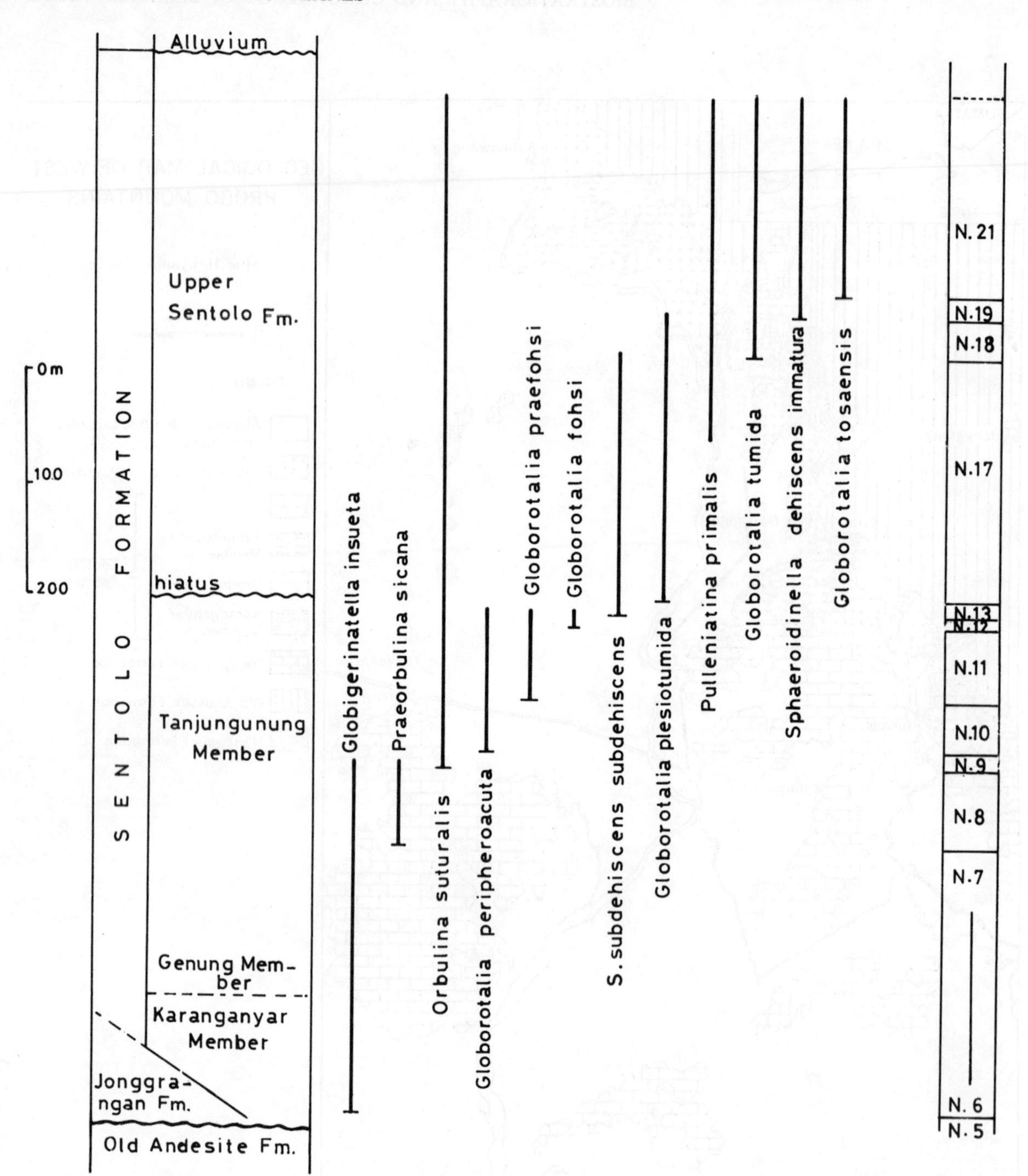

Fig. 5. Litho- and biostratigraphy of the Sentolo Formation, West Progo Mountains.

The planktonic foraminiferal biostratigraphy of the Cepu area was established by Bolli (1966) from the Bojonegoro Well-1 and by Wibisono (1971) through a surface section in the Kawengan area. A stratigraphic correlation was also made by Baumann (1975) between the Bojonegoro Well-1 with that of Blow's (1969) zonal scheme. In this report the litho- and planktonic foraminiferal biostratigraphy follows Soetantri *et al.* (1973), Bolli (1966), and Wibisono (1971).

The sedimentary sequence of the Cepu area encompasses a succession from Zone N8 to Zone N21. Eleven planktonic foraminiferal datum-planes have been recognized in the interval from the OK to Mundu Formations (Fig. 7). These are the first appearance datums of 1. *Praeorbulina glomerosa curva*, 2. *Orbulina*, 3. *Globorotalia praefohsi*, 4. *Globorotalia fohsi*, 5. *Neogloboquadrina acostaensis*, 6. *Globorotalia plesiotumida*, 7. *Pulleniatina primalis*, 8. *Globorotalia tumida*, 9. *Sphaeroidinella dehiscens immatura*, 10. *Globorotalia tosaensis*, and the last occurrence datum of 11. *Globigerinoides subquadratus*.

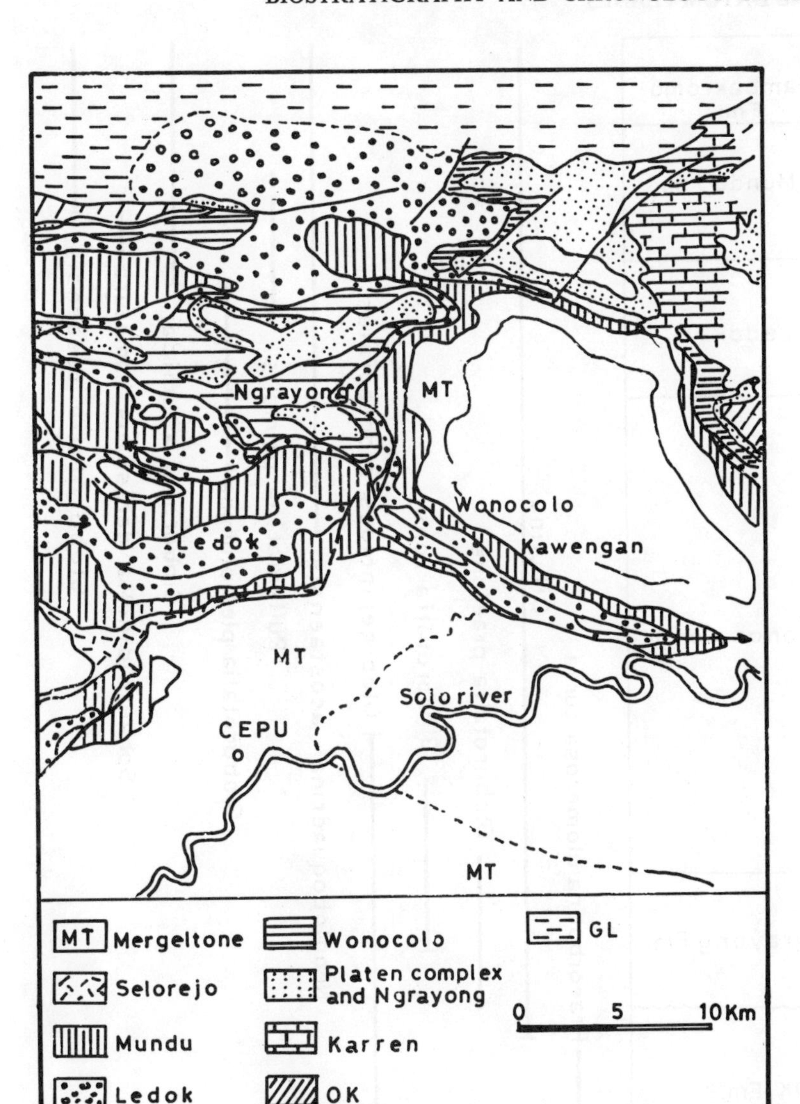

Fig. 6. Geological map of the Cepu area, after Soetantri *et al.* (1973).

Kutai Basin

The Kutai basin is located in the eastern part of Kalimantan and is fringed in the west by the Kuching High. It extends offshore and is bordered in the east by the trough of the Makassar Strait. Its northern boundary is the Mangkalihat Ridge, which is an east-southeast trending basement high. In the south, the Meratus High divides the basin into two parts known as the Barito and Pasir Basins (Fig. 8).

In the northern Kutai Basin and southern Tanjungmangkalihat area, the Late Oligocene is predominantly represented by shallow marine sediments redeposited as turbidites. This is followed by deeper marine sediments of Early Miocene age. The planktonic foraminiferal biostratigraphy of this part of the basin was estabilised by Djamas and Marks (1978) on drilled cutting samples and a surface section in the Keraitan area. The sedimentary sequence encompasses a succession from Zones N1 to N5. It is correlatable with an interval from Te 1 to Te 5 of Indo-Pacific letter stages. Five planktonic and six benthic larger foraminiferal datum levels are recognized (Fig. 9). These are the first appearance datums of *Globigerina angulisuturalis, Globigerinoides primordius, Globorotalia kugleri, Miogypsina kotoi, Spiroclypeus leupoldi, Heterostegina borneensis,* and the last occurrence of *G. kugleri, Globorotalia opima, H. borneensis,* and *Lepidocyclina papuaensis.*

In the offshore Kutai Basin, the Middle Miocene to Recent sediments consist prominently of shallow marine shelf to bathyal, deltaic sediments. Smaller benthic foraminifera are widely distributed and may eventually prove useful for correlation. Planktonic foraminifera and calcareous nannoplankton are absent from many parts of

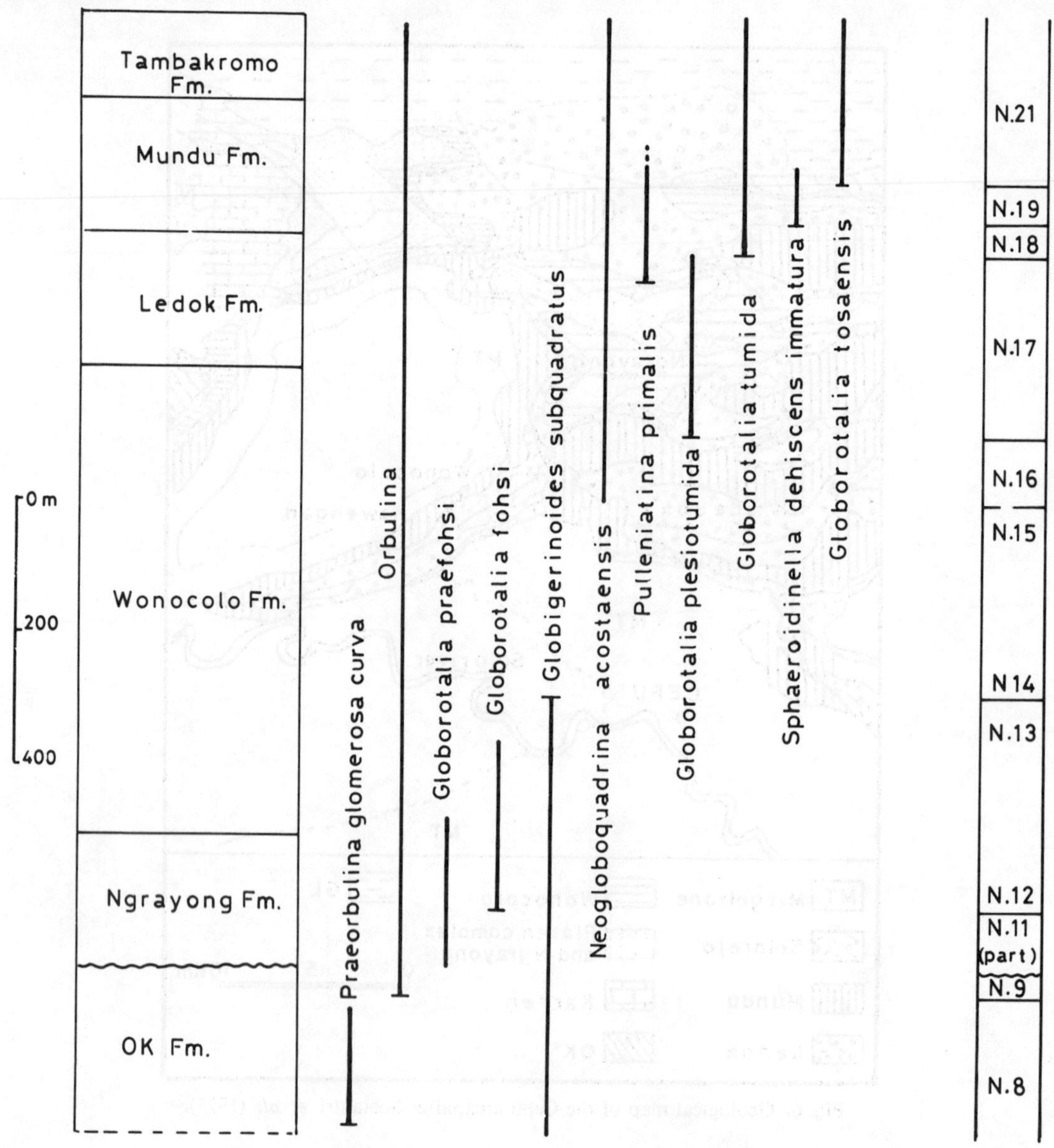

Fig. 7. Litho- and biostratigraphy of the Cepu area, East Java.

the section. However, they permit the smaller benthic foraminiferal zones established by Billman and Kartaadipura (1974), and later revised in nomenclature by Billman, Hottinger and Oesterle (1980), to be correlated with Blow's (1969) zonal scheme. The sedimentary sequence of this offshore area encompasses a succession from the *Asterorotalia yabei* to the *Calcarina* Zones of Billmann *et al.* (1980), an interval that is correlatable with Zones N13 to N23 of Blow. Five benthic foraminiferal datum levels are recognized. These are the last occurrence datums of *Asterorotalia yabei, Pseudorotalia catilliformis, Pseudorotalia globosa, Ammonia ikebei,* and *Ammonia pila* (Fig. 10).

References

Arpandi, D. and Padmosukismo, S., 1975: The Cibulakan Formation as one of the most prospective stratigraphic units in the Northwest Java basinal area. *Indonesian Petroleum Association Fourth Ann. Conv., Proc.*, v. **1**, p. 181–210.

Baumann, P., 1975: The Middle Miocene diastrophism, its influence to the sedimentary and faunal distribution of Java and the Java Sea basin. *Natl. Inst. Geology and Mining, Indonesia, Bull.*, v. **5**, no. 1, p. 13–28.

•Billman, H.G. and Kartaadipura, L.W. 1974: Late Tertiary biostratigraphic zonation, Kutai basin, offshore East Kalimantan, Indonesia. *Indonesian Petroleum Association Third Ann. Conv., Proc.*, p. 301–310.

Billman, H., Hottinger, L. and Oesterle, H., 1980: Neogene to Recent rotaliid foraminifera from the Indo-Pacific Ocean and their stratigraphic use. *Schweizerische Pal. Abh.*, v. **101**, p. 71–113.

Bolli, H.M. 1966: The planktonic foraminifera in well Bojonegoro-1 of East Java, *Eclogae Geol. Helv.*, v. **59**, no. 1, p. 449–465.

Djamas, Y.S. and Marks, E., 1978: Early Neogene fo-

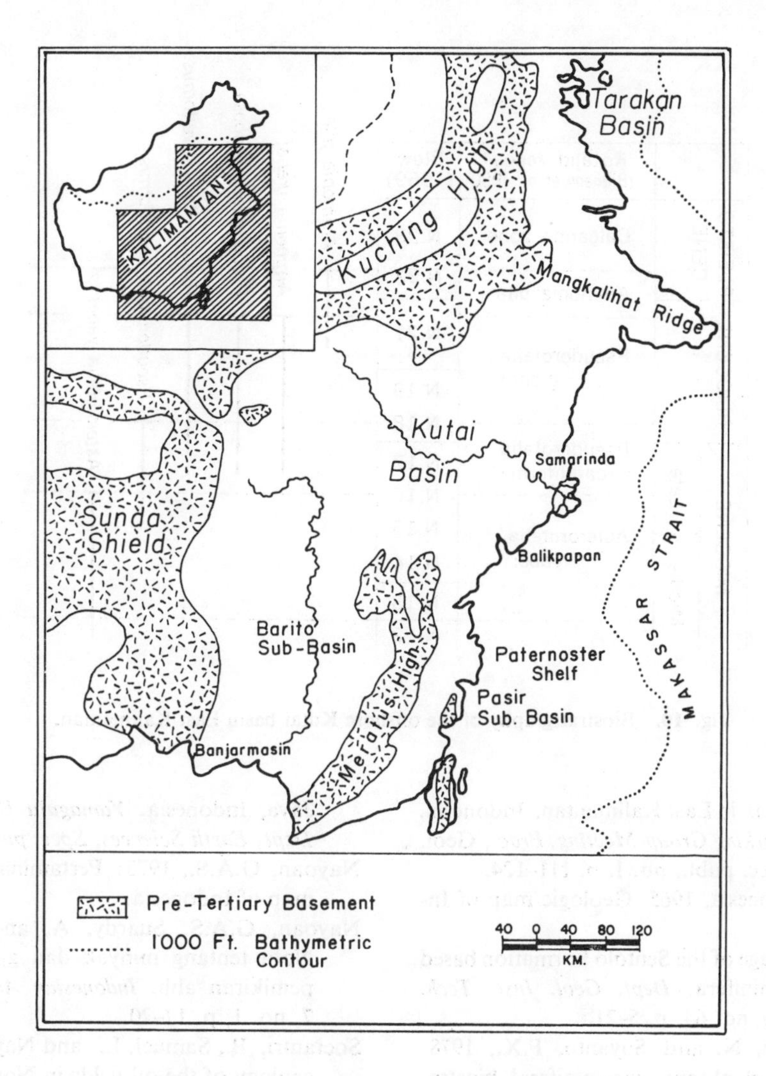

Fig. 8. Major structural elements of Kalimantan.

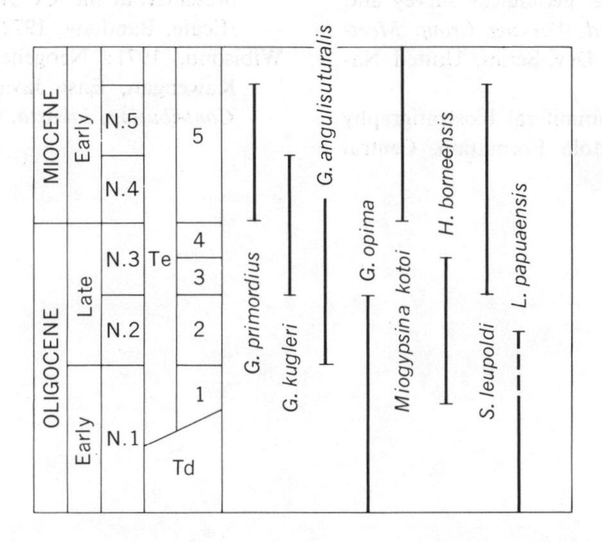

Fig. 9. Biostratigraphy of the northern Kutai basin and the Tanjungmangkalihat area.

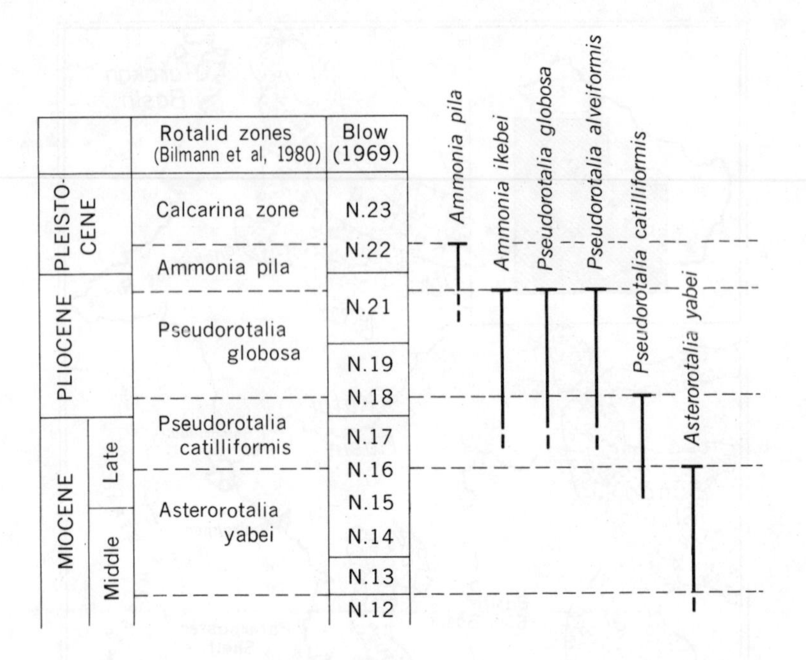

Fig. 10. Biostratigraphy of the offshore Kutai basin East Kalimantan.

raminiferal biohorizons in East Kalimantan, Indonesia. *IGCP 114 Second Working Group Meeting, Proc.*, Geol. Res. Dev. Center, Spec. publ., no. 1, p. 111–124.

Geological Survey of Indonesia, 1965: Geologic map of Indonesia 1:2,000,000.

Harsono, P., 1968: On the age of the Sentolo Formation based on planktonic foraminifera. *Dept. Geol. Inst. Tech. Bandung, Contribution*, no. 64, p. 5–21.

Harsono, P., Soeharsono, N. and Suyanto, F.X., 1978: Subsurface Neogene planktonic foraminiferal biostratigraphy of Northwest Java basin. *IGCP 114, Second Working Group Meeting, Proc.*, Geol. Res. Dev. Centre Spec. publ., no. 1, p. 125–136.

Kadar, D., 1979: Mapping by the geological survey and stratigraphic correlation. *Third Working Group Meeting, 1978, Proc.*, Mineral Res. Dev. Series, United Nations, Thailand, v. **6**, no. 45.

Kadar, D., 1981: Planktonic foraminiferal biostratigraphy of the Miocene-Pliocene Sentolo Formation, Central Java, Indonesia. *Yamagata University Fac. of Science Dept. Earth Sciences, Spec. publ.*, p. 35–47.

Nayoan, G.A.S., 1973: Pertamina preliminary oil and gas map of Indonesia.

Nayoan, G.A.S., Suardy, A. and Suyanto, F.X., 1980: Nota tentang minyak dan gas buim Indonesia suatu pemikiran ahli. *Indonesian Assoc. Geologists, Jour.* v. **7**, no. 1, p. 17–20.

Soetantri, B., Samuel, L., and Nayoan, G.A.S., 1973: The geology of the oil fields in Northeast Java. *Second Ann. Conv. Indonesian Petroleum Association, Proc.*, p. 149–175.

Udin, A.R., 1972: Lemigas stratigraphic studies, a paper presented at the CCOP 9th session /TAG 8 th session /Ecafe, Bandung, 1972.

Wibisono, 1971: Neogene planktonic foraminifera from Kawengan, East Java, Indonesia. *Lemigas Scientific Contribution, Jakarta*, v. **1**, no. 1, p. 1–69.

III-6

The Neogene of Andaman-Nicobar

M. S. Srinivasan

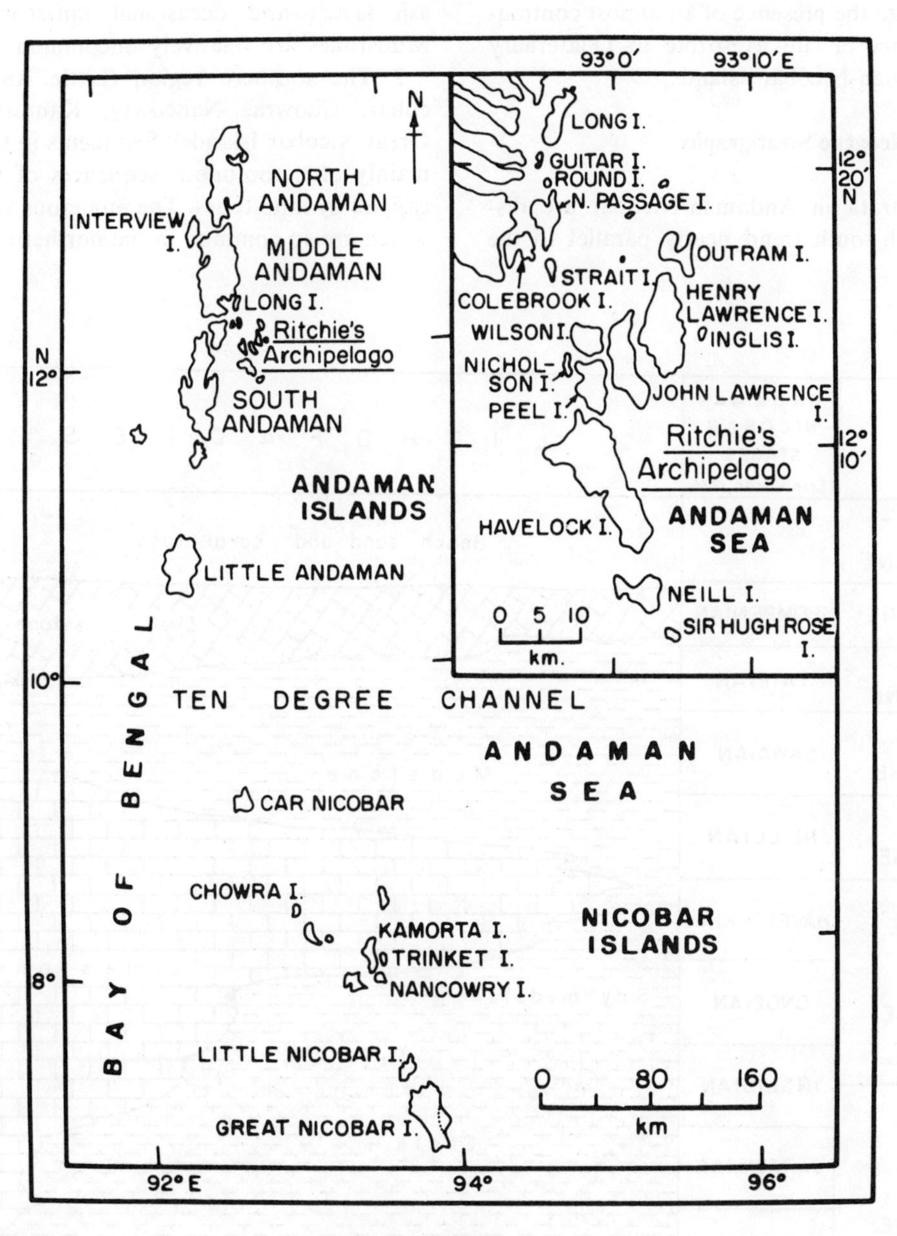

Fig. 1. Location map.

Department of Geology, Banaras Hindu University, Varanasi-221 005, India

Introduction

The Andaman-Nicobar group of islands on the eastern boundary of the Indian Plate in the northern Indian Ocean extends for about 850 km between latitude 6° 45′N and 13° 45′N (between longitude 92° 15′E and 94° 00′E) separating the Bay of Bengal from the Andaman Sea (Fig. 1). The islands are the subaerial expressions of a continuous ridge which connects the Arakan-Yoma Range of Western Burma to the festoon of islands south and west of Sumatra. Recent stratigraphic studies (Srinivasan, 1978, 1979; Srinivasan and Azmi, 1979) indicate the presence of an almost continuous marine sequence of late Mesozoic to Quaternary age on the Andaman-Nicobar Islands.

Neogene Stratigraphy

The Neogene strata in Andaman-Nicobar are distributed in a north-south trend nearly parallel to the axis of the islands and are referred to the Archipelago and Nicobar Series (Srinivasan, 1978). They are composed on the whole of sediments which were deposited under fluctuating shallow to deep-water conditions. The strata deposited in this basin varied greatly in thickness; nevertheless, thick continuous sequences are exposed on some islands. The Neogene strata of Andaman-Nicobar are broadly divisible into two regions:

1. The northern region (Ritchie's Archipelago, Round, Guitar, Long, Colebrook North Passage, Interview and Paget Islands). Here the sediments are mainly creamish-yellow chalk, including numerous volcanic ash layers and occasional limestone intercalations. Mudstones are relatively uncommon.

2. The southern region (Little Andaman, Car Nicobar, Chowra, Nancowry, Kamorta, Trinket and Great Nicobar Islands). Sediments in this region consist mainly of monotonous sequences of mudstones, often capped by limestones. The numerous volcanic ash layers which are so common in the northern region, especially

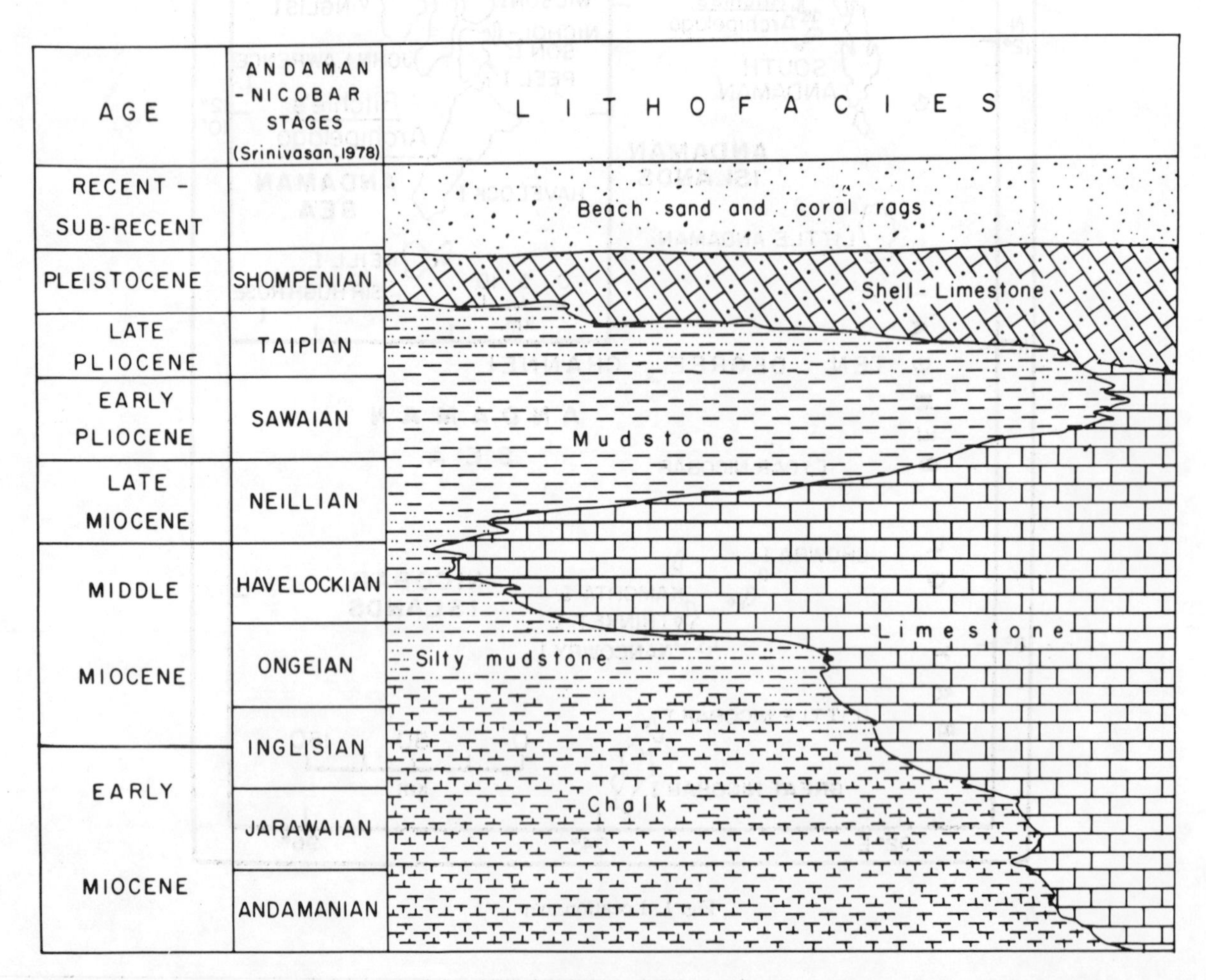

Fig. 2. Schematic diagram showing Neogene lithofacies variation in Andaman-Nicobar.

in Ritchie's Archipelago, are almost absent. The lithofacies variation through the Neogene is shown in Fig. 2.

Biostratigraphy

The sediments of the Archipelago (Miocene) and Nicobar (Pliocene-Pleistocene) Series are fairly rich in well-preserved foraminifera and other groups of microfossils, all of which are distinctly tropical Pacific in character. Investigations of a large number of overlapping sections exposed on the Andaman-Nicobar Islands have revealed the presence of an almost complete sequence of Neogene tropical planktonic foraminiferal zones (Srinivasan, 1977). Twenty-three biozones are recognized based on the known ranges of planktonic foraminifera. Excepting a few local breaks, the planktonic zones indicate the presence of an almost continuous sequence covering the interval from the early Miocene to Pleistocene (Fig. 3). The planktonic foraminifera of many sections have yet to be studied extensively from this region. Therefore, some of the zones, especially of the Pliocene, are tentative.

Major changes have been recorded in the Neogene planktonic foraminiferal assemblages, and reflecting these changes, nine chronostratigraphic divisions have been recognized (Srinivasan, 1978). In recent years, the datum concept, based on abrupt evolutionary appearance or last appearance of planktonic foraminifera, has been widely applied in correlating Neogene deep sea sections. Saito (1977) proposed 43 datums, most of them radiometrically dated, applicables for correlation of the Pacific Neogene marine sequences. Recently, Ikebe *et al.* (1981) listed 15 radiometrically dated important Neogene datum planes useful for regional correlation in the Western Pacific.

Stratigraphic ranges of planktonic foraminiferal species in the Andaman-Nicobar Neogene have allowed recognition of 25 datum levels useful for inter-basinal and inter-oceanic correlation. In the absence of radiometric dates and paleomagnetic records, the ages of

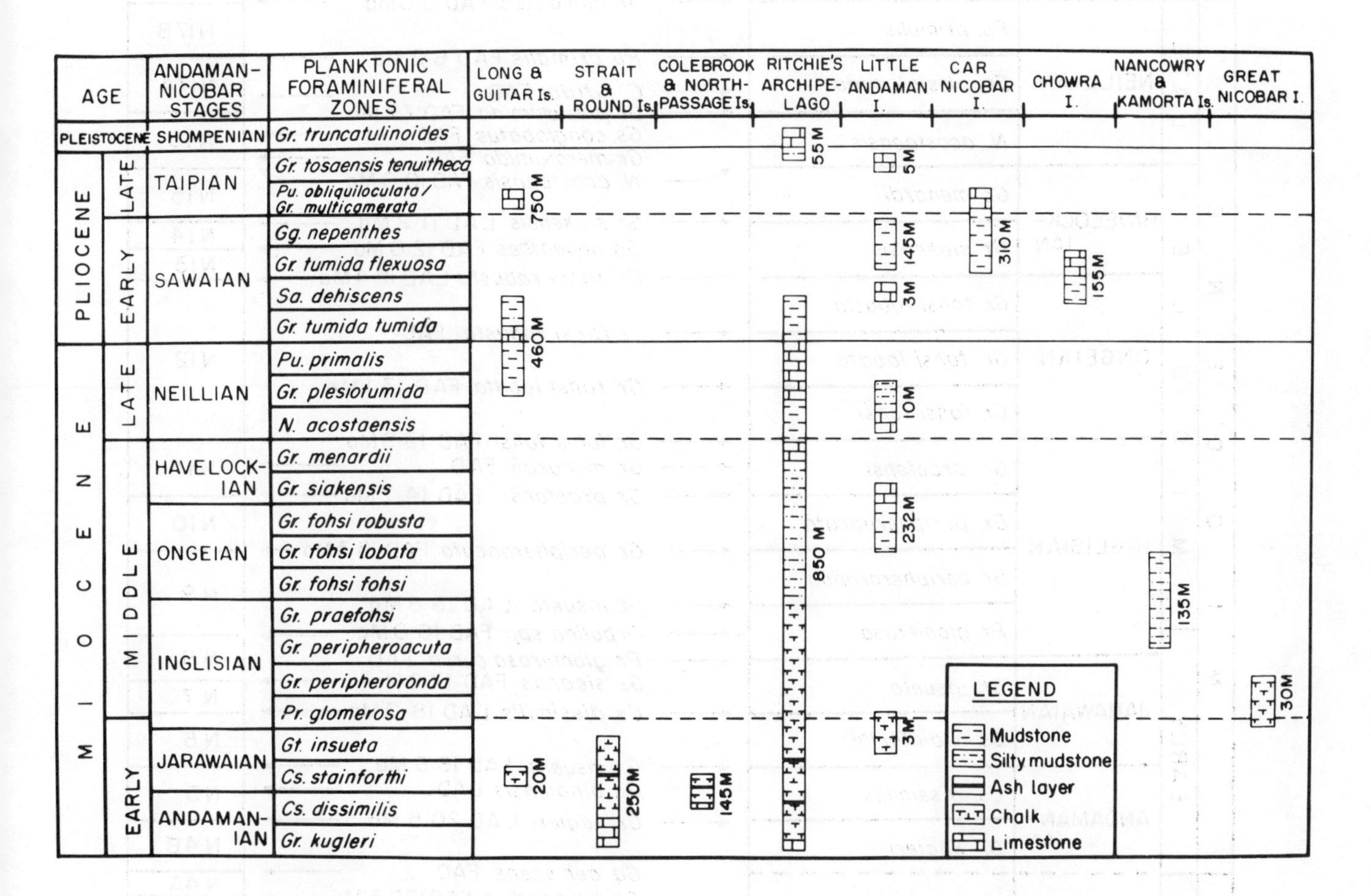

Fig. 3. Biostratigraphic correlation and chronostratigraphic extent of the sequences examined in Andaman-Nicobar Islands.

these datum levels in the Andaman-Nicobar have not been detected. However, the evolutionary bioseries as recorded in many planktonic foraminiferal lineages from the Neogene of Andaman-Nicobar are comparable with the equatorial Pacific DSDP Site 289, and therefore, the radiometric ages of Saito (1977) have been adopted (Fig. 4).

The datum levels in ascending stratigraphic order

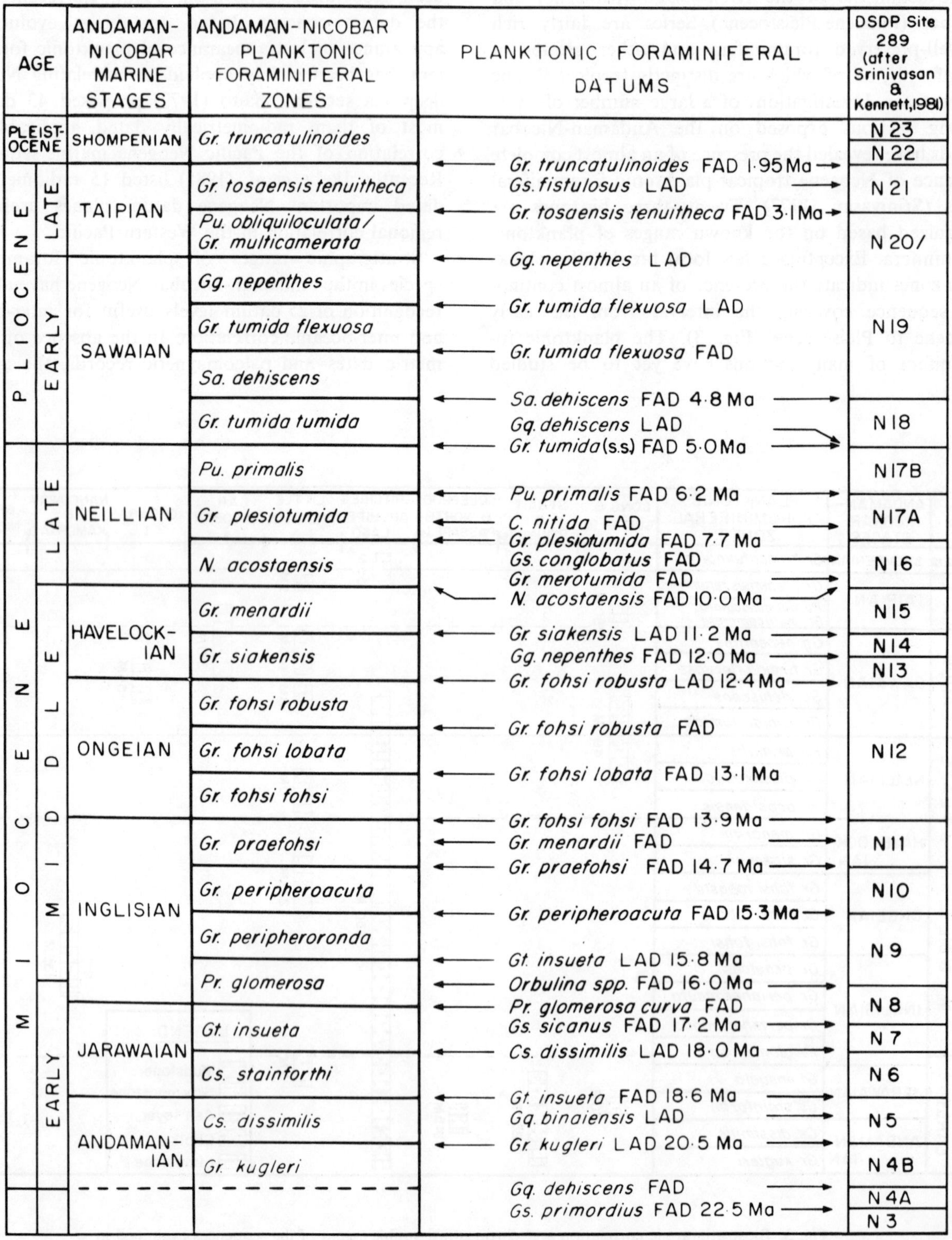

AGE: PLEISTOCENE; PLIOCENE (LATE, EARLY); MIOCENE (LATE, MIDDLE, EARLY)

ANDAMAN-NICOBAR MARINE STAGES: SHOMPENIAN; TAIPIAN; SAWAIAN; NEILLIAN; HAVELOCKIAN; ONGEIAN; INGLISIAN; JARAWAIAN; ANDAMANIAN

ANDAMAN-NICOBAR PLANKTONIC FORAMINIFERAL ZONES:
Gr. truncatulinoides
Gr. tosaensis tenuitheca
Pu. obliquiloculata / Gr. multicamerata
Gg. nepenthes
Gr. tumida flexuosa
Sa. dehiscens
Gr. tumida tumida
Pu. primalis
Gr. plesiotumida
N. acostaensis
Gr. menardii
Gr. siakensis
Gr. fohsi robusta
Gr. fohsi lobata
Gr. fohsi fohsi
Gr. praefohsi
Gr. peripheroacuta
Gr. peripheroronda
Pr. glomerosa
Gt. insueta
Cs. stainforthi
Cs. dissimilis
Gr. kugleri

PLANKTONIC FORAMINIFERAL DATUMS:
Gr. truncatulinoides FAD 1·95 Ma
Gs. fistulosus LAD
Gr. tosaensis tenuitheca FAD 3·1 Ma
Gg. nepenthes LAD
Gr. tumida flexuosa LAD
Gr. tumida flexuosa FAD
Sa. dehiscens FAD 4·8 Ma
Gq. dehiscens LAD
Gr. tumida (s.s.) FAD 5·0 Ma
Pu. primalis FAD 6·2 Ma
C. nitida FAD
Gr. plesiotumida FAD 7·7 Ma
Gs. conglobatus FAD
Gr. merotumida FAD
N. acostaensis FAD 10·0 Ma
Gr. siakensis LAD 11·2 Ma
Gg. nepenthes FAD 12·0 Ma
Gr. fohsi robusta LAD 12·4 Ma
Gr. fohsi robusta FAD
Gr. fohsi lobata FAD 13·1 Ma
Gr. fohsi fohsi FAD 13·9 Ma
Gr. menardii FAD
Gr. praefohsi FAD 14·7 Ma
Gr. peripheroacuta FAD 15·3 Ma
Gt. insueta LAD 15·8 Ma
Orbulina spp. FAD 16·0 Ma
Pr. glomerosa curva FAD
Gs. sicanus FAD 17·2 Ma
Cs. dissimilis LAD 18·0 Ma
Gt. insueta FAD 18·6 Ma
Gq. binaiensis LAD
Gr. kugleri LAD 20·5 Ma
Gq. dehiscens FAD
Gs. primordius FAD 22·5 Ma

DSDP Site 289 (after Srinivasan & Kennett, 1981): N 23; N 22; N 21; N 20/; N 19; N 18; N 17B; N 17A; N 16; N 15; N 14; N 13; N 12; N 11; N 10; N 9; N 8; N 7; N 6; N 5; N 4B; N 4A; N 3

Fig. 4. Comparison of planktonic foraminiferal datum levels between Andaman-Nicobar and DSDP site 289 (Chronology after Saito, 1977).

are as follows:

Early Miocene

(1) *Globorotalia kugleri* LAD
(2) *Globigerinatella insueta* FAD
(3) *Catapsydrax dissimilis* LAD
(4) *Praeorbulina glomerosa curva* FAD

Middle Miocene

(5) *Orbulina* spp. FAD
(6) *Globigerinatella insueta* LAD
(7) *Globorotalia peripheroacuta* FAD
(8) *Globorotalia praefohsi* FAD
(9) *Globorotalia menardii* FAD
(10) *Globorotalia fohsi fohsi* FAD
(11) *Globorotalia fohsi lobata* FAD
(12) *Globorotalia fohsi robusta* FAD
(13) *Globorotalia fohsi robusta* LAD
(14) *Globorotalia siakensis* LAD

Late Miocene

(15) *Neogloboquadrina acostaensis* FAD
(16) *Globorotalia plesiotumida* FAD
(17) *Candeina nitida* FAD
(18) *Pulleniatina primalis* FAD

Pliocene

(19) *Globorotalia tumida* FAD
(20) *Sphaeroidinella dehiscens* FAD
(21) *Globorotalia tumida flexuosa* FAD
(22) *Globorotalia tumida flexuosa* LAD
(23) *Globigerina nepenthes* LAD
(24) *Globorotalia tosaensis tenuitheca* FAD

Pleistocene

(25) *Globorotalia truncatulinoides* FAD

The integrated paleomagnetic, isotopic, and quantitative biostratigraphic studies currently planned would provide ages for these datum levels in Andaman-Nicobar. A comparison of these ages with the recorded radiometric ages from other areas in the Pacific would enhance the Neogene biochronology.

References

Ikebe, N., Chiji, M. and Huang, T.V., 1981: Important datum planes of the western Pacific Neogene. *Osaka Mus. Nat. Hist., Bull.*, no. 34, p. 79–86.

Saito, T., 1977: Late Cenozoic planktonic foraminiferal datum levels: the present state of knowledge toward accomplishing Pan-Pacific stratigraphic correlation. *1—CPNS, Tokyo 1976, Proc.*, p. 61–80.

Srinivasan, M.S., 1977: Standard planktonic foraminiferal zones of the Andaman-Nicobar Late Cenozoic: *Recent Res. Geol.,* Hindustan Publ. Corp., Delhi, v. **3**, p. 23–29.

Srinivasan, M.S., 1978: New chronostratigraphic divisions of the Andaman-Nicobar Late Cenozoic. *Recent Res. Geol.*, Hindustan Publ. Corp., Delhi, v. **4**, p. 22–36.

Srinivasan, M.S., 1979: Geology and mineral resources of Andaman-Nicobar Islands. *Andaman-Nicobar Information,* Govt. Press, Port Blair, p. 44–52.

Srinivasan, M.S. and Azmi, R.J., 1979: Correlation of Late Cenozoic marine sections in Andaman-Nicobar Northern Indian Ocean and equatorial Pacific. *Jour. Paleontology,* v. **53**, p. 1401–1415.

Srinivasan, M.S. and Kennett, J.P., 1981: A review of Neogene planktonic foraminiferal biostratigraphy: Applications in the equatorial and South Pacific, *SEPM Special Publ.*, no. 32, p. 395–432.

III-7

Neogene Biostratigraphy of Taiwan

Tunyow HUANG* and Ting-Chang HUANG**

Oligocene and Neogene sequences are well developed in the western foothills of the Hsuehshan Range and adjacent offshore regions of Taiwan (Fig. 1). Having been subjected to little or no metamorphism, these sequences are paleontologically the best known interval of the sedimentary column of the island.

From the viewpoint of basin analysis, sedimentation of the Oligocene and younger sequences occurred in three distinct provinces: 1) the Taihsi-Taichung Basin in northern Taiwan, 2) the Penghu platform in central Taiwan, and 3) the Tainan Basin in southern Taiwan (Sun, 1981). It is generally accepted that the Tertiary sediments were derived from the foreland north or northwest of the present northern part of Taiwan. The Taihsi-Taichung Basin in northern Taiwan is characterized by the rhythmic occurrence of progradational sequences, which range in environments from offshore marine to nonmarine (coal-bearing) conditions. The resulting lithofacies is generally highly quartzose. The standard succession of the Oligocene and Neogene is well exposed in an area from Miaoli northward to the coast.

The Penghu platform is a structurally stable element which divides the Taihsi-Taichung and the Tainan Basins. The platform runs NE–SW, with its northeastern extremity submerged beneath the present Chiayi-Yunlin coastal plain. Typical Neogene sequences lying above the platform do not outcrop but have been recovered from wildcat wells drilled by the Chinese Petroleum Corporation. The Kuoshing area, which is located northeast of Peikang, is renown for more or less continuous exposure of the Oligo-Miocene sequence. It provides, in the field, an opportunity to observe sedimentation under the marginal effects of the Penghu platform. The platform and its vicinity, being distal to sediment source, is characterized by deposition of suspended materials, mud, and siltstone, with an extremely low rate of accumulation. Thin-bedded limestone of shallow-shelf origin has a wide distribution on the platform. Additionally, minor hiatuses, caused by winnowing, non-deposition, or erosion were recorded in the Miocene in this province (Tang, 1977).

The Tainan Basin in southern Taiwan is located southeast of a distinct Neogene hinge fault zone which runs NE–SW and forms the southern limit of the Peikang platform. Development of this basin apparently occuured through the collapse of the passive Penghu platform. The rate of sedimentation is variable but generally exceedingly high because of extensive growth faulting. This province is now covered mostly by the Neogene sediments composed principally of mudstone. The lithostratigraphic units of each of the aforementioned provinces are listed in Figs. 2, 3 and 4. The succession of Tertiary calcareous nannofossils has been studied from many type sections in different regions since 1976. Important contributions include: T.C. Huang (1976, 1977, 1978 a, 1979) and Chen *et al.* (1977 a) for northern Taiwan; T.C. Huang and Ting (1979) and Chi (1979) for central Taiwan; and Chen *et al.* (1977b) for southern Taiwan.

Figure 3 is an effort to compile all of the nannofossil biostratigraphic events considered useful for the island-wide stratal correlation. A total of 18 datums (Datums A through R) can be recongnized as follows:

Datum A: First occurrence of *Sphenolithus ciperoensis* in the middle part of the Kankou Formation.

Datum B: Last occurrence of *Sphenolithus distentus* in the upper part of the Tsuku Sandstone and in the Lower part of the Shuichangliu Formation.

Datum C: Last occurrence of *Sphenolithus ciperoensis, Dictycoccites bisectus,* or *Zygrhablithus bijugatus* in the upper part of the following formations: Tatungshan, Chingtan, Wuchihshan, and

* *Office of Chief Geologist, Chinese Petroleum Corporation, 83, Chung Hwa Road, Taipei 100, Taiwan*
** *Offshore Petroleum Exploration Division, Chinese Petroleum Corporation, 2-7, Lane 129, Yen Ping S. Road, Taipei 100, Taiwan*

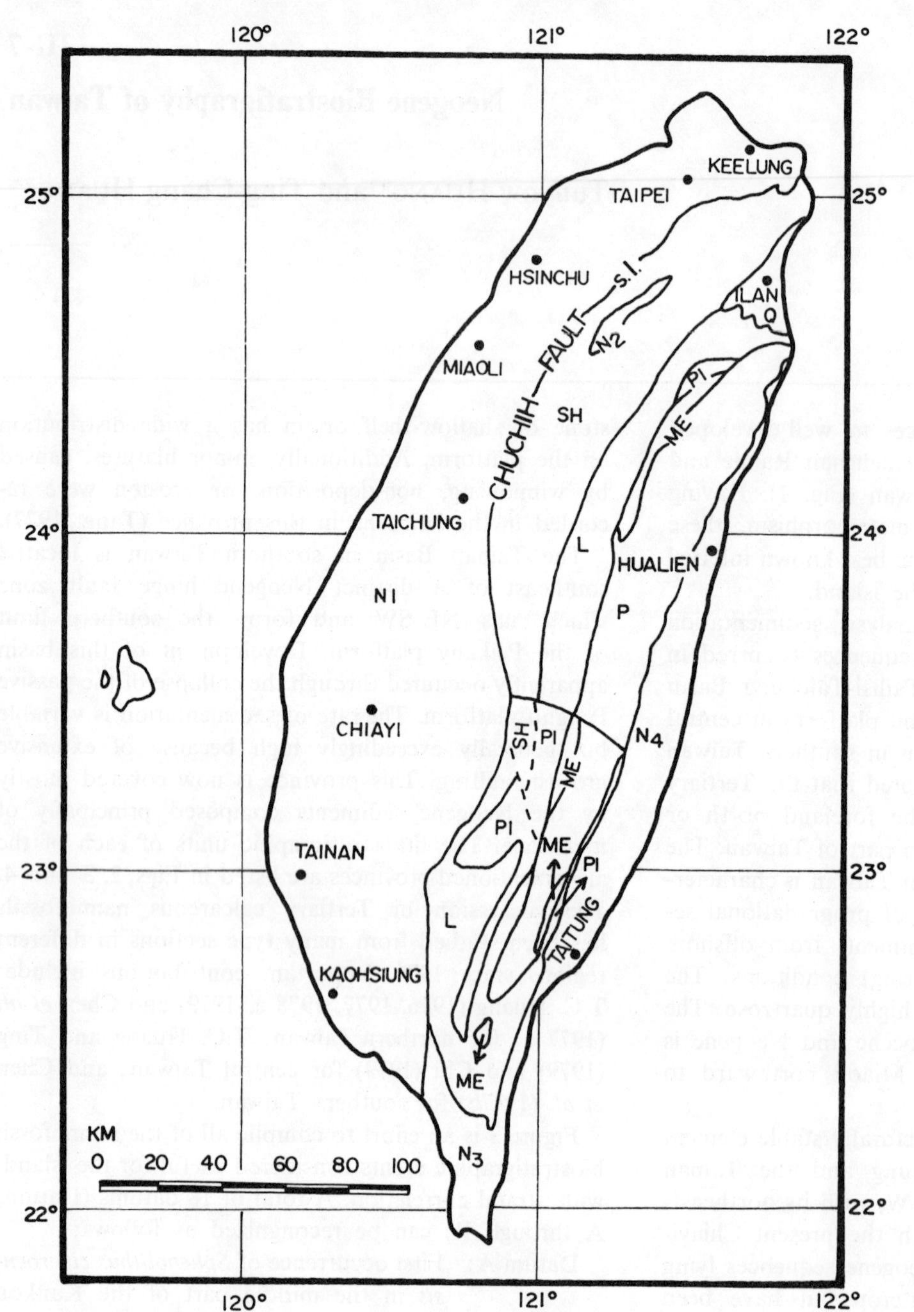

Fig. 1. Geological map of Taiwan. Western Foothills Region: N1, Neogene terrane, where Hsichih Group developed. Hsüehshan Range: SH, Paleogene terrane (or Shihtsaoan and Hsüehshanian undifferentiated, where Upper Wulai Group developed); N2, Miocene patch. Backbone Central Range including Hengchun Peninsula; P, Tananao Schist; ME, Pihou Formation and its equivalents; Pi, Pilushanian; L, Lushanian; N3, Neogene and Quaternary. Coastal Range: N4, Neogene, and Quaternary.

Shuichangliu.

Datum D: First occurrence of *Helicosphaera carteri* in the lower part of the Taliao Formation and the lower part of the Tanliaoti Member of the Takeng Formation.

Datum E: First occurrence of *Sphenolithus belemnos* in the lower part of the Peiliao Formation and in the middle part of the Shihmen Member of the Shuilikeng Formation.

Datum F: Last occurrence of *Sphenolithus belemnos* in the middle part of the Peiliao Formation and in the upper part of the Shihmen Member of the Shuilikeng Formation.

Datum G: Last occurrence of *Helicosphaera ampliaperta* in the lower part of the Talu Shale and in basal part of the Changhukeng Shale Member of the Shuilikeng Formation (?).

Datum H: Last occurrence of *Sphenolithus heteromorphus* in the middle part of Kuanyinshan Sandstone and in the upper part of the Changhukeng Shale Member of the Shuilikeng Formation.

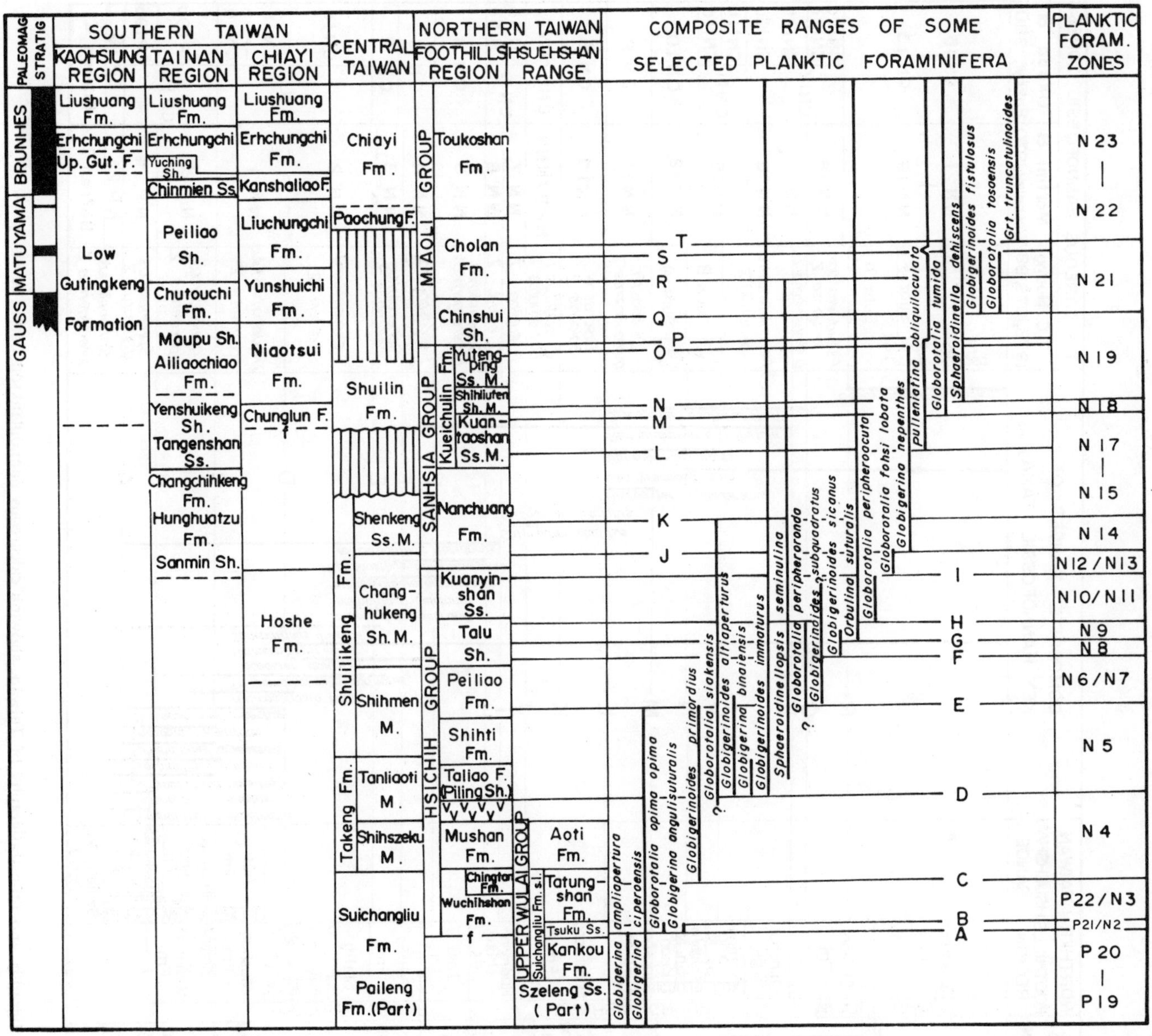

Fig. 2. Neogene biostratigraphy and lithostratigraphy of Taiwan, showing planktonic foraminiferal datum levels.

Datum I: Last occurrence of *Cyclicargolithus floridanus* in the upper part of the Changhukeng Shale Member of the Shuilikeng Formation.

Datum J: First occurrence of *Catinaster coalitus* in the uppermost part of the Changhukeng Shale Member of the Shuilikeng Formation and in the basal part of the Nanchuang Formation.

Datum K: Last occurrence of *Discoaster hamatus* in the subsurface equivalent of the Hunghuatzu Formation (?).

Datum L: First occurrence of *Discoaster quinqueramus* in the upper part of the Changchihkeng Formation.

Datum M: Last occurrence of *Discoaster quiquueramus* in the middle part of the Kuantaoshan Sandstone Member of the Kueichulin Formation and in the upper part of the Tangenshan Sandstone.

Datum N: First occurrence of *Gephyrocapsa* spp. in the Shihliufen Shale Member of the Kueichulin Formation and in the lower part of the Niaotsui Formation.

Datum O: Last occurrence of *Sphenolithus abies* in the lower part of the Yutengping

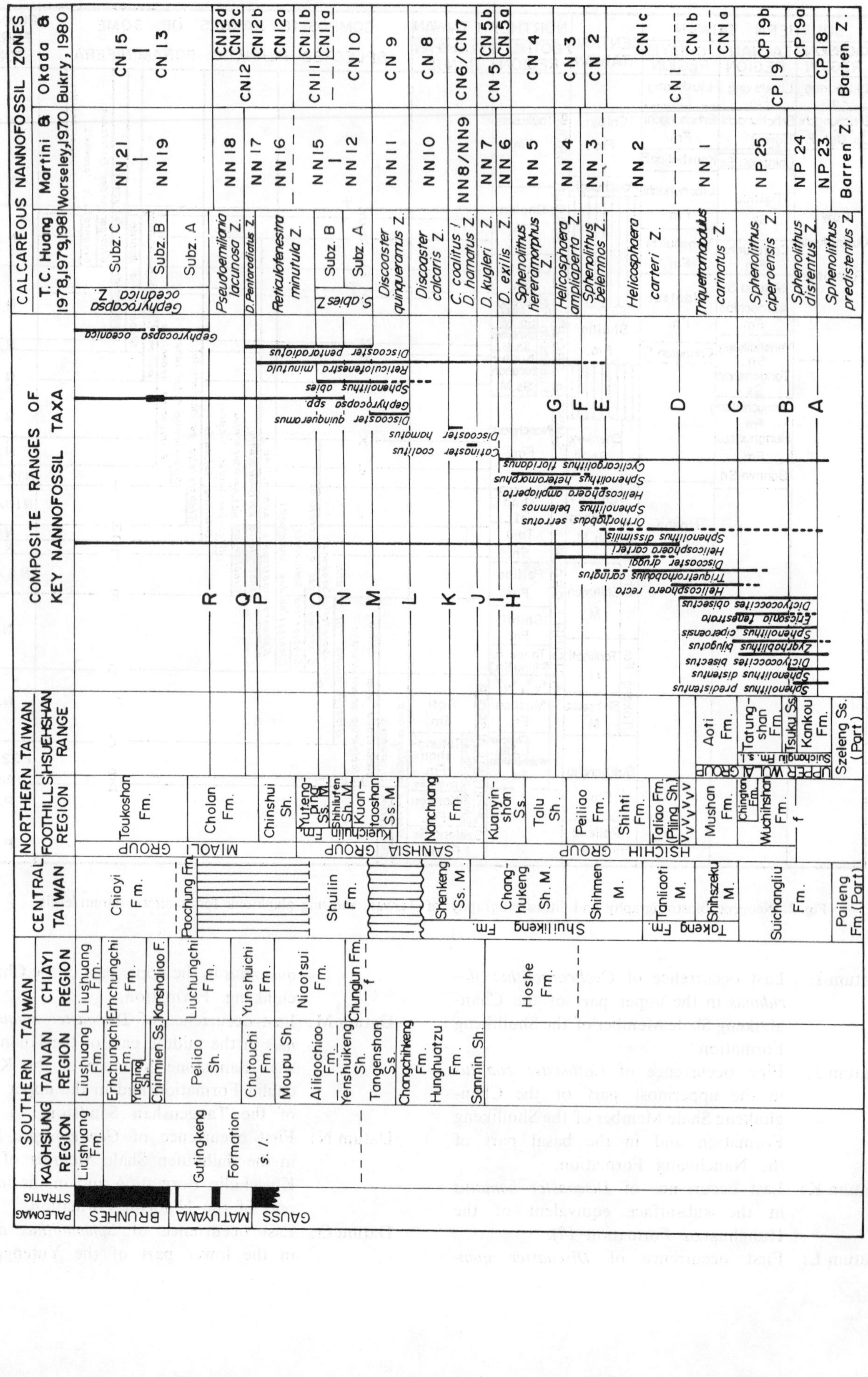

PALEOMAG	SOUTHERN TAIWAN: KAOHSIUNG REGION	SOUTHERN TAIWAN: TAINAN REGION	SOUTHERN TAIWAN: CHIAYI REGION	CENTRAL TAIWAN	NORTHERN TAIWAN: FOOTHILLS REGION	NORTHERN TAIWAN: HSUEHSHAN RANGE
BRUNHES	Liushuang Fm.	Liushuang Fm.	Liushuang Fm.	Chiayi Fm.	MIAOLI GROUP: Toukoshan Fm.	
	Gutingkeng Formation s. l.	Erhchungchi Fm.	Erhchungchi Fm.			
		Yuching Sh.				
		Chinmien Ss.	Kanshaliao F.			
MATUYAMA		Peiliao Sh.	Liuchungchi Fm.	Paochung Fm.	Cholan Fm.	
		Chutouchi Fm.	Yunshuichi Fm.		Chinshui Sh.	
GAUSS		Moupu Sh.	Niaotsui Fm.		SANHSIA GROUP: Kueichulin Fm.: Yutengping Ss. M.	
		Ailiaochiao Fm.		Shuilin Fm.	Shihliufen Sh. M.	
		Yenshuikeng Sh.	Chunglun Fm.		Kuantaoshan Ss. M.	
		Tangenshan Ss.				
		Changchihkeng Fm.				
		Hunghuatzu Fm.		Shuilikeng Fm.: Shenkeng Ss. M.	Nanchuang Fm.	
		Sanmin Sh.	Hoshe Fm.	Changhukeng Sh. M.	HSICHIH GROUP: Kuanyinshan Ss.	
					Talu Sh.	
				Shihmen M.	Peiliao Fm.	
					Shihti Fm.	
				Takeng Fm.: Tanliaoti M.	Taliao Fm. (Piling Sh.)	
				Shihszeku M.	Mushan Fm.	UPPER WULAI GROUP: Aoti Fm.
				Suichangliu Fm.	Chingtan Fm.	Suichangliu Fm. s.l.: Tatungshan Fm.
					Wuchihshan Fm.	Tsuku Ss.
						Kankou Fm.
				Paileng Fm. (Part)		Szeleng Ss. (Part)

COMPOSITE RANGES OF KEY NANNOFOSSIL TAXA

Datum levels (top to bottom): R, Q, P, O, N, M, L, K, J, I, H, G, F, E, D, C, B, A

Taxa: *Sphenolithus predistentus*, *Sphenolithus distentus*, *Dictyococcites bisectus*, *Zygrhablithus bijugatus*, *Sphenolithus ciperoensis*, *Ericsonia fenestrata*, *Dictyococcites abisectus*, *Helicosphaera recta*, *Triquetrorhabdulus carinatus*, *Discoaster druggi*, *Helicosphaera carteri*, *Sphenolithus dissimilis*, *Orthorhabdus serratus*, *Sphenolithus belemnos*, *Helicosphaera ampliaperta*, *Sphenolithus heteromorphus*, *Cyclicargolithus floridanus*, *Catinaster coalitus*, *Discoaster hamatus*, *Discoaster quinqueramus*, *Gephyrocapsa spp.*, *Sphenolithus abies*, *Reticulofenestra minutula*, *Discoaster pentaradiatus*, *Gephyrocapsa oceanica*

CALCAREOUS NANNOFOSSIL ZONES

T. C. Huang 1978, 1979, 1981		Martini & Worseley, 1970	Okada & Bukry, 1980	
Gephyrocapsa oceanica Z.	Subz. C	NN21	CN15	
	Subz. B	NN19	CN13	
	Subz. A			
Pseudoemiliania lacunosa Z.		NN18	CN12	CN12d
				CN12c
D. Pentaradiatus Z.		NN17		CN12b
Reticulofenestra minutula Z.		NN16		CN12a
S. abies Z.	Subz. B	NN15	CN11	CN11b
				CN11a
	Subz. A	NN12	CN10	
Discoaster quinqueramus Z.		NN11	CN9	
Discoaster calcaris Z.		NN10	CN8	
C. coalitus / *D. hamatus* Z.		NN8/NN9	CN6/CN7	
D. kugleri Z.		NN7	CN5	CN5b
D. exilis Z.		NN6		CN5a
Sphenolithus hereramorphus Z.		NN5	CN4	
Helicosphaera ampliaperta Z.		NN4	CN3	
Sphenolithus belemnos Z.		NN3	CN2	
Helicosphaera carteri Z.		NN2	CN1	CN1c
Triquetrorhabdulus carinatus Z.		NN1		CN1b
				CN1a
Sphenolithus ciperoensis Z.		NP25	CP19	CP19b
Sphenolithus distentus Z.		NP24		CP19a
Sphenolithus predistentus Z.		NP23	CP18	
		Barren Z.	Barren Z.	

Fig. 3. Late Cenozoic biostratigraphy and lithostratigraphy of Taiwan, showing calcareous nannoplankton datum levels.

Sandstone Member of the Kueichulin Formation, in the middle part of the Niaotsui Formation, and at the base of the Ailiaochiao Formation.

Datum P: Last occurrence of *Reticulofenestra minutula* on the top of the Chinshui Shale, in the lower part of the Chutouchi Formation, and in the basal part of the Yunshuichi Formation.

Datum Q: Last occurrence of *Discoaster pentaradiatus* in the lower part of the Cholan Formation and in the upper part of the Yunshuichi Formation.

Datum R: First occurrence of *Gephyrocapsa oceanica* in the upper part of the Cholan Formation, in the upper part of the Liuchungchi Formation, in the upper part of the Peiliao Shale, and in the Gutingkeng Formation.

Planktonic and benthonic foraminiferal biostratigraphy of the Neogene series in Taiwan has been studied by L. S. Chang (1959, 1960 a, b, 1962, 1972, 1975 a, b, 1976), T. Huang (1964, 1967, 1971 a, b, 1975, 1976, 1977, 1978a, b), T. Huang and Chiu (1973), and others (Oinomikado and Huang, 1957; Y. M. Chang, 1973). The regional biostratigraphic values of some benthonic foraminifers in western Taiwan were emphasized by Tunyow Huang (1964, 1971 b, 1978 b). Investigations of the Oligo-Miocene in northern Taiwan were done recently by C. Y. Huang (1981) and C. Y. Huang, *et al.* (1979).

Both planktonic and benthonic foraminiferal biostratigraphic events and the magnetozones defined by Chen, *et al.* (1977 a, b) are given in Figs. 2 and 4.

In the early Neogene series, 9 datum levels of planktonic foraminifers can be recognized (Fig. 2). The first occurrence of *Globigerinoides primordius* is in the upper Tatungshan Formation and Chingtan Formation. The lowest *Globigerinoides*-bearing horizon (L. S. Chang, 1962) is entirely within the calcareous nannoplankton Zone NP 25, and *Globigerinoides primordius* jointly occurs with Oligocene coccoliths (T. C. Huang, 1977). The first occurrence of *Globigerinoides quadrilobatus* and *Globigerinoides altiaperturus* is in the lower Taliao Formation, and the initial occurrence of *Globigerinoides subquadratus* is in the lower Peiliao Formation and Tsouho Formation. The first occurrence of *Globigerinoides sicanus* in the lower Talu Shale and middle Nankang Formation and the initial specimeus of *Praeorbulina glomerosa* are recognized in the lower Talu Shale. The first appearance datum of *Orbulina suturalis* is found in the middle part of the Talu Shale about 50 m higher than in the so-called Talu Sandstone in the Chukuangkeng section. The first occurrence of *Globorotalia peripheroacuta* is in the uppermost part of the Talu Shale. *Globorotalia fohsi lobata* is observed near the top of the Kuanyinshan Sandstone in the Nantou area (Huang and Chiu, 1973), and *Globorotalia fohsi robusta* from the Sanmin Shale in the Hunghuatzu section (T. Huang, 1967). The first occurrence of *Globigerina nepenthes* and the last occurrence of *Globorotalia siakensis* are recognized in the lower part of the Hunghuatzu Formation.

In the late Neogene series, 9 datum levels of planktonic foraminifers can be recognized (Fig. 2). The first occurrence of *Pulleniatina obliquiloculata* in the lower part of the Tangenshan Sandstone and the first occurrence of *Globorotalia tumida* are recognized in the upper part of the Tangenshan Sandstone. The first occurrence of *Sphaeroidinella dehiscens* was observed in the middle Shihliufeng Shale(SCS–63) of the Sanchienshan section, near Tapu in the Chiayi area (T. Huang, 1971a). The last occurrence of *Globigerina nepenthes* and the abrupt change in the coiling direction of the test of *Pulleniatina* from left to right are recognized in the Maupu Shale. The first appearance of both *Globorotalia tosaensis* and *Globigerinoides fistulosus* are in the same horizon in the lower part of the Chutouchi Formation (T. Huang, 1967). The last occurrences of *Sphaeroidinellopsis seminulina,* including *S. kochi,* and *Globoquadrina altispira* are observed in the lower part of the Chutouchi Formation and they are stratigraphically higher than the *Globorotalia tosaensis* datum. The abrupt change in coiling direction of the test of *Pulleniatina* from right to left and the last occurrence of *Globigerinoides fistulosus* in the same horizon in the lower Peiliao Shale are stratigraphically a little upper than the *Globorotalia truncatulinoides* base datum (T. Huang, 1967). The first appearance datum of *Globorotalia truncatulinoides* was observed in the upper part of the lower Gutingkeng Formation in southern Taiwan (T. Huang, 1971 a, 1977).

The Olduvai Event occurs after the first appearance of *Globorotalia truncatulinoides* and before the initial appearance of *Gephyrocapsa oceanica* and is located in the Cholan Formation and in the upper part of the lower Gutingkeng Formation. The Jaramillo Event and the Brunhes/Matuyama boundary in the magnetostratigraphy of the Chuhuangkeng section of northern Taiwan and the Chishan section of southern Taiwan (Chen, *et al.*, 1977 a, b) are within the Toukoshan Formation and the upper part of the lower Gutingkeng Formation.

In the Neogene series, some benthic species (Fig. 4) such as *Gaudryina, Textularia, Pseudorotalia, Asterorotalia,* and *Pararotalia* are valuable for regional stratigraphic correlations (T. Huang, 1964, 1971 b, 1978). The extinction of *Gaudryina hayasakai* and the first appearance

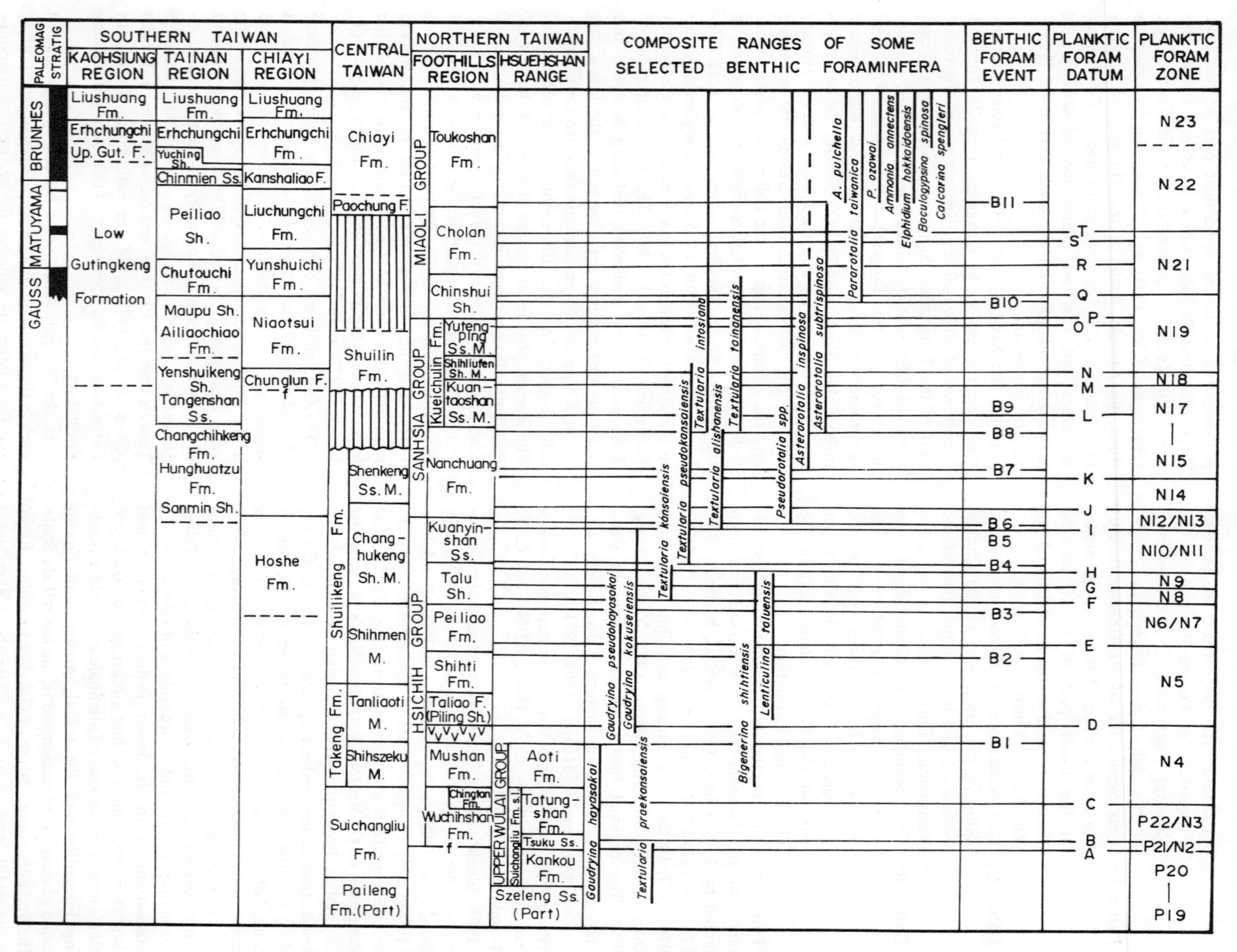

Fig. 4. Neogene biostratigraphy and lithostratigraphy of Taiwan, showing benthic foraminiferal events.

of *Gaudryina pseudohayasakai* are in the same horizon in the Kungkuan Tuff. An *Operculina bartschi multiseptata* concentration zone occurs in the uppermost part of the Shihti Formation (B2). The extinction of *Textularia kansaiensis* and the first appearance of *Textularia pseudokansaiensis* are recorded in the same horizon of the uppermost part of the Talu Shale, and this horizon lies just above the first occurrence datum of *Globorotalia peripheroacuta*. The first occurrence of *Pseudorotalia* is known from the base of the Nanchung Formation, and the first occurrence of *Asterorotalia inspinosa* from the middle Nanchung Formation, Tapu section of the Chiayi area. *Asterorotalia subtrispinosa* and *Asterorotalia multispinosa* first occur in the uppermost part of the Nanchung Formation, and *Pararotalia taiwanica* first occurs in the Chinshui Shale. The first occurrence of *Asterorotalia pulchella*, together with *Pararotalia ozawai*, *Ammonia annectence*, and *Elphidium hokkaidoensis* is recognized in the lower part of the Toukoshan Formation.

References

Chang, L.S., 1959: A biostratigraphic study of the Miocene in western Taiwan based on smaller foraminifera (part 1: Planktonics). *Geol. Soc. China, Proc.*, no. 2, p. 47–72.

Chang, L.S., 1960a: Tertiary biostratigraphy of Taiwan with special reference to smaller foraminifera and its bearing on the Tertiary geohistory of Taiwan. *Geol. Soc. China, Proc.*, no. 3, p. 7–30.

Chang, L.S., 1960b: Tertiary biostratigraphy study of the Miocene in western Taiwan based on smaller foraminifera (part 2: benthonics). *Geol. Surv. Taiwan, Bull.*, no. 12, p. 67–91.

Chang, L.S., 1962: A biostratigraphic study of Oligocene in northern Taiwan based on smaller foraminifera. *Geol. Soc. China, Proc.*, no. 5, p. 47–64.

Chang, L. S., 1972: A biostratigraphic study of the so-called Slate Formation on the east flank of the Central Range between Tawu and Taimali, southeastern Taiwan, based on smaller foraminifera. *Geol. Soc. China, Proc.*, no. 13, p. 129–142.

Chang, L.S., 1975a: Biostratigraphy of Taiwan. *In* Kobayashi, T. and Toriyama, R., eds., *Geol. Palaeont. Southeast Asia*, v. **15**, p. 337–361.

Chang, L.S., 1975b: Miocene/Pliocene boundary in Taiwan. *In* Saito. T. and Burckle, L. H. eds., *Late Neogene Epoch Boundaries*, Micropaleons. Spec. Public., no. 1, p. 106–114.

Chang, L.S., 1976: The Lushanian Stage in the Central Range of Taiwan and its fauna. *Progress in Micropaleontology*, Micropaleont. Press, p. 103–109.

Chang, Y. M., 1973: Biostratigraphic study of smaller foraminifera from the Wu-Chi section, Kuohsing, Nantou, Taiwan. *Petrol. Geol. Taiwan*, no. 2, p. 183–206.

Chen, P. H., Huang, T. C., Huang, C. Y., Jiang, M. J., Lo, L. L., and Kuo, C. L., 1977a: Paleomagnetic and cocolith stratigraphy of Plio-Pleistocene shallow marine sediments, Chuhuangkeng, Miaoli. *Petrol. Geol. Taiwan*, no. 14, p. 219–239.

Chen, P.H., Huang, C. Y., Huang, T. C., and Tsai, L. P., 1977b: A study of the late Neogene marine sediments of the Chishan area, Taiwan: paleomagnetic stratigraphy, biostratigraphy, and paleoclimate. *Geol. Soc. China, Mem.*, no. 2, p. 169–190.

Chi, W. R., 1979: Calcareous nonnoplankton biostratigraphy of the Nantou area, central Taiwan. *Petrol. Geol. Taiwan*, no. 16, p. 131–165.

Huang, C. Y., 1981: Smaller foraminiferal study on the Oligo-Miocene formations in northern Taiwan. National Taiwan Univ., *Dept. Geol.*, doctoral thesis, in Chinese.

Huang, C.Y., Cheng, Y. M. and Huang, Tunyow, 1978: Preliminary biostratigraphic study of the Nankang Formation in the Lilao section, near Taipei, northern Taiwan. *Ti-Chih*, v. **2**, p. 1–11. (in Chinese with English abstract)

Huang, T. C., 1976: Neogene calcareous nannoplankton biostratigraphy viewed from the Chuhuangkeng section, northwestern Taiwan *Geol. Soc. China, Proc.*, no. 19, p. 7–24.

Huang, T.C., 1977: Calcareous nannoplankton stratigraphy of the Upper Wulai Group (Oligocene) in northern Taiwan. *Petrol. Geol. Taiwan*, no. 14, p. 147–179.

Huang, T.C., 1978: Calcareous nannoplankton, paleoenvironment, age and correlation of the Upper Wulai Group and the Lower Hsichih Group (Oligocene to Miocene) in northern Taiwan. *Geol. Soc. China, Proc.*, v. **21**, no. 1, p. 128–150.

Huang, T.C., 1979: A supplementary note on the calcareous nannofossils, age and correlation of the Wuchihshan Formation. *Petrol. Geol. Taiwan*, no. 16, p. 85–93.

Huang, T.C., 1980: Oligocene to Pleistocene calcareous nannofossil biostratigraphy of the Hsuehshan Range and western foothills in Taiwan. *In* Kobayashi, T. and Toriyama, R., ed., *Geol. Paleontol. Southeast Asia*, v. **21**, p. 191–210.

Huang, T.C. and Ting, J. S., 1979: Calcareous nannofossil succession from the Oligo-Miocene Peikang-Chi section and revised stratigraphic correlation between northern and central Taiwan. *Geol. Soc. China, Proc.*, no. 22, p. 105–120.

Huang, T.C. and Ting, J.S., 1981: Calcareous nannofossil biostratigraphy of the Late Neogene shallow marine deposits in Taiwan. *Ti-Chih*, v. **3**, p. 105–119, (in Chinese with English abstract)

Huang, Tunyow, 1964: "*Rotalia*" group from the upper Cenozoic of Taiwan. *Micropaleontology*, v. **10**, no. 1, p. 49–62.

Huang, Tunyow, 1967: Late Tertiary planktonic foraminifera from southern Taiwan. *Tohoku Univ., Sci.Reports., 2nd ser.(Geoi.)*, v. **38**, no.2, p. 165–192.

Huang, Tunyow, 1971a: New developments in stratigraphic correlation of the Neogene sequence in western Taiwan. *Petrol. Geol. Taiwan*, no.9, p. 19–27.

Huang, Tunyow, 1971b: Some foraminiferal lineages in

Taiwan. *Geol. Soc. China, Proc.*, no. 14, p. 76–85.

Huang, Tunyow, 1975: Late Neogene foraminiferal zonation of southwestern Taiwan. *In* Saito, T. and Burckle, L.H., eds., *Late Neogene Epoch Boundaries,* Micropaleontology, Spec. Publ., no. 1, p. 106–114.

Huang, Tunyow, 1976: Some segnificant biostratigraphic events in the Neogene formations of Taiwan. *In* Takayanagi, Y. and Saito, T., eds., *Progress in Micropaleontology,* p. 103–109.

Huang, Tunyow, 1977: Late Neogene planktonic foraminiferal biostratigraphy of the Tainan foothills region, Tainan, Taiwan. *Petrol. Geol. Taiwan,* no. 14, p. 121–145.

Huang, Tunyow, 1978a: Foraminiferal biostratigraphy of the Hunghuatzu section, southern Taiwan. *Petrol. Geol. Taiwan,* no. 15, p. 35–48.

Huang, Tunyow, 1978b: Two important benthic foraminiferal datum in the Neogene of northern Taiwan. *Ti-Chih,* v. 2, p. 19–23. (in Chinese with English abstract).

Huang, Tunyow and Chiu, H.T., 1973: Some diagnostic Miocene planktonic foraminifera from west central Taiwan. *Geol. Soc. China, Proc.*, no. 16, p. 59–68.

Ikebe, N., Chiji, M. and Huang, Tunyow, 1981: Important datum-planes of the western Pacific Neogene. *Osaka Museum of National History, Bull.*, no. 34, p. 79–86.

Oinomikado, T. and Huang, Tunyow, 1957: Micropaleontological ivestigation of Kueitanchi section near Chutouchi oilfield. *Symposium on Petroleum Geology of Taiwan in the Celebration of the Tenth Anniversary,* Chinese Petroleum Coporation, p. 257–265.

Sun, C.C., 1981: The Tertiary basins in offshore Taiwan. *2nd ASCOPE Meeting, Manila, Abst.*

Tang, C.H., 1977: Late Miocene erosional unconformity on the subsurface Peikang High beneath the Chiayi-Yunlin coastal plain, Taiwan. *Geol. Soc. China, Mem.*, no. 2, p. 155–167.

III-8

Neogene Biostratigraphy of Korea

Bong Kyun KIM*, Sun YOON[2]* and Hyu Su YUN[3]*

Introduction

In southern Korea, the Tertiary deposits are distributed along the eastern coast in five small sedimentary basins called the Bugpyeong, Yeonghae, Pohang, Eoil, and Ulsan Basins from north to south (Fig. 1).

The Bugpyeong Basin contains the Bugpyeong Formation which unconformably underlies the Bugpyeong Conglomerate. The former consists mainly of mudstone and sandstone with the intercalation of a thin coal bed in this middle part. The lower and upper parts of the formation yield three species of planktonic and 17 species of benthic foraminifers and some molluscs.

The Tertiary strata in the Yeonghae Basin are divided into three formations, namely, the Dogogdong Formation, the Yeonghae Conglomerate, and the Yeongdong Formation in ascending order. In the former two formations the fossils are rare with the exception of some plant fossils suggesting a terrestrial environment for these formations. However, some foraminifers and molluscs were described from the Yeongdong Formation (Kim, 1965). On the basis of his foraminiferal research, Kim (1965) correlated it with the Eedong Formation of the Pohang Basin.

In the Pohang Basin, which is the largest of the five basins, sedimentary rocks can be divided into two groups

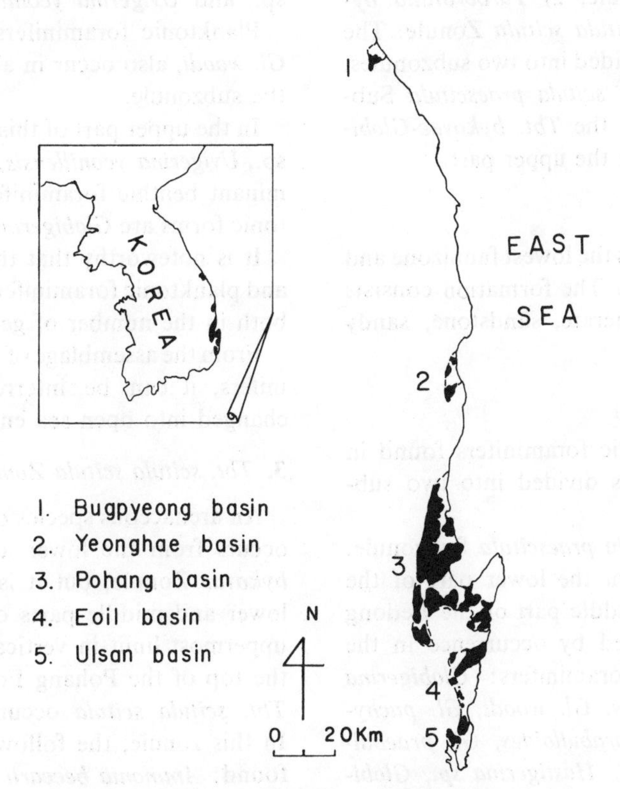

Fig. 1. Distribution of Tertiary basins in South Korea.

* *Seoul National University, Seoul,* 2* *Busan National University, Busan*
3* *Chungnam National University, Chungnam*

based on differences in lithology, namely, the Yeonil and Yangbuk Groups. The Yangbuk Group, which developed in the southeastern part of the basin, consists mainly of nonmarine sediments which are intercalated by lavas, tuffs, and lignite beds. Stratigraphically and lithologically, the Yeonil Group can be subdivided into the Seoam Conglomerate, Songhagdong, Daegock, Eedong, and Pohang Formations in ascending order. The lowermost part of this group is composed of terrestrial conglomerates, whereas the overlying formations, which consist mostly of mudstone, siltstone, and sandstone, yield rich fossil faunas such as foraminifers, diatoms, silicoflagellates, radilolarians, dinoflagellates, and molluscs. A biostratigraphic zonation is summarized in Table 1 and discussed below.

There is some controversy in regard to the ralationship between these two groups because their lithological contact is not exposed and the Yangbuk Group does not yield any marine fossil to be used for correlation. In spite of the lack of direct evidence for its relative age, the Yangbuk Group is believed to underlie the Yeonil Group because of their structural relationship.

Foraminiferal Zonation

A biostratigraphic zonation is established in upward sequence: 1) Radiolaria Zonule, 2) *Turborotalia bykovae* Zonule, and 3) *Tbt. scitula scitula* Zonule. The second zonule is further subdivided into two subzonules, namely the *Tbt. bykovae-Tbt. scitula praescitula* Subzonule in the lower part and the *Tbt. bykovae-Globigerina trilocularis* Subzonule in the upper part.

1. *Radiolaria Zonule*

The Radiolaria Zonule forms the lowest faunizone and does not yield any foraminifer. The formation consists of an alternation of conglomerate, sandstone, sandy shale, and shale.

2. *Turborotalia bykovae Zonule*

On the basis of characteristic foraminifers found in the *Tbt. bykovae* Zonule, it is divided into two subzonules.

2-a. *Tbt. bykovae-Tbt. scitula praescitula* Subzonule.

This subzonule extends from the lower part of the Daegock Formation to the middle part of the Eedong Formation and is characterized by occurrence in the lower part of planktonic foraminifers: *Globigerina angustiumbilicata, Gl. bulloides, Gl. woodi, Gl. pachyderma, Gl. trilocularis, Gl. parabulloides, Gl. praebulloides, Globigerinita glutinata., Hastigerina sp., Globigerinoides trilobus, Tbt. bykovae,* and *Tbt. scitula praescitula.* Among these, *Globigerina bulloides* and *Tbt. bykovae* are most dominant.

From the middle part are found *Bulimina elongata tenenera, Bolivina striata, Florilus kidoharaensis, Gyroidina hangukensis, Haplophragmoides* sp., *Sphaeroidina compacta, Uvigerina yeonillensis, Valvulineria asanoi,* and planktonics like *Globigerina bulloides. Gl. woodi, Tbt. bykovae,* and *Tbt. scitula praescitula* are reported. Especially, *Florilus kidoharaensis* is abundant and is confined to the middle part of this subzonule.

Abundant planktonic foraminifers in the lower part of the subzonule and the increase of benthic forms in the upper part seem to show that sedimentary conditions changed from an open sea to a lagoonal shallow marine environment.

2-b. *Tbt. bykovae-Globigerina trilocularis* Subzonule.

The lower limit of this subzonule is defined in the middle part of the gypsum beds, from where it extends upwards to the middle of the Eedong Formation.

In the lower part of this subzonule, the following benthic foraminifers are reported: *Bolivina koreanica, Bulimina exilis tenuata, B. inflata, Cancris indicus, Gyroidina hangukensis, Haplophragmoides* sp., *Melonis pompilioides, Uvigerina hootsi,* and *U. yeonillensis.* Among them, the most characteristic and abundant forms are *Bolivina koreanica* and *Globigerina trilocularis,* occurring in association with *Tbt. bykovae.* Other dominant species in this part are *Gyroidina hangukensis, Haplophragmoides* sp., and *Uvigerina yeonillensis.*

Planktonic foraminifers, *Globigerina trilocularis* and *Gl. woodi,* also occur in abundance in the lower part of the subzonule.

In the upper part of this subzonule, *Haplophragmoides* sp., *Uvigerina yeonillensis,* and *Uvigerina* sp. are the dominant benthic foraminifers, whereas dominant planktonic forms are *Globigerinita glutinata* and *Tbt. bykovae.*

It is noteworthy that the abundance of both benthic and planktonic foraminifers gradually decreases upwards both in the number of genera and species.

From the assemblage of benthic and planktonic foraminifers, it can be inferred that lagoonal conditions changed into open sea environments.

3. *Tbt. scitula scitula Zonule*

An arenaceous species of the genus *Haplophragmoides* occurs from the lower and middle parts of the *Tbt. bykovae* Zonule, but it is particulary abundant in the lower and middle parts of the Pohang Formation. Its uppermost limit in vertical distribution corresponds to the top of the Pohang Formation. On the other hand, *Tbt. scitula scitula* occurs only from this formation. In this zonule, the following benthic foraminifers are found: *Ammonia beccarii pohangensis, Bolivina striatula, Cibicides malloryi, Haplophragmoides* sp., *Martinottiella communis, Uvigerina hootsi,* and *Valvulineria yeonillensis.* Among these forms, *Ammonia beccarii pohangensis,*

Table 1. Important biostratigraphic zonation in the Pohang Basin.

Age		Formations			Foraminifera Kim (1965)	Mollusca Yoon (1975)	Diatom Lee (1977)		Silicoflagellata Lee (1979)	
Quart.		Yonghan F.								
Plio.		Seoguipo F.								
Miocene	Middle Miocene	Yeonil Group	Pohang F.		Tbt. scitula scitula zonule	Pa. — Mo. Co. — Ca. Pr.	?	?	?	
			Eedong F.		*Tbt. bykovae zonule*: *Tbt. bykovae—Glb. trilocularis* subzonule	Pr.	*Actinocyclus ingens* Zone	*Dtl. hustedtii—Cpd. rhombicus* subzono	*Distephanus crux crux* Zone	*Cnp. longispinus* subzone
			Daegock F.		*Tbt. bykovae zonule*: *Tbt. bykovae—Tbt. scitula praescitula* subzonule	Pr. (?)		*Dtl. hustedtii—Dtl. lauta* subzone		
			Songhagdong F.		Radioralia zonule	Pr. Ac. — Co.		*Rxa. peragalli* subzone		
			Seoam Cg.			Ac. — Co. Do. — Fe. Ve. — An.		*Csd. lewisianus* subzone		*Cbm. triacanth* subzone
	Early Miocene	Yangbug Group	Beomgogri F.							
			Janggi F.	Upper	Trachytic Basalt					
				Middle	Coal Measure					
				Lower	Keumkwang-dong Member					
					Janggi Cg.					

Ve. — An. : *Vicaryella—Anadara* assemblage
Do. — Fe. : *Dosinia—Felaniella* assemblage
Ac. — Co. : *Acesta—Conchocele* assemblage
Pr. : *Propeamussium* assemblage
Co. — Ca. : *Conchocele—Calyptogena* assemblage
Pa. — Mo. : *Patinopecten—Modiolus* assemblage
Tbt. : *Turborotalia*
Glb. : *Globigerina*
Cpd. : *Graspedodiscus*
Dtl. : *Denticula*
Rxa. : *Rouxia*
Csd. : *Coscinodiscus*
Cnp. : *Cannopilus*
Cbm. : *Corbisema*

Bolivina striatula, and *Valvulineria asanoi* are dominant and characteristic. The planktonic forms *Globigerina angustiumbilicata, Gl. bulloides, Gl. falconensis, Tbt. scitula scitula*, and *Tbt. lata?* are known. Among these the dominant and characteristic species are *Gl. bulloides* and *Tbt. scitula scitula*.

From the abundant occurrence of both planktonic and benthic forms in the lower part of the zonule, it is evident that sedimentary condition characterizing the underlying biostratigraphic unit continued into the lower part of the present formation.

Molluscan Assemblages

Six molluscan assemblages are distinguished in this basin (Yoon, 1975):

1. *Vicaryella-Anadara assemblage (Ve.-An.)*

This assemblage is found in the Seoam Conglomerate. Its main constituent species are *Anadara (Hataiarca) daitokudoensis, A. (H.) kakehataensis, Crassostrea gravitesta gravitesta, Megaxinus* (s. s.) *hataii, Trapezium* (*Neotrapezium*) *cheonbugense, Cyclina* (s. s) *japonica, Vicaryella ishiiana, Batillaria* (*Tateiwaia*) *tateiwai*, and *B. (T.) yamanarii*.

The elements of this assemblage are warm-water forms which lived in a shallow brackish embayment. This assemblage is the lowermost of the molluscan sequence of the Yeonil Group.

2. *Dosinia-Felaniella assemblage (Do.-Fe.)*

This assemblage is found in the upper part of the Seoam Conglomerate. Notably, specimens of *Pillucina (Sydlorina) yokoyamai* and *Felaniella* (s.s.) *ferruginata* are abundant. Other main constituent species are *Anadara* (s.s.) *makiyamai, Lithophaga (Leiosolenus) kimi, Chlamys* (s.s.) *akitana, Crassostrea gravitesta gravitesta, Saxidomus* cf. *nuttalli, Dosinia (Phacosoma) nomurai, Siratoria siratoriensis, Vicaryella tyosenica, Batillaria (Tateiwaia) yamanarii*, and *Crepidula isimotoi*.

3. *Acesta-Conchocele assemblage (Ac.-Co.)*

This assemblage is found in the Songhagdong Formation. Stratigraphically, the assemblage is also correlated with the horizon of the *Do.-Fe.* assemblage, but some of the elements occur in the overlying formations. Main representative species are *Acila* (s.s.) cf. *submirabilis, Solemya (Acharax) tokunagai, Chlamys* (s.s.) *akitana, Acesta* (s.s.) cf. *yagenensis, Lucinoma acutilineatum, Conchocele bisecta, Dentalium* cf. *yokoyamai*, and *Euspira meisensis*. Among these, *Solemya (Acharax) tokunagai, Lucinoma acutilineatum*, and *Conchocele bisecta* indicate water conditions deeper than those of the *Do.-Fe.* assemblage.

4. *Propeamussium assemblage (Pr.)*

Propeamussium (s.s.) *tateiwai* represents this assemblage. This species occurs randomly in a fine-grained sandy mudstone and siltstone facies of the upper part of the Songhagdong, Daegok, and Eedong, and the lower part of the Pohang Formations. The *Propeamussium* assemblage is found in horizons stratigraphically higher than the Do.-Fe. assemblage.

5. *Conchocele-Calyptogena assemblage (Co.-Ca.)*

Conchocele bisecta and *Calyptogena* cf. *elongata* constitute this assemblage. The fossils occur densely in a siltstone of the Pohang Formation. These species are to some extent deep-water and silty bottom dwellers.

6. *Patinopecten-Modiolus assemblage (Pa.-Mo.)*

This assemblage consists of *Yoldia (Orthoyoldia) sagittaria, Modiolus* (s.s.) *difficilis, Patinopecten (Mizuhopecten) kimurai ugoensis, Lucinoma acutilineatum*, and *Macoma* (s.s.) *optiva*, and occurs in the uppermost part of the Pohang Formation.

Diatom and Silicoflagellata Zones

Lee (1975, 1976) studied the Pohang Basin and established diatom and silicoflagellata zones. The diatom zonation comprises four subzones, namely, the *Coscinodiscus lewisianus, Rouxia peragalli, Denticula hustedtii-D. lauta*, and *D. hustedtii-Craspedodiscus rhombicus* Subzones. The silicoflagellate zones include the *Corbisema triacanth* and *Distephanus longispinus* Subzones. Table 1 shows the age of the zones and a correlation with foraminifer and mollusc zones.

Tertiary deposits of the Eoil Basin are divided into three formations, the Gampo Conglomerate, Hyodongri Volcanics, and Eoil Formation, in ascending order. They are mostly terrestrial sediments except for the middle part of the Eoil Formation which is interfingered with marine sediments. These marine sediments yield many molluscs such as *Vicarya callosa japonica, Anadara kakehataensis*, and *Crassostrea gravitesta eoilensis*. Recently some dinoflagellates are found from this bed (Yun, 1982 in print).

The Ulsan Basin is a very small one and is seperated from the Eoil Basin by high ridges composed of Mesozoic rocks. The only recognized formation, the Hwabongri Formation, yields many benthic foraminifers and some dinoflagellates. By means of foraminifers, the Hwabongri Formation can be correlated with the Eedong and Pohang Formations of the Pohang Basin.

The island of Jeju is situated at lat. 33° 35′ N., and long. 126° 10′ to 126° 55′ E. The Seogwipo Formation is distributed at Seogwipo village on the southern middle

Table 2. Correlation of the Tertiary deposits in southeastern Korea (Kim, 1970).

Geologic age		Pohang		Bugpyeong	Yeonghae		Eoil	Ulsan	Jeju
Pliocene				Bugpyeong Conglomerate					Seoguipo F.
Miocene	Mid. Miocene	Yeonil Group	Pohang F.	Bugpyeong Formation (?)	Yeonghae Group	Yeongdong Formation		Hwabongri Formation (?)	
			Eedong F.						
			Daegog F.						
			Songhagdong F.			Yeonghae Conglomerate			
			Seoam Conglomerate			Dogogdong Formation			
	Early Miocene	Yangbug Group	Beomgogri F.				Eoil F. and Basalt		
			Janggi – Upper: Trachytic Basalt				Hyodongri Volcanics		
			Janggi – Middle: Coal Measure						
			Janggi – Lower: Keumkwang-dong Member				Gampo Conglomerate		
			Janggi – Lower: Janggi Cg.						

coast of the island. The formation attains over 50 meters in thickness and consists mainly of light gray to brown fine to medium sandstone, sandy shale, and shale. The formation has three molluscan fossil beds and three diastems. Yokoyama (1923) proposed the age of the formation to be late Pliocene by his molluscan study, but Haraguchi (1931), who collected a number of fossil molluscs, brachiopods, echinoids, bryozoans, corals and tooth of fish, considered the formation to belong to the Pleistocene. Kim (1972) has studied this area paleontologically and found many foraminifers. The geological age of the formation may be more likely to be Pliocene by the planktonic foraminifers, *Turborotalia humerosa, Tbt. crassaformis, Tbt. acostaensis, Pulleniatina obliquiloculata, Orbulina universa,* and *Globigerinoides ruber ruber,* and sedimentary environments of the formation might have been a littoral zone in shallow open warm sea.

Formations in the above-mentioned six basins are correlated as shown in Table 2.

References

Chang,K.H. and Lee, Y.G., 1975: Diatom flora of Chonbuk (conglomerate) Formation in Chonbuk-myon, Wolsong-gun, Kyongsangbuk-do, Korea. *Kyongpook Natl. Univ., Res.Rev.*, v. **19**, p. 53–62 (in Korean with English abstract).

Kim, B.K., 1965: The stratigraphy and paleontologic studies on the Tertiary (Miocene) of the Pohang area, Korea. *Seoul Univ., Jour, Sci. Tech. Ser.*, v. **15**, p. 32–121.

Kim, B.K., 1970: A study on the Neogene Tertiary deposits in Korea. *Geol. Soc. Korea, Jour.*, v. **6**, no. 2, p.77–96. (in Korean with English abstract)

Kim, B.K., 1972: A stratigraphic and paleontologic study of the Seogwipo Formation, *Celeb. 60th Birth. Prof. Chi Moo Son, Mem.*, p. 169–182 (in Korean with English abstract).

Kim, B.K., 1979: Correlation of Miocene deposits in southeastern Korea. *Geol. Soc. China, Mem.*, no. 3, p. 93–101.

Lee,Y.G., 1975: Neogene diatoms of Pohang and Gampo areas, Kyogsangbug-do, Korea. *Geol. Soc. Korea, Jour.*, v. **11**, no. 2, p. 99–114.

Lee, Y.G., 1976: Fossil diatoms in the upper part of the Foil Formation, Eoil area, Gyeongsangbuk-do, Korea. *Korean Inst. Min. Geol., Jour.*, v. **9**, no. 2, p. 77–83.

Yoon, S., 1975: Geology and paleontology of the Tertiary Pohang Basin, Pohang district, Korea. Part 1, Geology. *Geol. Soc. Korea, Jour.*, v. **11**, no. 4, p. 187–214.

Yoon, S., 1976: The Tertiary deposits of the Ulsan Basin. No. 1, Tertiay deposits in the eastern block. *Coll. Lib. Arts and Sci.*, v. **15**, Nat. Sci. Ser., p. 67–71.

Yoon, S., 1979: Neogene molluscan fauna of Korea. *Geol. Soc. China, Mem.*, no. 3, p. 125–130.

Yun, H.S., 1981: Dinoflagellates from Pohang Tertiary Basin, Korea. *Geosci. Miner. Resources Kier, Rept.*, v. **11**, p. 5–15.

III-9
Neogene Biostratigraphy and Chronology of Japan

Ryuichi TSUCHI

Introduction

Neogene marine sequences in Japan are distributed more extensively over the islands than those of any other period and include abundant pyroclastic materials (Fig.1). These deposits commonly contain planktonic flora and fauna as well as benthic molluscs and larger foraminifers.

Neogene chronostratigraphy in Japan has been much advanced in recent years by incorporating various methods of detailed biostratigraphy using planktonic microfossils, radiometric dating, and magnetostratigraphy. During the Negogene, a warm-water fauna dominated southern Japan, whereas a cold-water fauna prevailed in northern Japan and the Sea of Japan region. Thus, Neogene faunal distributions reflect similar oceanographic and climatic conditions to those of the present. In turn, these distributional trends make precise interregional correlations of sedimentary sequences in southern Japan with those in northern Japan and the Sea of Japan side difficult.

Correlation of Sedimentary Sequences in Southern Japan with those in Northern Japan

A summary of biostratigraphic and chronostratigraphic correlation of the Neogene marine sequences of Japan is presented on Fig.2. Selected Neogene sections from south to north are arranged from left to right in the figure. Correlations of sections in southern Japan with those in northern Japan and the Sea of Japan side based upon planktonic foraminifera and diatom zonations are also shown in this figure.

In sections of southern Japan, 25 early Miocene to early Plesistocene planktonic foraminiferal datum levels can be recognized. Most of these datums are traceable from tropical to middle latitude areas including southern Japan. The presence of tropical planktonic foraminfers has allowed recognition of Blow's Neogene planktonic foraminiferal zones N.6 to N.22, even though some of the *Globortalia foshi* group are absent in this region. Magnetostratigraphy as well as K-Ar and fission track dating of volcanic and pyroclastic rocks have allowed independent calibration of foraminiferal biohorizons in a number of sequences.

In sections along the Sea of Japan, only a few datum levels of tropical planktonic foraminifers are traceable from southern Japan during part of the early Neogene. Alternately, datum levels of temperate planktonic foraminifers such as *Globoquadrina asanoi* can be traced throughout southern and northern Japan. These taxa are useful for the correlation of sections in northern Japan and also on the Sea of Japan side. Radiolarians and calcareous nannoplankton are also important in this region. Thus, Neogene sections in northern Japan and along the Sea of Japan coast are not easily correlated with tropical planktonic foraminferal zones of southern Japan.

Chronologic Framework of Microplanktonic Biohorizons

Planktonic foraminiferal datum levels applicable to the correlation of land-based Neogene sequences in Japan and their estimated ages are arranged on the radiometric time scale on the left side of Fig.2. This chart represents an attempt to construct a new microplanktonic time scale incorporating revised values of K-Ar dating and a revised geomagnetic polarity time scale. The geomagnetic polarity time scale for the interval 0–5 Ma used here is that given by Mankinen and Dalrymple (1979). For the interval 5–6 Ma, a revision of the scale originally proposed by McDougall *et al.* (1977) in accordance with the new decay constants is used.

Planktonic foraminiferal datum levels for the interval 0–6 Ma are calibrated against the geomagnetic polarity time scale based upon biostratigraphies and magnetostratigraphies of deep sea cores V24–59 (Hays *et al.*, 1969) and RC 12–66 (Saito *et al.*, 1975) from the equa-

Geoscience Institute, Faculty of Science, Shizuoka University, Shizuoka 422, Japan

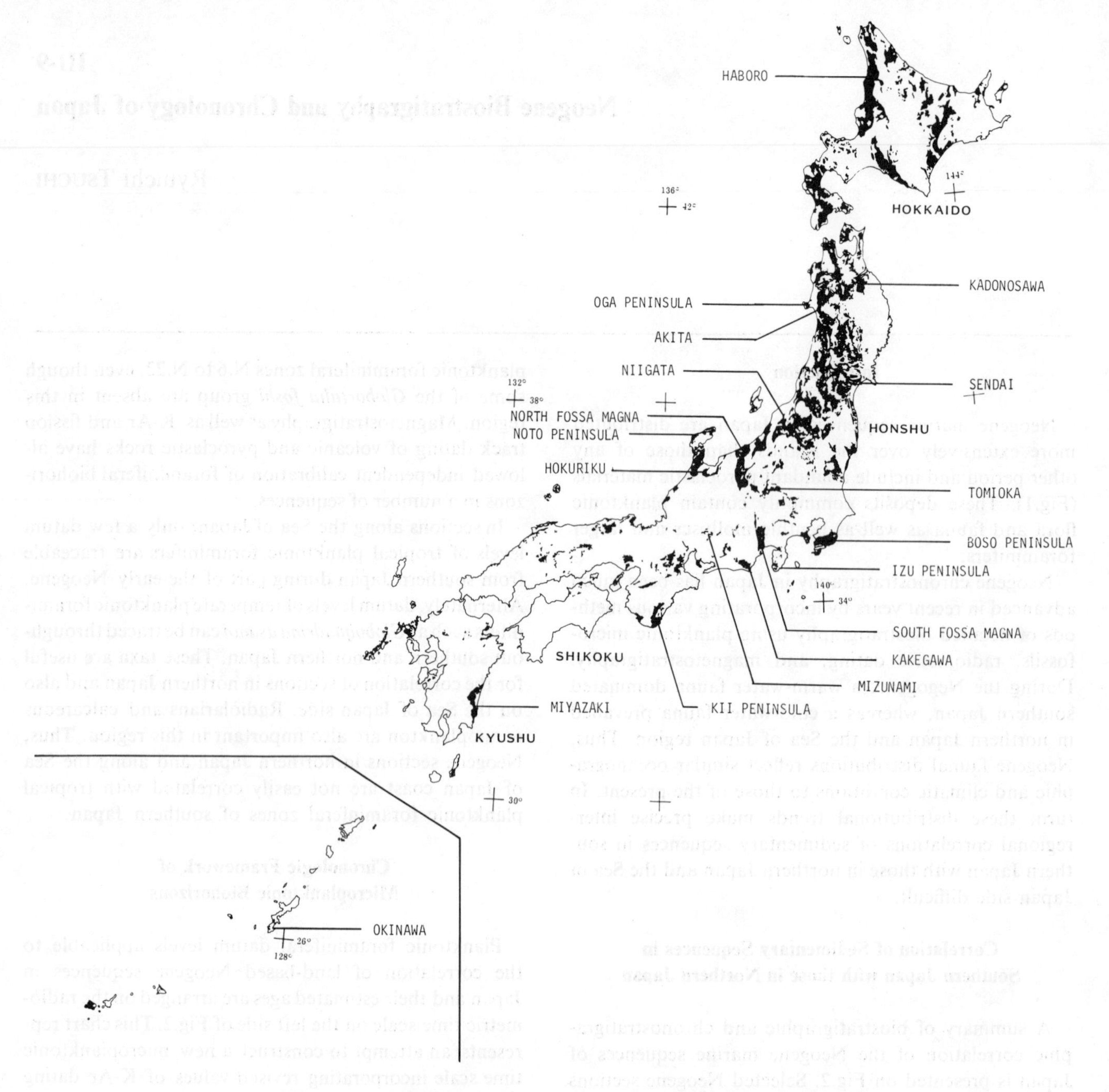

Fig. 1. Distribution of Neogene marine strata and Neogene sedimentary basins in Japan, compiled from Geological Map of Japan (1978).

torial Pacific, V20–119 (Maiya *et al.*, 1976) from the North Pacific, and a subareal Neogene section of the Kakegawa area (Tsuchi and Ibaraki, 1978). For the interval prior to 6 Ma, chronological calibrations are made primarily on the basis of the biostratigraphy of DSDP Site 289 in the equatorial Pacific (Srinivasan and Kennett, 1981) by assuming a constant rate of sediment accumulation between radiometrically estimated ages for the *Orbulina* datum at 15.5 Ma and the *Pulleniatina* datum at 5.8 Ma.

Here, the *Orbulina suturalis* base datum at 15.5 Ma is estimated from an interpolation of K-Ar ages in section of the Kii Peninsula and the Hokuriku area (Ikebe *et al.*, 1972; Tsuchi *et al.*, 1981). The *Globigerina nepenthes* base datum at 11. 6 Ma is derived from K-Ar ages of a tuff just below the datum level in sections of Tomioka area (Shibata *et al.*, 1979). The *Pulleniatina primalis* base datum at 5.8 Ma is estimated from the geomagnetic polarity time scale on the basis of biostratigraphy in the equatorial Pacific given by Saito *et al.* (1975). Planktonic foraminiferal zones referred to are thoes of Blow(1969).

Diatom bio-events for the interval 0–6 Ma are chronologically calibrated against the geomagnetic polarity time scale based upon biostratigraphies given by Donahue (1970), Koizumi (1975), and Burckle and Opdyke (1977). For the interval prior to 6 Ma, chronological calibrations are made according to biostratigraphic data available from land-exposed sections and K-Ar and fission track ages, and also estimated rates of sediment accumulation in the Northwest Pacific presented by Barron (1980), Harper (1980), and Koizumi (1973, 1979).

It is noteworthy that the estimated ages of planktonic foraminiferal and diatom datum levels have been made independently by different procedures. The *Orbulina* datum estimated at 15.5 Ma is, however, considered to be nearly synchronous with the *Denticulopsis lauta* base datum on the basis of field examinations (Morozumi and Koizumi, 1981).

Besides planktonic foraminifers and diatoms, some datum levels of radiolarians and calcareous nannoplankton have also been utilized for the correlation of the Neogene sequences. Chronological calibrations of datum levels of both taxa are shown in Fig.3. Radiolarian events for the interval 0–6 Ma are chronologically calibrated against the geomagnetic polarity time scale based upon biostratigraphies of the North Pacific and Japanese subareal sections (Hays, 1970; Hays and Schackleton, 1976; Sakai *et al.*, 1981). For the interval 6–17 Ma, chronological calibrations are made in the same manner as for planktonic foraminifera based on the biostratigraphy of DSDP Site 289, partly combining biostratigraphic data available from DSDP Site 436 in the Northwest Pacific. For the interval prior to 17 Ma, some revisions of the calibration given by Theyer *et al.* (1978) are made. Radiolarian zones are those proposed by Riedel and Sanfilippo (1978) and Sakai (1980).

Calcareeous nannoplankton bio-events for the interval 0–6 Ma are chronologically calibrated against the geomagnetic polarity time scale based upon biostratigraphies of deep sea cores (Haq *et al.*, 1977; Haq and Berggren, 1978; Gartner, 1979; Haq *et al.*, 1980). For the interval prior to 6 Ma, chronological calibrations are also made according to the biostratigraphy of DSDP Site 289 in

---

Abbreviations for generic and specific names used in Figs. 2,3, and 4.

(Planktonic foraminifera)
Gds., Globigerinoides;
Gla., Globigerinatella;
Gna., Globigerina;
Gqa., Globoquadrina;
Grt., Globorotalia;
Gta., Globigerinita;
Spa., Sphaeroidinella;
Sps., Sphaeroidinellopsis.

(Diatom zones)
D. s., Denticulopsis seminae;
R. c., Rhizosolenia curvirostris;
A. o., Actinocyclus oculatus;
D. s. v., Denticulopsis seminae var. *fossilis;*
D. k., Denticulopsis kamtschatica;
D. h., Denticulopsis hustedtii;
D. l., Denticulopsis lauta;
A. i., Actinocyclus ingens;
K. c., Kisseleviella carina.

(Diatom)
A., Actinocylus;
C., Coscinodiscus;
Cm., Cosmiodiscs;
H., Hemidiscus;
N., Nitzschia;
P., Pseudoeunotia;
R., Rhizosolenia;
T., Thalassiosira

(Radiolaria)
A., Axoprunum angelinum;
B., Batryostrobus aquilonaris;
B., Buccinosphaera invaginata;
C., Calocycletta costata;
C., Cannartus petterssoni, C. sp. B, *C.* sp. A;
C., Collosphaera tuberosa;
C., Cyrtocapsella tetrapera, C. japonica, C. cornuta;
D., Dorcadospyris alata;
E., Eucyrtidium matuyamai, E. yatuoense, E. inflatum;
L., Lamprocyrtis heteroporos;
L., Lithopera neotera, L. bacca, L. renzae;
L., Lychnocanoma nipponica magnacornuta;
O., Ommatartus antepenultimus, O. penultimus, O. hughesii
P., Pterocanium prismatium;
S., Stichocorys peregrina, S. delmontensis, S. wolffi, S. almata;
S., Stylacontarium acquilonium;
S., Sphaeropyle langii;
T., Thecosphaera japonica;

(Calcareous nannoplankton)
A., Amaurolithus primus;
C., Calcidiscus macintyrei;
C., Catinaster coalitus;
C., Cyclicargolithus floridanus;
D., Discoaster brouweri, D. tamalis, D. quinqueramus, D. berggrenii, D. deflandrei;
E., Emiliania huxleyi;
G., Gephyrocapsa oceanica;
H., Helicosphaera sellii;
P., Pseudoemiliania lacunosa;
R., Reticulofenestra pseudoumbilica;
S., Sphenolithus heteromorphus;

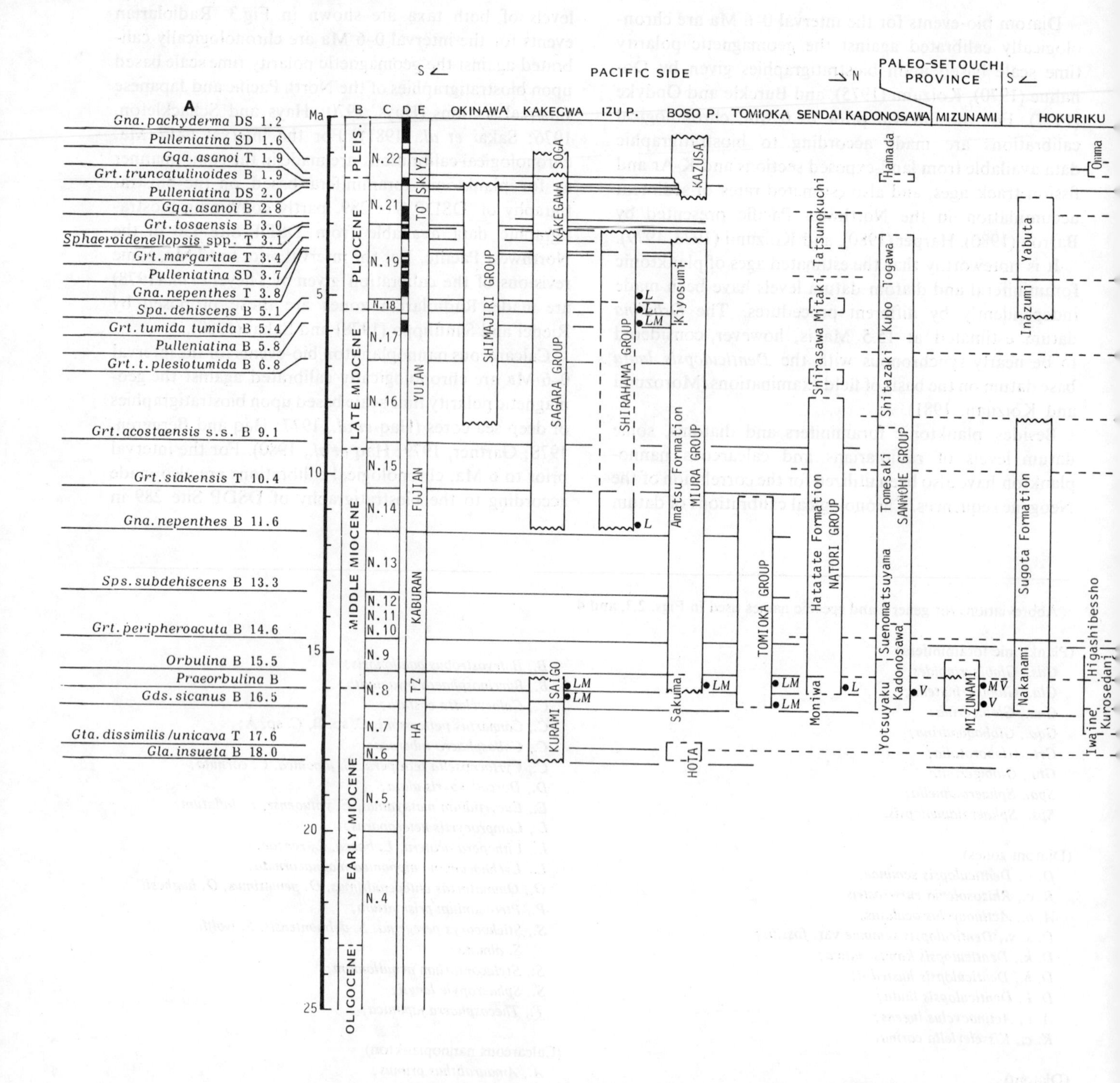

Fig. 2. Bio-stratigraphic and chronostratigraphic correlation of selected Neogene sequences in Japan. (Slightly modified from Tsuchi *et al.*, 1981)

A: Planktonic foraminiferal datum levels and their estimated ages (Ma), B: Geologic age, C: Planktonic foraminiferal zones after Blow(1969), D: Geomagnetic polarity time scale, E: Japanese Neogene Stages (HA: Haranoyan, TZ: Tozawan, TO: Totomian, S: Suchian, K: Kechienjian, YZ: Yuzanjian), F: Diatom zones after Donahue (1970) and Koizumi(1973, 1979), G: Datum levels and

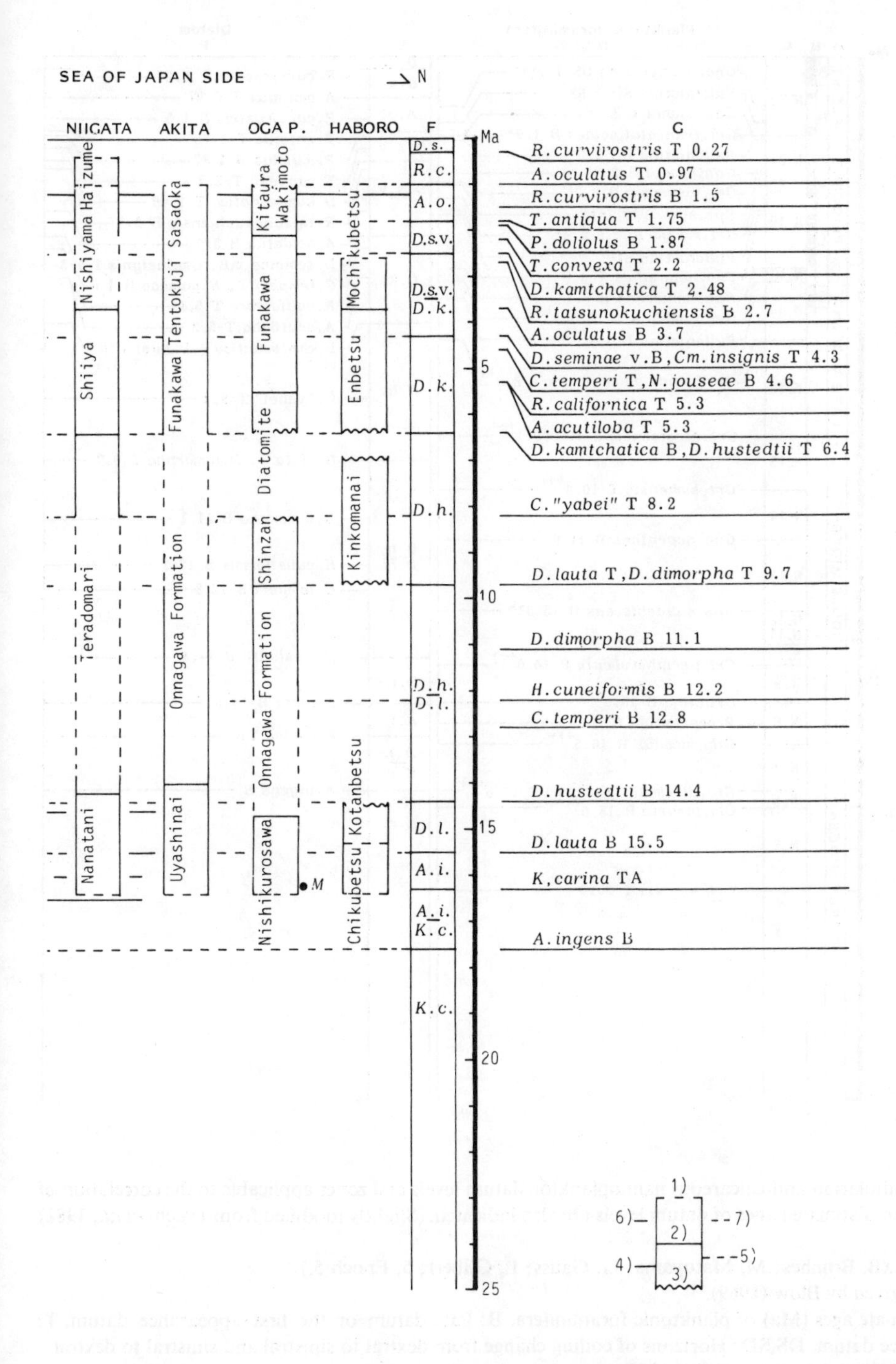

their approximate ages (Ma) of diatoms, B: Base datum, T: Top datum, TA: Top abundant or the highest level of the acme of the species, DS,SD: Horizon of coiling change from dextral to sinistral or sinistral to dextral in planktonic foraminifera.

1) Presumed biostratigraphic position of upper or lower limit of the strata, 2) Conformity, 3) Unconformity, 4) Planktonic foraminiferal datum level confirmed, 5) Diatom datum level confirmed, 6) Planktonic foraminiferal datum level inferred, 7) Diatom datum level inferred.

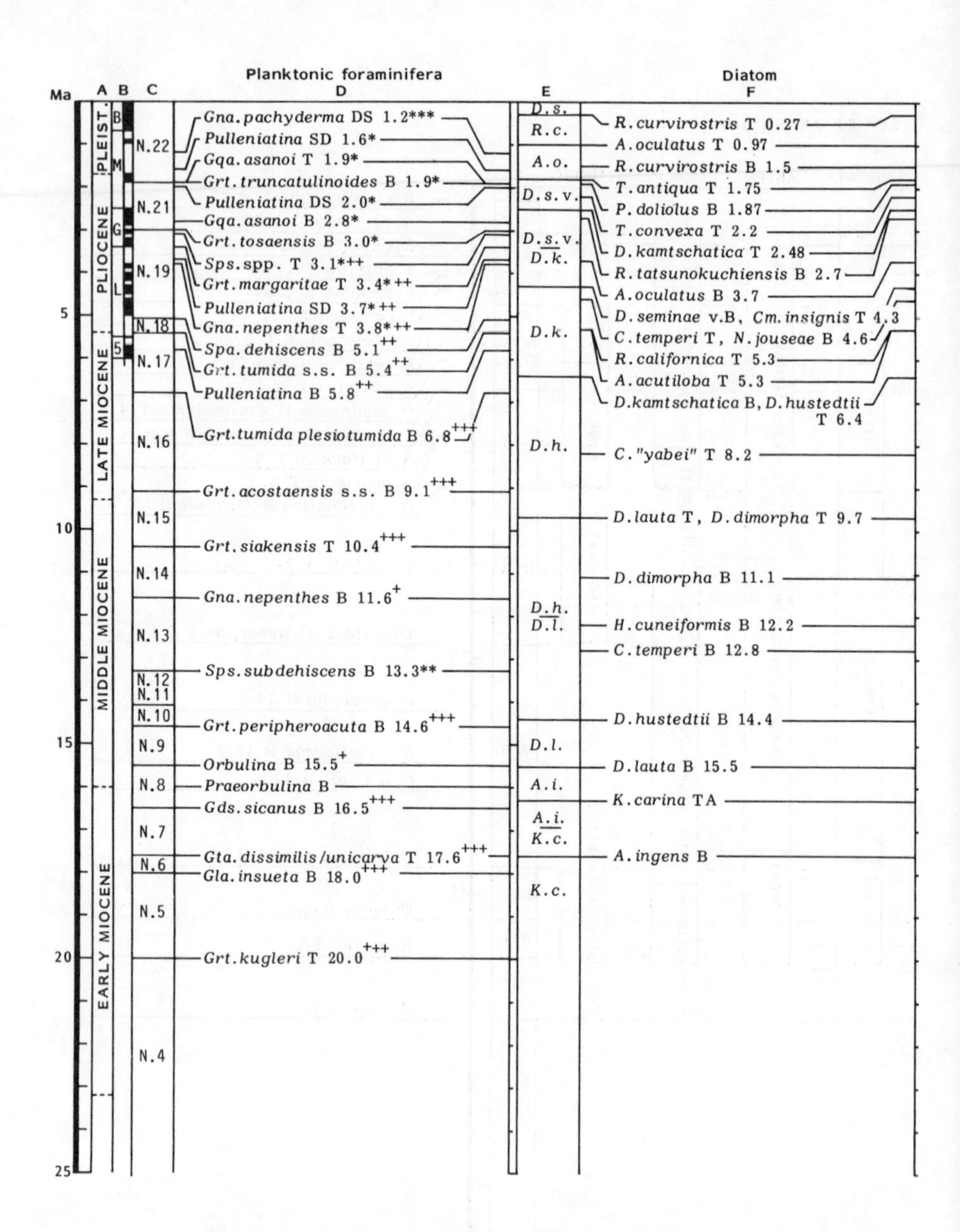

Fig. 3. Planktonic foraminiferal, diatom, radiolarian and calcareous nannoplankton datum levels and zones applicable to the correlation of Neogene land-based sections in Japan. Estimated ages of datum levels are also indicated. (Slightly modified from Tsuchi *et al.*, 1981)

A: Geologic age.
B: Geomagnetic polarity time scale.(B, Brunhes; M, Matuyama; G, Gauss; L, Gilbert; 5, Epoch 5.)
C: Planktonic foraminiferal zones given by Blow (1969).
D: Datum levels and their approximate ages (Ma) of planktonic foraminifera. B: Base datum or the first appearance datum. T: Top datum or the last appearance datum. DS,SD: Horizons of coiling change from dextral to sinistral and sinistral to dextral.
E: Diatom zones after Donahue (1970) and Koizumi (1973, 1979).
F: Datum levels and their approximate ages (Ma) of diatom species. TA: Top abundant or the highest level of the acme of the species.
G: Radiolarian zones given by Riedel and Sanfillippo (1978) and Sakai (1980).
H: Datum levels and their approximate ages (Ma) of radiolarians.
I: Calcareous nannoplankton zones by Martini (1971).
J: Zones by Okada and Bukry (1980).
K: Datum levels and their approximate ages (Ma) of calcareous nannoplankton.

Published sources of age estimations of the respective datum levels are marked by symbols as follows:

Planktonic foraminifera ——

+: Estimate from K-Ar ages (Kawai and Hirooka, 1966; Shibata and Nozawa, 1968; Ikebe *et al.*, 1972; Shibata, 1973; Shibata *et al.*, 1979).

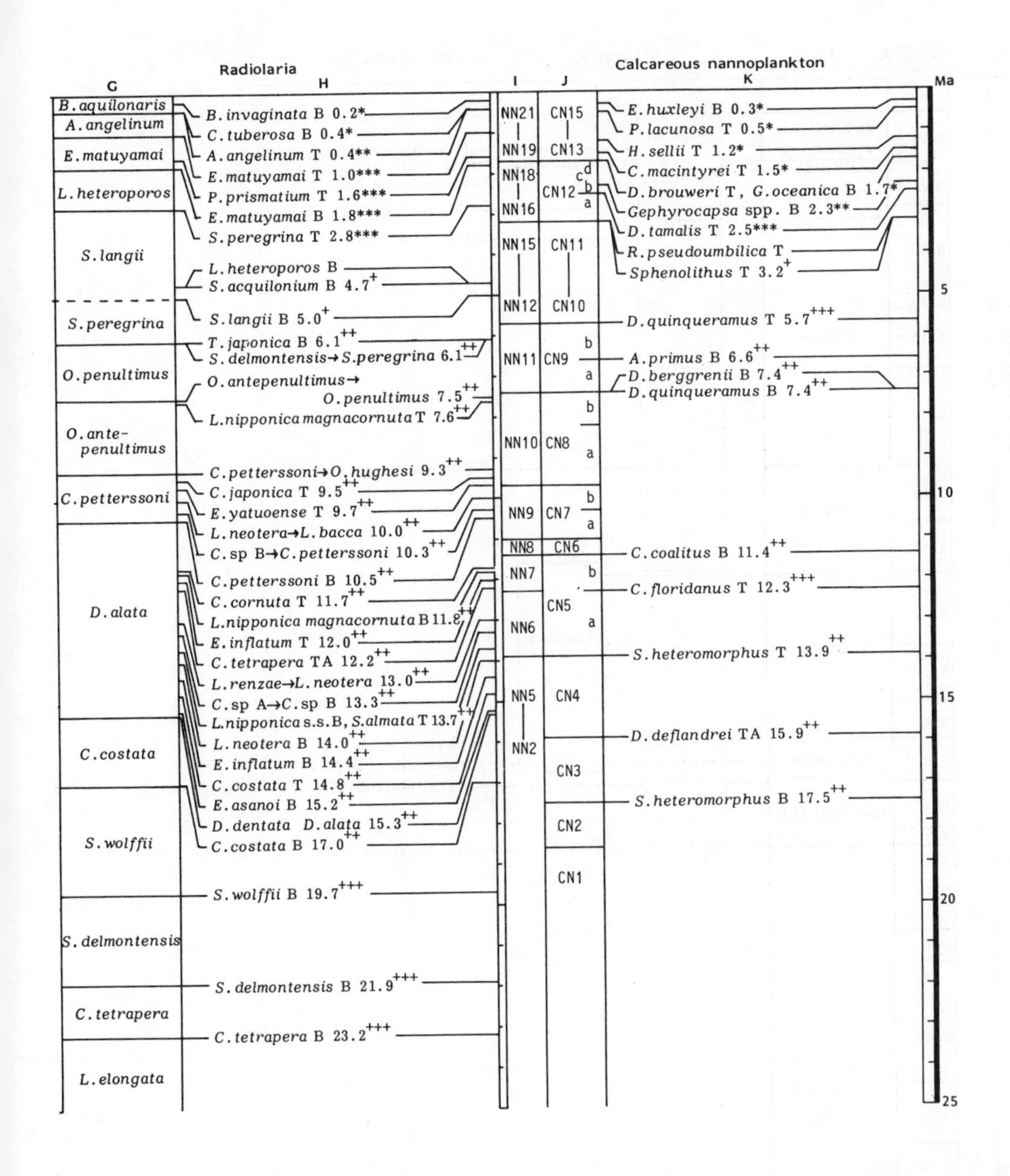

++: Estimate from biostratigraphy and magnetostratigraphy of core V24-59 and RC12-66 (Hays *et al.*, 1969; Saito *et al.*, 1975).
+++: Estimate from biostratigraphy of DSDP Site 289 (Andrews and Packam *et al.*, 1975; Srinivasan and Kennett, 1981)
*: Estimate from biostratigraphic and magnetostratigraphy of the Neogene Kakegawa section (Tsuchi and Ibaraki, 1978).
**: Estimate from biostratigraphy of the Neogene section of Boso Peninsula (Oda, 1977)
***: Estimate from biostratigraphic and magnetostratigraphy of core V20–119 (Maiya *et al.*, 1976).

Radiolaria——
*: Estimate based on Sakai *et al.* (1981).
**: Estimate from that given by Hays and Shackleton (1976).
***: Estimate from that given by Hays (1970).
+: Estimate based on unpublished data by T. Sakai.
++: Estimate from biostratigraphy of DSDP Site 289, partly combining biostratigraphic data available from Site 436.
+++: Estimate from that given by Theyer *et al.* (1978).

Calcareous nannoplankton——
*: Estimate from that given by Gartner (1979).
**: Estimate from that given by Haq *et al.* (1977).
***: Estimate from that given by Okada and Bukry (1980).
+: Estimate from that proposed by Haq and Berggren (1978).
++: Estimate from biostratigraphy of DSDP Site 289 (Unpublished data by H. Okada)
+++: Estimate from that given by Haq *et al.* (1980).

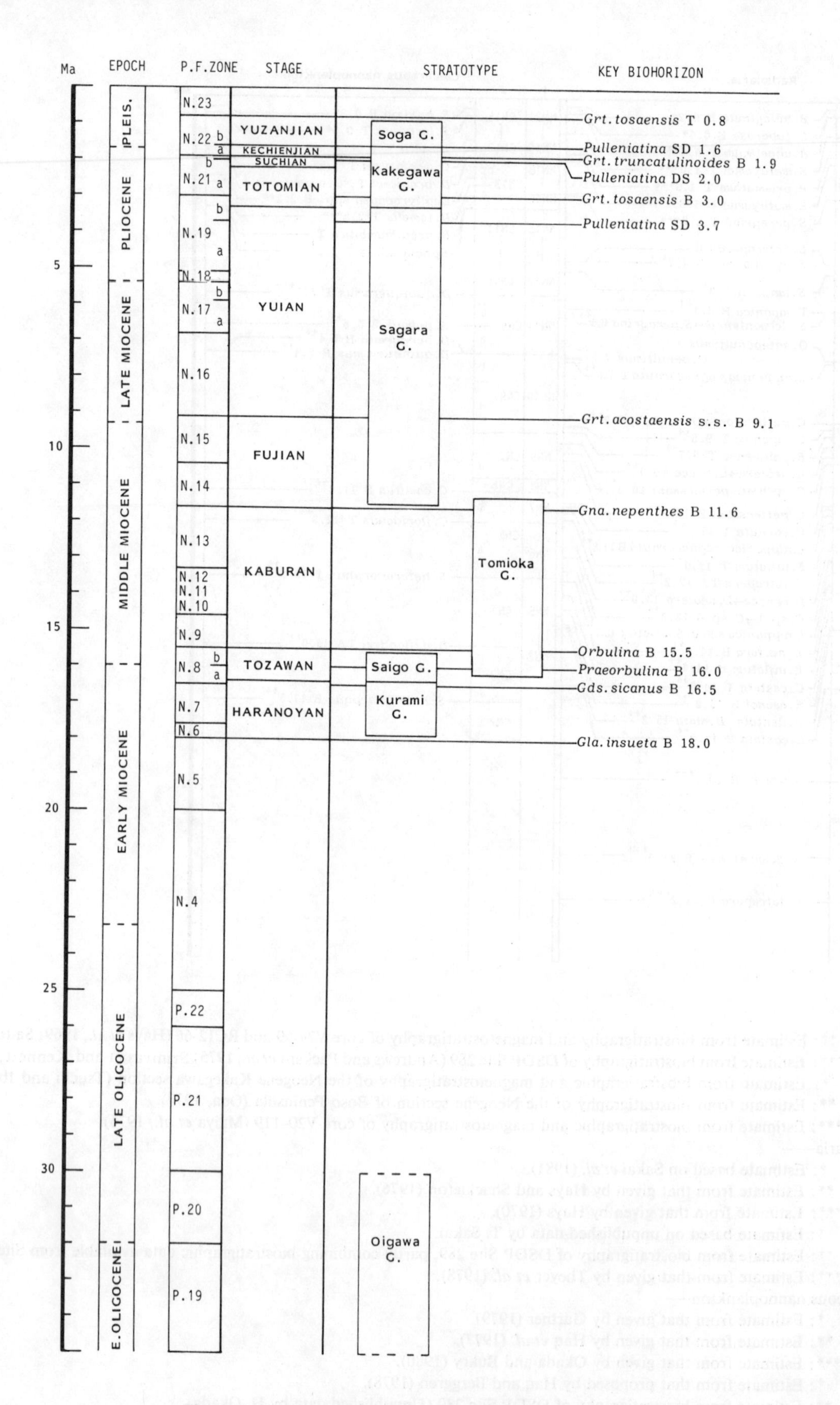

Fig. 4. Japanese Neogene stages, stratotypes and key bio-horizons.

the equatorial Pacific. Calcareous nannoplankton zones are those proposed by Martini (1971), and Okada and Bukry (1980).

Japanese Neogene Stages

Japanese Neogene stages were first proposed by Makiyama (1939) on the basis of lithostratigraphies of central Japan and their benthic faunal characters. Those included the Asagaian, Oigawan, Togarian, Yuian, Dainitian, and Kechienjian stages, in ascending order. Later, the "middle Miocene" Tozawan and the "late Pliocene" Yuzanjian were added (Makiyama, 1947), and the Dainitian was replaced by the Suchian (Makiyama and Sakamoto, 1957). Recently, Ikebe (1978) and Ikebe and Chiji (1981) proposed two middle Miocene stages, the Kaburan and the Fujian, and also two early Miocene stages provisionally called LM-1 and LM-2.

According to recent biostratigraphic studies on planktonic foraminifers, the Tozawan, Togarian, and a part of the Oigawan Stages are all of latest early Miocene age, and another part of the Oigawan is included in the late Oligocene. The Asagaian is, at present, considered to be of late Oligocene age. Here, the author proposes the new name Haranoyan for the LM-2 stage on the basis of recent studies of the stratotype, the Kurami Group in the Kakegawa area (Ibaraki *et al.*, 1983), and uses Fujian as corresponding to a lower part of the Yuian, which was originally assigned to the entire section of the Sagara Group extending from middle Miocene to early Pliocene time. Chronostratigraphic positions of stages lower than the Haranoyan are not yet been fully understood. The Haranoyan, Kaburan, and Fujian have been defined by planktonic foraminiferal biohorizons. The stratotypes of the Kaburan and Fujian are, respectively, the Tomioka Group and a part of the Nishiyatsushiro Group in central Japan. The Tozawan was defined originally with the *Lepidocyclina-Miogypsina*-bearing Saigo Group in the Kakegawa area as its type. The Yuian and later stages have been defined by molluscan characteristics in the Kakegawa area (Tsuchi, 1961) and recently also by planktonic foraminiferal events (Tsuchi and Ibaraki, 1978). The biostratigraphic and chronostratigraphic positions of respective stages and stratotypes based on planktonic biostratigraphies are shown in Fig. 4.

Transitions in Epoch Boundaries

Sequences assignable to the early part of the early Miocene have only rarely been identified in Japan. In fact, there may have been a large hiatus in the lower Miocene of many Neogene sequences in Japan and a continuous section which can define the Oligocene/Miocene or the Paleogene/Neogene boundary has yet to be ascertained. Recently, however, some early Miocene sequences have been examined in the so-called "Paleogene" complex on the Pacific coast of southwestern Japan (Shuto, 1963; Saito, 1980).

The Miocene/Pliocene boundary defined at the base of zone N.18 can be drawn in the Kakegawa area at a horizon within the Sagara Alternating Sand and Silt of the Sagara Group (Tsuchi and Ibaraki, 1978) and in Oga Peninsula at a black shale horizon of the lower part of the Funakawa Formation (Koizumi and Kanaya, 1977). The Miocene-Pliocene transition in marine sequences has, therefore, been recognized in a continuous sequence consisting of a similar lithology.

The Pliocene/Pleistocene or Neogene/Quaternary boundary defined by a horizon within a continuous section including closely successive datum levels of the base *Globortalia truncatuinoides* at 1.9 Ma, top *Globoquadrina asanoi,* top *Discoaster broweri,* and the SD horizon of *Pulleniatina* at 1.6 Ma, a horizon of abrupt coiling change in the test of *Pulleniatina* spp. from sinistral to dextral can be examined in the Kakegawa area, Boso Peninsula, and other areas where no marked change of the paleoenvironment has been noticed. However, some climatic fluctuations are suggested by the disappearance of tropical elements in the Pliocene Kakegawa molluscan fauna in the Kakegawa section near the SD horizon of *Pulleniatina* at about 1.6 Ma.

Acknowledgements

I wish to express my sincere thanks to Prof. Nobuo Ikebe, leader of the International Geological Correlation Programme Project 114–Evaluation of biostratigraphic datum planes of the Pacific Neogene, for his leadership in this project. The deepest gratitude is also due to all the members of the national working group of Japan for their helpful cooperation toward the accomplishment of our research project.

References

Andreus, J.E., Packam, G., *et al.*, 1975: Site 289. *DSDP, Init. Repts.*, v. **30**, p. 231-398,

Barron, J.A., 1980: Lower Miocene to Quaternary diatom biostratigraphy of Leg 57, off northwestern Japan, Deep Sea Drilling Project. *DSDP, Init. Repts.*, v. **56–57**, pt. 2, p. 641–685.

Blow, W.H., 1969: Late Middle Eocene to Recent planktonic foraminiferal biostratigraphy. *In* Bronnimann, P. and Renz, H.H., eds., *lst Internat. Conf. Plankt. Microfossils, Genova, 1967, Proc.*, v. **1**, p. 119–421.

Burckle, L.H. and Opdyke,N.D., 1977: Late Neogene diatom correlations in the circum-Pacific. *1-CPNS, Tokyo, 1976, Proc.*, p. 255–284.

Donahue, J.G., 1970: Pleistocene diatoms as climatic indicators in North Pacifc sediments. *In* Hays, J.D. ed., *Geol. Invest. North Pacific, Geol. Soc. Amer., Mem.*, no. 126, p. 121–138.

Gartner, S., 1977: Calcareous nannofossil biostratigraphy and revised zonation of the Pleistocene. *Marine Micropal.*, v. **2**, no. 1, p. 1–25.

Haq, B.U., Berggren, W.A. and van Couvering, J.A., 1977: Corrected age of the Pliocene/Pleistocene boundary. *Nature*, v. **269**, p. 483–488.

Haq, B. U. and Berggren, W.A., 1978: Late Neogene calcareous plankton biochronology of the Rio Grande Rise (South Atlantic Ocean). *Jour.Pal.*, v. **52**, p. 1167–1194.

Haq, B.U. *et al.*, 1980: Late Miocene marine carbon-isotopic shift and synchroneity of some phytoplanktonic biostratigraphic events. *Geology,* v. **8**, p. 427–431.

Harper, H.E. Jr., 1980: Diatom biostratigraphy of Sites 434,435 and 436, north-western Pacific, Leg 56, Deep Sea Drilling Project. *DSDP, Init. Repts.*, v. **56-57**, pt. 2, p. 633–639.

Harper, H.E., Jr. and Schackleton, N.J., 1976: Globally synchronous extinction of the radiolarian *Stylatractus universus. Geology*, v. **4**, p. 649–652.

Hays, J.D., 1970: Stratigraphy and evolutionary trends of Radiolaria in North Pacific deep-sea sediments. *In* Heys J.D., ed., *Geol. Inv. North Pacific., Geol. Soc. Am., Mem.*, v. **126**, p. 185–218.

Hays, J.D., Saito, T., Opdyke, N.D., and Burckle, L.H., 1969: Pliocene-Pleistocene sediments of the equatorial Pacific; Their paleomagnetic, biostratigraphic and climatic record. *Geol. Soc. Amer., Bull.*, v. **80**, p. 1481–1513.

Ibaraki, M., Tsuchi, R. and Takayanagi, T., 1983: Early Neogene planktonic foraminiferal biostratigraphy in the Kakegawa area, the Pacific coast of central Japan. *Shizuoka Univ. Fac. Sci., Repts.*, v. **17**, p. 101–116.

Ikebe,N., 1978: Bio- and chronostratigraphy of Japanese Neogene, with remarks on paleogeography. *In* Hujita, K. *et al.*, eds., *Cenozoic Geology of Japan, Prof. N. Ikebe Mem., Vol.*, p. 13–34.

Ikebe, N. and Chiji, M., 1981: Important datum planes of the western Pacific Neogene (revised) with remarks on the Neogene stages in Japan. *In* Tsuchi, R., ed., *Neogene of Japan*, IGCP-114 Natl. Working Group of Japan, p. 1–14.

Ikebe, N., Takayanagi, Y., Chiji, M., and Chinzei, K., 1972: Neogene biostratigraphy and radiometric time scale of Japan– An attempt at international correlation. *Pacific Geology*, no. 4, p. 39–78.

Kawai, N. and Hirooka, K., 1966: Radiometric dates of some Cenozoic igneous rocks in Southwest Japan. *Sympo. Age of the acidic rocks in Japan, Preprint*, no. 5, Geol. Soc. Japan.

Koizumi, I., 1973: The late Cenozoic diatoms of Sites 183–193, Leg. 19, Deep Sea Drilling Project. *In* Creager, J.S., Scholl, D.W., *et al.*, eds., *DSDP, Init. Repts.*, v. **19**, p. 805–855.

Koizumi, I., 1975: Late Cenozoic diatom biostratigraphy in the circum-Pacific region. *Geol. Soc. Japan, Jour,* v. **81**, p. 611–627.

Koizumi, I., 1979: A note on microbiostratigraphy of deep-sea cores and landbased sections. *Chikyu,* v. **1**, p. 226–229.

Koizumi, I. and Kanaya,T., 1977: Correlation of late Neogene sections on the Oga Peninsula and Akita City, Northeast Japan. *In* Takayasu, T., *et al.*, eds., *Prof.K. Huzioka Mem. Vol.*, p. 401–412.

Maiya, S., Saito, T. and Sato, T., 1976: Late Cenozoic planktonic foraminiferal biostratigraphy of northwest Pacific sedimentary sequences. *In* Takayanagi, Y. and Saito, T., eds., *Progress in Micropalaeontology*, Spec. Pub. Micropal. Press, p. 395–422.

Makiyama, J., 1939: The Neogenic stratigraphy of the Japan Islands. *6th Pac. Sci. Cong., Proc.*, v. **2**, p. 641–649.

Makiyama, J., 1947: Two stages of Middle Miocene in Japan. *Kyoto Univ. Coll. Sci., Mem., ser. B*, v. **19**, pt. 1, p. 33–36.

Makiyama, J. and Sakamoto,T., 1957: Geological map and explanatory text of the geological map of Japan, Mitsuke and Kakegawa, 1:50,000. Geol. Surv. Japan.

Mankinen, E.A. and Dalrymple,G.B., 1979: Revised geomagnetic polarity time scale for the interval 0–5 m.y.B.P. *Jour. Geophys. Res.*, v. **84**, p. 615–626.

Martini, E., 1971: Standard Tertiary and Quaternary calcareous nannoplankton zonation. *2nd Conf. Plankt. Microfossils, Proc.*, v. **2**, p. 739–786.

McDougall, I., Saemundsson, K., Johannesson, H., Watkins, N.D., and Kristjansson,L., 1977: Extension of the geomagnetic polarity time scale to 6.5 m.y.: K-Ar dating, geological and paleomagnetic study of a 3,500m lava succession in western Iceland. *Geol. Soc. Amer., Bull.*, v. **88**, p. 1–15.

Morozumi, Y. and Koizumi,I., 1981: Himi and Yatsuo areas. *In* Tsuchi, R. ed., *Neogene of Japan*, IGCP Natl. Working Group of Japan, p. 65–67.

Oda, M., 1977: Planktonic foraminiferal biostratigraphy of the late Cenozoic sedimentary sequence, central Honshu, Japan. *Tohoku Univ., Sci Repts., 2nd ser. (Geol.)*, v. **48**, no. 1, p. 1–72.

Okada, H. and Bukry, D., 1980: Supplementary modification and introduction of code numbers to the low-latitude coccolith biostratigraphic zonation. *Marine Micropaleont.*, v. **5**, p. 321–325.

Riedel, W.R. and Sanfillippo, A., 1978: Stratigraphy and evolution of tropical Cenozoic radiolarians. *Micropaleontology*, v. **24**, p. 61–96.

Saito,T., 1980: An Early Miocene (Aquitanian) planktonic foraminiferal fauna from the Tsuro Formation, the younger part of the Shimanto Supergroup, Shikoku, Japan. *In* Taira, A. and Tashiro, M. eds., *Geology and Paleontology of the Shimanto Belt*, p. 227–234.

Saito, T., Burckle, L.H. and Hays, J.D., 1975: Late Miocene to Pleistocene biostratigraphy of equatorial Pacific sediments. *In* Saito, T and Burckle, L.H. eds., *Late Neogene Epoch Boundaries*, Amer. Mus. Nat. Hist., p. 226–244.

Sakai, T., 1980: Radiolarians from sites 434,435 and 436, Northwest Pacific, Leg. 56, Deep Sea Drilling Project. *DSDP, Init. Repts.*, v. **56–57**, p. 695–733.

Sakai, T. *et al.*, 1981: Micropaleontological studies of cored materials. *Hakuho-Maru Cruise KH-80-3, Prilim. Rept.*, Ocean Res. Inst. Univ. Tokyo.

Shibata, K., 1973: K-Ar ages of volcanic rocks from the Hokuriku Group. *Geol. Soc. Japan, Mem.*, no.8, p. 143–149.

Shibata, K. *et al.*, 1979: K-Ar age results-1 *Geol. Surv. Japan, Bull.*, v. **30**, p. 675–686.

Shibata, K. *et al.* and Nozawa, T., 1968: K-Ar age of granitic rocks from the outer zone of Southwest Japan *Geochem Jour.*, v. **1**, p. 131–137.

Shuto, T., 1963: Geology of the Nichinan area, with special reference to the Takachiho disturbance. *Kyushu Univ. Fac. Sci., Mem., ser. D*, v. **6**, p. 135–166.

Srinivasan, M.S. and Kennett, J.P., 1981: A review of Neogene planktonic foraminiferal biostratigraphy: Application in the equatorial and south Pacific. *In* Warme, J.E. *et al.* eds., *The Deep Sea Drilling Project: A Decade of Progress*, SEPM Spec. Pub., no. 32, p. 395–432.

Theyer, F., Mato, C.Y. and Hammond, S.R., 1978: Paleomagnetic and geochronologic calibration of latest Oligocene to Pliocene radiolarian events, equatorial Pacific. *Mar. Micropal.*, v. **3**, p. 377–395.

Tsuchi, R., 1961: On the late Neogene sediments and molluscs in the Tokai region, with notes on the geologic history of the Pacific coast of Southwest Japan. *Japanese Jour. Geol. Geog.*, v. **32**, p. 437–456.

Tsuchi, R. and Ibaraki, M., 1978: Late Neogene succession of molluscan fauna on the Pacific coast of southwestern Japan, with reference to planktonic foraminiferal sequence. *The Veliger*, Calif. Malacozool. Soc., v. **21**, p. 216–222.

Tsuchi, R. and Ibaraki, M., 1978b: Definition and faunal characteristics of late Neogene stages on the Pacific coast of southwestern Japan. *2nd. Internat. Meeting IGCP-114, Bandung 1977, Proc.*, p. 53–62.

Tsuchi, R., Takayanagi, Y. and Shibata, K., 1981: Neogene bio-events in the Japanese Islands. *In* Tsuchi, R. ed., *Neogene of Japan*, IGCP-114, Natl. Working of Japan, p. 15–32.

III-10
Neogene Stratigraphy of Northeast Asia (Kamchatka, Sakhalin)

Yuri B. GLADENKOV

Neogene marine deposits are widely distributed in northeastern Asia including the Kamchatka Peninsula and Sakhalin Island. The northernmost deposits occur in Chukotka; farther south they are found in the Koryak Upland, Kamchatka, Sakhalin, and the Kuril Islands. This region stretches from the Bering Strait to the southern Kuril Islands, a distance of more than 3,500 km (Fig.1).

Previous studies have shown that the Neogene of Northeast Asia, or the Far East subregion of the North Pacific boreal region, contains three principal provinces of the Chukotsk-Koryak, Kamchatka (and Kuril), and Sakhalin. Each province should have its stages correlated with those of other provinces (Fig. 2). For example, a general correlation has been effected beween regional stages of Kamchatka and Sakhalin. But it has been revealed that their boundaries often do not coincide, which reflects peculiar geological development of the provinces.

In subdividing the Tertiary deposits of northeast Asia, a geologist experiences the following difficulties (Gladenkov, Menner *et al.*, 1980):

1) The Chukotka-Sakhalin region embraces climatic facies ranging from polar to subboreal. Thus, correlation is difficult because synchronous paleontological assemblages at various latitudes may show strong differences in faunal and floral composition.

2) Heterochronous tectonic structures of various oders can be distinguished within the region. Thus, we must distinguish regional and local stratigraphic subdivisions in each region. The uniqueness of these units often makes correlation difficult. The tectonic history of Kamchatka, for instance, differs in many respects from Sakhalin and the Koryak Upland. There are certain structural-facies zones in each that contrast strongly in their geological histories.

3) Geosynclinal Neogene sequences in this region are characterized by very great thicknesses (as much as 3–4 kilometers or more). There are also considerable volumes of volcanogenic rocks, frequent changes of facies, several hiatuses and unconformities, and units up to several meters thick that are "barren" of fossils.

4) There are also purely "paleontological" difficulties in age determination. The occurrence and distribution of fossil groups so frequently used in the correlation of these sections are largely controlled by facies composition of these stratigraphic sequences and calcareous planktonic organisms are rare. Moreover, planktonic microfossils in this region are mostly of boreal lineages which cannot be satisfactorily related to global warm-water zonations. Only some poor assemblages occur in the Neogene of Kamchatka and Sakhalin. Knowledge of other paleontological groups is still relatively poor.

5) Finally, much of the correlation of sedimentary rocks of these various regions has been carried out on the basis of lithostratigraphy, and these purely lithostratigraphic correlations have frequently resulted in significant errors.

The follwing is a brief characterization of the distribution of Neogene marine deposits (from north to south) in the Chukotka to Sakhalin region. Marine deposits on Chukotka have penetrated in the Anadyr depression. Here a thick (up to 3 km) Paleogene-Neogene section is developed. The thickest formations (more than 1,000 m thick) are located on the Bering Sea coast in the Opukh-Pekulneisky trough of early Miocene-Pliocene age (Volobueva, 1979).

Three structural-facies zones are usually distinguished in Kamchata: the western, central, and eastern zones. Each of these is characterized by unique stratigraphic sections. Marine deposits are most widely distributed in the western zones (the Western Kamchatka trough). The Tochilinsky section is considerd the stratotype for the Paleogene and Neogene of Kamchatka. It is the most continuous and most fossiliferous section in the western zone. The thickness of the Neogene is up

Geological Institute, U.S.S.R. Academy of Sciences, Pyzhevsky Per. 7, Moscow 109017, U.S.S.R.

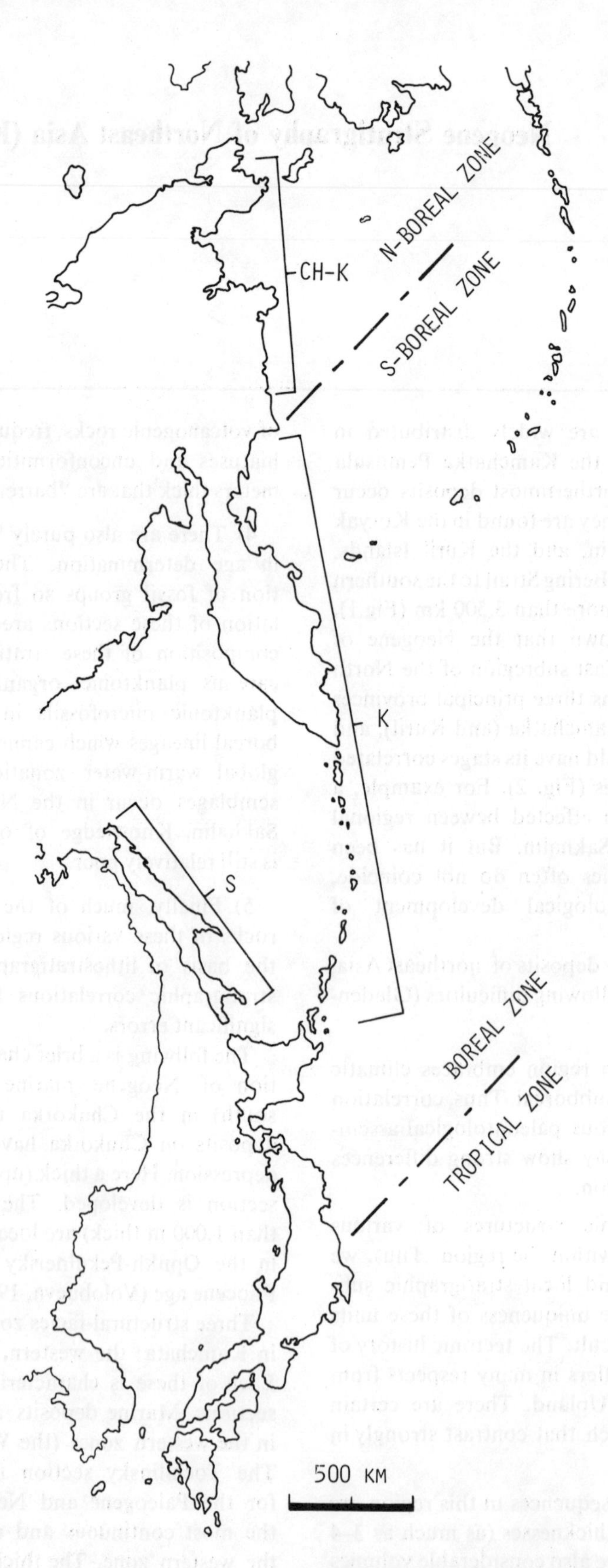

Fig. 1. Biogeographic regions of Northeast Asia during the Neogene. CH-K. Chukotka-Korjanian province, K. Kamchatka and Kurilian province, S. Sakhalin.

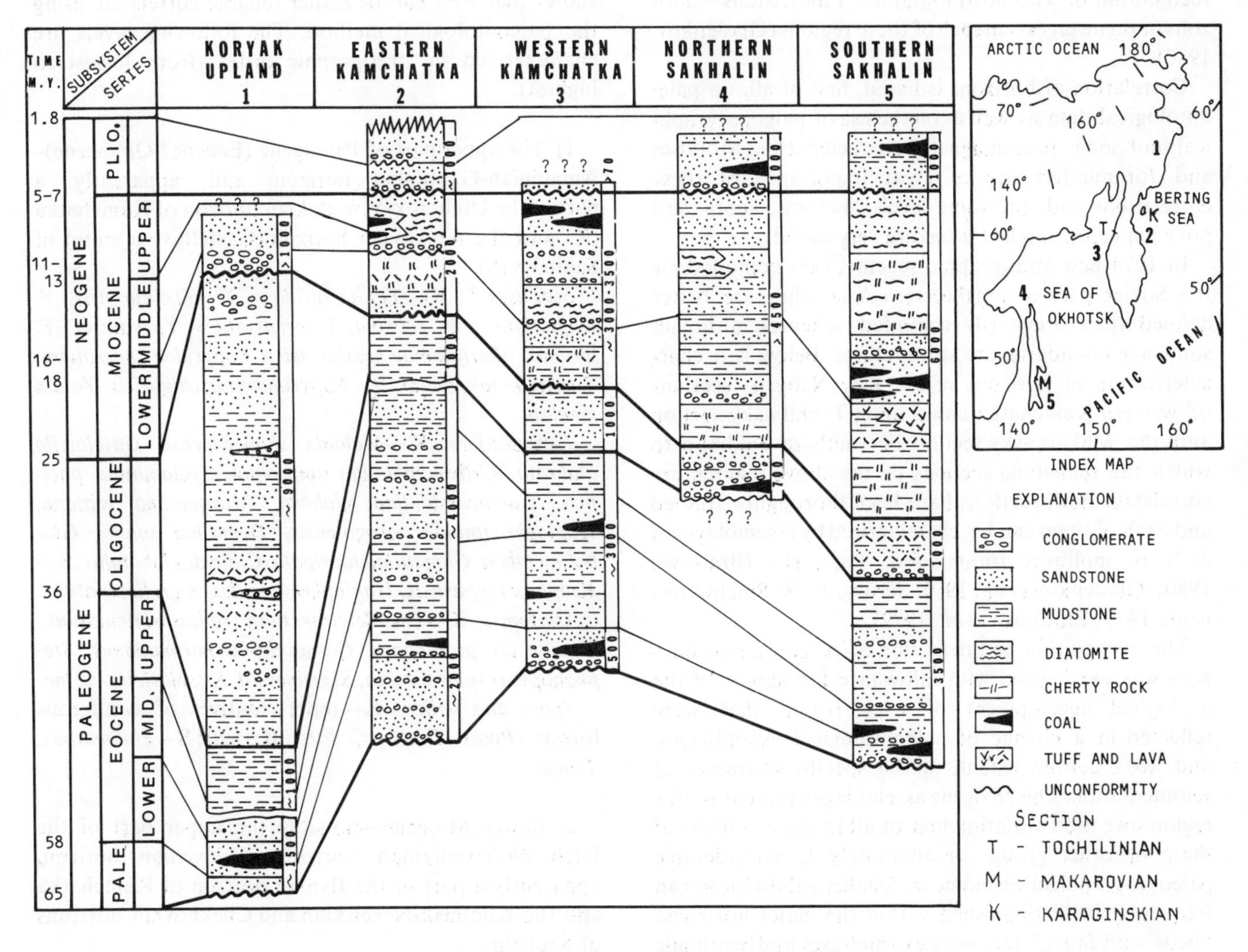

Fig. 2. Lithostratigraphy of the Neogene sequences of Northeast Asia.

to 3.5 km. Volcanogenic rocks reaching 2–3 km or more in thickness comprise the central zone (the Central Kamchatka anticlinorium). The most complicated geological structure occurs in the eastern zone where several troughs and anticlinal structures can be recognized. This structural heterogeneity is responsible for the diversity of stratigraphic sections throughout this zone. Locally, marine beds of the eastern zone grade laterally into volcanogenic and coal-bearing facies. The key sections are on Karaginsky Island and in the Korf region in the northeastern part of the zone. The Neogene sections range from 1.5 to 3 km in thickness (Gladenkov, 1978).

Three large structural zones are distinguished within Sakhalin: the north-northeastern, western, and eastern zones (Zhidkova *et al.*, 1974). Neogene deposits are widely distributed in the northeastern zone; their thickness ranges from 5 to 9 km. The most complete Cenozoic section occurs in the western zone (the West Sakhalin trough) where the thickness of Neogene strata is about 4 to 7 km. Neogene deposits, up to 5 km thick, are also fully exposed in the eastern part of Sakhalin.

In previous years, lithostratigraphic subdivisions—series and suites—were distinguished in each of the above-mentioned regions (Koryak Upland, Kamchatka, and Sakhalin). A series is a subdivision corresponding to the most significant stage of sedimentation whereas suites are units of distinctive lithological composition. Each series usually includes some suites that appear to be similar to North American formations. Initial correlations beween these regions were based on suites and series. Intensive investigations in this region showed that the boundaries of the suites are frequently diachronous. Therefore, geologists working in Kamchatka and Sakhalin have placed emphasis on the biostratigraphic method of correlation. The study of changes in paleontological assemblages in these sections has enabled recognition of chronostratigraphic subdivisions—horizons or regiostages—in each of these regions (Gladenkov, 1980).

Correlation of horizons is based, first of all, on paleontological data as well as on the use of paleogeographical and some paleomagnetic characteristics. Molluscs and foraminifers are of great importance of these correlations and, for some levels, diatoms, spore and pollen spectra, and leaf floras are very useful.

In 1974 new stratigraphic schemes were compiled for the Soviet part of northeastern Asia. They are better defined than previously compiled schemes, although some age boundaries are still tentaive. Below is a characterization of horizons of the Key Neogene sections of western Kamchatka based upon Tochilinsky section and the Makarovsky section of southern Sakhalin to which the remaining sections of the above regions are correlated. Both sections have been thoroughly studied and each of them can be characterized by assemblages of diatoms, molluscs, foraminifers, flora, etc. (Bratseva, 1980; Gladenkov *et al.*, 1980; Serova, 1978; Sinelnikova *et al.*, 1979; Zhidkova *et al.*, 1979).

The recognition and definition of paleontologic horizons was carried out with allowance for stages of the geological development of each region that were reflected in a change of paleontological assemblages, and, to a certain extent, in the specific character of sedimentation. The Neogene assemblages present in this region owe their variation first of all to the evolution of these or other groups or alternately to considerable paleogeographical phenomena. Smaller subdivisions can frequently be distinguished within the major horizons: "beds with fauna" (according to molluscs and benthonic foraminifers) and zones (by means of diatoms). If diatomaceous zones are of a great correlative importance, the "beds with fauna" are now used as local subdivisions only.

Appreciable numbers of species within benthonic assemblages are endemic or forms with lengthy age ranges. Therefore, only a relatively small portion of the benthonic assemblages can be used for correlation.

A brief summary of the characteristics of sections in western and eastern Kamchatka is given in Fig.3, which shows the general position of mollusc and diatom assemblages. Figure 4 illustrate the relationship of subdivisions distinguished by means of various paleontological groups of western Kamchatka and southern Sakhalin. Neogene climatic fluctuations were also taken into account in recognition of correlative units and assemblages (Fig.5).

The analysis of horizons on Kamchatka and Sakhalin shows that they can be rather reliably correlated using the paleontological method. The following levels are characterized in stratigraphic order (from lowest to highest).

1) The upper part of Paleogene (Eocene?-Oligocene)–Amaninian-Gakhinian horizon, and, apparently, a part of the Utcholok-Viventekian horizon of Kamchatka (K) and the Arakayian horizon (Gastellovian suite) of Sakhalin (S).

Molluscs *: K—*Yoldia nitida* = *Y. yotsukurensis, Y. longissima* = *Y. sobrina, Y. cerussata* = *Y. watasei, Y. biremis, Periploma besshoensis, Papyridea harrimani, Turritella tokunagai*; S—*Papyridea matschigarica, Yoldia caudata.*

Foraminifers: K—*Melonis shimokinense, Gavelinella glabrata, Cribroelphidium sumitomoi, Cyclammina pacifica, Globocassidulina globosa, Marginulina eratana, Haplophragmoides asagaiensis, Dentalina soluta, Globulina gibba, Glandulina laevigata, Cibicides lobatulus*; S—*Reophax tappuensis, Haplophragmoides* ex gr. *laminatus.*

Diatoms: K—*Pseudotriceratium radiosoreticulatum, Pyxilla* aff. *prolungata, Cosmiodiscus normanianus, Stephanopyxis spinosissima, S. grunovelii, Kisseleviella carina.*

Spore and pollen: K—representatives of coniferous forests (*Pinus, Tsuga, Cedrus, Myrica*); S—*Podocarpus, Tsuga.*

2) Lower Miocene—possibly the upper part of the Utcholok-Viventekian horizon, Kuluvenian horizon, apparently a part of the Ilynian horizon of Kamchatka and the Kholmsk-Nevelskian and Chekhovian horizons of Sakhalin.

Molluscs: K—*Mytilus tichanovitchi, Spisula equilateralis, Peronidea t-matsumotoi*; C—*Mytilus tichanovitchi, Chlamys kaneharai.*

Foraminifers: K—*Pseudoelphidiella subcarinata, Pseudononion kishimaense, Melonis soldanii, Asanospira carinata, Elphidium kushiroensis, Guttulina pacifica*; S—*Pseudoelphidiella subcarinata, Cyclammina obesa, Gyroidina orbicularis.*

Spore and pollen: K—representatives of coniferous forests mixed with broad-leaved ones (*Pinus, Tsuga*, Taxodiaceae, *Fagus*): S—*Tsuga, Taxodium, Trapa.*

3) The upper part of Lower and the lower part of Middle Miocene—the upper part of the Ilynian—and the lower part of the Kakertian horizons of Kamchatka, the Upperdujan horizon of Sakhalin.

Molluscs: K—*Chlamys kaneharai, Penitella kotakae, Panomya elongata, Lucinoma hannibali, Polinices ramon-*

* These and the following are mainly species with widely distributed and common forms.

ensis, Yoldia thraciaeformis (appearance), *Margarites costalis.*

Foraminifers: K—*Ammonia takanabensis, Pseudoelphidiella problematica*; S—*Miliammina complanata, Asteroammonia borovlevae, Ammodiscus macilentus.*

Diatoms: K—*Denticula lauta, Actinocyclus ingens, Coscinodiscus endoi, C. symbolophorus, Stephanopyxis schenckii, S. corona.*

Leaf flora: S—*Acer ezonium, Fagus antipofii, Quercus castanea.*

Spore and pollen: K—coniferous-broad-leaved forests with *Taxodium, Sequoia, Cryptomeria, Fagus* (up to 36%); S—Taxodiaceae, Juglandaceae, *Trapa, Parthenocissus.*

4) Middle Miocene—Kakertian (the major part) horizon of Kamchatka and the Ausinian horizon, likely the lower part of the Kurussian horizon of Sakhalin.

Molluscs: K—*Chlamys cosibensis, Securella ensifera, Macoma optiva* (appearance); S—*Yoldia thraciaeformis* (appearance), *Macoma optiva* (appearance), *Mizuhopecten subyessoensis.*

Foraminifers: S—*Asteroammonia katangliensis, Cyclammina praecancellata.*

Diatoms: K—*Denticlua lauta, D. hustedtii, Thalassiosira manifesta, Coscinodiscus gracilis*; S—*Denticula lauta, Stephanopyxis schenckii, Goniothecium tenue.*

Radiolarians: S—*Lichnocanium nipponicum.*

5) Middle Miocene (perhaps, Upper Miocene, in part)—Etolonian horizon of Kamchatka and, likely, a part of the Kurassian and Maruyamian horizons of Sakhalin.

Molluscs: —*Chlamys daishakaensis, C. cosibensis cosibensis, C. cosibensis heteroglypta, Securella ensifera chehalisensis, Anadara tsudai, A. ninohensis, Neptunea pluricostulata*; S—*Thyasira disjuncta alta.*

Foraminifers: K—*Anomalinoides altamiraensis, Cibicidoides malloryi, Trifarina kokozuraensis, Elphidiella tenera, Discorbis opercularis, D. nagaoi, Cribroelphidium subarcticum, Sigmomorphina setanaensis, Perfectononion subgranulosus*; C—*Epistominella pacifica, Islandiella laticamerata.*

Diatoms: K—*Denticula lauta, D. hustedtii, Thalassiosira undulosa, T. kryophila, T. eccentrica, Stephanopyxis miocenica*; S—*D. hustedtii.*

Spore and pollen (for horizons 4–5): K—coniferous forests admixed with broad-leaved ones (*Taxodium, Tsuga, Picea, Juglans, Ulmus, Fagus*), in places up to 12%; S—Taxodiaceae, *Pinus, Alnus.*

6) Upper Miocene—the Ermanovian horizon of Kamchatka and the Maruyamian horizon of Sakhalin.

Molluscs: K—*Anadara obispoana, Acila blancoensis, Mulinea densata, Turritella fortilirata habei, Septifer margaritanus, Mytilus kewi, Glycymeris coalingensis*; S—*Spisula voyi, Mya truncata.*

Diatoms: K—*Melosira praedistans, M. praegranulata, M. praeislandica.*

Leaf flora: K—coniferous-broad-leaved forests admixed with southern boreal elements (*Osmunda sachalinensis, Ginkgo* ex gr. *adiantoides, Salix alaskana, Populus sambonsugii, Juglans, Pterocarya, Carpinus*).

7) Pliocene (lower part)—Enemtenian horizon of Kamchatka and the lower part of the Pomyrian horizon of Sakhalin.

Molluscs: K—*Fortipecten takahashii, F. kenyoshiensis, Anadara trilineata trilineata, Securella securis, Chlamys piltunensis, Swiftopecten swiftii* kindled; S—*F.-takahashii, Spisula densata.*

Foraminifers: K—*Pseudoelphidiella oregonensis;* S—*P. oregonensis, Saidovella nagaoi.*

Diatoms: K—*Denticula kamtschatica, Paralia sulcata, Thalassiosira zabelinae, T. gravida* f., *Coscinodiscus marginatus, Stephanopyxis nipponica*; S—*Thalassiosira zabelinae.*

Spore and pollen (for horizons 6–7): K—coniferous forests admixed with amentaceous plants (*Picea, Pinus, Abies, Alnus, Betula, Salix,* Ericaceae); S—*Pinus, Betula,* Ericaceae, *Tsuga, Alnaster.*

8) Pliocene (upper parts)—Liminitevayam horizon of Kamchatka.

Molluscs: *Astarte wortonensis, A. diversa, A. actis.*

Diatoms: *D. seminae f.* Zone, *Thalassiosira gravida, Bacterosira fragilis, Melosira albicans.*

Spore and pollen: *Picea. Eupicea, Omorica, Betula, Alnus, Tsuga.*

Some of the above horizons are good markers not only within the Sakhalin and Kamchatka areas, but within many regions of the North Pacific as well. This enables us to outline some important stratigraphic levels for the region concerned. The following are the most distinct levels:

1. Oligocene (horizon 1); traced in North Japan (Momijiyama) and North America ("Blakeley").
2. The upper part of lower Miocene (horizon 2). North Japan (Asahi).
3. The upper part of the lower beginning of the middle Miocene (horizon 3) representing the Neogene climatic optimum and the *Denticula lauta* Zone; traced in North Japan (the lowermost part of Takinoue=Daijima-Nishikurosawa) and North America (a part of "Temblor").
4. Middle Miocene-the major part (horizons 4–5) including the *Denticula hustedtii-D. lauta* Zone; traced in Japan, North America.

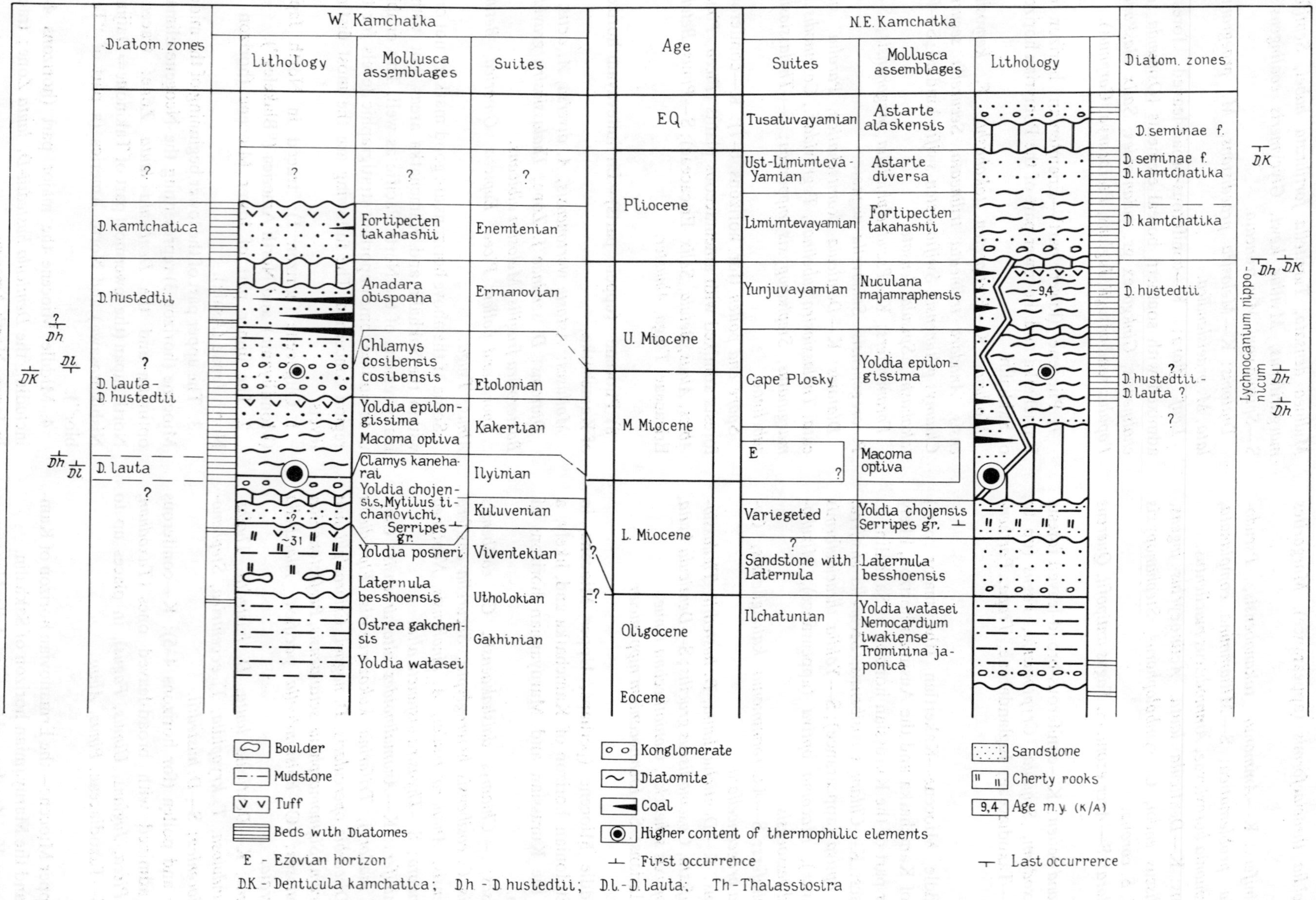

	W. Kamchatka: Diatom zones	W. Kamchatka: Mollusca assemblages	W. Kamchatka: Suites	Age	N.E. Kamchatka: Suites	N.E. Kamchatka: Mollusca assemblages	N.E. Kamchatka: Diatom zones	
				EQ	Tusatuvayamian	Astarte alaskensis	D. seminae f.	
	?	?	?	Pliocene	Ust-Limimteva-Yamian	Astarte diversa	D. seminae f. D. kamtchatika	DK (last occurrence)
	D. kamtchatica	Fortipecten takahashii	Enemtenian	Pliocene	Limimtevayamian	Fortipecten takahashii	D. kamtchatika	
	D. hustedtii	Anadara obispoana	Ermanovian	U. Miocene	Yunjuvayamian	Nuculana majamraphensis	D. hustedtii	Dh, DK; Lychnocanium nipponicum
? Dh; Dl; DK	? D. lauta - D. hustedtii	Chlamys cosibensis cosibensis	Etolonian	U. Miocene	Cape Plosky	Yoldia epilongissima	? D. hustedtii - D. lauta ?	Dh; Dh
		Yoldia epilongissima; Macoma optiva	Kakertian	M. Miocene			?	
Dh; Dl	D. lauta	Clamys kaneharai	Ilyinian	M. Miocene	E ?	Macoma optiva		
	?	Yoldia chojensis, Mytilus tichanovichi, Serripes gr.	Kuluvenian	L. Miocene	Variegeted	Yoldia chojensis; Serripes gr.		
		Yoldia posneri	Viventekian ?	L. Miocene	? Sandstone with Laternula	Laternula besshoensis		
		Laternula besshoensis	Utholokian -?	Oligocene	Ilchatunian	Yoldia watasei; Nemocardium iwakiense; Trominina japonica		
		Ostrea gakchensis	Gakhinian	Oligocene				
		Yoldia watasei		Eocene				

Lithology columns: W. Kamchatka ~31 (age m.y.); N.E. Kamchatka 9.4 (age m.y.).

Legend:
- Boulder
- Mudstone
- Tuff
- Beds with Diatomes
- Konglomerate
- Diatomite
- Coal
- Higher content of thermophilic elements
- Sandstone
- Cherty rooks
- 9,4 Age m.y. (K/A)
- E - Ezovian horizon
- ⊥ First occurrence
- ⊤ Last occurrerce

DK - Denticula kamchatica; Dh - D. hustedtii; Dl - D. lauta; Th - Thalassiosira

Fig. 3. Correlation of the Neogene sequences of West and East Kamchatka. (Tochilian and Karaginskian sequences). (above).
Fig. 4. Neogene key sequences of Sachalin and Kamchatka. (below).

S. SAKHALIAN (MAKAROVIAN SEQUENCE)

AGE		HORIZON	SUITE	M.	Mollusca	Foraminifera	Polynoflora	Diatoms
Pliocene		Pomyrian	Pomyrian	1700 - 1800	Spisula voyi, Mya truncata: Fortipecten takahashii, Spisula (Pseudocardium) densata	Elhidiella oregonensis, Saidovella nagaoi	III: Pinus pumila (ex gr. sibirica), Tsuga, Betula sect. nana, Sphagnum, Alnaster	Actinocyclus splendens; Thalassiosira zabelinae
Miocene	Upper	Marujamian	Marujamian	1700 - 1800	Acila marujamensis, Liocyma praefluctuosa, Echina rachniius sakhalinensis; —; Conchocele disjuncta alta	—; Epistominella pacifica, Islandiella laticamerata; —	III: Taxodiaceae, Pinus, Persticarioipollis pliocenicus, Alnus	?; Denticula hustedtii, Chaetoceros sp.
		Kurassian	Kurassian	150 - 200	Megayoldia thraciaeformis, Macoma optiva: —	Cyclammina praecancellata, Budashevaella laevigata		Stephanopyxis schenckii, Goniothecium tenue, Denticula lauta
	Middle	Uglegorskian	Ausinian	350 - 400	Mizuhopecten subyessoensis	Asteroammonia katangliensis, Pseudoelphidiella problematica		Lychnocanium nipponicum
			Upper duja	700	Corbicula gabbiana; Flora: Acer ezoanum, Fagus antipofii ●; Quercus, Castanea	Miliammina complanata, Asteroammonia borovlevae; —	II: Taxodiaceae, Juglans, Parthenocissus sp., Trapa comitantiborealis	?
	Lower	Nevelskian	Chekhov, Nevelsk	100 - 500; 800	Mytilus ochotensis, Chlamys kaneharai; Neilo multidentata, Nuculana crassatelloides	Pseudoelphidiella aff. subcarinata; Cyclammina pilvoensis; Cyclammina obesa	Tsuga sauerae, Taxodium, Trapa sp.	
		Kholmskian	Kholmskian	150 - 360		Pseudoelphidiella subcarinata	I: Podocarpus aff. totara, Tsuga parva	
Oligocene		Arakaiyan ("Machigarian")	Gastellovian	700 - 1000	Papyridea matschigarica, Mya grewingki	Reophax tappuensis, Haplophragmoides ex gr. laminatus, Budashevaella deserta		

K_2

W. KAMCHATKA (TOCHILINIAN SEQUENCE)

HORIZON	SUITE	M.	Mollusca	Foraminifera	Polynoflora	Diatoms
?						
Enemtenian	Enemtenian	>70	Fortipecten takahashii, Anadara trilineata trilineata	Elphidiella oregonensis	Picea, Alnus, Betula, Salix, Ericaceae	Denticula kamtschatica
Ermanovian	Ermanovian	400	Flora: Osmunda sachalinensis, Gingko ex.gr. adiantoides; Anadara obispoana, Septifer margaritanus	—	V: Taxodium, Tsuga, Picea	? Melosira praedistans, M. praegranulata
Etolonian	Etolonian	340	Securella ensifera chehalisensis, Chlamys cosibensis cosibensis	Cribroelphidium subarcticum, Elphidiella nagaoi, Anomalinoides altamiraensis	IV: ● Fagus 12%; Pinus, Ulmus, Pterocarya, Juglans	Denticula hustedtii; Denticula Lauta – D. hustedtii
Kakertian	Kakertian	505	Macoma optiva, Panomya elongata, Glycymeris wishkahaensis	—; Ammonia takanabensis	III: Taxodium, Sequoia, Criptomeria, ● Fagus (36%)	D. Lauta
Ilyinian	Ilyinian	156	Chlamys kaneharai, Penitella kotakae	Cribroelphidium micrum	Pinus, Tsuga, Taxodiaceae, Fagus (1%)	?
Kuluvenian	Kuluvenian	290	Mytilus tichanovichi, Spisula equilateralis	Pseudononion kishimaense, Pseudoelphidiella subcarinata	II	
Utcholok-Viventekian	Viventekian	210	Yoldia posneri, Delectopecten pedroanus	Haplophragmoides spadix	I: Pinus, Tsuga, Cedrus, Mirica	
	Utcholokian	270	Laternula besshoensis	Astrononion hamadaensis, Cribroelphidium aracaicum		
Amanin-Gakchinian	Gakchinian	740 - 780	Yoldia watasei, Y. longissima, Ostrea gakchensis	Asanospira carinata, Haplophragmoides asagaiensis, Quinqueloculina lamarkia		
	Amaninian	200 - 225	Laternula besshoensis, Papiridea harrimani	Galvinella glabrata, Melonis simokinense		

AGE			NORTH AMERICA	KORYAK UPLAND	EASTERN KAMCHATKA	WESTERN KAMCHATKA	NORTHERN SAKHALIN	SOUTHERN SAKHALIN	NORTHERN JAPAN
NEOGENE	PLIOCENE		"San-Joaquin" "Etchegoin"	Horizons not yet determined	Ust-Limimtevayamian Limimtevayamian	? ? ? Enemtenian	Pomyrian	Pomyrian	Setana "Takikawa"
	MIOCENE	MIDDLE	"Jacalitos" "Margaritan"		Classician Medvezkinian	Ermanovian Etolonian Kakertian Ilynian	Nutovian Okobykayian	Maruyamian Kurassian Uglegorskian	"Wakkanai" Kawabata Takinoue
		LOWER	"Temblor" "Vaqueros"	?—?—? Undal-umenian	II Pakhachinian I	Kuluvenian ? Viventekian Utholokian ?	Daginian Uininian Daekhuriinian	Chekhovian Nevelskian Kholmskian	Asahi ?
	OLIGOCENE		"Blakeley"	Mallenian	Aluginian	Gakhinian Amaninian ?	Machigarian	"Arakayian" Lesogorian	Momijiama

Fig. 6. Correlation of the Neogene stages in Northeast Asia with those of North Japan and North America.

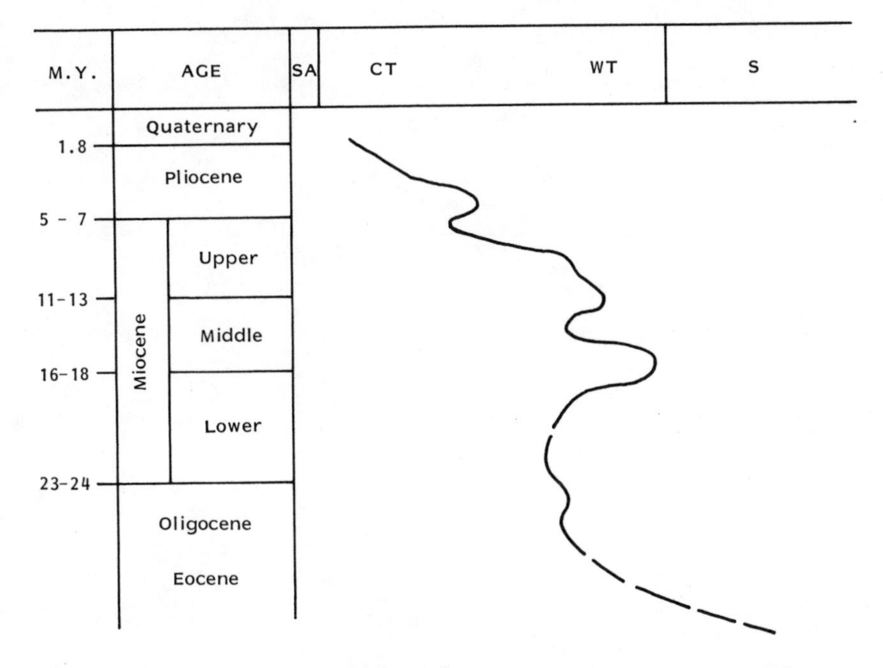

Fig. 5. Neogene climatic fluctuations in regions of Kamchatka. Climate: SA. Subarctic, CT. Cool temperate, WT. Warm temperate

5. Pliocene (horizons 7–8). The *Denticula kamtschatica—D. seminae f.*, Zones, horizon with *Fortipecten takahashii—Elphidiella oregonensis.* North Japan (Takikawa); Horizon with *Astarte diversa* (Alaska).

The poorest represented are levels of the lower and upper Miocene due to a relatively poor paleontological characterization. The problem of the exact position of the lower boundary of the Miocene in the Kamchatka sections has not been solved yet.

Comparison of Neogene deposits of Sakhalin and Kamchatka to synchronous deposits of North Japan (Hokkaido) allows tentative correlation of Neogene marine deposits in both regions (Fig. 6).

References

Bratseva, G.M., 1980: Palynological characteristic of Neogene deposits of Kamchatka. *In* "Palynology in the USSR" (1976–1980). "*Nauka*", p. 91–92 (in Russian).

Gladenkov, Yu.B., 1978: Marine Upper Cenozoic of the northern regions. *Trudy GIN*, v. **313**, p. 194 (in Russian).

Gladenkov, Yu.B., 1980: Stratigraphy of Marine Paleogene and Neogene of northeastern Asia (Chukotka, Kamchatka, Sakhalin). *Amer. Assoc. Petrol. Geol., Bull.*, v. **64**, No. 7, p. 1087–1093.

Gladenkov, Yu.B., Menner, V.V., Serova, M.Ya., and Sinelnikova, V.N., 1980: Upper Cenozoic of the Arcto-Boreal area. *In* "Stratigraphy in investigations of the Geological Institute of the USSR Academy of Sciences". "*Nauka*", p. 208–223 (in Russian).

Serova, M.Ya., 1978: Stratigraphy and foraminifers of the Neogene of Kamchatka. *Trudy GIM*, v. **323**, p. 173 (in Russian).

Sinelnikova, V.N., Serova, M.Ya., Bratseva, G.M., Giterman, R.E., Gladenkov Yu.B., Gladikova V.N., Dolmatova L.M., *et al.*, 1979: The Neogene key section of Western Kamchatka. *In* XIV Pacific Scientific Congress, section B III, p. 111–112 (in Russian).

Volobueva, V.I., 1979: Paleogene and Neogene biostratigraphy of the eastern part of the Koryak Upland, In: "XIV Pacific Scientific Congress", section B III, p. 36–38 (in Russian).

Zhidkova, L.S., Mishakov, G.S., Neverova, T.I., Popova, L.A., Salnikov, B.A., Salnikova, N.B., and Sheremetieva, G.N., 1974: Biofacies peculiarities of Intercenozoic basins of Sakhalin and Kuril Islands, "Nauka", Novosibirsk, p. 252 (in Russian).

Zhidkova, L.S., Salnikov, B.A., Brutman, N.Ya., Saklinskaya, E.D., Kuznetsova, V.N., Moiseeva, A.I., Popova, L.A., Fotjanova, L.I., and Shanyan, S.Kh., 1979: Makarovsky key stratigraphic section of Paleogene-Neogene deposits of Sakhalin. *In: "XIV Pacific Scientific Congress", section B III, Abst.*, p.52–53 (in Russian).

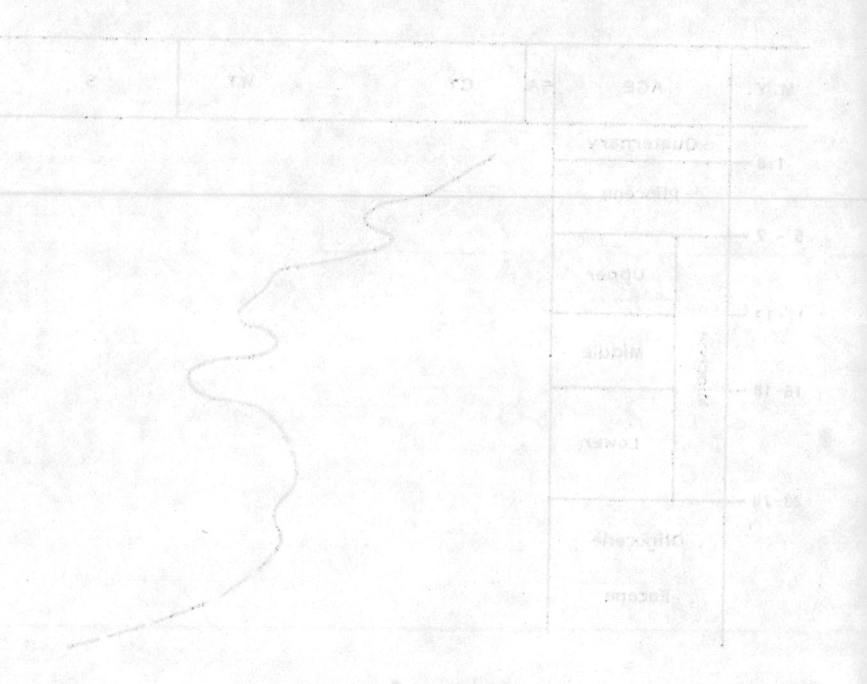

III-11

Correlation of North Pacific Neogene Molluscan Biostratigraphic Frameworks

John M. ARMENTROUT*, Kiyotaka CHINZEI[2]* and Yuri B. GLADENKOV[3]*

Abstract

Molluscan fossils have historically played an important role in establishing correlations, particularly of rocks deposited in neritic environments. A new correlation chart has been constructed showing the relationship of North Pacific Neogene molluscan biostratigraphic frameworks with oceanic microfossil zones and subzones, and the worldwide geologic time scale. The provincial molluscan biostratigraphic frameworks include those for Japan, Korea, South Sakhalin, Kamchatka, southern Alaska, Washington, Oregon, and California.

Calibration of the North Pacific Neogene provincial molluscan framework to each other and to the worldwide geologic time scale permits refinement of correlations and recognition of synchronous North Pacific Neogene events. Additionally, precise correlation of first and last appearances of widely distributed molluscan taxa will allow definition of provincial molluscan datum levels calibrated with microfossil datum levels and correlated to the worldwide geologic time scale.

The correlation chart was constructed by the cooperative effort of the "Working Group for North Pacific Neogene Correlations" of the International Geological Correlation Programme Project 114—"Evaluation of the biostratigraphic datum-planes of the Pacific Neogene for the purpose of global-scale correlation," meeting in Osaka, Japan, 25–29 November 1981.

Introduction

The main goal of stratigraphical geology is the attainment of ever more precise correlations (Berggren and van Couvering, 1974, p. IX), correlation being the determination of the equivalence in geologic age and stratigraphic position of rock units in geographically separate areas. The ultimate purpose of rock-unit correlation is to correctly place events in the worldwide geologic time scale and thus relate the events to coeval events elsewhere (Glaessner, 1967, p. 1). From the correlation of these events, it is possible to reconstruct the geologic history of the earth.

Molluscan fossils have historically played an important role in establishing correlations, particularly of rocks deposited in neritic environments. This paper reviews the historical background and current status of Neogene molluscan biostratigraphic frameworks for the North Pacific; more specifically for Japan, Korea, Kamchatka, South Sakhalin, Alaska, Oregon, Washington, and California.

The cooperative research effort that led to this report was initiated in 1976 by the International Geological Correlation Programme (IGCP) Project 114, "Evaluation of the biostratigraphic datum-planes of the Pacific Neogene for the purpose of global-scale correlation." During the IGCP-114 First International Congress on Pacific Neogene Stratigraphy in 1976, the "Working Group for North Pacific Neogne Correlations" was formed. Members of the working group met at each subsequent IGCP-114 conference, working principally on correlations based on molluscan fossils and associated fossil groups. At the 1981 Osaka, Japan meeting of IGCP-114, the "Working Group for North Pacific Neogene Correlations" summarized the present status of provincial molluscan biostratigraphic frameworks and correlated them to each other, to microfossil biostratigraphic frameworks, and to the worldwide geologic time scale. That summary is the basis of this report.

Molluscan Biostratigraphy

In the mid-1830s, marine rocks were discovered by Townsend (1839) near the mouth of the Columbia River, which separates the states of Oregon and Wash-

* *Mobil Exploration and Producing Services, Inc., P.O. Box 900, Dallas, Texas 75221, U.S.A.*

[2]* *Geological Institute, University of Tokyo, Hongo, Tokyo 113, Japan*

[3]* *Geological Institule, USSR Academy of Sciences, Pyzhevsky Per. 7, Moscow 109017, U.S.S.R.*

ington. Townsend collected fossil molluscs from the rocks and sent them to T. A. Conrad for study. Conrad named and described the fossils and correctly correlated them with the Miocene of Europe, thus beginning the history of molluscan biostratigraphy of the North Pacific margin.

Molluscan assemblages continued to serve as the principal tool by which the marine Neogene formations of the North Pacific were correlated during the 1800s and on into the 1930s. During the late 1930s large invertebrate fossils began to be replaced by foraminifera for purposes of provincial age determination and correlation along the Pacific margin of the United States. This change to microfossils was brought about by the utility of foraminifera in establishing correlations for rock-units sampled during oil-well drilling (Addicott, 1972, p. 630). The use of microfossil groups for correlation was further accelerated by the Deep Sea Drilling Program and the application of planktonic microfossil biostratigraphy for oceanwide and worldwide correlations. As a consequence, molluscan biostratigraphy of ocean margin stratigraphic units received less attention. However, the abundance of molluscan fossils in rocks deposited in some middle- and inner-neritic environments is seldom equaled by any microfossil group. Thus, the correlation of shallow marine rocks is in large part still dependent on the use of molluscan biostratigraphy.

Neogene molluscan faunas of the North Pacific margin are characterized by their high degree of endeminsm (Addicott, 1976; Chinzei, 1978). The endemism in Neogene faunas reflects the strong latitudinal thermal gradient and distinct water masses of the North Pacific. As a consequence of these and other environmental variables, the Neogene molluscan faunas have been segregated into several faunal provinces (Valentine, 1966; Allison, 1978; Chinzei, 1978). As with the boundaries of Holocene and recent faunas (Hall, 1964; Valentine, 1966), these Neogene provinces are characterized by relatively high percentages of endemic species and their boundaries are marked by rapid faunal change (Addicott, 1976).

Five North Pacific Neogene and Holocene molluscan provinces are considered in this report (Fig. 1). Chinzei (1978) emphasizes that the molluscan faunas of Japan represent two different water systems, one warm and the other cold. The warm-water fauna is represented by assemblages from southwest Japan and Korea; the cold-water fauna is represented by assemblages from northeast Japan. Menner and Gladenkov (1979) recognize two distinct molluscan sequences along the Pacific Coast of Russia: one represented by the regiostage sequence of Kamchatka, the second represented by the regiostage sequence of South Sakhalin. Addicott (1976) recognizes three faunal provinces along the eastern side of the North Pacific. These provinces are referred to as the California, Pacific Northwest (Oregon-Washington-British Columbia), and Gulf of Alaska provinces.

Separate biochronologies are needed for each of the North Pacific Neogene provinces. Each provincial biochronology is useful for correlations within that province, but correlation between provinces is difficult due to faunal endemism. Intraprovincial correlation is facilitated, however, by the use of less endemic microfossil assemblages that co-occur with the molluscan fossil assemblages.

Of the microfossil groups studied, diatoms are the most biostratigraphically useful in North Pacific Neogene sequences. Diatoms are diverse and abundant at higher latitudes and occur in many Neogene rock units. Recent studies of diatom biostratigraphy have improved correlations between high and low latitudes and permit

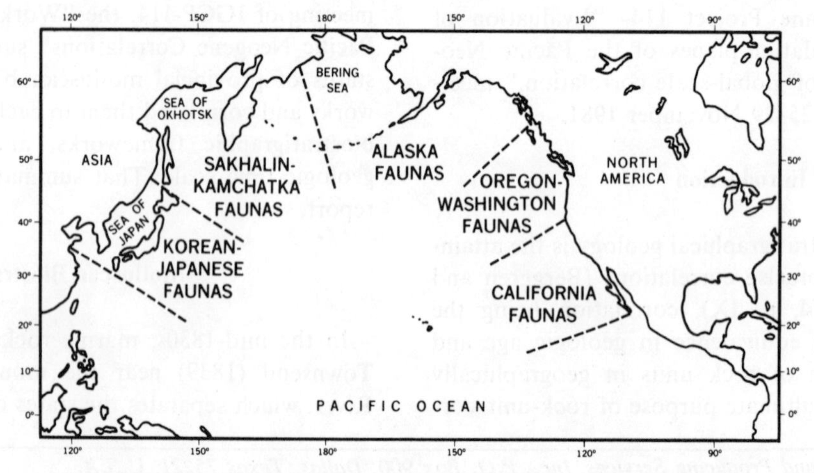

Fig. 1. Map showing the study area with North Pacific faunal provinces of Neogene molluscan assemblages.

correlation of the diatom biochronologic framework to the worldwide geologic time scale (Schrader, 1973; Koizumi, 1977; Burckle, 1978; Barron, 1981b). Diatoms co-occur with mollusks in many Neogene rock units (Armentrout, 1981; Barron, 1981a; Barron and Armentrout, 1980; Ishizaki and Takayanagi, 1981; Chinzei, 1981; Morozumi and Koizumi, 1981; Maiya and others, 1981a, 1981b; Saito, 1981). By using the co-occurrence of diatoms with mollusks, it has been possible to correlate the North Pacific provincial Neogene molluscan biochronologies to the diatom biochronologic framework and thus to the worldwide geological time scale.

Cross-checks of the correlation by diatoms of the provincial molluscan biochronologies to the worldwide time scale are possible where other chronologically significant indices occur with molluscan assemblages. Such cross-checks have been made using planktonic foraminifera, magnetostratigraphy, and radiometric dates (Morrison and Sarna-Wojciki, 1981; Tsuchi and Ibaraki, 1981); planktonic foraminifera and calcareous nannoplankton (Ibaraki, 1981; Okamoto and Huang, 1981); and magnetostratigraphy (Burckle, Dodd, and Stanton, 1980).

Correlation of Molluscan Biostratigraphic Frameworks

The correlation of North Pacific Neogene molluscan biostratigraphic frameworks with the worldwide geologic time scale and oceanic zones and subzones is shown in Fig. 2. This correlation chart was constructed by the cooperative effort of molluscan working group participants at the 1981 Osaka, Japan meeting of IGCP-114. Contributors are listed under Acknowledgement. Development of the worldwide geologic time scale and correlation of oceanic zones and subzones to that time scale, as used in Fig. 2, is discussed by Armentrout, Echols, and Ingle (this volume).

The molluscan working group arranged the provincial molluscan biostratigraphic units according to their correlation with the diatom zonal schemes following Barron (1981b) for the northeast Pacific margin and Koizumi (1981) for the northwest Pacific margin. Calcareous nannofossil calibrations (Berggren, 1981) were also used for the early Miocene interval not zoned by diatoms (see also Poore, Barron, and Addicott, 1981).

Japanese-Korean faunas

The molluscan faunal sequence from Japan is from Chinzei (1978, 1981), whereas the Korean faunal distribution is from Kim (1981), Yoon (1981), and Yun (1981). Both Chinzei and Yoon participated in construction of the correlation chart (Fig. 2). Provincial stages are not currently used by Japanese or Korean molluscan workers.

The geographical and stratigraphical distribution of the benthic molluscan faunas of Japan and Korea reflect control by the history of sedimentation and water mass conditions. Ecologically analogous associations, or fossil communities, occur at distinct stratigraphic levels where similar environmental conditions repeatedly appeared. Based on this repetition, the Neogene molluscan faunas can be grouped into five faunas of different age. Faunas from southwest Japan are predominantly warm-water faunas, whereas those from Korea and Hokkaido are predominantly cool-water faunas (Chinzei, 1978).

The latest Early Miocene Kadonosawa and Chenogogsa faunas contain large numbers of tropical and subtropical taxa (Chinzei, 1978; Tsuchi, Takayanagi, and Shibata, 1981; Yoon, 1981). These widely distributed warm-water faunas represent a climatically controlled bio-event that serves as a correlation horizon at or near the beginning of the *Actinocyclus ingens* Diatom Zone.

Soviet Union regiostages

The molluscan regiostages (provincial stages) for Kamchatka and South Sakhalin are from Menner and Gladenkov (1979), the latter of whom participated in construction of the correlation chart (Fig. 2). Reviews of stratigraphic work along the Soviet Pacific margin are provided by Gladenkov (1981); Gladenkov, Vitukin, and Oreshkina (1979); and Margulis, Shpetalenko, and Gritesendo (1981).

The presence of planktonic foraminifera and diatoms in some horizons containing molluscan assemblages facilitates correlation of the regiostages of South Sakhalin and Kamchatka to the worldwide time scale and to the provincial zones and stages of North America and Japan (Gladenkov, 1980). The resulting correlation shows that the sedimentological and biostratigraphical history of Kamchatka, South Sakhalin, Japan, and Korea are similar. In particular, the latest Early Miocene warm bio-event occurs throughout the northwest Pacific margin Neogene record. Additionally, the horizon of *Fortipecten takahashii* is recognized as being correlative in Hokkaido, South Sakhalin, and Kamchatka. Differences in the duration of the bio-events may be real or may reflect differences in the age resolution of the biostratigraphic data used to calibrate the event. Further research on these bio-events is planned.

Alaska faunas

Molluscan faunas of the Alaskan faunal province have not received detailed investigation as have the

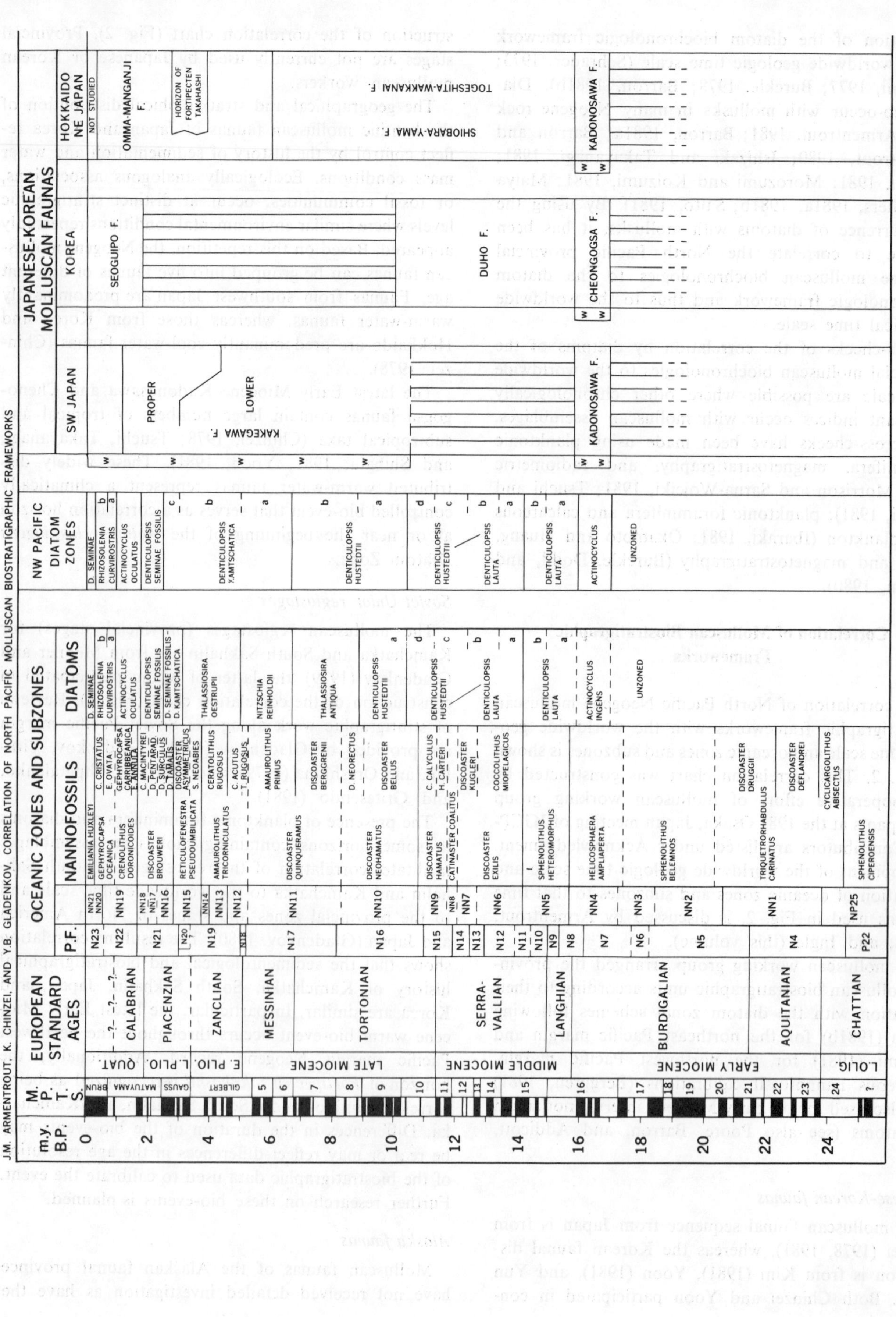

J.M. ARMENTROUT, K. CHINZEI, AND Y.B. GLADENKOV, CORRELATION OF NORTH PACIFIC MOLLUSCAN BIOSTRATIGRAPHIC FRAMEWORKS

m.y. B.P.
M. P. T. S.
EUROPEAN STANDARD AGES
OCEANIC ZONES AND SUBZONES
P.F.
NANNOFOSSILS
DIATOMS
NW PACIFIC DIATOM ZONES
JAPANESE-KOREAN MOLLUSCAN FAUNAS
SW JAPAN
KOREA
HOKKAIDO NE JAPAN

0 2 4 6 8 10 12 14 16 18 20 22 24

BRUN MATUYAMA GAUSS GILBERT 5 6 7 8 9 10 11 12 13 14 15 16 17 18 19 20 21 22 23 24 ?

QUAT. L. PLIO. E. PLIO. LATE MIOCENE MIDDLE MIOCENE EARLY MIOCENE L. OLIG.

-?- -?- -?-
CALABRIAN
PIACENZIAN
ZANCLIAN
MESSINIAN
TORTONIAN
SERRA-VALLIAN
LANGHIAN
BURDIGALIAN
AQUITANIAN
CHATTIAN

N23 N22 N21 N20 N19 N18 N17 N16 N15 N14 N13 N12 N11 N10 N9 N8 N7 N6 N5 N4 P22

NN21 NN20 NN19 NN18 NN17 NN16 NN15 NN14 NN13 NN12 NN11 NN10 NN9 NN8 NN7 NN6 NN5 NN4 NN3 NN2 NN1 NP25

EMILIANIA HUXLEYI
GEPHYROCAPSA OCEANICA
C. CRISTATUS
E. OVATA
CRENOLITHUS DORONICOIDES
GEPHYROCAPSA CARRIBBEANICA
E. ANNULA
C. MACINTYREI
DISCOASTER BROUWERI
D. PENTARAD
D. SURCULUS
D. TAMALIS
RETICULOFENESTRA PSEUDOUMBILICATA
DISCOASTER ASYMMETRICUS
S. NEOABIES
AMAUROLITHUS TRICORNICULATUS
CERATOLITHUS RUGOSUS
C. ACUTUS
T. RUGOSUS
AMAUROLITHUS PRIMUS
DISCOASTER QUINQUERAMUS
DISCOASTER BERGGRENII
D. NEORECTUS
DISCOASTER NEOHAMATUS
DISCOASTER BELLUS
DISCOASTER HAMATUS
C. CALYCULUS
H. CARTERI
CATINASTER COALITUS
DISCOASTER KUGLERI
DISCOASTER EXILIS
COCCOLITHUS MIOPELAGICUS
SPHENOLITHUS HETEROMORPHUS
HELICOSPHAERA AMPLIAPERTA
SPHENOLITHUS BELEMNOS
DISCOASTER DRUGGII
TRIQUETRORHABDULUS CARINATUS
DISCOASTER DEFLANDREI
CYCLICARGOLITHUS ABISECTUS
SPHENOLITHUS CIPEROENSIS

D. SEMINAE
RHIZOSOLENIA CURVIROSTRIS b a
ACTINOCYCLUS OCULATUS
DENTICULOPSIS SEMINAE FOSSILIS
D. SEMINAE FOSSILIS - D. KAMTSCHATICA
THALASSIOSIRA OESTRUPII
NITZSCHIA REINHOLDII
THALASSIOSIRA ANTIQUA
DENTICULOPSIS HUSTEDTII a
DENTICULOPSIS HUSTEDTII d c b
DENTICULOPSIS LAUTA a
DENTICULOPSIS LAUTA b a
ACTINOCYCLUS INGENS
UNZONED

D. SEMINAE
RHIZOSOLENIA CURVIROSTRIS b a
ACTINOCYCLUS OCULATUS
DENTICULOPSIS SEMINAE FOSSILIS
DENTICULOPSIS KAMTSCHATICA c b a
DENTICULOPSIS HUSTEDTII b a
DENTICULOPSIS HUSTEDTII d c
DENTICULOPSIS LAUTA b a
DENTICULOPSIS LAUTA b a
ACTINOCYCLUS
UNZONED

W W W W W W W W
KAKEGAWA F.
PROPER
LOWER
KADONOSAWA F.

SEOGUIPO F.
DUHO F.
CHEONGOGSA F.

NOT STUDIED
OMMA-MANGANJI F.
HORIZON OF FORTIPECTEN TAKAHASHI
SHIOBARA-YAMA F.
TOGESHITA-WAKKANAI F.
KADONOSAWA F.

Fig. 2. Correlation of North Pacific molluscan biostratigraphic frameworks to both microfossil biostratigraphic frameworks and to the worldwide geologic time scale. F = Fauna; W = Warm Fauna; C = Cool Fauna.

SOVIET UNION MOLLUSCAN REGIOSTAGES

S. SAKHALIN
NOT STUDIED
POMYRIAN
FORTIPECTEN TAKAHASHI HORIZON
MARUJAMIAN
KURASSIAN
AUSINIAN
UPPER DUJIAN
CHEKHOVIAN
NEVELSKIAN
"ARAKAIAN"

KAMCHATKA
KARAGINSKIAN
OLCHOVSKIAN
TSUATUVJAMIAN
USTLIMIMTEVAMIAN
ENEMTENIAN
FORTIPECTEN TAKAHASHI HORIZON
ERMANOVIAN
ETOLONIAN
KAKERTIAN
ILYINIAN
KULUVIAN
VIVENTEKIAN
UTCHOLOKIAN
GAKLINIAN

ALASKAN MOLLUSCAN FAUNAS

ALASKAN PENINSULA
TERRACES
TACHILNI FORMATION FAUNAS (MILKY RIVER BEDS)
BEAR LAKE FORMATION FAUNAS
UNGA CONGLOMERATE FAUNA

YAKATAGA DISTRICT
TERRACES
YAKATAGA FORMATION FAUNAS
DIFFERENT INTERPRETATIONS OF BOUNDARY AGE
POUL CREEK FORMATION FAUNAS

NORTH AMERICAN MOLLUSCAN STAGES

OREGON-WASHINGTON
ASSEM. OF WILDCAT GROUP
MOCLIPSIAN
GRAYSIAN
WISHKAHAN
NEWPORTIAN
PILLARIAN
JUANIAN

CALIFORNIA
NOT STUDIED
"SAN JOAQUIN"
"ETCHEGOIN"
"JACALITOS"
"MARGARITAN"
"TEMBLOR"
"VAQUEROS"

CALIFORNIA FORAMINIFERAL STAGES
HALLIAN
WHEELERIAN
VENTURIAN
REPPETIAN
"DELMONTIAN"
MOHNIAN
UPPER "TYPE" DELMONTIAN
LUISIAN
RELIZIAN
SAUCESIAN
ZEMORRIAN

NE PACIFIC DIATOM ZONES
D. SEMINAE
RHIZOSOLENIA CURYIROSTRIS b a
ACTINOCYCLUS OCULATUS
DENTICULOPSIS SEMINAE FOSSILIS
D. SEMINAE FOSSILIS
D. KAMCHATICA
THALASSIOSIRA OESTRUPII
NITZSCHIA REINHOLDII
THALASSIOSIRA ANTIQUA
DENTICULOPSIS HUSTEDTII
DENTICULOPSIS HUSTEDTII d c
DENTICULOPSIS LAUTA b a
DENTICULOPSIS LAUTA b a
ACTINOCYCLUS INGENS
UNZONED

BRUN MATUYAMA GAUSS GILBERT 5 6 7 8 9 10 11 12 13 14 15 16 17 18 19 20 21 22 23 24 ?

m.y. B.P.
0 2 4 6 8 10 12 14 16 18 20 22 24

faunas of the other North Pacific provinces. The correlation of formational faunas presented on Fig. 2 is based on faunal lists for each formation assembled and reviewed by Allison (1976, 1978). Principal faunal studies include those on the Poul Creek Formation and Yakataga Formation faunas by Addicott, Kanno, Sakamoto, and Miller (1971), Ariey (1978 a, 1978 b), and Kanno (1971); the Topsey Formation fauna by Marincovich (1980); and the Unga Conglomerate Member fauna of the Bear Lake Formation by MacNeil (1973). Allison and Marincovich (1981) have described a late Oligocene or earliest Miocene fauna from Sitkinak Island, Kodiak Archipelago, which in part may be a correlative of the Gaklinian Regiostage of Kamchatka and Juanian Stage of Oregon and Washington.

Age relationships of Alaskan faunas (Allison, 1978) are based on correlation with the molluscan stages of Oregon and Washington (Armentrout 1975, 1981; Addicott, 1976). Differences of opinion exist for some critical boundaries, one example being the Poul Creek/Yakataga formational boundary (Fig. 2). Allison (1978), using data on molluscs, correlates that formational boundary with the Newportian/Pillarian Molluscan Stage boundary at about 19 Ma. Armentrout, Echols, and Nash (1978), using molluscs, benthic and planktonic foraminifera, and glauconite K-Ar age dates, correlate the Poul Creek/Yakataga boundary at about 6.5 Ma. Resolution of such different interpretations and refinement of correlations related to Alaskan faunas will require considerable new research. Recognition of bio-events, either evolutionary or climatic, must await more detailed data on local faunas, identification of marker species common to Alaskan, Asiatic and North American provinces, and associated studies of planktonic microfossil groups, magnetostratigraphy, and radiometric dating of biostratigraphically significant rocks. Such research should permit definition of provincial molluscan stages for the Alaskan province and a more precise correlation of provincial Alaskan units to the worldwide time scale (Allison, 1978; Allison and Marincovich, 1979; McCoy and Ariey, 1979).

North American west coast stages

The molluscan biostratigraphic framework of Oregon-Washington and California has undergone considerable evolution in recent years. Many publications follow the correlation of the provincial molluscan stages with the European periods, epochs, and stages as proposed by Weaver *et al.* (1944). Current usage follows Addicott (1976) and Armentrout (1975, 1979, 1981) for Oregon-Washington, and Addicott (1972) for California. Both Addicott and Armentrout participated in the construction of Fig. 2.

The Neogene molluscan faunas of the Pacific Coast of North America show a strong faunal endemism reflecting the influence of geographically separate basins along a broad latitudinal belt. The resultant problems in establishing correlations within and between faunal provinces has led to the definition of separate provinicial stage sequences for Oregon-Washington and California (Addicott, 1977). The Oregon-Washington Neogene province includes British Columbia (Addicott, 1976). The boundaries between the Oregon-Washington and California molluscan provinces fluctuated during the Neogene, but a practical boundary may be drawn at Cape Mendocino, California (Addicott, 1970).

The provincial stages of Oregon-Washington and California are based on range zones of individual taxa and concurrent range zones as determined in the stratotypes and from reference sections (Addicott, 1976). Because of faunal endemism these stages can be recognized with certainty only within one province. Correlation with other provincial sequences is based upon the few chronostratigraphically significant taxa that range outside a single province (see Allison, 1978) or by correlation using microfossil groups which are less endemic. The provincial molluscan stages of Oregon-Washington and California have both been correlated with the California benthic foraminiferal stage sequence based on the co-occurrence of molluscs and foraminifers within the same stratigraphic sections (Armentrout, 1981; Armentrout and Echols, 1981). The California foraminiferal stage sequence and its correlation to the northeast Pacific diatom zones follows Armentrout, Echols, and Ingle (this volume).

One note of importance concerns the concept of the Wishkahan and Graysian Stages. Addicott (1976) defined these two stages as successive chronostratigraphic units based on the molluscan assemblages of the Montesano Formation with stratotypes along the Wishkah River, Grays Harbor Basin, Washington. The concept of the Wishkahan Stage is supplemented by the fauna of the Empire Formation at Coos Bay, Oregon (Addicott, 1976). However, the diatom florules of the Empire Formation are younger than those of the lower part of the Graysian stratotype. Diatom florules of the Empire Formation are assigned to the *Thalassiosira antiqua* and *Nitzschia reinholdii* Zones (Barron, 1981 a; Barron and Armentrout, 1980). Diatom florules of the lower Graysian stratotype are assigned to the *Denticulopsis hustedtii–D. lauta* and *D. lauta* Zones (Barron, 1981 a). No age diagnostic diatoms have been recovered from the Wishkahan stratotype. If the Empire Formation fauna is Wishkahan in age, then the Wishkahan and Graysian Stages are in part coeval, an interpretation followed by Armentrout (1981) and Armentrout and Echols (1981). However, Addicott (1981, 1982, personal communications) suggests that the Empire Formation fauna be reassigned to the Graysian Stage and that the faunal assemblages of the Wishkahan and Graysian Stages be redefined to reflect that change. We have followed Addicott's recommendation thus retaining the original concept of the Wishkahan and Graysian Stages as successive non-overlapping chronostratigraphic units (see Fig. 2).

The late Early to early Middle Miocene Newportian Stage and "Temblor" Stage molluscan faunas and age-equivalent latest Saucesian, Relizian, and earliest Luisian Stage foraminiferal faunas contain large numbers of tropical and subtropical taxa (Addicott, 1972, 1976, 1977). The late Early to early Middle Miocene warm-water fauna occurs as far north as the Narrow Cape Formation of Kodiak Island in the western part of the Gulf of Alaska province (Addicott, 1976; Allison, 1978). These widely distributed warm-water faunas reflect a warm-water bio-event at or near the beginning of the *Actinocyclus ingens* Diatom Zone.

Future Work

The correlation of North Pacific Neogene molluscan biostratigraphic frameworks to the worldwide time scale provides a refined chronostratigraphic framework for interpreting the Pacific rim geologic record.

The refined chronostratigraphic framework also makes it possible to re-evaluate the vertical distribution of molluscan marker species. Precise correlation of first and last appearances of molluscan taxa that occur over broad geographic areas will allow for definition of provincial molluscan datum levels calibrated with microfossil datum levels and correlated to the worldwide geologic time scale. Such supplementary datum levels have been proposed for Japan by Tsuchi, Takayanagi, and Shibata (1981). Studies of molluscan genera offering potential for supplementary datum levels include those on *Mya* (MacNeil, 1965), *Yoldia* (Gladenkov, 1970, 1974), the Pectinidae (Addicott, 1974, 1978; Masuda, 1980; MacNeil, 1967; and Sinelnikova, 1975), the Arcidae (Noda, 1966), the Cassididae (Kanno, 1973), the Naticidae (Marinocovich, 1977), *Neptunea* (Nelson, 1974, 1978), and the Turritellidae (Kotaka, 1978).

Definition of provincial molluscan datum levels calibrated with oceanic microfossil datum levels will permit correlation of regionally significant events recorded within stratigraphic sequences deposited in marginal-marine, open shelf, or deep sea basins. Such correlations will allow the testing of hypotheses on regionally correlative unconformities (Amano, 1981; Keller and Barron, 1981; Martinez-Pardo, 1981; Ujiie, 1981) and synchronous climatic events such as that occurring at or near the beginning of the *Actinocyclus ingens* Diatom Zone in Japan, Korea, South Sakhalin, Kamchatka, on Kodiak Island, and in Oregon, Washington, and California (Addicott 1976, 1977; Gladenkov, 1980; Tsuchi, Takayanagi, and Shibata, 1981).

Concluding Remarks

Precise correlation requires the use of as many chronostratigraphic parameters as possible. Molluscan biostratigraphy is the most useful chronostratigraphic tool in shallow-marine sequences. Calibration of the North Pacific Neogene provincial molluscan biostratigraphic frameworks to each other and to the worldwide geologic time scale has already permitted refinement of correlations and recognition of some synchronous North Pacific Neogene events. The addition of molluscan datum levels, calibrated by planktonic biostratigraphy, magnetostratigraphy, and radiometric dates, will result in even more precise correlations, and will ultimately provide the basis for Pan-Pacific and global correlation of Neogene stratigraphy.

Acknowledgements

The authors wish to acknowledge our debt of gratitude to those colleagues whose unselfish sharing of data and ideas has led to this report. In particular, we wish to recognize W. O. Addicott, R. C. Allison, R. J. Echols, J. C. Ingle, S. McCoy, and L. Marincovich of the United States of America; V. V. Menner of the Union of Soviet Socialist Republics; S. Yoon of the Republic of Korea; and S. Kanno, T. Kotaka, K. Masuda, H. Noda, K. Ogasawara, T. Shuto, T. Tanai, and R. Tsuchi of Japan. Most importantly, we thank N. Ikebe and M. Chiji for their organization of IGCP-114 which provided the opportunity to develop this summary of correlations of North Pacific molluscan biostratigraphic frameworks.

References

Addicott, W. O., 1970: Latitudinal gradients in Tertiary molluscan faunas of the Pacific coast. *Palaeogeog. Palaeoclimatol. Palaeoecol.*, v. **8**, p. 287–312.

Addicott, W. O., 1972: Provincial middle and late Tertiary molluscan stages, Temblor Range, California. *In Symposium on Miocene biostratigraphy of California*, SEPM Pacific Section, Bakersfield, California, p. 1–26.

Addicott, W.O., 1974: Giant pectinids of the eastern North Pacific margin: Significance in Neogene zoogeography and chronostratigraphy. *Jour. Pal.*, v. **48**, p. 180–194.

Addicott, W.O., 1976: Neogene molluscan stages of Oregon and Washington. *In* Wornardt, W. W., ed., *Symposium on the Neogene of the Pacific Coast*. SEPM Pacific Section, San Francisco, California, p. 95–115.

Addicott, W. O., 1977: Neogene chronostratigraphy of nearshore marine basins of the eastern North Pacific:

northwestern Mexico to Canada. *In* Saito, T. and Ujiie, H., eds., *I-CPNS, Tokyo 1976, Proc.*, p. 151–175.

Addicott, W. O., 1978: Pectinids as biochronologic indicies in the Neogene of the eastern North Pacific. *Second Working Group Meeting, Biostrat. Datum-Planes of the Pacific Neogene, IGCP Project 114, Proc., Republic of Indonesia, Geological Research and Development Centre, Special Publication*, no. 1, p. 11–23.

Addicott, W. O., Kanno, S., Sakamoto, K., and Miller, D. J., 1971: Clark's Tertiary molluscan types from the Yakataga district, Gulf of Alaska. *In* U.S. Geological Survey Research 1971. *U.S. Geol. Survey Professional Paper*, 750-C, p. C18–C33, 6 figs.

Allison, R. C., 1977: Late Oligocene through Pleistocene molluscan faunas of theG ulf of Alaska region. *In* Saito, T., and Ujiie, H., eds., *I-CPNS, Tokyo 1976, Proc.*, p. 313–316.

Allison, R. C., 1978: Late Oligocene through Pleistocene molluscan faunas in the Gulf of Alaska region. *The Veliger*, v. **21**, no. 2, p. 171–188.

Allison, R. C. and Marincovich, L., Jr., 1979: Asiatic Molluscs in late Paleogene strata of the western Gulf of Alaska: Prospects for Circum-North Pacific correlation. *In* Shilo, N. A., ed., *Stratigraphy and Paleobiogeography of the Pacific Ring's Cenozoic (Section BIII), XIV Pacific Sci. Cong., Moscow 1979, Abst.*, v. **2**, p. 15–17.

Allison, R. C. and Marincovich, L., Jr., 1981: A late Oligocene or earliest Miocene molluscan fauna from Sitkinak Island, Alaska. *U.S. Geol. Survey Professional Paper*, 1233, 10, p.

Amano, K., 1981: The stratigraphy of Miocene series in northeast Honshu, Japan and eustatic sea-level changes. *In* Ikebe, N., and others, eds., *IGCP-114 Internatl. Workshop on Pacific Neogene Biostrat., Osaka, Japan, 1981, Proc.*, p. 109.

Ariey, C. A., 1978: Molluscan biostratigraphy of the upper Poul Creek and lower Yakataga Formations, Yakataga district, Gulf of Alaska. *University of Alaska, Fairbanks, M. S. thesis*, 249 p., 6 pls., 13 figs.

Ariey, C. A., 1978b: Molluscan biostratigraphy of the upper Poul Creek and lower Yakataga Formations, Yakataga district, Gulf of Alaska. *Stanford University Publications, Geol. Sci.*, v. **14**, p. 1–2.

Armentrout, J.M., 1975: Molluscan biostratigraphy of the Lincoln Creek Formation, southwest Washington. *In* Weaver, D. E., and others, eds., *Future energy horizons of the Pacific Coast; Paleogene symposium and selected technical papers*, AAPG, SEPM, SEG, Pacific Sections, Annual Meeting, Long Beach, California, p. 14–28.

Armentrout, J. M., 1979: Progress report on Cenozoic correlations, west coast North America. *In* Shilo, N. A., ed., *Stratigraphy and Paleobiogeography of the Pacific Ring's Cenozoic (Section BIII), XIV Pacific Sci. Cong., Moscow, 1979, Abst.*, v. **2**, p. 19–20.

Armentrout, J. M., 1981: Correlation and ages of Cenozoic chronostratigraphic units in Oregon and Washington. *Geol. Soc. Amer., Special Paper*, 184, p. 137–148.

Amentrout, J.M. and Echols, R. J., 1981: Biostratigraphic-chronostratigraphic scale of the northeastern Pacific Neogene. *In* Ikebe, N. and others, eds., *IGCP-114 Int. Workshop of Pacific Neogene Biostrati. Osaka, Japan, 1981, Proc.*, p. 7–27.

Armentrout, J. M., Echols, R. J. and Ingle, J. C., Jr., 1984: A Neogene biostratigraphic-chronostratigraphic scale for the northeastern Pacific margin. *In* Ikebe, N., and Tsuchi, R., eds., *Pacific Neogene Datum Planes*, Univ. Tokyo Press, p. 171–177.

Armentrout, J. M., Echols, R. J. and Nash, K. W., 1978: Late Neogene climatic cycles of the Yakataga Formation, Gulf of Alaska. *In* Addicott, W. O., and Ingle, J. C., eds., *Correlation of tropical through high latitude marine Neogene deposits of the Pacific basin, Stanford, California, Stanford Univ. Pub. Geol. Sci.*, v. **4**, p. 3–4.

Barron, J. A., 1981a: Marine diatom biostratigraphy of the Montesano Formation near Aberdeen, Washington. *Geol. Soc. America, Special Paper*, 184, p. 113–126.

Barron, J. A., 1981b: Late Cenozoic diatom biostratigraphy and paleoceanography of the middle-latitude eastern North Pacific, DSDP Leg 63. *In* Yeats, R. S., Haq, B. U., and others, eds., *DSDP, Initial Reports*, Washington, D. C., U. S. Government Printing Office, v. **63**, p. 507–538.

Barron, J. A. and Armentrout, J. M., 1980: Late Miocene diatom florules of the Empire Formation, Coos Bay, Oregon. *Geol. Soc. America, Abst.*, v. **12**, p. 95–96.

Berggren, W. A., 1981: Correlation of Atlantic, Mediterranean and Indo-Pacific Neogene stratigraphies: Geochronology and chronostratigraphy. *In* Ikebe, N., and others, eds., *IGCP-114 Int. Workshop on Pacific Neogene Biostrat. Osaka, Japan, 1981, Proc.*, p. 29–59.

Berggren, W. A. and van Couvering, J. A., 1974: Biostratigraphy, geochronology and paleoclimatology of the last 15 million years in marine and continental sequences. *Palaeogeog. Paleoclimatol. Paleoecol.*, v. **16**, p. 1–216.

Burckle, L. H., 1978: Early Miocene to Pliocene diatom datum levels for the equatorial Pacific. *Second Working Group Meeting, Biostrat. Datum-Planes of the Pacific Neogene, IGCP Project 114, Proc., Geol. Research and Development Centre, Republic of Indonesia, Special Publication*, no. 1, p. 25–44.

Burckle, L. H., Dodd, J. R. and Stanton, R. J., Jr., 1980: Diatom biostratigraphy and its relationship to paleomagnetic stratigraphy and molluscan distribution in the Neogene Centerville Beach section, California. *Jour. Pal.*, v. **54**, no. 4, p. 644–674.

Chinzei, K., 1978: Neogene molluscan faunas in the Japanese Islands. An ecologic and zoogeographic synthesis. *The Veliger*, v. **21**, no. 2, p. 155–170.

Chinzei, K., 1981: Kadonosawa Area-Stratigraphic Outline, *In* Tsuchi, R., ed., *Neogene of Japan–Its Biostratigraphy and Chronology*, IGCP-114 National Working Group Japan, Shizuoka, Japan, p. 57–61.

Gladenkov, Y. B., 1970: *Yoldia* in the Paleogene and Neogene of the North Pacific area. *Akademiia nauk SSR, Izvestiia, Seriia Geologicheskaia*, v. **2**, p. 112–122 (in Russian).

Gladenkov, Y. B., 1974: The Neogene Period in the subarctic sector of the Pacific. *In* Herman, Y., ed., *Marine Geology and Oceanography of the Arctic Seas*, Springer-Verlag,

New York, p. 271–281.

Gladendov, Y. B., 1980: Stratigraphy and marine Paleogene and Neogene of Northeast Asia (Chukotka, Kamchatka, Sakhalin): *AAPG, Bull.*, v. **64**, no. 7, p. 1087–1093.

Gladenkov, Y. B., 1981: Results and prospects of stratigraphic work on the Cenozoic sequence of the Boreal regions. *International Geology Reviews*, v. **23**, no. 12, p. 1379–1385.

Gladenkov, Y. B., Vitukin, D. I. and Oreshkina, T. V., 1979: Correlation of eastern Kamchatka Cenozoic and oceanic deposits, *In* Shilo, N. A., ed., *Stratigraphy and Paleobiogeography of the Pacific Ring's Cenozoic (Section BIII): 14th Pacific Sci. Cong. Moscow, 1979, Abst.*, v. **2**, p. 15–17.

Glaessner, M. F., 1967: Time scales and Tertiary correlations. *In* Hatai, K., ed., *Tertiary correlations and climatic changes in the Pacific, 11th Pacific Sci. Cong. Tokyo, 1966*, p. 1–5.

Hall, C. A., Jr., 1964: Shallow-water marine climates and molluscan provinces. *Ecology*, v. **45**, no. 2, p. 226–234.

Ibaraki, M., 1981: Okinawa Island-Stratigraphic outline, *In* Tsuchi, R., ed., *Neogene of Japan–Its Biostratigraphy and Chronology*, IGCP-114 National Working Group of Japan, Shizuoka, Japan, p. 34–36.

Ishizaki, K. and Takayanagi, Y., 1981: Boso Peninsula-Stratigraphic outline, *In* Tsuchi, R., ed., *Neogene of Japan–Its Biostratigraphy and Chronology*, IGCP-114 National Working Group of Japan, Shizuoka. Japan, p. 46–49.

Kanno, S., 1971: Tertiary molluscan fauna from the Yakataga district and adjacent area of southern Alaska. *Paleont. Soc. Japan. Spec. Paper*, 16, 154 p., 18 pls.

Kanno, S., 1973: Japanese Tertiary cassidids (Gastropoda) and their related mollusks from the west coast of North America. *Tohoku Univ. Sendai, Japan, Sci. Rep. Ser., 2*, Spec. Vol., no. 6, p. 217–233, pls. 19–22.

Keller, G. and Barron, J. A., 1981: Integrated planktic foraminiferal and diatom biochronolgy for the northeast Pacific and Monterey Formation. *In* Garrison, R. E., and Douglas, R. G., eds., *The Monterey Formation and related siliceous rocks in California*. SEPM, Pacific Section, Los Angeles, California, p. 43–54.

Kim, B. K., 1981: A micropaleontological study (silicoflagellate, ebridian and nannofossil) on Neogene Tertiary in the Pohang Basin (Korea). *In* Ikebe, N. and others, eds., *IGCP-114 International Workshop on Pacific Neogene Biostratigraphy, Osaka, Japan, 1981, Proc.*, p. 122.

Koizumi, I., 1977: Diatom biostratigraphy in the North Pacific region. *In* Saito, T. and Ujiie, H., eds., *I-CPNS, Tokyo 1976, Proc.*, p. 235–253.

Kotaka, T., 1978: World-wide biostratigraphic correlation based on Turritellid phylogeny. *The Veliger*, v. **21**, no. 2, p. 189–196.

MacNeil, F. S., 1965: Evolution and distribution of the genus *Mya*, and Tertiary migrations of Mollusca. *U.S. Geol. Survey Professional Paper, 483-G*, p. G1–G49, 11 pls.

MacNeil, F. S., 1967: Cenozoic pectinids of Alaska, Iceland, and other northern regions. *U. S. Geol. Survey Professional Paper*, 553, p. 1–53, 25 pls.

MacNeil, F. S., 1973: Marine fossils from the Unga Conglomerate Member of the Bear Lake Formation, Cape Aliaskin, Alaska Peninsula, Alaska. *Tokohu Univ. Sendai, Japan, Sci. Rep. Ser. 2*, Spec. Vol. no. 6., p. 117–123, 2 pls.

Maiya, S., Ichinoseki, T. and Akiba, F., 1981a: Oshima Peninsula-Stratigraphic outline, *In* Tsuchi, R., ed., *Neogene of Japan–Its Biostratitraphy and Chronology*, IGCP-114 National Working Group of Japan, Shizuoka, Japan, p. 76–80.

Maiya, S., Ichinoseki, T. and Akiba, F., 1981b: Hidaka Area-Stratigraphic outline, *In* Tsuchi, R., ed., *Neogene of Japan–Its Biostratigraphy and Chronology.*, IGCP-114 National Working Group of Japan, Shizuoka, Japan, p. 85–89.

Margulis, L. S., Shpetalenko, M. A., Gritesenko, I. I., and Boldyreva, V. P., 1981: Stratigraphic position of upper Cenozoic deposits of the Tartar Strait. *International Geology Reviews*, v. **23**, no. 11, p. 1347–1354.

Marincovich, L., Jr., 1977: Cenozoic Naticidae (Mollusca: Gastropoda) of the northeastern Pacific. *Amer. Pal. Bull.*, v. **70**, no. 294, 494 p.

Marincovich, L., Jr., 1980: Miocene mollusks of the Topsey Formation, Lituya District, Gulf of Alaska Tertiary Province, Alaska. *U. S. Geol. Survey Professional Paper*, 1125-C, p. C1–C14.

Martinez-Pardo, R., 1981: An unknown upper Miocene-lower Pliocene regional hiatus along the marginal northeast Pacific?. *In* Ikebe, N. and others, eds., *IGCP-114 Inter. Workshop on Pacific Neogene Biostrat. Osaka, Japan, 1981, Proc.*, p. 124–126.

Masuda, K., 1980: Pliocene biostratigraphy in Japan based on pectinids. *Saito Ho-on Kai Museum, Research Bull.*, no. 48, p. 9–23.

McCoy, S., Jr. and Ariey, C., 1979: Molluscan biostratigraphy of the Poul Creek and Yakataga Formations, Yakataga district, Alaska. *In* Shilo, N. A., ed., *Stratigraphy and Paleobiogeography of the Pacific Ring's Cenozoic (Section BIII), XIV Pacific Science Congress, Moscow, 1979, Abst.*, v. **2**, p. 86–87.

Menner, V. V. and Gladenkov, Y. B., 1979: On the creation of a correlation scheme for the Neogene of the northern part of the Circum-Pacific Belt, *In* Shilo, N. A., ed., *Stratigraphy and Paleobiogeography of the Pacific Ring's Cenozoic (Section BIII), XIV Pacific Science Congress, Moscow, 1979, Abst.*, v. **2**, p. 93–94.

Morozumi, Y. and Koizumi, I., 1981: Himi and Yatsuo Areas-Stratigraphic outline. *In* Tsuchi, R., ed., *Neogene of Japan–Its Biostratigraphy and Chronology*, IGCP-114 National Working Group of Japan, Shizuoka, Japan, p. 65–67.

Morrison, S. and Sarna-Wojcicki, A., 1981: Time equivalent bay and outer shelf faunas of the Neogene Humboldt Basin, California and correlation to the North Pacific microfossil zones of DSDP 173, *In* Ikebe, N., and others, eds., *IGCP-114 Internatl. Workshop on Pacific Neogene Biostrat. Osaka, Japan, 1981, Proc.*, p. 130–131.

Nelson, C. M., Jr., 1974: Evolution of the late Cenozoic gastropod *Neptunea* (Gastropoda: Buccinacea). University of California, Berkeley, Ph.D. thesis., 802 p., 66 pls., 17 figs.

Nelson, C. M., 1978: *Neptunea* (Gastropoda: Buccinacea) in the Neogene of the North Pacific and Adjacent Bering Sea, *The Veliger*, v. **21**, no. 2, p. 203–215.

Noda, H., 1966: The Cenozoic Arcidae of Japan. *Tohoku Univ., Sci. Rep., Series 2*, v. **38**, no. 1, 161 p.

Okamoto, K. and Huang, T., 1981: Early Pleistocene mollusca and nannofossils from the acoustic C formation in the southwestern Japan sea, *In* Ikebe, N., and others, eds., *IGCP-114 Internatl. Workshop on Pacific Neogene Biostrat. Osaka, Japan, 1981, Proc.*, p. 134.

Poore, R. Z., Barron, J. A. and Addicott, W. O., 1981: Biochronology of the Northern Pacific Miocene, *In* Ikebe, N., and others, eds., *IGCP-114 Internatl. Workshop on Pacific Neogene Biostrat. Osaka, Japan, 1981, Proc.*, p. 134.

Saito, T., 1981: Haboro-Embetsu Area-Stratigraphic Outline. *In* Tsuchi, R., ed., *Neogene of Japan–Its Biostratigraphy and Chronology*, IGCP-114 National Working Group of Japan, Shizuoka, Japan, p. 81–84.

Schrader, H. J., 1973: Cenozoic diatoms from the northeast Pacific, Leg 18. *In* Kulm, L. D., von Huene, R., and others, eds., *DSDP, Init. Repts*, Washington, D. C., U. S. Government Printing Office, v. **18**, p. 673–798.

Sinelnikova, V. N., 1975: Mio-Pliocene Pectinidae of Kamchatka. *Academy of Sciences of the USSR, Geol. Inst., Transactions*, v. **229**, 140 p., 25 pls. (in Russian).

Townsend, J. I., 1839: Narrative of a journey across the Rocky Mountains to the Columbia River, and a visit to the Sandwich Islands, Chili, etc., with a scientific appendix. Philadelphia, Henry Perkins.

Tsuchi, R. and Ibaraki, M., 1981: Kakegawa Area-Stratigraphic outline. *In* Tsuchi, R., ed., *Neogene of Japan–Its Biostratigraphy and Chronology*, IGCP-114 National Working Group of Japan, Shizuoka, Japan, p. 37–41.

Tsuchi, R., Takayanagi, Y. and Shibata, K., 1981: Neogene bioevents in the Japanese Islands. *In* Tsuchi, R., ed., *Neogene of Japan–Its Biostratigraphy and Chronology*, IGCP-114 National Working Group of Japan, Shizuoka, Japan, p. 81–84.

Ujiie, H., 1981: "North Pacific Middle Miocene Hiatus" and its significance in the Pacific Neogene stratigraphy. *In* Ikebe, N., and others, eds., *IGCP-114 Internatl. Workshop on Pacific Neogene Biostrat. Osaka, Japan, 1981, Proc.*, p. 145-146.

Valentine, J.W., 1966: Numerical analysis of marine molluscan ranges on the extratropical northeastern Pacific shelf. *Limnology and Oceanography*, v. **11**, no. 2, p. 198–211.

Weaver, C. E., chairman, and others, 1944: Correlation of the marine Cenozoic formations of western North America. *Geol. Soc. Amer., Bull.*, v. **55**, p. 569–598.

Yoon, S., 1981: The Seoguipo Fauna (Mollusca) of the Jeju Island, Korea. *In* Ikebe, N., and others, eds., *IGCP-114 Internatl. Workshop on Pacific Neogene Biostrat. Osaka, Japan, 1981, Proc.*, p. 149.

Yun, H., 1981: Dinoflagellates from Pohang Tertiary Basin Korea. *In* Ikebe N. and others, eds. *IGCP-114 Internatl. Workshop on Pacific Neogene Biostrat. Osaka, Japan, 1981, Proc.*, p. 150.

IV

IGCP-114: Its Activities and Main Achievements

IV-1

IGCP-114: Its Activities and Main Achievements

Nobuo IKEBE* and Manzo CHIJI**

IGCP-114

The purpose of IGCP project No. 114, "Evaluation of the biostratigraphic datum-planes of the Pacific Neogene for the purpose of global scale correlation," is to evaluate various biostratigraphic datum planes (mainly of planktonic microorganisms) of the Pacific Neogene by means of stratigraphical and volcanostratigraphical field surveys, and through reference to radiometrically dated horizons or magnetostratigraphical data obtained both from land and deep-sea bottom (DSDP, etc.).

Through these processes of research and synthesis we expect to obtain a sequence of valuable chronostratigraphic and biostratigraphic horizons relevant to the establishment of correlation of the Pacific Neogene and with the Mediterranean and American Neogene. Such a sequence of reliable key biohorizons would be valuable not only for scientific purposes but also for the exploration and exploitation of mineral and energy resources in the Pacific Neogene. The circum-Pacific regions, especially those "island arcs" along the western Pacific, are Neogene terrains composed of marine sediments with abundant micro- and megafossils, which are also intercalated with such volcanic beds as lavas or tephra layers. These volcanic materials serve as a very useful key-bed for intrabasinal correlation and may be useful in confirming the position of important bio-events in a given stratigraphical sequence. Further, these volcanic materials provide very good material for radiometric dating and for establishing magnetostratigraphy.

The project began in September 1975. It was affiliated with IGCP Project No. 1, "Accuracy in time" ("Accuracy and precision in stratigraphic and time correlations, calibration of correlation methods"; Leader: C.W. Drooger, Utrecht) as a Pacific or Japanese working team.

The present project was accepted by the IGCP Board on 19–24 March 1976, as the project No. 114 (Division I, Priority area 1, Category B) which would complete its activities in 1982. IGCP-114 would intimately be connected with IGCP-1 (above cited), IGCP-32 (Stratigraphic correlations between the sedimentary basins of the ESCAP regions, organized and led by the Mineral Resources Section, ESCAP at Bangkok) and IGCP-25 (Stratigraphic correlation of the Tethys-Paratethys Neogene, Leader: J. Senes, Bratislava).

On the other side, the project would intimately be related to the more comprehensive work of the Regional Committee on Pacific Neogene Stratigraphy, or RCPNS (N. Ikebe, convener-chairman, 1973–1976–1980 July; N. de B. Hornibrook, New Zealand, chairman 1980 July–) of the Subcommission on Neogene Stratigraphy (Chairman: R. Selli, 1973–1979; J. Senes, 1979–) of the Commission on Stratigraphy of IUGS. Thus, the First Meeting of the Working Group of Project 114 was held in Tokyo in May 1976, in conjunction with the First International Congress on Pacific Neogene Stratigraphy (1-CPNS) of IUGS, sponsored by the Science Council of Japan (Saito, T. and Ujiie, H., eds., Proc. 1-CPNS, 1977). In this first meeting, the International Working Group was established and working methods were discussed (Chiji, M., 1977; Ikebe, N., 1977).

Working Group Activities

During 1977–1979 international meetings of IGCP-114 were held according to the following schedule and with the following items of discussion.

The Second Working Group Meeting of IGCP-114 was held at the Geological Research and Development Centre, Bandung, Indonesia, under the theme "Critical revisions on important biostratigraphic datum planes of the Pacific Neogene, with special emphasis on the regions of Southeast Asia", organized by the

* *(Tezukayama University) 6–4, Hama-cho, Ashiya 659, Japan*
** *Osaka Museum of Natural History, Nagai Park, Higashi-sumiyoshi-ku, Osaka 546, Japan*

Organizing Committee chaired by Darwin Kadar and sponsored by the Geological Research and Development Centre of Indonesia (H. M. S. Hartono, Director). The meeting was followed by excursions or field-discussions in West Java (2 June) and in Central Java (3–5 June) (Kadar, D., ed., Proc. 2-IGCP-114, 1978; Ikebe, N., 1978).

The following thirteen papers were presented at this meeting:

Addicott, W.O.: Pectinids as biochronologic indices in the Neogene of the eastern North Pacific.

Burckle, L.H.: Early Miocene to Pliocene diatom datum levels for the equatorial Pacific.

Djamas, Y.S. and Marks, E.: Early Neogene foraminiferal biohorizons in E.Kalimantan, Indonesia.

Ikebe, N. and Chiji, M.: Neogene datum-planes of western Pacific, a proposal for discussion.

Kadar, D.: Upper Pliocene and Pleistocene planktonic foraminiferal zonation of Ambengan drill hole, southern part of Bali Island.

Kim, B.K.: Preliminary benthonic foraminiferal zonation and faunal analysis based on the quantitative method in the Tertiary Pohang Basin, Korea.

Matsumaru, K.: Biostratigraphy and paleoecological transition of larger foraminifiera from the Minamizaki Limestone, Chichi-jima, Japan.

McGowran, B.: Australian Neogene sequences and events.

Nishimura, S., Sasajima,S., Thio, K.H. and Hehuwat, F.: The 2nd report of fission track dating in the Sunda Arc.

Pringgoprawiro, H., Soeharsono,N. and Sujanto, F.X.: Subsurface Neogene planktonic foraminiferal biostratigraphy of North—West Java Basin.

Shuto, T.: Molluscan biohorizons in the Indo—West Pacific Neogene.

Theyer, F., Mato, C.Y. and Hammond, S.R.: Neogene radiolarian events of the Tropical Pacific: A catalog of magnetostratigraphic calibrations.

Tsuchi, R. and Ibaraki, M.: Definition and faunal characteristics of late Neogene stages on the Pacific coast of southwestern Japan.

The Third Working Group Meeting was held at Stanford University, California, on 26–28, June, 1978, under the theme "Correlation of tropical through high latitude marine Neogene deposits of the Pacific basin," organized by the Local Committee co-chaired by W.O. Addicott and J.C. Ingle. The meeting was preceded by a four-day field trip (21–24 June) through the San Joaquin Hills, south of Low Angeles—La Panza Range—Santa Cruz—San Francisco—Stanford, observing Neogene biostratigraphy of selected areas in the California Coast Ranges. During the session at Stanford, a meeting on the correlation of northeast Pacific Neogene was convened by J.M. Armentrout (Addicott, W.O. and Ingle, J.C., Jr., eds., Abstract and Program, 3-IGCP-114, Stanford, 1978; Addicott, W.O., ed., 1978; Ikebe, N., 1979).

The following thirty-seven papers were presented at this meeting:

Ariey,C.: Molluscan biostratigraphy of the upper Poul Creek and lower Yakataga Formation, Yakataga District, Gulf of Alaska.

Armentrout,J.M., Echols,R.J. and Nash,K.W.: Late Neogene climatic cycles of the Yakataga Formation, Robinson Mountains, Gulf of Alaska area.

Barron,J.A., von Huene,R. and Nasu,N.: Correlation of high and low latitude upper Miocene diatom datum levels at DSDP Site 438 in the northwest Pacific.

Burckle,L.H., Dodd,J.R. and Stanton,R.J.: Diatom biostratigraphy of the Centerville Beach section, California.

Burckle,L.H., Gartner,S., Opdyke,N.D., Sciarrillo,J.R., and Shakleton,N.J.: Paleomagnetics, oxygen isotopes and biostratigraphy of a late Pliocene section from the central Pacific.

Crouch,R.W. and Poag,C.W.: *Amphistegina gibbosa* (d'Orbigny) from the California borderlands: the Caribbean connection.

Gladenkov,Y.B.: Mollusks and Neogene correlation of the North Pacific.

Harper, H.E.: Diatom biostratigraphy of the Miocene/Pliocene boundary in the North Pacific.

Hashimoto,W. and Matsumaru,K.: Consideration of the stratigraphy of the Caraballo Range, northern Luzon, checking with the larger foraminiferal ranges from the Cenozoic sediments of the Philippines.

Huang,T.: Significant new look on the Neogene stratigraphy of Taiwan.

Hughes,G.W.: Planktonic foraminiferal biostratigraphic datum planes for Plio-Pleistocene sedimentary rocks from the Solomon Islands.

Ikebe,N. and Chiji,M.: Evaluation of some important datum-planes of the Pacific Neogene.

Ishida,S.: Paleoecology and stratigraphy of Neogene in some areas of Japan and Pacific.

Kanno,S. and Noda,K.: Biostratigraphical and paleogeographical distribution of the gastropod genus *Vicarya*.

Kassab, I.I.M.: The Paleogene/Neogene boundary in Iraq.

Keigwin,L.D.,Jr.: Middle Miocene to Pliocene stable isotopic datums in east equatorial and north central Pacific deep sea drilling (DSDP) sites.

Keller,G.: Late Neogene paleoceanography and planktonic foraminiferal datum levels of mid latitudes of the North Pacific.

Kennett,J.: Cenozoic microfossil datums in Antarctic-Subantarctic deep-sea sedimentary sequences and the evolution of southern ocean planktonic biogeography.

Kurihara,K.: Foraminiferal datum levels recognized in the Neogene sections of the Kanto District, central Japan.

Lagoe,M.B.: Foraminifera from the uppermost Poul Creek and lowermost Yakataga Formation, Yakataga District, Alaska.

Masuda,K.: Pectinid biostratigraphy of the Neogene in central to south Japan.

Noda,H.: Neogene anadaran distribution in Japan and Southeast Asia.

Obradovich,J.D., Naeser,C.W. and Izett,G.A.: Geochronology of late Neogene marine strata in California.

Poore,R.A. and McDougall,K.: Calcareous microfossil biostratigraphy and paleoecology of the type section of the Luisian Stage of California.

Quilty,P.G.: Comparison of tropical-Antarctic deep marine and land based late Tertiary sections, sothwest Pacific.

Rau,W.W., Plafker,G. and Winkler,G.R.: Foraminiferal biostratigraphy in the Gulf of Alaska Tertiary province.

Sancetta,C.A.: Neogene planktonic provinces: a synthesis of DSDP material.

Serova,M.Y.: Foraminiferal datum planes and correlative assemblages in the northwest Pacific Neogene.

Shuto,T.: Molluscan succession in the Oligo-Miocene of southwest Japan—with special reference to the Oligocene/Miocene boundary.

Soeka,S., Suminta, and Sudjaah,T.: Neogene benthonic foraminiferal biostratigraphy and datum planes of the East Java Basin, Indonesia.

Srinivasan,M.S. and Azmi,R.J.: Correlation of late Cenozoic-marine sections in Andaman-Nicober and the equatorial Pacific.

Stoll,S.J., Wonfor,J.S., Shaffer,B.L., Smith,D.J., and Park, N.Y.: Biostratigraphic problems of the Neogene sequencefrom the Shimane Peninsula to the Noto Peninsula, southwest Honshu, Japan.

Taylor,D.J. and Deighton,I.: The two *Sphaeroidinella* datums in the southwest Pacific.

Theyer,F.: Geochronology and stratigraphic reliability of Neogene radiolarian events, tropical Pacific.

Thompson,P.R.: Late Tertiary planktonic foraminiferal datum biostratigraphy of the western North Pacific.

Tsuchi,R. and Ibaraki,M.: Notes on correlation of late Neogene sediments on the southern Japan with those on the northern Japan.

Zinsmeister,W.J.: Review of the Neogene of the Pacific margin of Antarctica.

The Fourth Working Group Meeting was held in Khabarovsk, USSR, on 22–26 August 1979, convened simultaneously with the scientific session of the 2nd International Congress of Pacific Neogene Stratigraphy (2-CPNS) of RCPNS, in conjunction with the 14th Pacific Science Congress of the Pacific Science Association. The meeting was held on the theme "Miocene and Pliocene marker horizons in the Pacific", convened by V.V. Menner and Y.B. Gladenkov, and sponsored by the Academy of Sciences, USSR. The meeting was followed by a filed trip to see the Makarov key section (Oligocene-Pliocene) in Sakhalin, which was held 2–8 September (Shilo, ed. 1979; Menner, ed. 1979; Ikebe, N., 1980).

The following thirty papers were orally presented at the meeting:

Ablayeva,A.G. and Schmidt,I.N.: The Neogene stage in the development of Sikhote-Alin florae and vegetation.

Addicott, W.O. and Poore, R.A.: Paleogene-Neogene boundary in the marine sequences of the eastern North Pacific.

Akiba,F.: Revised Japanese Neogene diatom biostratigraphy.

Allison,R.C. and Marincovich,L.,Jr.: Asiatic mollusks in Late Paleogene and Neogene strata of the western Gulf of Alaska—Prospects for Circum-North Pacific correlations.

Baranova,Y.P.: Regional Neogene horizons in Northeast Asia.

Beu,A.G.: *Hartungia typica* Bronn, 1861 (Gastropoda, Family Janthinidae)—a late Pliocene pelagic macrofossil datum plane.

Biske,S.F., Baranova,Y.P. and Shvareva,N.Ya.: Climate of Paleogene and Neogene of the north-eastern Asia.

Brutman,N.Ya.: Palynozones and pollen of Genus *Tsuga* from the Neogene of Sakhalin.

Fotyanova,L.I.: Correlative floral levels of the Upper Oligocene—Lower Miocene in Europe and the Far East.

Fradkina,A.F.: Correlation of Neogene deposits in the north-east Asia by aid of palynological data.

Gladenkov,Y.B., Vitukhin,D.N. and Oreshkina,T.V.: Correlation between Cenozoic of eastern Kamchatka and oceanic deposits.

Heath,R.: Correlation of Indo-Pacific Neogene planktonic foraminiferal events.(read by N.Ikebe)

Hornibrook,N. de.B.: *Globorotalia crassaformis* and *G.crassula* in the Pliocene of New Zealand and a Late Pliocene cool marine phase in the Lower Nukumaruan Stage. (read by A.G.Beu)

Ikebe,N.: Evaluation of biostratigraphic datum-planes of the Pacific Neogene—Introductory to the Symposium on "Miocene and Pliocene marker horizons in the Pacific."

Kanno,S.: Molluscan faunal sequences of the Neogene System in Hokkaido, northern Japan.

Linkova,T.I.: Correlation of Cenozoic deep sea bottom sediments based on paleomagnetic and biostratigraphic data.

Margulis, L.S., Savitsky,V.O. and Tyutrin,I.I.: The Cenozoic sediments of south Sakhalin and adjacent sea regions.

Masuda,K. Early Pliocene biostratigraphy in Japan based on Pactinids.

McCoy,S., Jr. and Ariey,C.: Molluscan biostratigraphy of the Poul Creek and Yakataga Formations, Yakataga District, Alaska.

Menner,V.V. and Gladenkov,Y.B.: Elaboration of correlation scheme of the Neogene in the northern part of the Pacific Ring.

Moiseyeva,A.I., Sheshukova,V.S., Poretskaya, and Boldyreva, V.P.: Neogene diatom assemblages in the Makarov key section, Sakhalin, and their implication for stratigraphy and correlation.

Ogasawara,K.: Geological and paleogeographical significances of the Omma-Mangajian Fauna of Japan Sea borderland.

Sasa,Y.: Comment on the geological maps of Sakhalin issued by Japanese geologists, 1960.

Savitsky,V.O., Boldyreva,V.P., Mitrophanova,L.I., and Taboyakova,L.A.: Lower and Upper Miocene boundary in marine Cenozoic section of Sakhalin.

Serova,M.Y.: Biostratigraphy and correlation of marine Neogene in northern Pacific.

Sinelnikova,V.N., Serova,M.Y., Bratseva,G.M., Giterman, R.E., Gladenkov,Y.B., Gladikova,V.M., Dolmatova, L.M., Kafanov,A.N., Korobkov,A.I., Krishtofovich, L.V., Kuklina,T.G., Konova,L.A., Lupikina,E.G., Popov, S.V., Vitukhin, D.I., Fotyanova, L.I., Grechin, V.I., and Shmidt,O.I.: Neogene key section in western Kamchatka.

Tochilina,S.V.: Neogene correlation in the Sea of Japan.

Tsuchi,R. and Working Group on Neogene biostratigraphy and radiometric dating of Japan. Recent progress in bio- and chronostratigraphy of the Japanese Neogene—1979 Report.

Volobuyeva, V.I.: Paleogene and Neogene biostratigraphy of the eastern part of the Koryak upland, and their correlation with respective formations in the north of the Pacific belt.

Zhidkova, L.S., Salinikov,B.A., Brutman,N.J., Zaklinskaya, E.D., Kuznetsova, V.N., Moiseyeva,A.E., Popova,L.A., Fotyanova,L.I. and Shainyan, S.H.: Makarov key stratigraphic section of the Paelogene-Neogene in Sakhalin, and their role in stratigraphy and correlation.

Fifth Working Group Meeting: For the year 1980, the activities of IGCP-114 were chiefly concentrated on two special meetings under the framework of the 26th IGC in Paris: S 04.2.12 "Stratigraphic correlations in the Neogene" (J. Senes, convener, sponsored by IGCP) and Global Neogene business meeting (of IUGS bodies and IGCP porjects on the Neogene problems). At the former meeting, a summary report of the progress of IGCP-114 was introduced by Ikebe and others and Tsuchi and others on 12 July 1980, as mentioned below. At the business meeting a plan for the 1981 Osaka Meeting, including a provisional plan for compiling the final report of the project, was introduced by Ikebe and Chiji (Ikebe, N., 1981).

Ikebe,N., Chiji,M. and Huang,T.Y.: Important datum-planes of the western Pacific Neogene.

McCoy,S.,Jr.: Neogene biostratigraphy and correlations of the eastern gulf of Alaska.

Tsuchi, R., Takayanagi, Y. and Shibata, K.: Stratigraphic succession of Japanese Neogene events.

Ujiie,H.: Relationship between a stratigraphic hiatus above the lowermost middle Miocene and the spreading of the Sea of Japan.

The Japan-Indonesia cooperative research on Neogene—Palaeogene biostratigraphy in Java was conducted in the Sentolo and Nanggulan sections in Central Java, under the leadership of T. Saito and D. Kadar from June to August of 1979 (Saito, ed., 1981), and from July to September of 1981, sponsored by the Ministry of Education of the Government of Japan.

The Sixth Working Group Meeting, the final meeting, was held in Osaka and Kobe between 25 and 29 November 1981 under the title "International Workshop on Pacific Neogene Biostratigraphy (IWPNB)". The meeting was followed by an excursion to southern Korea from 29 Nov. to 3 Dec., 1981. The purpose of the meeting was to summarize the work accomplished during the preceding 6 years.

The meeting consisted of two parts. The first part was the general session which was held at the Osaka Museum of Natural History on 25 and 26 Nov. Twenty-one overseas (Chile; 1, Indonesia; 3, Korea; 3, New Zealand; 1, Taiwan; 1, U.K.; 2, U.S.A.; 9, U.S.S.R.; 1) and 70 domestic participants took part in this session, and 28 contributions were presented and discussed. The second part was a semi-closed workshop which was held at the Kansai Inter-University Seminar House, Kobe, between 27 and 29 November. For this latter half of the meeting, 21 overseas and 28 domestic participants attended. In this workshop, methods as well as procedures for drawing conclusions from our work were discussed. In particular, sectional meetings were organized to discuss datum planes of individual fossil group including Planktonic Foraminifera, Calcareous Nannoplankton, Diatom, Radiolaria, Larger Foraminifera and Molluscs. A group of specialists also compiled a combined paleomagnetic and radiometric time scale.

The proceedings of this meeting were published, and IGCP-114 National Working Group of Japan (Leader; R. Tsuchi) also issued a commemorative volume on this meeting. These two publications were distributed to the meeting participants (Ikebe *et al.*, eds., 1981; Tsuchi, ed., 1981). The post-meeting excursion (Nov. 29 — Dec. 3) visited the Pohang and Yeonil areas and Cheju Is. of southern Korea for stratigraphical and paleontological checks, under the leadership of Profs. B. K. Kim, S. Yoon and H. Yun. Thirty colleagues (Chile; 1, Indonesia; 2, Japan; 19, Korea; 3, Taiwan; 1, U.S.A.; 4) took part in the excursion (Yoon, ed., 1981). The 6th business meeting of the IGCP-114 was held on 27 November 1981, and compilation and publication of the final report was discussed (Ikebe, N., 1982).

The following twenty nine papers were presented in the general session:

Adams,C.G.: Neogene larger foraminifera, evolutionary and geological events in the context of planktonic datum planes.

Amano, K.: The stratigraphy of Miocene Series in northeast Honshu, Japan and eustatic sea-level changes.

Armentrout, J.M. and Echols, R.J.: Biostratigraphic-chronostratigraphic scale of the northeastern Pacific Neogene.

Backlman, J. and Shackleton, N.J.: Quantitative biochronology of calcareous plankton of the Pliocene and Early Pleistocene of the Pacific region.

Berggren,W.A.: Correlation of Atlantic, Mediterranean and Indo-Pacific Neogene stratigraphies: Geochronology and chronostratigraphy.

Djamas,Y.: Biostratigraphy of Early Neogene time in South Sumatra Basin, Indonesia.

Gladenkov, Y.B.: Neogene horizons (Regiostages) of Sakhalin and Kamchatka and the problem of their correlation.

Haq,B.U.: Calibration of Pacific Neogene nannofossil datums to magnetostratigraphy.

Hirooka,K., Tsuda,S., Nishimura,S., Sasajima,S., Thio,K. H., and Hehuwat, F.: Paleomagnetic study of Miocene sediments in Karansanbung area, Central Java, Indonesia.

Hornibrook,N. de B.: Evaluation of Pacific southern mid latitude and equatorial latitude planktonic foraminiferal datum levels.

Huang,T.: Biostratigraphy of the Late Miocene in western Taiwan.

Ibaraki,M.: Geologic ages of "*Lepidocyclina*" and *Miogypsina* horizons in Japan as determined by planktonic foraminifera.

Ikebe,N. and Chiji,M.: IGCP-114: Biostratigraphic datum-planes of the Pacific Neogene—purpose and activities.

Ingle,J.C., Jr. Neogene reference sections of the Pacific Coast of North America; their regional significance and correlation.

Kennett, J.P. and Srinivasan,M.S.: Neogene equatorial to subantarctic planktonic foraminiferal biostratigraphy and datum levels: South Pacific.

Kim, B.K.: A micropaleontological study (silicoflagellate, ebridian and nannofossil) on Neogene Tertiary in the Pohang basin, Korea.

Koizumi, I.: Evolutionary trends of the marine species of the genus *Denticulopsis*.

Marincovich,L.: Late Miocene Mollusks of the Tachilni Formation, Alasaka Peninsula, Alaska.

Martinez-Pardo,R.: An unknown Upper Miocene-Lower Pliocene regional hiatus along the marginal Northeast Pacific?

Matsuoka, K.: Late Cenozoic dinoflagellate cyst assemblage in the Niigata region, Central Japan.

Morrison,S. and Sarna-Wojcicki,A.: Time equivalent bay and outer shelf faunas of the Neogene Humboldt basin, California and correlation to the North Pacific microfossil zones of DSDP 173.

Poore, R.Z., Barron, J.A. and Addicott, W.O.: Biochronology of the northern Pacific Miocene.

Shibata, K.: Radiometric dating of Neogene igneous activity in Japan with reference to biostratigraphy.

Shuto,T.: Biogeography and correlation of Late Neogene gastropod faunas of Southeast and East Asia.

Soeka,S., Suminta and Siswoyo, The Miocene/Pliocene boundary in the North-East Java basin.

Tsuchi,R. and IGCP-114, National Working Group of Japan Recent progress in bio- and chronostratigraphy of Japanese Neogene.

Ujiie,H.: "North Pacific Middel Mocene Hiatus" and its significance in the Pacific Neogene stratigraphy.

Yoon,S.: The Seoguipo fauna (mollusca) of the Jeju Island, Korea.

Yun,H.: Dinoflagellates from Pohang Tertiary basin, Korea.

Main Achievements

This project attempted to make some significant progress in the following fields of research concerning the Pacific Neogene stratigraphy over a pcriod of seven years:

1) Establishment of regional biostratigraphic successions based on appropriate micro- and/or megafossil taxa (ref.: chapter III of this book).

2) Cross-correlation of sequences of bioevents defined by various microfossil taxa mainly with comparative examination of magnetostratigraphic data.

3) To increase the number of reliable radiometric data which will help determine geochronological position of important bio-events especially the base- or top-datum of selected microorganisms.

Many members of a number of national working groups of the project have contributed new paleontological, radiometric and magnetostratigraphic data. Particularly, biostratigraphic data covering such varied taxa as planktonic foraminifers, diatoms, radiolarians, larger foraminifers and molluscs enabled the recognition of many reliable datum planes defined by these taxonomic groups. Evaluation of selected datum planes is materialized through the following processes: critical examination of the parallelization between bio-events defined by these taxa, comparison with those obtained from deep-sea core sequences, geochronological checking of selected bio-events with the use of radiometric and magnetostratigraphic methods, and geochronological comparison of sequences of bio-events recognized in various regions.

Revision of chronological positions of some selected planktonic foraminiferal datums in the western Pacific region during the term of the project are shown in Fig. 1.

A correlation between the Pacific and Japan Sea sides of Japan for the uppermost part of the Neogene has been accomplished by means of the top and base datums of *Globoquadrina asanoi*. For correlating the Lower/Middle Miocene boundary, the *Orbulina* base datum is useful. In the Japan Sea side region of Japan *Orbulina* occurs very rarely. Thus, the base datum of *Globigerina bulloides* is useful in this region in place of the *Orbulina* datum. The importance of *G. bulloides* datam has been proven in such sections of Japan as Kanazawa, Noto Peninsula and Tomioka (Ikebe and Chiji, 1973, Chiji,and Konda, 1978; 1981, Ikeda, 1982). In the central Paratethys, the same relation of the *G.bulloides* datum to the *Orbulina* datum has been reported by Rögl (1975).

An example of the evaluation of planktonic foraminiferal bio-events relative to the geomagnetic polarity events in the equatorial latitudes and south mid-latitudes (New Zealand) of the Pacific is the work by Hornibrook (1981). Between these two areas, time-lag as well as synchroneity of bio-events are recognized in various degrees going back to the Polarity Zone 7. Hornibrook also compared a sequential order of significant planktonic foraminiferal events inthe Neogene section lower than Polarity Zone 7 in the same area.

In the Japanese Neogene, planktonic foraminiferal datum planes between 0 and 6 Ma are chronologically calibrated against the geomagnetic polarity time scale based upon biostratigraphies and magnetostratigraphies of some deep-sea cores and of a land exposed Neogene section of the Kakegawa area on the Pacific side of Central Japan. For the interval prior to 6 Ma, calibration was done primarily on the basis of biostratigraphy of DSDP Site 289 by assuming a constant sedimentation rate at the site between the base datum of *Orbulina* at 15.5 Ma and the base datum of *Pulleniatina* at 5.8 Ma.

Diatom bio-events for the 0 and 5 Ma interval are chronologically calibrated against the geomagnetic polarity time scale based on biostratigraphy given by

1 Ikebe et al. (1972) Ikebe (1973)
- Grt.truncatulinoides B 2±0.2
- Spa.dehiscens B (5.5)
- Pulleniatina B (7.5)
- Grt.acostaensis B (9)
- Grt.siakensis T (10)
- Gna.nepenthes B (11)
- Grt.fohsi foshsi s.Bolli B (14)
- Orbulina B 15±1
- (Praeorbulina) B (16)
- Gds.sicanus B (16.5)
- Csx.dissimilis/unicava T (18)
- Globigerinoides B (23)

2 Ikebe et al. (1977)
- Grt.tosaensis T 0.5-0.6
- Gqa.asanoi T 1.8
- Grt.truncatulinoides B 1.8-2.0
- Gqa.asanoi B 3.3-3.4
- Grt.tosaensis B 4-4.5
- Spa.dehiscens B (5)
- Grt.tumida B (5.3)
- Pulleniatina B 7.0-7.5
- Grt.acostaensis B 9.5-10
- Grt.siakensis T 10.8
- Gna.nepenthes B 11.3±0.4
- Sps.subdehiscens B 12.5
- Grt.peripheroacuta B (14)
- Orbulina B 15±0.5
- Praeorbulina B (16)
- Gds.sicanus B 16.5-17
- Csx.dissimilis/unicava T (18)
- Gtl.insueta B (19.5)
- Globigerinoides B (23)

3 Ikebe (1978)
- Grt.tosaensis T 0.5-0.6
- Gqa. asanoi T 1.8
- Grt.trun'noides B 1.9
- Gqa.asanoi B 2.8
- Grt.tosaensis B 3.3-3.5
- Spa.dehiscens B 4.8
- Grt.tumida B 5.3
- Pulleniatina B 6.2
- Grt.acostaensis B 10.0
- Grt.siakensis T 11
- Gna.nepenthes B 11.3±0.4
- Sps.subdehiscens B (12.0)
- Grt.peripheroacuta B 14
- Orbulina B 15±0.5
- Praeorbulina B (16)
- Gds.sicanus B 16.5-17
- Csx.dissimilis/unicava T (18)
- Gtl.insueta B (18.6-19.5)
- Globigerinoides B (23)

4 Ikebe (1978) (IGCP Spec.publ.)
- Grt.tosaensis T 0.5-0.6
- Gqa.asanoi T 1.6-1.8
- Grt.trun'noides B 1.8-2.0
- Gqa.asanoi B (2.5)
- Grt.tosaensis B 3.3-3.5
- Spa.dehiscens B 4.5-4.8
- Grt.tumida B 5.0-5.5
- Pulleniatina B 6.2-7.0
- Grt.acostaensis B 10.0-10.5
- Grt.siakensis T 11.0-11.2
- Gna.nepenthes B 11.3-11.7
- Sps.subdehiscens B (12.0)
- Grt.peripheroacuta B 14.5
- Orbulina B 15.0-15.5
- Praeorbulina B (16.0)
- Gds.sicanus B 16.5-17.0
- Csx.dissimilis/unicava T 17.5-18.5
- Gtl.insueta B (19.0-20.0)
- Globigerinoides B (22.5-23.5)
- Grt.kugleri B (25)

5 Ikebe et al. (1980) IGC
- Grt.trun'noides B 1.8-2.0
- Grt.tosaensis B 3.1-3.5
- Spa.dehiscens B 4.5-4.8
- Grt.acostaensis B 10-10.5
- Gna.nepenthes B 11.3-11.7
- Grt.peripheroacuta B 14-14.5
- Orbulina B 15-15.5
- Praeorbulina B (16)
- Gds.sicanus B 16.5-17.2
- Csx.dissimilis T 17.5-18
- Gtl.insueta B (18.6-20)
- Globigerinoides B (22.5-23.5)

6 Ikebe et al. (1981) Tsuchi et al. (1981)
- Grt.trun'noides B 1.9
- Grt.tosaensis B 3.0
- Spa.dehiscens B 5.1
- Grt.tumida B 5.4
- Pulleniatina B 5.8
- Grt.acostaensis B 9.1
- Grt.siakensis T 10.4
- Gna.nepenthes B 11.6
- Grt.peripheroacuta B 14.6
- Orbulina B 15.5
- Praeorbulina B (16)
- Gds.sicanus B 16.5
- Csx.dissimilis/unicava T 17.6
- Gtl.insueta B (18)
- Globigerinoides B (22-24)

7 IGCP-114, IWPNB (Nov. 1981)
- Grt.trun'noides B 1.9
- Grt.tosaensis B 3.0
- Gna.nepenthes T 3.9
- Spa.dehiscens B 5.0
- Grt.tumida B 5.3
- Pulleniatina B 5.9
- Grt.t.plesiotumida B 8.2
- Grt.acostaensis B 11.2
- Grt.siakensis T 12.0
- Gna.nepenthes B 12.4
- Grt.peri'acuta B 15.0
- Orbulina B 15.5 or 15.2
- Praeorbulina B 15.8
- Gds.sicanus B 16.4
- Csx.dissimilis T 17.3
- Grt.kugleri T 21.4
- Gqa.dehiscens B 22.8
- Grt.kugleri B 23.8
- Gds.primordius B 25.1

Ma: 5, 10, 15, 20, 25

Fig. 1. Successions of key planktonic foraminiferal datums in the western Pacific Neogene. (1-6, Western Pacific data mainly from Japan; 7, Standard for the Pacific Neogene, data plotted against the magnetic anomaly scale of Ness *et al.*, 1980).

IMPORTANT BIOSTRATIGRAPHIC DATUMS	Southern mid latitude (New Zealand)	Equatorial	Northern mid latitude (Japan)
B *Globorotalia truncatulinoides*	2.0	1.9	1.9
B *Globoquadrina asanoi*			2.8
B *Globorotalia tosaensis*	2.5	3.0	3.0
T *G. margaritae*	3.2	(3.3)	3.4
B *G. inflata*	3.8	?	2.8
B *Sphaeroidinella dehiscens*		5.0	(5.1)
B *Globorotalia crassaformis*	5.1	?	(5.2)
B *G. tumida*	?	5.3	(5.4)
B *Pulleniatina primalis*		5.9	(5.8)
B *Globorotalia conomiozea*	5.8	(6.3)	
B *G. acostaensis*		11.2	(9.1)
T *G. siakensis*	12.0	12.0	10.4
B *Globigerina nepenthes*	12.7	12.4	11.6
B *Globorotalia peripheroacuta*		15.0	(14.6)
B *Globigerina bulloides*		(10.0)	15.5
B *Orbulina suturalis*	16.0	15.5	15.5
B *Praeorbulina glomerosa curva*	16.4	15.8	(16.0)
B *Globigerinoides sicanus*	16.6	16.4	(16.5)
B *Catapsydrax dissimilis*	19.5	17.3	(17.6)
T *C. stainforthi*		(19.5)	
B *Globigerinatella insueta*	?	18.5	(18.0)
B *Globorotalia incongnita*	19.5	(20.5)	
T *G. kugleri*	?	21.4	(20.0)
B *Globoquadrina dehiscens*	24.0	22.8	?
B *Globorotalia kugleri*	?	23.8	

Fig. 2. Selected important planktonic foraminiferal datum scale and its applicable areas in the western Pacific.

Donahue (1970), Koizumi (1975), Burckle and Opdyke (1977). For the interval prior to 5 Ma, chronological calibration relied on biostratigraphic data available from land-exposed sections in Japan, published K-Ar or fission track ages, and also estimated sedimentation rates in the Northwest Pacific.

It is worth while to mention here that, through the effort of the members of the molluscan subgroup of the working group, especially Armentrout, Chinzei and Gladenkov, correlation of molluscan biostratigraphies of the Northern Pacific Neogene in the light of the diatom biostratigraphy has lately become possible.

As a magnetostratigraphical and chronostratigraphical standard time scale, we adopted, at the International Workshop held in 1981 at Osaka and Kobe, the time scale given by Ness, Levi and Couch (1980). We also have attempted to use all the relevant data available from the Atlantic and Italian sections as well as from the equatorial Pacific. At present, there are insufficient data available to cover te entire Neogene sequence based only on the Pacific data.

Through the comparison between the standard planktonic datum-scale of the equatorial (E) and the north mid latitude (N) as well as the south mid-latitude (S) (Fig. 2), the following twelve planktonic foraminiferal datum planes are selected as the key datums for long distance correlation (S — E — N):

Base *Globorotalia truncatulinoides*	(1.9 — 2.0)
Base *Globorotalia tosaensis*	(2.5 — 3.0)
Top *Globorotalia margaritae*	(3.2 — 3.4)
Base *Globorotalia inflata*	(2.8 — 3.8)
Base *Globorotalia crassaformis*	(5.1 — 5.2)
Base *Globorotalia tumida*	(5.3 — 5.4)
Base *Globigerina nepenthes*	(11.6 — 12.7)
Base *Orbulina suturalis*	(15.5 — 16.0)
Base *Praeorbulina glomerosa curva*	(15.8 — 16.4)
Base *Globigerinoides sicanus*	(16.4 — 16.6)
Base *Globigerinatella insueta*	(18.0 — 18.5)

Top *Globorotalia kugleri* *	(20.0 — 21.4)
Base *Globoquadrina dehiscens*	(22.8 — 24.0)
Base *Globorotalia kugleri*	(23.8 ±)

The following seven datums are applicable to establish correlation in somewhat restricted areas:

(S — E)	
Base *Globorotalia conomiozea*	(5.8 — 6.3)
Base *Globorotalia incongnita*	(19.5 — 20.5)
(E — N)	
Base *Sphaeroidinella dehiscens*	(5.0 — 5.1)
Base *Pulleniatina primalis*	(5.8 — 5.9)
Base *Globorotalia periopheroacuta*	(14.6 — 15.0)
Top *Catapsydrax stainforthi*	(19.5 ±)
(N)	
Base *Globoquadrina asanoi*	(2.8)

In northern temperate regions, the base datums of *Globoquadrina asanoi* and *Globigerina bulloides* would be useful as the key biohorizons for areal correlation, as mentioned before.

The larger foraminifera are an important taxonomic group for Tertiary biostratigraphic correlation in the Indo-Pacific Region. It is another important achievement that occurences of larger foraminifera were made clear by Adams with respect to Blow's planktonic foraminiferal zones.

We express our sincere gratitude to all the members as well as collaborators of IGCP Project 114 for their kind cooperation and for presenting valuable contributions during the term of the Project from 1976 to 1982.

Our acknowledgement also goes to the secretariat of IGCP-UNESCO and IUGS, for their continuous support of this project.

References

(References will be seen in IV–2 Bibliopgraphy related to IGCP-114; those not included in the Bibliography are listed here.)

Donahue, J.G., 1970: Pleistocene diatoms as climatic indicators in North Pacific sediments, *Geol. Soc. Amer., Mem.*, no. 126, p. 121–138.

Ikebe, N. and Chiji, M., 1973: Some problems on the geological age of "*Lepidocyclina-Miogypsina* Zone" and "*Miogypsina-Operculina* Zone" in Japan (Japanese w. English abstract), *Geol. Soc. Japan, Mem.*, no. 8, p. 74–84.

Koizumi, I., 1975: Late Cenozoic diatom biostratigraphy in the circum-Pacific region. *Geol. Soc. Japan, Jour.*, v. **81**, p. 611–627.

Ness, G., Levi, S. and Couch, R., 1980: Marine magnetic anomaly timescales for the Cenozoic and Late Cretaceous: a précis, critique and synthesis. *Geophy. Space Physics, Review*, v. **18**, p. 753–770.

Nishi, H., 1983: Planktonic foraminiferal biostratigraphy of the Nichinan Group (in Japanese). *Geol. Soc. Japan, 90th Annual Meet., Abstr.* Kagoshima, 1983, p. 121.

Rögl, F., 1975: Die plankonischen Foraminiferen der zentralen Paratethys. *VI-CMNS, Bratislava, Proc.*, v. **1**, p. 113–120.

* Lately, fairly reliable occurrence of *Globigerinoides primordius, Globorotalia kugleri* and *Globoquadrina dehiscens* have been reported from some sections of the Takigahirayama Formation of the Nichinan Group in Kyushu (Nishi, 1983).

Appendix

Members of International Working Group, IGCP-114

ADAMS, G. C. — British Museum (Natural History), Cromwell Road, London SW7, 5 BD, U.K.

ADDICOTT, W.O. — U. S. Geological Survey, 345 Middlefield Road, MS 15, Menlo Park, Calif. 94025, U. S. A.

ARMENTROUT, J. M. — Mobil Exploration and Producing Services Inc., P.O. Box 900, Dallas, Texas 75221, U.S.A.

BANNER, F. T. — Dept. Geology & Oceanography, University College of Wales, Swansea, England

BERGGREN, W. A. — Woods Hole Oceanographic Inst., Woods Hole, MA 02543, U.S.A.

BUKRY, D. — U.S. Geological Survey, c/o Scripps Inst. Oceanography, La Jolla, Calif. U.S.A.

BURCKLE, L. B. — Lamont-Doherty Geological Observatory, Columbia Univ., Parisades, N.Y. 10984, U.S.A.

CAMERON, B. E. B. — Geological Survey of Canada, 100 West Pender, Vancouver, B.C., Canada

CHIJI, M. — Osaka Museum of Natural History, Nagai Park, Higashisumiyoshi-ku, Osaka 546, Japan

CHOU, J. T. — Chinese Petroleum Co., 3 Tun-Hwa South Road, Taipei 100, Taiwan

GARTNER, S. — Dept. Oceanography, Texas A. & M. University, College Station, Texas 77843, U.S.A.

GLADENKOV, Y. B. — Geological Inst., Academy of Science, Pyzhevsky per 7, Moscow 109017, U.S.S.R.

HAQ, W. — Woods Hole Oceanographic Inst., Woods Hole, MA 02543, U.S.A.

HARTONO, H. M. S. — Geol. Research and Develop. Centre, Jalan Diponegolo 57, Bandung, Indonesia

HORNIBROOK, N. de B. — N. Z. Geological Survey, P.O. Box 30368, Lower Hutt, New Zealand

HUANG, T. Y. — Chinese Petroleum Co., 83 Chung Hwa Road, Taipei 100, Taiwan

IKEBE, N. — (Tezukayama University) 6–4, Hamacho, Ashiya 659, Japan

INGLE, J. — School of Earth Sciences, Stanford Univ., Stanford, CA. 94305, U.S.A.

JENKINS, D. J. — The Open University, Milton Keynes, England

KADAR, D. — Geol. Research and Develop. Centre, Jalan Diponegolo 57, Bandung, Indonesia

KENNETT, J. P. — Graduate School, Oceanography, Univ., Rhode Is., Kingston, Rhode Island 02881, U.S.A.

KIM, B. K. — College of Natural Sciences, Seoul National Univ., Seoul 151, Korea

KOIZUMI, I. — Dept. General Education, Osaka University, Machikaneyama, Toyonaka 560, Japan

LEE, Y. — Dept. Geology, Kyungpook National Univ., Daegu, Korea

MARTINEZ-PARDO, R. — Dept. Geology, University of Chile, Casilla 13581, Correo 21, Santiago, Chile

MATHUR, U. B. — Geological Survey of India, Western Region, E-180, Subhash Marg., Jaipur 302001, India

MCGOWRAN, B. — Dept. Geology, University of Adelaide, Adelaide, SA 5001, Australia

NAKAGAWA, H. — Inst. Geology & Paleontology, Tohoku University, Aobayama, Sendai 980, Japan

MCCOY, S. — Amoco Production Company, Denver, Col., U.S.A.

OBRADOVICH, J. — U.S. Geological Survey, Federal Center, Denver, Colorado 80225, U.S.A.

QUILTY, P. G. — School of Earth Sciences, Macquarie University, Sydney, N.S.W., Austraila

SAITO, T. — Dept. Earth Sciences, Yamagata University, Koshirakawa, Yamagata 990, Japan

SEROVA, M. Y. — Geological Inst., Academy of Science, Pyzhevsky per 7, 109017 Moscow, U.S.S.R.

SHACKLETON, N. J. — Sub-department Quaternary Re-

	search, Univ. Cambridge, Cambridge, England
SHIBATA, K.	Geological Survey of Japan Higashi 1–1–3, Yatabe-cho, Tsukuba-gun, Ibaraki 305, Japan
SRINIVASAN, M. S.	Dept. Geology, Banaras Hindu University, Varanasi 221005, India
TAKAYAMA, T.	Dept. Geol., Coll. Lib. Ar., Kanazawa University, Marunouchi, Kanazawa 920, Japan
TAKAYANAGI, Y.	Inst. Geology & Paleontology, Tohoku University, Aobayama, Sendai 980, Japan
THEYER, F.	Geophysics Inst., University of Hawaii, Honolulu, Hawaii 96822, U.S.A.
TSUCHI, R.	Geoscience Inst., Faculty of Science, Shizuoka University, Shizuoka 422, Japan
UJIIÉ, H.	Oceanography Inst., Ryukyu University, Naha, Okinawa 903, Japan
VELLA, P.	Dept. Geology, Victoria University, Wellington, New Zealand
YOON, S.	Dept. Geology, Busan National University, Busan 607, Korea

Secretariat, IGCP-114

IKEBE, N., Leader; CHIJI, M., Secretary

Osaka Museum of Natural History, Nagai Park, Higashi-sumiyoshi-ku, Osaka 546, Japan

Explanation of picture:
Participants of the 6th (the final) Working Group Meeting of IGCP-114, International Workshop on Pacific Neogene Biostratigraphy (IWPNB) on 25, Nov., 1981, in front of the Osaka Museum of Natural History, Osaka, Japan.

IV-2

Bibliography Related to the Activities of IGCP-114

Compiled by Nobuo IKEBE

Ablayev, A. G. and Schmidt, I. N., 1979: The Neogene stage in the development of Sikhote Alin Florae and vegetation [abstr.]. *In* Shilo, N. A. ed., *4-IGCP-114 in XIV-PSC, Khabarovsk 1979, Abstr.*, Sect. B-III, v. **2**, p. 3–5.

Adams, C. G., 1981: Neogene larger foraminifera, evolutionary and geologic events in the context of planktonic datum planes [abstr.]. *In* Ikebe, N. *et al.*, eds., *6-IGCP-114, IWPNB Osaka 1981, Proc.* p. 5–6.

Addicott, W. O., 1976: Neogene molluscan stages of Oregon and Washington. *In* Fritsche, A. E. and others, eds., *Neogene Symposium*: Society of Economic Paleontologists and Mineralogists, Pacific Section, San Francisco, California, April 1976, p. 95–115.

Addicott, W. O., 1976: New molluscan assemblages from the upper Twin River Formation, western Washington: Significance in Neogene biostratigraphy. *U.S. Geological Servey Journal of Research*, v. **4**, p. 437–447.

Addicott, W. O., 1976: Molluscan paleontology of the early Miocene Clallam Formation, northwestern Washington. *U.S. Geological Survey Professional Paper*, 976, 44 p., 9 pls.

Addicott, W. O., 1977: Neogene chronostratigraphy of near-shore marine basins of the eastern North Pacific. *In* Saito, T. and Ujiie, H., eds., *1-CPNS, Tokyo 1976, Proc.*, p. 151–175.

Addicott, W. O., 1978: Pectinids as biochronologic indices in the Neogene of eastern north Pacific. *In* Kadar, D., ed., *2-IGCP-114, Bandung 1977, Proc.* p. 11–22.

Addicott, W. O., ed., 1978: Neogene biostratigraphy of selected areas in the California Coast Ranges. *Field Conf., 3-IGCP-114, U.S.G.S. Open-file Report*, 78–446, p. 1–110.

Addicott, W. O., 1978: Marine paleogeography and paleontology of the Salinas Basin during the latest part of the Miocene with notes on microfossils from near San Lucas, California. *In* Addicott, W. O., ed., *Neog. Biostratigr. of selected areas in Cal. Coast Ranges, Field Conf., 3-IGCP-114, Stanford 1978, U.S.G.S. Open-file Report*, 78–446, p. 83–90.

Addicott, W. O., 1978: Notes on the geology of point Lobos State Reserve, Monterey County, California. *In* Addicott, W. O., ed., *Neog. Biostratigr. of selected areas in Cal. Coast Ranges, Field Conf., 3-IGCP-114, Stanford 1978, U.S.G.S. Open-file Report*, 78–446, p. 91–96.

Addicott, W. O., ed., 1978: Papers on Neogene Mollusks of the North Pacific Margin, read before the 1-CPNS (1-IGCP-114) 1976, Tokyo. *The Veliger*, Berkley, California v. **21**, p. 155–235.

Addicott, W. O., 1980: Miocene stratigraphy and fossils, Cape Blanco, Oregon. *Orgon Geology*, v. **42**, p. 87–97.

Addicott, W. O., 1980: Project 114 biostratigraphic datum-planes of the Pacific Neogene. *U.S. Contribution to the IGCP*, U. S. Nat. Comm. IGCP, Washington, D.C. p. 15–17.

Addicott, W. O., 1981: Significance of pectinids in Tertiary biochronology of the Pacific Northwest. *Geological Society of America, Special Paper*, 184, p. 17–37.

Addicott, W. O., Barron, J. A. and Miller, J. W., 1978: Marine late Neogene sequence near Santa Cruz, California. *In* Addicott, W. O., ed., *Neog. biostratigr. of selected areas in Cal. Coast Ranges, Field Conf., 3-IGCP-114, Stanford 1978, U.S.G.S. Open-file Report*, 78–446, p. 97–109.

Addicott, W. O. and Ingle, J. C. *et al.*, ed., 1978: *Third Working Group Meeting, IGCP-114 (3-IGCP-114), Stanford, Abstracts, 1978.* Stanford University Publication, California, p. 1–65.

Addicott, W. O. and Poore, R. Z., 1979: The Paleogene-Neogene boundary in eastern North Pacific marine sequences [abstr.]. *In* Shilo, N. A., ed., *4-IGCP-114 in XIV-PSC, Khabarovsk 1979, Abstr.*, Sect. B-III, v. **2**, p. 5–6.

Addicott, W. O. and Poore, R. Z., 1980: Paleogene/Neogene boundary in western North American sequences. *5-IGCP-114 in 26-IGC, Paris 1980, Abstracts*, v. **1**, p. 194.

Addicott, W. O., Poore, R. Z., Barron, J. A., Gower, H. D., and McDougall, K., 1978: Neogene biostratigraphy of the Indian Creek-Shell Creek area, northern La Panza Range, California. *In* Addicott, W. O., ed., *Neogene biostrat. of selected areas in Cal. Coast Ranges, Field Conf., 3-IGCP-114, Stanford 1978, U.S. G.S. Open-file Report*, 78–446, p. 49–81.

Akhmetyev, M. A. and Chelebayeva, A. I., 1979: Main stages of development of Neogene flora in the southern Soviet Far East and Kamchatka and the problems of interregional correlation [abstr.], *In* Shilo, N. A., ed., *4-IGCP in XIV-PSC, Khabarovsk 1979, Abstr.*, Sect. B-III, v. **2**, p. 6–9.

(Tezukayama University) 6–4, Hama-cho, Ashiya 659, Japan

Akiba, F., 1979: Revised Japanese Neogene diatom biostratigraphy [abstr.]. *In* Shilo, N. A. ed., *4-IGCP-114 in XIV-PSC, Khabarovsk 1979, Abstr.*, Sect. B-III, v. **2**, p. 9–10.

Akiba, F., 1981: Kamiiso Area, Hokkaido (Additional data). *In* Tsuchi, R., ed., *Fundam. data on Japan. Neogene, supplement*, Shizuoka, p. 80.

Akiba, F. and Ichinoseki, T., 1983: The Neogene micro- and chronostratigraphies in Hokkaido- Special reference to those of the southwestern part of the Kushiro Coal Field area, eastern Hokkaido, Japan—. *Japan. Assoc. Petrl. Techn., Jour.*, v. **48**, no.1, p. 49–61.

Aleksandrova, A. N., 1979: The lower boundary of the Quaternary System in Sakhalin [abstr.]. *In* Shilo, N. A., ed., *4-IGCP-114 in XIV-PSC, Khabarovsk 1979, Abstr.*, Sect. B-III, v. **2**, p. 10–12.

Alekseev, M., 1977: An attempt of correlation of the East Siberian and Japanese Pliocene and lower Quaternary sequences [abstr.]. *In* Saito, T. and Ujiie, H., eds., *I-CPNS Tokyo 1976, Proc.*, p. 311–312.

Allison, R. C., 1977: Late Oligocene through Pleistocene molluscan faunas in the Gulf of Alaska region [abstr.]. *In* Saito, T. and Ujiie, H. eds., *I-CPNS, Tokyo 1976, Proc.*, p. 313–316.

Allison, R. C., 1978: Late Oligocene through Pleistocene molluscan faunas in the Gulf of Alaska region. *The Veliger*, v. **21**, p. 171–188.

Allison, R. C. and Marincovich, L., Jr., 1979: Asiatic molluscs in late Paleogene and Neogene strata of the western Gulf of Alaska; prospects for circum-North Pacific correlations [abstr.]. *In* Shilo, N. A., ed., *4-IGCP-114 in XIV-PSC, Khabarovsk 1979, Abstr.*, Sect. B-III, v. **2**, no. 2, p. 15–17.

Amano, K., 1981: The stratigraphy of Miocene Series in northeast Honshu, Japan and eustatic sea-level changes [abstr.]. *In* Ikebe, N. *et al.*, eds., *6-IGCP-114, IWPNB Osaka 1981, Proc.*, p. 109.

Ariey, C., 1978: Molluscan biostratigraphy of the upper Poul Creek and lower Yakataga Formation, Yakataga district, Gulf of Alaska [abstr.]. In Addicott, W. O. and Ingle, J. C., eds. *3-IGCP-114, Stanford 1978, Stanford Univ. Publ. Geol. Sci.*, v. **14**, p. 1–2.

Armentrout, J. M., 1979: Progress report on Cenozoic correlations, west coast of North America [abstr.]. *In* Shilo, N.A., ed., *4-IGCP-114, in XIV-PSC, Khabarovsk 1979, Abstr.*, Sect. B-III, v. **2**, p. 19–20.

Armentrout, J. M., 1980: A new late Oligocene unit, Sitkinak Island Trinity Islands Group, Kodiak Island Archipelago, Alaska. *Geological Society of America, Abstracts with Programs*, v. **12**, no. 3, p. 94.

Armentrout, J. M., 1980: Field trip road log for Cenozoic stratigraphy of Coos Bay and Cape Blanco, southwestern Oregon. *Oregon Department of Geology and Mineral Industries, Bulletin*, v. **101**, p. 175–216.

Armentrout, J. M., 1981: Correlation and ages of Cenozoic chronostratigraphic units in Oregon and Washington. *In* Armentrout, J. M., ed., *Pacific Northwest Cenozoic Biostratigraphy: Geological Society of America, Special Paper*, 184, p. 137–148.

Armentrout, J. M. and Echols, R. J., 1981: Biostratigraphic-chronostratigraphic scale of the northeastern Pacific Neogene. *In* Ikebe, N. *et al.*, eds., *6-IGCP-114, IWPNB Osaka 1981, Proc.*, p. 7–27.

Armentrout, J. M., Echols, R. J. and Nash, K. W., 1978: Late Neogene climatic cycles of the Yakataga Formation, Robinson Mountains, Gulf of Alaska Area. *In* Addicott, W.O. and Ingle, J.C., eds., *3-IGCP-114, Stanford 1978, Stanford Univ. Publ., Geol. Sci.*, v. **14**, p. 3–4.

Arsanov, A. S., 1979: Tertiary sequence of the Kronotsky region (eastern Kamchatka) [abstr.]. *In* Shilo, N. A., ed., *IGCP-114 in XIV-PSC, Khabarovsk 1979, Abstr.*, Sect. B-III, v. **2**, p. 21–23.

Backman, J. and Shackleton, N. J., 1981: Quantitative biochronology of calcareous plankton of the Pliocene and early Pleistocene of the Pacific region [abstr.]. *In* Ikebe, N. et al., eds., *6-IGCP-114, IWPNB Osaka 1981, Proc.* p. 110–112.

Baldauf, J. G. and Barron, J. A., 1980: *Actinocyclus ingens* var. *nodus*: a new, stratigraphically useful diatom of the circum-North Pacific. *Micropaleontology*, v. **26**, no. 1, p. 103–110.

Baranova, Yu. P., 1979: Regional Neogene horizons in Northeast Asia [abstr.]. *In* Shilo, N. A., *4-IGCP-114 in XIV-PSC, Khabarovsk 1979, Abstr.*, Sect. B-III, v. **2**, p. 23–24.

Baranova, Yu. P. and Biske, S. F., 1979: Tertiary climates in Northeast Asia [abstr.]. *In* Shilo, N. A., ed., *4-IGCP-114 in XIV-PSC, Khabarovsk 1979, Abstr.*, Sect. B-III, v. **2**, p. 25–26.

Barbieri, F., Iaccarino, S., and Rossi, U., 1977: Paleoecology of the Pliocene in the sites 62 and 63 DSDP (Leg 7-equatorial West Pacific) [abstr.]. *In* Saito, T. and Ujiie, H., eds., *I-CPNS, Tokyo 1976, Proc.*, p. 316–318.

Barron, J. A., 1976: Marine diatom and silicoflagellate biostratigraphy of the type Delmontian Stage and the type *Bolivina obliqua* Zone. *U.S. Geological Survey, Journal of Research*, v. **4**, p. 339–351.

Barron, J. A., 1976: Revised Miocene and Pliocene diatom biostratigraphy of upper Newport Bay, Newport Beach, California. *Marine Micropaleontology*, v. **1**, p. 27–63.

Barron, J. A., 1980: Lower Miocene to Quaternary diatom biostratigraphy of Leg 57, off northeastern Japan, Deep Sea Drilling Project. *DSDP, Initial Reports*, Washington, U.S. Government Printing Office, v. **56**, **57**, p. 641–685.

Barron, J. A., 1981: Marine diatom biostratigraphy of the Montesano Formation near Aberdeen, Washington. *In* Armentrout, J. M., ed., *Pacific Northwest Cenozoic Biostratigraphy: Geological Society of America Special Paper*, 184, p. 113–126.

Barron, J. A., 1981: Late Cenozoic diatom biostratigraphy and paleoceanography of the middle-latitude eastern North Pacific, DSDP Leg 63. *DSDP, Initial Reports*, U.S. Government Printing Office, Washington, D.C., v. **63**, p. 507–538.

Barron, J. A. and Armentrout, J. M., 1980: Late Miocene diatom florules of the Empire Formation, Coos Bay,

Oregon. *Geological Society of America Abstracts with Programs*, v. **12**, p. 95–96.

Barron, J. A. and Ingle, J. C., 1978: Neogene section at the Mission Hills. *In* Addicott, W.O., ed., *Neog, biostratigr. of selected areas in Cal. Coast Ranges. Field Conf., 3-IGCP-114, Stanford 1978, U.S.G.S. Open-file Rep.*, 78–446, Menlo Park, 1978, p.29–36.

Barron, J.A., Poore, R.Z. and Reinhard, W., 1981: Biostratigraphic summary, Deep Sea Drilling Project Leg 63. *DSDP. Initial Reports.*, v. **63**, p. 927–941.

Barron, J.A., von Huene, R. and Nasu, N., 1978: Correlation of high and low latitude upper Miocene diatom datum levels at DSDP Site 438 in the northwest Pacific [abstr.]. *In* Addicott, W.O. and Ingle, J.C., eds., *3-IGCP-114, Stanford 1978, Stanford Univ. Publ., Geol. Sci.*, v. **14**, p.5–7.

Baskakova, L.A., 1979: Biostratigraphic subdivision of the Miocene in the southern continental part of the Soviet Far East [abstr.] *In* Shilo, N.A., ed., *4-IGCP-114 in XIV-PSC, Khabarovsk 1979, Abstr.*, Sect. B-III, v. **2**, p. 27–29.

Becker, H.F., 1977: Oligocene/Miocene plant fossils of southwestern Montana (USA) in correlation with floras of Japan [abstr.] *In* Saito, T. and Ujiie, H. *et al.*, eds., *1-CPNS, Tokyo 1976, Proc.*, p. 318–320.

Belaya, B.V., Kysterova, I.B., Terekhova, V.E., and Narkhinova, V.E., 1979: History of Tertiary flora of Northeast Asia by palynological research [abstr.]. *In* Shilo, N.A., ed., *4-IGCP-114 in XIV-PSC, Khabarovsk 1979, Abstr.*, Sect. B-III, v. **2**, p. 29–31.

Belova, V.A., 1979: The main stages in Neogenic flora formation in intermontane depressions of the Baikal type [abstr.]. *In* Shilo, N.A., ed., *4-IGCP-114 in XIV-PSC, Khabarovsk 1979, Abstr.*, Sect. B-III, v. **2**, p. 31–33.

Berggeren, W.A., 1981: Correlation of Atlantic, Mediterranean and Indo-Pacific Neogene stratigraphies: Geochronology and chronostratigraphy. *In* Ikebe, N. *et al.*, eds., *6-IGCP-114, IWPNB Osaka 1981, Proc.* p. 29–60.

Beu, A.G., 1979: *Hartungia typica* Bronn, 1861 (Gastropoda, Family Janthinidae); a late Pliocene pelagic macrofossil datum plane [abstr.]. *In* Shilo, N. A., ed., *4-IGCP-114 in XIV-PSC, Khabarovsk 1979, Abstr.*, Sect. B-III, v. **2**, p. 35–36.

Bhatia, S. B. and Mathur, A. K., 1977: The Neogene charophyte flora of the Siwalik Group, India and its biostratigraphical significance [abstr.]. *In* Saito, T. and Ujiie, H., eds., *1-CPNS, Tokyo 1976, Proc.*, p. 320.

Brutman, H. Ya., 1979: Palinozones and pollen of *Tsuga* genus from the Neogene of Sakhalin [abstr.]. *In* Shilo, N. A., ed., *4-IGCP-114 in XIV-PSC, Khabarovsk 1979, Abstr.*, Sect. B-III, v. **2**, p. 40.

Burckle, L. H., 1978: Early Miocene to Pliocene Diatom levels for the equatorial Pacific. *In* Kadar, D., ed., *2-IGCP-114, Bandung 1977, Proc.*, p. 25–44.

Burckle, L. H., Dodd, J. R. and Stanton, R. J., 1978: Diatom biostratigraphy of the Centerville Beach section, California [abstr.]. *In* Addicott, W.O. and Ingle, J.C., eds., *3-IGCP-114, Stanford 1978, Stanford Univ. Publ. Geol. Sci.*, v. **14**, p. 8–9.

Burckle, L. H., Dodd, J. R. and Stanton, R. J., 1980: Diatom biostratigraphy and its relationship to paleomagnetic stratigraphy and molluscan distribution in the Neogene Centerville Beach Section, California. *Journal of Paleontology*, v. **54**, p. 664–674.

Burckle, L. H., Gartner, S., Opdyke, N. D., Sciarrillo, J. R., and Shackleton, N. J., 1978: Paleomagnetics, oxygen isotopes and biostratigraphy of a late Pliocene section from the central Pacific [abstr.]. *In* Addicott, W.O. and Ingle, J.C., eds., *3-IGCP-114, Stanford 1978, Stanford Univ. Publ., Geol. Sci.*, v. **14**, p. 10–11.

Burckle, L. H. and Opdyke, N. D., 1977: Late Neogene diatom correlations in the Circum-Pacific. *In* Saito, T. and Ujiie, H., eds., *1-CPNS Tokyo 1976, Proc.*, p. 255–284.

Chihara, K., 1977: Volcanostratigraphy in the inner part of the Honshu Arc on basis of basaltic activity [abstr.]. *In* Saito, T. and Ujiie, H., eds., *1-CPNS, Tokyo 1976, Proc.*, p. 321–323.

Chiji, M., 1977: IGCP Project 114, Minute of the First Business Meeting of Working Group. *In* Saito, T. and Ujiie, H., eds., *1–CPNS, Tokyo 1976, Proc.*, p. 45–48.

Chiji, M. and Konda, I., 1977: On Japanese middle Miocene stages [abstr.]. *In* Saito, T. and Ujiie, H., eds., *1-CPNS, Tokyo 1976, Proc.*, p. 323–327.

Chiji, M. and Konda, I., 1978: Planktonic foraminiferal biostratigraphy of the Tomioka Group and the Nishiyatsushiro and Shizukawa Groups, central Japan, with some considerations on the Kaburan Stage (Middle Miocene) (Japanese w. English abstr.). *In* Huzita, K. *et al.*, eds., *Cenozoic Geology of Japan*, p. 73–92.

Chiji, M. and Konda, I., 1979: Shimobe area, Yamanashi Prefecture. *In* Tsuchi, R., ed., *Fundam. data on Japan. Neogene*, Shizuoka, p. 15–16.

Chiji, M. and Konda, I., 1979: Tomioka area, Gunma Prefecture; 1. *In* Tsuchi, R., ed., *Fundam. data on Japan. Neogene* Shizuoka, p. 30–31.

Chiji, M. and Konda, I., 1981: Shimobe Area, Yamanashi Prefecture. *In* Tsuchi, R., ed., *Fundam. data on Japan. Neogene, supplement*, Shizuoka, p. 7–8.

Chiji, M. and Konda, I., 1981: Tomioka Area, Gumma Prefecture. *In* Tsuchi, R., ed., *Fundam. data on Japan. Neogene, supplement*, Shizuoka, p. 9–10.

Chiji, M. and Konda, I., 1981: Stratigraphic outlines of selected Neogene sequences, 3. Shimobe area. *In* Tsuchi, R., ed., *Neogene of Japan*, p. 42–45.

Chiji, M. and Konda, I., 1981: Stratigraphic outlines of selected Neogene sepuences, 5. Tomioka area. *In* Tsuchi, R., ed., *Neogene of Japan*, p. 50–52.

Chinzei, K., 1977: Neogene molluscan faunas of Japan; a paleoecological and paleobiogeographical synopsis [abstr.]. *In* Saito, T. and Ujiie, H., eds., *1-CPNS, Tokyo 1976, Proc.*, p. 327–330.

Chinzei, K., 1978: Neogene molluscan faunas in the Japanese Islands: an ecologic and zoogeographic synthesis. *The Veliger*, v. **21**, no. 2, p. 155–170.

Chinzei, K., 1979: Kadonosawa-Sannohe area; 1. *In* Tsuchi, R., ed., *Fundam. data on Japan. Neogene*, Shizuoka 1979,

p. 50–52.

Chinzei, K., 1981: Stratigraphic outlines of selected Neogene sequences, 7. Kadonosawa area. *In* Tsuchi, R. ed., *Neogene of Japan*, p. 57–61.

Chou, J. T., 1977: Neogene stratigraphy in western Taiwan [abstr.]. *In* Saito, T. and Ujiie, H., eds., *1-CPNS, Tokyo 1976, Proc.*, p. 330–331.

Crouch, R. W. and Poag, C. W., 1978: *Amphistegina gibbosa* (d'Orbigny) from the California borderland: the Caribbean connection [abstr.]. In Addicott, W. O. and Ingle, J. C., eds., *3-IGCP-114, Stanford 1978, Stanford Univ. Publ., Geol. Sci.*, v. **14**, p. 12–13.

Devyatilov, A. D., 1979: Biostratigraphy and correlation of the Paleogene deposits in the Penzhina-Anadyr region [abstr.]. *In* Shilo, N. A., ed., *4-IGCP-114 in XIV-PSC, 1979, Abstr.*, Sect. B-III, v. **2**, p. 41–43.

Djamas, Y. S., 1981: Biostratigraphy of early Neogene time in South Sumatra Basin, Indonesia [abstr.]. *In* Ikebe, N. *et al.*, eds., *6-IGCP-114, IWPNB Osaka 1981, Proc.* p. 113.

Djamas, Y. S. and Marks, E., 1978: Early Neogene foraminiferal biohorizons in E. Kalimantan, Indonesia. *In* Kadar, D., ed., *2-IGCP-114, Bandung 1977, Proc.* p. 111–124.

Dolmatova, L. M., 1979: Marine diatoms from the Neogene deposits of Karaginsky Island [abstr.]. *In* Shilo, N. A., ed., *4-IGCP-114 in XIV-PSC, Khabarovsk 1979, Abstr.*, Sect. B-III, v. **2**, p. 43–45.

Drooger, C. W., 1976: IGCP-1, Accuracy in time. *Geol. Correlation, IGCP*, no. 4, p. 16–17.

Edwards, A. R. and Hornibrook, N. de B., 1980: An integrated bio- and magneto-stratigraphy of the Late Miocene to Pleistocene of the New Zealand Region. [abstr.]. *Geological Society of New Zealand, Christchurch Conference 1980*, p. 35.

Fotyanova, L. I., 1979: Correlative floral levels of the upper Oligocene-Lower Miocene in Europe and the Far East [abstr.]. *In* Shilo, N. A., ed., *4-IGCP-114 in XIV-PSC, Khabarovsk 1979, Abstr.*, Sect. B-III, v. **2**, p. 45–48.

Fradkina, A. F., 1979: Correlation of Neogene deposits of Northeast Asia by the aid of palynological data [abstr.]. *In* Shilo, N. A., ed., *4-IGCP-114 in XIV-PSC, Khabarovsk 1979, Abstr.*, Sect. B-III, v. **2**, p. 49–51.

Fujita, T., Matsumoto, Y., Shimazu, M., and Wadatsumi, K., 1978: Volcanostratigraphy of the Neogene formations in Southwest Japan and the Fossa Magna region. *In* Huzita, K. *et al.*, ed., *Cenozoic Geology of Japan*, p. 121–134.

Fukuta, O., 1977: On the iodine-gas deposits of dissolved-in-water type with special reference to late Cenozoic stratigraphy of Japan [abstr.]. *In* Saito, T. and Ujiie, H., eds., *1-CPNS, Tokyo 1976, Proc.*, p. 332–334.

Ganeshin, G. S., 1979: On the lower boundary of the Quaternary System in the Far East [abstr.]. *In* Shilo, N. A., ed., *4-IGCP-114 in XIV-PSC, Khabarovsk 1979, Abstr.*, Sect. B-III, v. **2**, p. 52–53.

Gartner, S., 1977: Nannofossil biostratigraphy of Pleistocene deep-sea sediments of the western Pacific [abstr.]. *In* Saito, T. and Ujiie, H., eds., *1-CPNS, Tokyo 1976, Proc.*, p. 335.

Giterman, R. E. and Kartashova, G. G., 1979: On Pliocene vegetation of the Far Northeast USSR [abstr.]. *In* Shilo, N. A., ed., *4-IGCP-114 in XIV-PSC, Khabarovsk 1979, Abstr.*, Sect. B-III v. **2**, p. 54–55.

Gladenkov, Y. B., 1977: Stages in the evolution of mollusks and subdivisions of the North Pacific Neogene. *In* Saito, T. and Ujiie, H., eds., *1-CPNS, Tokyo 1976, Proc.*, p. 89–91.

Gladenkov, Y. B., 1977: Malacofaunal assemblages of upper Cenozoic in the Pacific and Atlantic [abstr.]. *In* Saito, T. and Ujiie, H., eds., *1-CPNS, Tokyo 1976, Proc.*, p. 336.

Glandenkov, Y. B., 1978: Correlation of upper Cenozoic marine deposits in Boreal regions (based on mollusks). *Inter. Geology Rev.*, v. **20**, no. 1.

Gladenkov, Y. B., 1978: Mollusks and Neogene correlation of the North Pacific [abstr.]. *In* Addicott, W.O. and Ingle, J.C., eds., *3-IGCP-114, Stanford 1978, Stanford Univ. Publ., Geol. Sci.*, v. **14**, p. 14.

Gladenkov, Y. B., 1978: Premises of Neogene correlation in the northern part of the Circum-Pacific. *The Veliger*, v. **21**, no. 2, p. 225–226.

Gladenkov, Y. B., 1980: Marine upper Cenozoic of boreal regions of two ecosystems, Atlantic and Pacific. *26-IGC, Abstract*, v. **1**, p. 231.

Gladenkov, Y. B., 1981: Neogene horizons (Regiostages) of Sakhalin and Kamchatka and the problem of their correlation [abstr.]. *In* Ikebe, N. *et al.*, eds., *6-IGCP-114, IWPNB Osaka 1981, Proc.*, p. 113A.

Gladenkov, Y. B., Vitukhin, D. I., and Oreshkina, T. V., 1979: Correlation of eastern Kamchatka Cenozoic and oceanic deposits [abstr.]. *In* Shilo, N. A., ed., *4-IGCP-114 in XIV-PSC, Khabarovsk 1979, Abstr.*, Sect. B-III v. **2**, p. 55–57.

Grechin, V. I., 1979: Siliceous rocks in Paleogene-Neogene deposits of the northwestern margin of the Pacific Ocean [abstr.]. *In* Shilo, N. A., ed., *4-IGCP-114 in XIV-PSC, Khabarovsk 1979, Abstr.*, Sect. B-III, v. **2**, p. 57–59.

Grinenko, O. V. and Sergeyenko, A. I., 1979: Paleogene near-shore marine deposits in the northwestern part of the Circum-Pacific mobile belt [abstr.]. *In* Shilo, N. A., ed., *4-IGCP-114 in XIV-PSC, Khabarovsk 1979, Abstr.*, Sect. B-III, v. **2**, p. 59–61.

Hadiwisastra, S. and Martodjojo, S., 1977: "*West Java*," *Excursion guidebook, 2-IGCP 114*, Bandung 1977, p. 1–10.

Haq, B. U., 1981: Calibration of Pacific Neogene nannofossil datums to magnetostratigraphy [abstr.]. *In* Ikebe, N. *et al.*, eds., *6-IGCP-114, IWPNB Osaka 1981, Proc.*, p. 114.

Harada, K., 1977: Mn-nodules as deep-sea environmental indicator of late Cenozoic [abstr.]. *In* Saito, T. and Ujiie, H., eds., *1-CPNS, Tokyo 1976, Proc.*, p. 337.

Harada, K. and Yamamoto, K., 1977: Recent dinoflagellate cyst distribution in the Northwest Pacific off Japan [abstr.]. *In* Saito, T. and Ujiie, H., eds., *1-CPNS, Tokyo 1976, Proc.*, p. 337–340.

Harata, T. and Tokuoka, T., 1978: A consideration on the Paleogene paleogeography in southwest Japan (Japanese w. English abstr.). *In* Huzita, K. *et al.*, ed., *Cenozoic Geololgy of Japan,* p. 1–12.

Harper, H. E., 1978: Diatom biostratigraphy of the Miocene/ Pliocene boundary in the North Pacific [abstr.]. *In* Addicott, W.O. and Ingle, J.C., eds., *3-IGCP-114, Stanford 1978, Stanford Univ. Publ., Geol. Sci.*, v. **14**, p. 15.

Hasegawa, S., 1979: Foraminifera of the Himi Group, Hokuriku Province, central Japan. *Tohoku Univ., Sci, Rep., 2nd ser.*, v. **49**, p. 89–163, pls. 3–10.

Hasegawa, S., 1979: Southern Noto Peninsula; Asahiyama, Himi. *In* Tsuchi, R., ed., *Fundam. data on Japan. Neogene.* Shizuoka, p. 85–86.

Hasegawa, Y., 1977: Early Neogene birds from Honshu and Kyushu, Japan [abstr.]. *In* Saito, T. and Ujiie, H., eds., *1-CPNS, Tokyo 1976, Proc.*, p. 340–341.

Hashimoto, W. and Balce, G. R., 1977: A new correlation scheme for the Philippine Cenozoic formations. *In* Saito, T. and Ujiie, H., eds., *1-CPNS, Tokyo 1976, Proc.*, p. 119–132.

Hashimoto, W. and Matsumaru, K., 1978: Consideration of the stratigraphy of the Caragallo Range, northern Luzon, checking with the larger foraminiferal ranges from the Cenozoic sediments of the Philippines [abstr.]. *In* Addicott, W.O. and Ingle, J.C., eds., *3-IGCP- 114, Stanford 1978, Stanford Univ. Publ., Geol. Sci.*, v. **14**, p. 16–17.

Hayasaka, S., 1979: Minamitanecho, Tanegashima Island *In* Tsuchi, R., ed., *Fundam. data on Japan. Neogene*, Shizuoka, p. 5–6.

Hayasaka, S., Fukuda, Y., and Hayama, A., 1980: Discovery of molluscan fossils and the paleoenvironmental aspects of the Kumage Group in Tane-ga-Shima, south Kyushu, Japan. *In* Ito, H. and Noda, H., eds., *Prof. S. Kanno Mem. Vol.*, p. 59–70, pl. 7.

Hirooka, K., Sasajima, S., Nishimura, S., *et al.*, 1977: Preliminary report of the paleomagnetic study in Sunda Arc [abstr.]. *In* Saito, T. and Ujiie, H., eds., *1-CPNS, Tokyo 1976, Proc.*, p. 341–342.

Hirooka, K., Tsuda, S., Nishimura, S., Sasajima, S., Thio, K. H., and Hehuwatt, F., 1981: Paleomagnetic study of Miocene sediments in Karansanbung area, Central Java, Indonesia [abstr.]. *In* Ikebe, N. *et al.*, eds., *6-IGCP-114, IWPNB Osaka 1981, Proc.*, p. 115.

Ho, C. S., 1977: Melanges in the Neogene sequence of Taiwan [abstr.]. *In* Saito, T. and Ujiie, H., eds., *1-CPNS, Tokyo 1976, Proc.*, p. 342–344.

Honda, N., 1981: Kakegawa Area (2), *In* Tsuchi, R., ed., *Fundam. data on Japan. Neogene, supplement*, Shizuoka 1981, p. 5–6.

Hornibrook, N. de B., 1977: The Neogene (Miocene-Pliocene) of New Zealand. *In* Saito, T. and Ujiie, H., eds., *1-CPNS, Tokyo 1976, Proc.*, p. 145–150.

Hornibrook, N. de B., 1979: *Globorotalia crassaformis* and *G. crassula* in the Pliocene of New Zealand and a late Pliocene cool marine phase in the lower Nukumaruan stage [abstr.]. *In* Shilo, N. A., ed., *4-IGCP-114 in XIV-PSC, Khabarovsk 1979, Abstr.*, Sect. B-III, v. **2**, p. 62–63.

Hornibrook, N. de B., 1980: Correlation of Pliocene biostratigraphy, magnetostratigraphy and ^{18}O fluctuations in New Zealand and DSDP Site 284. *Newsletters on Stratigraphy*, 9, p. 114–120.

Hornibrook, N. de B., 1981: *Globorotalia* (Planktonic Foraminiferida) in the Late Pliocene and Early Pleistocene of New Zealand. *N. Z. Journal of Geology and Geophysics*, v. **24**, p. 263–292.

Hornibrook, N. de B., 1981: Evaluation of Pacific southern mid latitude and equatorial latitude planktonic foraminiferal datum levels. *In* Ikebe, N. *et al.*, eds., *6-IGCP-114, IWPNB Osaka 1981, Proc.*, p. 61–71.

Hornibrook, N. de B., 1982: Late Miocene to Pleistocene *Globorotalia* (Forminiferida) from DSDP Leg 29, Site 284, S. W. Pacific.

Hoskins, R. H., 1978: New Zealand Middle Miocene Foraminifera: The Waiauan Stage. *Ph. D. thesis, University of Exeter*, U. K., 2 vols. 388 p.

Howell, D. G. and McLean, H., 1977: Basin analysis of provincial lower and middle Miocene strata, Santa Rosa and Santa Cruz Islands, California [abstr.]. *In* Saito, T. and Ujiie, H., eds., *1-CPNS, Tokyo 1976, Proc.*, p. 344–345.

Huang, H. and Okamoto, K., 1979: Planktonic foraminiferal assemblages form the Kawai Formation and its correlations in the San-in areas, Japan. *Mizunami Fossil Mus., Bull.*, no. 6 p. 101–110, pls. 17–18.

Huang, T. C. and Okamoto, K., 1980: Calcareous nannofossils from the Miocene formations in Yuya and Iki, Southwest Japan. *Mizunami Fossil Mus., Bull.*, no. 7, p. 69–72, pl. 4.

Huang, T. Y., 1978: Significant new look on the Neogene stratigraphy of Taiwan [abstr.]. *In* Addicott, W.O. and Ingle, J.C., eds., *3-IGCP-114, Stanford 1978, Stanford Univ. Publ., Geol. Sci.*, v. **14**, p. 18.

Huang, T. Y., 1981: Biostratigraphy of the Late Miocene in western Taiwan, ROC [abstr.]. *In* Ikebe, N. *et al.*, eds., *6-IGCP-114, IWPNB Osaka 1981, Proc.*, p. 116–117.

Hughes, G. W., 1978: Planktonic foraminiferal biostratigraphic datum planes for Plio-Pleistocene sedimentary rocks from the Solomon Islands [abstr.]. *In* Addicott, W.O. and Ingle, J.C., eds., *3-IGCP-114, Stanford 1978, Stanford Univ., Geol. Sci.*, v. **14**, p. 19–20.

Huzita, K., 1978: Crustal movements and sea-level changes since Miocene in southwest Japan in relation to sedimentation and topographic surfaces. *In* Huzita, K. *et al.*, ed., *Cenozoic Geology of Japan*, p. 169–186.

Huzita, K. *et al.* (ed.), 1978: *Cenozoic Geology of Japan–* Prof. Ikebe Memorial Volume [in Japan. w. English abstr.], Osaka, p. 1–230.

Ibaraki, M., 1976: Notes on planktonic foraminifera from the Harada Formation, Shirahama Group, the southern Izu Peninsula. *Shizuoka Univ., Geosci. Rep.*, no. 2, p. 1–7.

Ibaraki, M., 1978: Notes on planktonic foraminifera from the "Nishikoiso" and "Oiso" Formations, Kanagawa Pref. *Shizuoka Univ., Geosci. Rep.*, no. 3, p. 1–8.

Ibaraki, M., 1979: Izu Peninsula. *In* Tsuchi, R., ed., *Fundam. data on Jap. Neogene*, Shizuoka, p. 17–18.

Ibaraki, M., 1979: Coast of Nishikoiso, Kanagawa Profecture. *In* Tsuchi, R., ed., *Fundam. data on Japan. Neogene*, Shizuoka, p. 19.

Ibaraki, M., 1979: Okinawa Island; 2. *In* Tsuchi, R., ed., *Fundam. data on Japan. Neogene*, Shizuoka, p. 3–4.

Ibaraki, M., 1981: Kakegawa Area (1). *In* Tsuchi, R., ed., *Fundam. data on Japan. Neogene, supplement*, Shizuoka, p. 3–4.

Ibaraki, M., 1981: South Fossa Magna (Additional data). *In* Tsuchi, R., ed., *Fundam. data on Japan. Neogene, supplement*, Sizuoka, p. 71–74.

Ibaraki, M., 1981: Chichibu Area (Additional data). *In* Tsuchi, R., ed., *Fundam. data on Japan. Neogene, supplement*, Shizuoka, p. 75.

Ibaraki, M., 1981: Mizunami Area, Gifu Prefecture (Additional data), *In* Tsuchi, R., ed., *Fundam. data on Japan. Neogene, supplement*, Shizuoka, p. 81.

Ibaraki, M., 1981: Yatsuo Area, Toyama Prefecture (Additional data), *In* Tsuchi, R., ed., *Fundam. data on Jap. Neogene, supplement*, Shizuoka, p. 81.

Ibaraki, M., 1981: Stratigraphic outlines of selected Neogene sequences, 1. Okinawa Island. *In* Tsuchi, R., ed., *Neogene of Japan*, p. 34–36.

Ibaraki, M., 1981: Geologic ages of "*Lepidocyclina*" and *Miogypsina* horizons in Japan as determined by planktonic foraminifera [abstr.]. *In* Ikebe, N. *et al.*, eds., *6-IGCP-114, IWPNB Osaka 1981, Proc.*, p. 118–119.

Ibaraki, M. and Tsuchi, R., 1976: Planktonic foraminifera from the lower part of the Kakegawa Group, Shizuoka Pref., Japan. *Shizuoka Univ. Fac. Sci., Rep.*, v. **11**, p. 161–178.

Ibaraki, M. and Tsuchi, R., 1980: Planktonic foraminifera from mollusca-bearing horizons of the Neogene sequence on the west coast of Boso Peninsula, Japan. *Shizuoka Univ., Fac. Sci., Rep.*, v. **14**, p. 86–101.

Ichikawa, K. and Kitamura, N., 1978: Late Cenozoic sedimentary basins of Japan in relation to the basement structure. *In* Huzita, K. *et al.*, ed., *Cenozoic Geology of Japan*, p. 187–204.

Ichikura, M. and Ujiie, H., 1977: Stratigraphy of the sediments in the Sea of Japan and its paleoenvironmental significance [abstr.]. *In* Saito, T. and Ujiie, H., eds., *1-CPNS, Tokyo 1976, Proc.*, p. 346–348.

IGCP., 1980: IGCP-114 (a biblioraphy). *IGCP Catalogue 1973–1979*, p. 133–145.

IGCP-114 National Working Group of Japan, 1981: Paleogeographic map of the Japanese Islands during 16–15 Ma, the earliest Middle Miocene (with explanation by Chinzei, K. and Itoigawa, J.). *In* Tsuchi, R., ed., *Neogene of Japan*, p. 105–109.

Ikebe, N., 1973: Neogene biostratigraphy and radiometric time scale. *Osaka City Univ., Geosci. Jour.*, v. **16**, p. 51–67.

Ikebe, N., 1976: Stratigraphic position of the Kakegawa Series in the Japanese Neogene. *In* Tsuchi, R., ed., *Guide Book for excursion 3, Kakegawa District, 1-CPNS, Tokyo, Japan*, p. 22–27.

Ikebe, N., 1977: Report of Progress IGCP-114 for the year 1976. *Geol. Correl., IGCP*, no. 5, p. 67–68.

Ikebe, N., 1978: Bio- and chronostratigraphy of Japanese Neogene, with remarks on paleogeography. *In* Huzita, K. *et al.*, ed., *Cenozoic Geology of Japan*, p. 13–34.

Ikebe, N., 1978: Biostratigraphic datum-planes of the Pacific Neogene, Project 114, *Geol. Correl., IGCP*, spec. issue, p. 65–66.

Ikebe, N., 1978: Report of Progress IGCP-114 for the year 1977. *Geol. Correl., IGCP*, no. 6, p. 70–71.

Ikebe, N., 1979: Report of Progress IGCP-114 for the year 1978. *Geol. Correl., IGCP*, no. 7, p. 98–99.

Ikebe, N., 1980: Report of Progress IGCP-114 for the year 1979. *Geol. Correl., IGCP*, no. 8, p. 137–139.

Ikebe, N., 1981: Report of Progress IGCP-114 for the year 1980. *Geol. Correl., IGCP*, no. 9, p. 46.

Ikebe, N., 1982: Report of Progress IGCP-114 for the year 1981. *Geol. Correl., IGCP*, no. 10, p. 37–38.

Ikebe, N. and Chiji, M., 1978: Neogene datum-planes of western Pacific, a proposal for discussion. *In* Kadar, D., ed., *2-IGCP-114, Bandung 1977, Proc.* p. 159–161.

Ikebe, N. and Chiji, M., 1978: Evaluation of some important datum-planes of the Pacific Neogene [abstr.]. *In* Addicott, W.O. and Ingle, J. C., eds., *3-IGCP- 114, Stanford 1978, Stanford Univ. Publ., Geol. Sci.*, v. **14**, p. 21.

Ikebe, N. and Chiji, M., 1981: Important datum-planes of the western Pacific Neogene (revised) with remarks on the Neogene stages in Japan. *In* Tsuchi, R., ed., *Neogene of Japan*, p. 1–14.

Ikebe, N. and Chiji, M., 1981: IGCP-114: Biostratigraphic datum-planes of the Pacific Neogene-purpose and activities. *In* Ikebe, N. *et al.*, eds., *6-IGCP-114, IWPNB Osaka 1981, Proc.*, p. 1–4.

Ikebe, N., Chiji, M. and Huang, T. Y., 1980: Important datum-planes of the western Pacific Neogene. *5-IGCP-114 in 26-IGC, Paris 1980, Abstracts*, v. **1**, p. 241.

Ikebe, N., Chiji, M. and Huang, T.Y., 1981: Important datum-planes of the western Pacific Neogene. *Osaka Mus. Nat. History, Bull.*, no. 34, p.79–86.

Ikebe, N., Takayanagi, Y., Chiji, M., and Chinzei, K., 1972: Neogene biostratigraphy and radiometric time scale of Japan–An attempt at intercontinental correlation. *Pacific Geology*, no. 4, p. 39–78.

Ikebe,N., Chiji,M., Tsuchi,R., and Morozumi,Y. (eds.), 1981: *Proceedings of IGCP-114 Internatinal Workshop on Pacific Neogene Biostratigraphy (IWPNB), 6th Working Group Meeting of IGCP-114 (6 IGCP-114), Osaka 1981.* IGCP-114, Osaka Museum of Natural History, Osaka. p.i-iv; 1–150.

Ikebe, N. and Working Group, 1977: Summary of bio- and chronostratigraphy of the Japanese Neogene. *In* Saito, T. and Ujiie, H., eds., *1-CPNS, Tokyo 1976, Proc.*, p. 93–114.

Ikebe, Y., Katahira, T. and Miyazaki, H., 1978: Some problems on petroleum geology in Japan. *In* Huzita, K. *et al.*, ed., *Cenozoic Geology of Japan*, p. 205–216.

Ikebe, Y. and Maiya, S., 1981: Stratigraphic outlines of selected Neogene sequences, 10, Akita and Niigata areas. *In* Tsuchi, R., ed., *Neogene of Japan*, p. 68–75.

Ikeda, T., 1982: Miocene planktonic foraminiferal biostratigraphy in the northeastern part of the Noto Peninsula, Central Japan. *Earth Science* (*Chikyu Kagaku*), v. **36**, no. 1, p. 1–9, 2pls.

Ingle, J. C., Jr., 1977: Summary of late Neogene planktic forminiferal biofacies, biostratigraphy, and paleoceanography of the marginal North Pacific Ocean [abstr.]. *In* Saito, T. and Ujiie, H., eds., *1-CPNS, Tokyo 1976, Proc.*, p. 177–182.

Ingle, J. C., Jr., 1978: Neogene biostratigraphy and paleoenvironments of the western Ventura Basin with special reference to the Balcom Canyon Section. *In* Addicott, W. A., ed., *Neog. biostratigr. of selected areas in Cal. Coast Ranges, Field Conf., 3-IGCP-114, Stanford 1978, U.S.G.S. Open-file Rep.*, 78–446, Menlo Park, p. 37–47.

Ingle, J. C., Jr., 1979: Cenozoic climatic history and marine biostratigraphy of the Pacific coast of North America [abstr.]. *In* Shilo, N. A., ed., *4-IGCP-114 in XIV-PSC, Khabarovsk 1979, Abstr.*, Sect. B-III, v. **2**, p. 63–65.

Ingle, J. C., Jr., 1979: The Humboldt (Eel River) Basin of northern California; a high latitude Pacific Neogene reference section [abstr.]. *In* Shilo, N. A., ed., *4-IGCP-114 in XIV-PSC, Khabarovsk 1979, Abstr.*, Sect. B-III v. **2**, p. 65–67.

Ingle, J. C., Jr., 1980: Depositional history and significance of Neogene diatomites from the Pacific rim. *26-IGC., Abstracts*, v. **2**, p. 487.

Ingle, J. C., Jr., 1981: Origin of Neogene diatomites around the North Pacific Rim. *In* Garrison, R. E. and Douglas, R. G., eds., *The Monterey Formation and related siliceous rocks in California*, Society of Economic Paleontologists and Mineralogists, Pacific Section, Los Angeles, California, p. 159–179.

Ingle, J. C., Jr., 1981: Neogene reference sections of the Pacific coast of North America; their regional significance and correlation [abstr.]. *In* Ikebe, N. *et al.*, eds., *6-IGCP-114, IWPNB Osaka 1981, Proc.*, p. 120–121.

Ingle, J. C. and Barron, J. A., 1978: Neogene biostratigraphy and paleoenvironments of the San Joaquin Hills and Newport Bay areas, California. *In* Addicott, W. O., ed., *Neog. biostratigr. of selected areas in Cal. Coast Ranges, Field Conf., 3-IGCP-114, Stanford 1978, U.S.G. S. Open-file Rep.* 78–446, Menlo Park, 1978, p. 3–27.

Ingle, J. C., Jr. and Garrison, R. E., 1977: Origin, distribution, and diagenesis of Neogene diatomites around the North Pacific rim [abstr.]. *In* Saito, T. and Ujiie, H., eds., *1-CPNS, Tokyo 1976, Proc.*, p. 348–350.

Ishida, S., 1978: Paleoecology and stratigraphy of Neogene in some areas of Japan and Pacific [abstr.]. *In* Addicott, W.O. and Ingle, J.C., eds., *3-IGCP-114, Stanford 1978, Stanford Univ. Publ., Geol. Sci.*, v. **14**, p. 22–23.

Ishida, S., 1979: Oga Peninsula; 2; F. T. Ages. *In* Tsuchi, R., ed., *Fundam. data on Japan. Neogene*, Shizuoka, p. 71–73.

Ishida, S., Maenaka, K. and Yokoyama, T., 1977: Magnetostratigraphy and biostratigraphy of Plio-Pleistocene in Kinki district, Japan [abstr.]. *In* Saito, T. and Ujiie, H., eds., *1-CPNS, Tokyo 1976, Proc.*, p. 350–352.

Ishizaki, K. and Takayanagi, Y., 1981: Stratigraphic outlines of selected Neogene sequences, 4. Boso Peninsula. *In* Tsuchi, R., ed., *Neogene of Japan*, p. 46–49.

Ishizaki, K. and Takayanagi, Y., 1981: Stratigraphic outlines of selected Neogene sequences, 6. Sendai area. *In* Tsuchi, R., ed., *Neogene of Japan*, p. 53–56.

Itihara, M., 1979: Minakuchi area, Shiga Prefecture. *In* Tsuchi, R., ed., *Fundam. data on Japan. Neogene*, Shizuoka, p. 110–111.

Itihara, M. (ed.), Chiji, M., Hayashi, T., *et al.*, 1976: *Guide Book for excursion 4, south of Osaka, 1976, 1-CPNS, Tokyo*, 21p.

Itihara, M., Shimoda, C., and Itihara, Y., 1978: Chemical composition of glasses in volcanic ash beds of the Osaka Group. *In* Huzita, K. *et al.*, eds., *Cenozoic Geology of Japan*, p. 113–120.

Itihara, M. and Yoshikawa, S., 1976: Geology of the Plio-Pleistocene Osaka Group: *In* Itihara, M., ed., *Guide Book for excursion 4, south of Osaka, 1-CPNS, Tokyo*, p. 9–17.

Itoigawa, J., 1979: Mizunami area, Gifu Prefecture. *In* Tsuchi, R., ed., *Fundam. data on Japan. Neogene*, Shizuoka, p. 106–107.

Itoigawa, J., 1981: Kawakami Area, Okayama Prefecture (Additional data). *In* Tsuchi, R., ed., *Fundam. data on Japan. Neogene, supplement*, Shizuoka, p. 82.

Itoigawa, J., 1981: Stratigraphic outlines of selected Neogene sequences, 8. Mizunami area. *In* Tsuchi, R., ed., *Neogene of Japan*, p. 62–64.

Itoigawa, J., Chinzei, K. and Kokawa, S., 1978: Paleoecology and paleoenvironmentology, with examples from the Japanese Cenozoic. *In* Huzita, K. *et al.*, eds., *Cenozoic Geology of Japan*, 1978, p. 155–168.

Itoigawa, J., Shibata, H., Nishimoto, H., and Okumura, Y., 1981–1982: Miocene fossils of the Mizunami Group, central Japan, 2. Mollusca (Japanese with English captions). *Mizunami Fossil Museum, Monograph*, no. 3–A, p. 1–53, pls. 1–52; no. 3–B, p. 1–330.

Jenkins, G. D., 1980: Planktonic foraminifera and the Oligocene-Miocene boundary in the Southwest Pacific. *26-IGC. Abstract*, v. **1**, p. 243.

Kadar, D. (ed.), 1977: *Collected abstracts, 2-IGCP-114, Bandung* 1–73, (mimeographed).

Kadar, D. (ed.), 1977: *Excursion Guide-books, "West Java" and "Central Java"*, 2-IGCP-114. (mimeographed).

Kadar, D., 1977: Planktonic foraminifera biostratigraphy of the Sentolo Formation, central Java, Indonesia [abstr.]. *In* Saito, T. and Ujiie, H., eds., *1-CPNS, Tokyo 1976, Proc.* p. 353.

Kadar, D. (ed.), 1978: *Proceedings of the Second Working Group Meeting of IGCP Project 114* (2-IGCP-114), *Bandung, 1977.* Geol. Research and Development Center, Indonesia, p. 1–171.

Kadar, D., 1978: Upper Pliocene and Pleistocene planktonic foraminiferal zonation of Ambengan drill hole, southern part of Bali Island. *In* Kadar, D., ed., *2-IGCP-114, Bandung 1977, Proc.*, p. 137–158.

Kadar, D., 1981: Planktonic foraminiferal biostratigraphy of the Miocene-Pliocene Sentolo Formation, Central Java,

Indonesia. *In* Saito, T., ed., *Micropaleontology, petrology and lithostratigraphy of Cenozoic rocks of the Yogyakarta region, Central Java.*, Yamagata Univ., p. 35–47.

Kadar, D., Raharjo, W. and Suradi, T., 1977: "Central Java" *Excursion guidebook, 2-IGCP-114, Bandung 1977*, p. 1–19.

Kamada, Y., 1979: Joban area; 1. *In* Tsuchi, R., ed., *Fundam. data on Japan. Neogene*, Shizuoka, p. 34–35.

Kamei, T. and Okazaki, Y., 1977: Mammalian fauna of the Miocene Mizunami Group and the Neogene mammals in Japan [abstr.]. *In* Saito, T. and Ujiie, H., eds., *1-CPNS, Tokyo 1976, Proc.*, p. 353–354.

Kanno, S., 1977: Migratory patterns of some Cenozoic mollusks in the northern Pacific regions; a basic study for the correlation of the Pacific molluscan faunas [abstr.]. *In* Saito, T. and Ujiie, H., eds., *1-CPNS, Tokyo 1976, Proc.*, p. 355–356.

Kanno, S., 1979: Molluscan faunal sequences of the Neogene System in Hokkaido, northern Japan [abstr.]. *In* Shilo, N. A., ed., *4-IGCP-114 in XIV-PSC, Khabarovsk 1979, Abstr.*, Sect. B-III, v. **2**, p. 70–71.

Kanno, S. and Noda, K., 1978: Biostratigraphical and paleogeographical distribution of the gastropod Genus *Vicarya* [abstr.]. *In* Addicott, W.O. and Ingle, J.C., eds., *3-IGCP-114, Stanford 1978, Stanford Univ. Publ., Geol. Sci.*, v. **14**, p. 24.

Kassab, I. I. M., 1978: The Paleogene/Neogene boundary in Iraq [abstr.]. *In* Addicott, W.O. and Ingle, J.C., eds., *3-IGCP-114, Stanford 1978, Stanford Univ. Publ., Geol. Sci.*, v. **14**, p. 25.

Kato, M., 1979: Joban area; 2. *In* Tsuchi, R., ed., *Fundam. data on Japan. Neogene*, Shizuoka, p. 36–43.

Kato, M., 1980: Planktonic foraminiferal biostratigraphy of the Takaku and Taga Groups in the Joban coal field, northeast Honshu, Japan. *Tohoku Univ., Sci. Rep., 2nd ser.*, v. **50**, nos. 2–3, p. 35–95.

Keigwin, L. D., Jr., 1978: Middle Miocene to Pliocene stable isotopic datums in east equatorial and north central Pacific deep sea drilling (DSDP) sites [abstr.]. *In* Addicott, W.O. and Ingle, J.C., eds., *3-IGCP-114, Stanford 1978, Stanford Univ. Publ., Geol. Sci.*, v. **14**, p. 26–27.

Keller, G., 1978: Late Neogene paleooceanography and planktonic foraminiferal datum levels of mid latitudes of the North Pacific [abstr.]. *In* Addicott, W.O. and Ingle, J.C., eds., *3-IGCP-114, Stanford 1978, Stanford Univ. Publ., Geol. Sci.*, v. **14**, p. 28–29.

Kennett, J. P., 1977: Late Neogene paleoceanography of the South Pacific [abstr.]. *In* Saito, T. and Ujiie, H., eds., *1-CPNS, Tokyo 1976, Proc.*, p. 81–82.

Kennett, J. P., 1978: The development of planktonic biogeography in the southern ocean during the Cenozoic. *Mar. Micropal.*, v. 3, p. 301–345.

Kennett, J. P., 1978: Cenozoic microfossil datums in Antarctic-Subantarctic deep-sea sedimentary sequences and the evolution of southern ocean planktonic biogeography [abstr.]. *In* Addicott, W.O. and Ingle, J.C., eds., *3-IGCP-114, Stanford 1978, Stanford Univ. Publ., Geol. Sci.*, v. **14**, p. 30–31.

Kennett, J. P., 1979: Late Cenozoic paleoceanography and planktonic foraminiferal biostratigraphy of the South Pacific; tropics to poles [abstr.]. *In* Shilo, N. A., ed., *4-IGCP-114 in XIV-PSC, Khabarovsk 1979, Abstr.*, Sect. B-III, v. **2**, p. 72–74.

Kennett, J.P., 1980: Paleoceanographic and biogeographic evolution of the southern ocean during the Cenozoic, and Cenozoic microfossil datums. *Palaeogeogr., Palaeoclimat., Palaeoecol.*, Elsevier, v. **31**, p.123–152.

Kenett, J. P. and Srinivasan, M. S., 1981: Neogene equatorial to subantarctic planktonic foraminiferal biostratigraphy and datum levels. *In* Ikebe, N. *et al.*, eds., *6-IGCP-114, IWPNB Osaka 1981, Proc.*, p. 73–90.

Kilmer, F. H., 1978: History of the Pliocene molluscan fauna of northern Japan. *The Veliger*, v. **21**, p. 227–231.

Kim, B. K., 1977: On the Neogene Tertiary deposits in southern Korea. *In* Saito, T. and Ujiie, H., eds., *1-CPNS, Tokyo 1976, Proc.*, p. 115–118.

Kim, B. K., 1978: Preliminary benthonic foraminiferal zonation and faunal analysis based on the quantitative method in the Tertiary Pohang basin, Korea. *In* Kadar, D., ed., *2-IGCP-114, Bandung 1977, Proc.*, p. 45–51.

Kim, B. K., 1981: A micropaleontological study (silicoflagellate, ebridian, and nannofossil) on Neogene Tertiary in the Pohang basin [abstr.]. *In* Ikebe, N. *et al.*, eds., *6-IGCP-114, IWPNB Osaka 1981, Proc.*, p. 122.

Kitamura, N., 1977: Neogene tectonic development of the Northeast Japan Arc [abstr.]. *In* Saito, T. and Ujiie, H., eds., *1-CPNS, Tokyo 1976, Proc.*, p. 356–358.

Kitamura, H. and Takayanagi, Y., 1977: Problem on reconstruction of the Neogene history. *In* Takayasu, T. *et al.*, eds., *Prof. Huzioka Mem. Vol.*, p. 193–222.

Kitazato, H., 1979: Oga Peninsula; 1. *In* Tsuchi, R., ed., *Fundam. data on Japan. Neogene*, Shizuoka, p. 69–70.

Koizumi, I., 1977: Diatom biostratigraphy in the North Pacific region. *In* Saito, T. and Ujiie, H., eds., *1-CPNS, Tokyo 1976, Proc.*, p. 235–253.

Koizumi, I., 1979: Kadonosawa-Sannohe area; 2. *In* Tsuchi, R., ed., *Fundam. data on Japan. Neogene*, Shizuoka, p. 53–55.

Koizumi, I., 1979: Coast of Shosanbetsu, Rumoi, Hokkaido; 1. *In* Tsuchi, R., ed., *Fundam. data on Japan. Neogene*, Shizuoka, p. 59–60.

Koizumi, I., 1979: Coast of Morai, Ishikari, Hokkaido. *In* Tsuchi, R., ed., *Fundam. data on Japan. Neogene*, Shizuoka, p. 63–64.

Koizumi, I., 1979: Tate area in Oshima Peninsula, Hokkaido. *In* Tsuchi, R., ed., *Fundam. data on Japan. Neogene*, Shizuoka, p. 65–66.

Koizumi, I., 1979: Kamiiso area in Oshima Peninsula, Hokkaido. *In* Tsuchi, R., ed., *Fundam. data on Japan. Neogene*, Shizuoka, p. 67–68.

Koizumi, I., 1979: Southern Noto Peninsula; Ao-Mitakubo, Himi. *In* Tsuchi, R., ed., *Fundam. data on Japan. Neogene*, Shizuoka, p. 89–90.

Koizumi, I., 1979: Southern Noto Peninsula, Coast of Nadaura. *In* Tsuchi, R., ed., *Fundam. data on Japan. Neogene*, Shizuoka, p. 91–92.

Koizumi, I., 1979: Northern Noto Peninsula; Suzu. *In* Tsuchi, R., ed., *Fundam. data on Japan. Neogene*, Shizuoka, p. 93–94.

Koizumi, I., 1979: DSDP Site 299, the Sea of Japan. *In* Tsuchi, R., ed., *Fundam. data on Japan. Neogene*, Shizuoka, p. 118–119.

Koizumi, I., 1979: DSDP Site 301, the Sea of Japan. *In* Tsuchi, R., ed., *Fundam. data on Japan. Neogene*, Shizuoka, p. 120–121.

Koizumi, I., 1979: DSDP Site 302, the Sea of Japan. *In* Tsuchi, R., ed., *Fundam. data on Japan. Neogene*, Shizuoka, p. 122–133.

Koizumi, I., 1979: A composite scheme of paleontological events in northern Japan [abstr.]. *In* Shilo, N. A., ed., *4-IGCP-114 in XIV-PSC, Khabarovsk 1979, Abstr.*, Sect. B-III, v. **2**, p. 75–76.

Koizumi, I., 1979: Pre-Quaternary sediments in the Sea of Japan and its paleogeographical significance [abstr.]. *In* Shilo, N. A., ed., *4-IGCP-114 in XIV-PSC, Khabarovsk 1979, Abstr.*, Sect. B-III, v. **2**, p. 77–78.

Koizumi, I., 1981: Joban Area (1)–(5), *In* Tsuchi, R., ed., *Fundam. data on Japan. Neogene, supplement*, Shizuoka, p. 18–27.

Koizumi, I., 1981: Mizunami Area, Gifu Prefecture, *In* Tsuchi, R., ed., *Fundam. data on Japan. Neogene, supplement*, Shizuoka, p. 68–69.

Koizumi, I., 1981: Evolutionary trends of the marine species of the genus *Denticulopsis* [abstr.]. *In* Ikebe, N. *et al.*, eds., *6-IGCP-114, IWPNB Osaka 1981, Proc.*, p. 123.

Koizumi, I., Barron, J. A., and Harper, H. E., Jr., 1980: Diatom correlation of Legs 56 and 57 with onshore sequences in Japan. *DSDP, Init. Rep.*, v. **56, 57**, p. 687–693.

Koizumi, I. and Kanaya, T., 1977: Correlation of late Neogene section on the Oga Peninsula, Akita Pref. Japan. *In* Takayasu, T. *et al.*, eds., *Prof. Huzioka Mem. Vol.*, p. 401–412.

Koizumi, I. and Ujiie, H., 1976: On the age of the Nobori Formation, Shikoku, southwest Japan–particularly based on diatoms. *Nat. Sci. Mus., Tokyo, Mem.*, no. 9, p. 61–68.

Konda, I., 1980: Benthonic foraminifera biostratigraphy of the standard areas of Middle Miocene in the Pacific side province, central Japan. *Kyoto Univ. Fac. Sci. Geol. Min., Mem.*, v. **47**, p. 1–42.

Konda, T. and Taguchi, K., 1977: Volcanostratigraphy of the Neogene Tertiary of Tohoku area, Japan [abstr.]. *In* Saito, T. and Ujiie, H., eds., *1-CPNS, Tokyo 1976, Proc.*, p. 358–360.

Kotaka, T., 1977: World-wide correlation based on turritellid phylogeny [abstr.]. *In* Saito, T. and Ujiie, H., eds., *1-CPNS, Tokyo 1976, Proc.*, p. 360–363.

Kotaka, T., 1978: World-wide biostratigraphic correlation based on Turritellid phylogeny. *The Veliger*, v. **21**, no. 2, p. 189–196.

Kotaka, T. and Kato, H., 1979: Additional fossil shells from the Utsutoge Formation, Yamagata Pref., northeast Honshu, Japan. *Saito Ho-on Kai Mus. Nat. Hist., Res. Bull.*, no. 47, p. 13–18, pls. 2–3.

Kotaka, T. and Ogasawara, K., 1977: Turritellid zone along the Japan Sea borderland, Japan. *In* Takayasu, T. *et al.*, eds., *Prof. Huzioka Mem. Vol.*, p. 345-354, pl. 1.

Kotaka, T., Ogasawara, K., and Tanaka, T., 1979: Boso Pen.-insula; central area. *In* Tsuchi, R., ed., *Fundam. data on Japan. Neogene*, Shizuoka, p. 28–29.

Kovalenko, F. Ya. and Kuptsova, I. A., 1979: Cenozoic Marine deposits, East Siberian Sea [abstr.]. *In* Shilo, N. A., ed., *4-IGCP-114 in XIV-PSC, Khabarovsk 1979, Abstr.*, Sect. B-III, v. **2**, p. 80–82.

Krasilov, V. A., 1979: Systems of communities and stratigraphic hypotheses [abstr.]. *In* Shilo, N. A., ed., *4-IGCP-114 in XIV-PSC, Khabarovsk 1979, Abstr.*, Sect. B-III v. **2**, p. 82.

Kurihara, K., 1977: Stratigraphic occurrence of planktonic foraminifera in the sections from lower to upper Miocene of the Kanto District, central Japan [abstr.]. *In* Saito, T. and Ujiie, H., eds., *1-CPNS, Tokyo 1976, Proc.*, p. 364–365.

Kurihara, K., 1978: Foraminiferal datum levels recognized in the Neogene sections of the Kanto District, central Japan [abstr.]. *In* Addicott, W.O. and Ingle, J.C., eds., *3-IGCP-114, Stanford 1978, Stanford Univ. Publ. Geol. Sci.*, v. **14**, p. 32–33.

Lagoe, M. B., 1978: Foraminifera from the uppermost Poul Creek and lowermost Yakataga Formations, Yakataga district, Alaska [abstr.]. *In* Addicott, W.O. and Ingle, J.C., eds., *3-IGCP-114, Stanford 1978, Stanford Univ. Publ., Geol. Sci.*, v. **14**, p. 34–35.

Ling, H. Y., 1977: Late Cenozoic silicoflagellates and ebridians from the eastern North Pacific region. *In* Saito, T. and Ujiie, H., eds., *1-CPNS, Tokyo 1976, Proc.*, p. 205–233.

Ling, H. Y., 1977: Miocene radiolarian and silicoflagellate zonation from the Aleutian abyssal plain (Site 183, DSDP Leg 19) [abstr.]. *In* Saito, T. and Ujiie, H., eds., *1-CPNS, Tokyo 1976, Proc.*, p. 365.

Linkova, T. I., 1 979: Correlation of Cenozoic deep sea bottom sediments with the aid of paleomagnetic and biostratigraphic data [abstr.]. *In* Shilo, N. A., ed., *4-IGCP-114 in XIV-PSC, Khabarovsk 1979, Abstr.*, Sect. B-III p. 85–86.

Maiya, S., 1978: Late Cenozoic planktonic foraminiferal biostratigraphy of the oil-field region of northeast Japan. *In* Huzita, K., *et al.*, ed., *Cenozoic Geology of Japan*, p. 35–60.

Maiya, S., 1979: Wakkanai area, Hokkaido. *In* Tsuchi, R., ed., *Fundam. data on Japan. Neogene*, Shizuoka, p. 56–57.

Maiya, S., 1979: Embetsu area, Hokkaido. *In* Tsuchi, R., ed., *Fundam. data on Japan. Neogene*, Shizuoka, p. 58.

Maiya, S., 1979: Coast of Shosanbetsu, Rumoi, Hokkaido; 2. *In* Tsuchi, R., ed., *Fundam. data on Japan. Neogene*, Shizuoka, p. 61–62.

Maiya, S., 1979: Shatsky Rise, the Northwest Pacific. *In* Tsuchi, R., ed., *Fundam. data on Japan. Neogene*, Shizuoka, p. 117.

Maiya, S., Akiba, F. and Ichinoseki, T., 1981: Hidaka Area (1)–(2). *In* Tsuchi, R., ed., *Fundam. data on Japan.*

Neogene, supplement, Shizuoka, p. 34–37.

Maiya, S., Akiba, F. and Ichinoseki, T., 1981: Kushiro Area (1)–(4). *In* Tsuchi, R., ed., *Fundam. data on Japan. Neogene, supplement*, Shizuoka, p. 38–46.

Maiya, S., Akiba, F., Ichinoseki, T., and Kodato, T., 1981: Tempoku Area (1)–(2). *In* Tsuchi, R., ed., *Fundam. data on Japan. Neogene, supplement*, Shizuoka, p. 47–48.

Maiya, S., Ichinoseki, T. and Akiba, F., 1981: Stratigraphic outlines of selected Neogene sequences, 11. Oshima Peninsula. *In* Tsuchi, R. ed., *Neogene of Japan*, p. 76–80.

Maiya, S., Ichinoseki, T. and Akiba, F., 1981: Stratigraphic outlines of selected Neogene sequences, 13. Hidaka area. *In* Tsuchi, R., ed., *Neogene of Japan*, p. 85–89.

Maiya, S., Inoue, Y. and Akiba, F., 1981: Sea of Kashima, *In* Tsuchi, R., ed., *Fundam. data on Japan. Neogene, supplement*, Shizuoka, p. 13–17.

Maiya, S., Inoue, Y. and Ogata, H., 1980: Paleoenvironment and organic matter in sediments. Part 1. Evolutionary changes of paleoenvironment and paleogeography in Niigata Neogene basin. *Japan Assoc. Petrol. Tech., Jour.*, v. **45**, p. 323–336.

Maiya, S., Kodato, T. and Ichinoseki, T., 1981: Haboro Area, *In* Tsuchi, R., ed., *Fundam. data on Japan. Neogene, supplement*, Shizuoka, p. 49–51.

Maiya, S., Saito, T. and Sato, T., 1976: Late Cenozoic planktonic foraminiferal biostratigraphy of Northwest Pacific sedimentary sequences. *In* Takayanagi, Y. and Saito, T., eds., *Progress in Micropaleontology*, Micropal. Press, N.Y., p. 395–422.

Mamontova, I. B., 1979: The role of gymnosperm pollen in regional correlation [abstr.]. *In* Shilo, N. A., ed., *4-IGCP-114 in XIV-PSC, Khabarovsk 1979, Abstr.*, Sect. B-III, v. **2**, p. 87–88.

Margulis, L. S., Savitsky, V. O. and Tyutrin, I. I., 1979: The Cenozoic sediments of South Sakhalin and adjacent sea regions [abstr.]. *In* Shilo, N. A., ed., *4-IGCP-114 in XIV-PSC, Khabarovsk 1979, Abstr.*, Sect. B-III, v. **2**, p. 88–91.

Marincovich, L., Jr., 1979: Miocene mollusks of the Topsy Formation, Lituya District, Gulf of Alaska Tertiary Province, Alaska. *U.S. Geological Survey Shorter Contributions to Paleontology*, p. Cl-C14.

Marincovich, L., Jr., 1981: Late Miocene mollusks of the Tachilni Formation, Alaska Peninsula, Alaska [abstr.]. *In* Ikebe, N. *et al.*, eds., *6-IGCP-114, IWPNB Osaka 1981, Proc.*, p. 123A.

Martinez-Pardo, R., 1978: Discovery of Miocene marine strata on the Mejillones Peninsula, Antofagasta Province, Chile [abstr.]. *In* Addicott, W.O. and Ingle, J.C., eds., *3-IGCP-114, Stanford 1978, Stanford Univ. Publ., Geol. Sci.*, v. **14**, p. 65.

Martinez-Pardo, R., 1981: An unknown upper Miocene-lower Pliocene regional hiatus along the marginal Northeast Pacific? [abstr.]. *In* Ikebe, N. *et al.*, eds., *6-IGCP-114, IWPNB Osaka 1981, Proc.*, p. 124–126.

Maruyama, T., 1981: Ichinoseki Area, Iwate Prefecture, *In* Tsuchi, R., ed., *Fundam. data on Japan. Neogene, supplement*, Shizuoka, p. 32–33.

Maruyama, T., 1981: Sendai Area. *In* Tsuchi, R., ed., *Fundam. data on Japan. Neogene, supplement*, Shizuoka, p. 28–29.

Masuda, K., 1977: Miocene molluscs from the Shimokurosawa Formation, Ichinoseki City, Iwate Pref. Northeast Honshu, Japan. *Saito Ho-on Kai Mus. Nat. Hist., Research Bull.*, v. **45**, p. 3–9, pl. 1.

Masuda, K., 1977: Neogene Pectinidae of the northern Pacific [abstr.]. *In* Saito, T. and Ujiie, H., eds., *1-CPNS, Tokyo 1976, Proc.*, p. 366–368.

Masuda, K., 1978: Pectinid biostratigraphy of the Neogene in central to south Japan [abstr.]. *In* Addicott, W.O. and Ingle, J.C., eds., *3-IGCP-114, Stanford 1978, Stanford Univ. Publ., Geol. Sci.*, v. **14**, p. 36–37.

Masuda, K., 1978: Neogene Pectinidae of the northern Pacific. *The Veliger*, v. **21**, no. 2, p. 197–202.

Masuda, K., 1979: Early Pliocene pectinid biostratigraphy in Japan [abstr.]. *In* Shilo, N. A., ed., *4-IGCP-114 in XIV-PSC, Khabarovsk 1979, Abstr.*, Sect. B-III, v. **2**, p. 91–92.

Masuda, K., 1980: Pliocene biostratigraphy in Japan based on Pectinids. *Saito Ho-on Kai Mus. Nat. Hist., Research Bull.*, no. 48, p. 9–19, pls. 2–3.

Masuda, K. and Chiyoda, A., 1981: Izumi Area, Miyagi Prefecture. *In* Tsuchi, R., ed., *Fundam. data on Japan. Neogene, supplement*, Shizuoka, 1981, p. 30–31.

Matoba, Y. and Oda, M., 1982: Late Pliocene to Holocene planktonic foraminifers of the Guaymas Basin, Gulf of California, Sites 477 through 481. *DSDP, Init. Rep.*, v. 64, p. 1003–1026.

Matsumaru, K., 1977: Miocene larger foraminiferal biostratigraphy of Japan and interregional correlation in the West Pacific Province [abstr.]. *In* Saito, T. and Ujiie, H., eds., *1-CPNS, Tokyo 1976, Proc.*, p. 368–371.

Matsumaru, K., 1978: Biostratigraphy and paleoecological transition of larger foraminifera from the Minamizaki Limestone, Chichi-jima, Japan. *In* Kadar, D., ed., *2-IGCP-114, Bandung 1977, Proc.*, p. 63–88.

Matsumaru, K., 1979: Fossil localities of larger foraminifera. *In* Tsuchi, R., ed., *Fundam. data on Japan. Neogene*, Shizuoka, p. 124–133.

Matsumaru, K., 1980: Notes on a species of *Miogypsina* from Japan. *In* Igo, H. and Noda, H., eds., *Prof. S. Kanno Mem. Vol.*, p. 213–219, pl. 25.

Matsumaru, K., 1980: Cenozoic larger foraminiferal assemblages of Japan. Part 1. Comparison with Southeast Asia. *In* Kobayashi, T. *et al.*, eds., *Geol. Paleont. SE. Asia.*, v. **21**, p. 211–224.

Matsumaru, K., 1981: Fossil localities for larger foraminifera (Supplement). *In* Tsuchi, R., ed., *Fundam. data on Japan. Neogene, supplement*, Shizuoka, p. 84–86.

Matsumaru, K., Azuma, Y. and Takeyama, K., 1979: Discovery of *Miogypsina* and *Operculina* rfom the Miocene sediments of the mountains, Fukui Pref. and its significance. *Geol. Soc. Japan, Jour.*, v. **85**, p. 771–774[J].

Matsumaru, K. and Hayashi, A., 1980: Neogene stratigraphy of the eastern margin areas of Kwanto Mountains, central Japan. *Geol. Soc. Japan, Jour.*, v. **86**, p. 225–242.

Matsumaru, K. and Matsuo, Y., 1981: Eastern end of Kanto Mountains (Additional data). *In* Tsuchi, R., ed., *Fun-*

dam. data on Japan. Neogene, supplement, Shizuoka, p. 76.

Matsumaru, K. and Matsuo, Y., 1981: Some records of larger foraminifera in comparison with planktonic foraminiferal zone from the Kanto region, Japan [abstr.]. *In* Ikebe, N. *et al.*, eds., *6-IGCP-114, IWPNB Osaka 1981, Proc.*, p. 127.

Matsuoka, K., 1979: Brackish water molluscan fauna from the Miocene Bihoku Group, Hiroshima Pref., Southwest Japan. *Mizunami Fossil Mus., Bull.*, no. 6, p. 27–39, pls. 5–6.

Matsuoka, K., 1981: Niigata area; 2. *In* Tsuchi, R., ed., *Fundam. data on Japan. Neogene, supplement*, Shizuoka 1981, p. 79–80.

Matsuoka, K., 1981: Dinoflagellate cysts and *Pediastrum* from the Nanggulan and Sentolo Formations in the middle Java Island, Indonesia. *In* Saito, T., ed., *Micropaleontology, petrology and lithostratigraphy of Cenozoic rocks of the Jogyakarta region, Central Java.* Yamagata University.

Matsuoka, K., 1981: Late Cenozoic dinoflagellate cyst assemblage in the Niigata region, Central Japan [abstr.]. *In* Ikebe, N. *et al.*, eds., *6-IGCP-114, IWPNB Osaka 1981, Proc.*, p. 128–129.

McCoy, S., Jr., 1980: Neogene biostratigraphy and correlations of the eastern Gulf of Alaska. *5-IGCP-114 in 26-IGC, Paris 1980, Abstracts*, v. **1**, p. 253.

McCoy, S., Jr. and Ariey, C., 1979: Molluscan biostratigraphy of the Poul Creek and Yakataga Formations, Yakataga district, Alaska [abstr.]. *In* Shilo, N. A., ed., *4-IGCP-114 in XIV-PSC, Khabarovsk 1979, Abstr.*, Sect. B-III, v. **2**, p. 86–87.

McGrowran, B., 1978: Australian Neogene sequences and events. *In* Kadar, D., ed., *2-IGCP-114, Bandung 1977, Proc.*, p. 165–167.

Menner, V. V. (ed.) 1979: The Makarovsky key section of Paleogene and Neogene deposits in Sakhalin. *Guidebook of Tour VI, XIV-PSC, Khabarovsk*, p. 1–32.

Menner, V. V., Baranova, Yu. P. and Zhidkova, L.S., 1977: Neogene of the northeastern USSR (Kolyma region, Kamchatka, and Sakhalin). *In* Shilo, N. A., ed., *4-IGCP-114 in XIV-PSC, Khabarovsk 1979, Abstr.*, Sect. B-III, v. **2**, p. 83–88.

Menner, V. V. and Glandenkov, Y. B., 1979: On the creation of a correlation scheme for the Neogene of the northern part of the Circumpacific belt [abstr.]. *In* Shilo, N. A., ed., *4-IGCP-114, XIV-PSC, Khabarovsk 1979, Abstr.*, Sect. B-III, v. **2**, p. 93–94.

Mikhailova, N. P., Glevaskaya, A. M., Tsykora, V. N., *et al.*, 1979: Comparison of paleomagnetic characteristics of Cenozoic volcanites of the Pacific and Alpine volcanic belts [abstr.]. *In* Shilo, N. A., ed., *4-IGCP-114 in XIV-PSC, Khabarovsk 1979, Abstr.*, Sect. B-III, v. **2**, p. 94–95.

Mitsunashi, T., Nakagawa, H., Suzuki, Y. *et al.*, eds., 1976: *Guide book for excursion 2, Boso Peninsula. 1-CPNS, Tokyo 1976*, p. 1–82.

Moiseyeva, A. I., Sheshukova-Poretskaya, V. S. and Boldyreva, V. P., 1979: Neogene diatom assemblages in the Makarov reference section on Sakhalin and their implication for stratigraphy and correlation [abstr.]. *In* Shilo, N. A., ed., *4-IGCP-114 in XIV-PSC, Khabarovsk 1979, Abstr.*, Sect. B-III, v. **2**, p. 96–97.

Morishita, A., 1977: Biostratigraphy of the Japanese Neogene by fossil echinoids [abstr.]. *In* Saito, T. and Ujiie, H., eds., *1-CPNS, Tokyo 1976, Proc.*, p. 371–372.

Morozumi, Y. and Koizumi, I., 1981: Stratigraphic outlines of selected Neogene sequences, 9. Himi and Yatsuo areas. *In* Tsuchi, R., ed., *Neogene of Japan*, p. 65–67.

Morozumi, Y., Ishigaki, T., Koizumi, I., *et al.*, 1979: Southern Noto Peninsula; Ao-Unami, Himi. *In* Tsuchi, R., ed., *Fundam. data on Japan. Neogene*, Shizuoka, p. 87–88.

Morrison, S. and Sarna-Wojcicki, A., 1981: Time equivalent bay and outer shelf faunas of the Neogene Humboldt basin, California and correlation to the North Pacific microfossil zones of DSDP 173 [abstr.]. *In* Ikebe, N. *et al.*, eds., *6-IGCP-114, IWPNB Osaka 1981, Proc.*, p. 130–131.

Nakagawa, H., 1981: Data on Japanese Neogene Magnetostratigraphy. *In* Tsuchi, R., ed., *Fundam. data on Japan. Neogene, supplement*, Shizuoka, p. 87–100.

Nakagawa, H., Kitamura, N., Takayanagi, Y., *et al.*, 1977: Magnetostratigraphic correlation of Neogene and Pleistocene between the Japanese Islands, central Pacific, and Mediterranean regions. *In* Saito, T. and Ujiie, H., eds., *1-CPNS, Tokyo 1976, Proc.*, p. 285–310.

Nakagawa, H. and Niitsuma, H., 1977: Magnetostratigraphy of the late Cenozoic of the Boso Peninsula, central Japan. *Quat. Research*, v. **7**, p. 294–301.

Natori, H., 1976: Planktonic foraminiferal biostratigraphy and datum planes in the late Cenozoic sedimentary sequence in Okinawa-jima, Japan. *In* Takayanagi, Y. and Saito, T., eds., *Progress in Micropaleontology*, Micropal. Press, N.Y., p.214–243, pls. 1–6.

Natori, H., 1977: Late Cenozoic planktonic foraminiferal biostratigraphy in southwestern Japan [abstr.]. *In* Saito, T. and Ujiie, H., eds., *1-CPNS, Tokyo 1976, Proc.*, p. 372–374.

Natori, H., 1979: Okinawa Island; 1. *In* Tsuchi, R., ed., *Fundam. data on Japan. Neogene*, Shizuoka, p. 1–2.

Natori, H., 1979: Miyazaki area. *In* Tsuchi, R., ed., *Fundam. data on Japan. Neogene*, Shizuoka, p. 7–9.

Natori, K., Nakamura, R. and Matoba, Y., 1980: Paleomagnetic study of late Cenozoic sediments in Akita Pref., Japan. *Akita Univ. Mining Coll., Jour., ser. A*, v. **5**, no. 4, p. 1–13.

Nelson, C. M., 1977: The gastropod *Neptunea* (Prosobranchia: Buccinacea) in the North Pacific Neogene [abstr.]. *In* Saito, T. and Ujiie, H., eds., *1-CPNS, Tokyo 1976, Proc.*, p. 374–376.

Nelson, C. M., 1978: *Neptunea* (Gastropoda: Buccinacea) in the Neogene of the North Pacific and adjacent Bering Sea. *The Veliger*, v. **21**, no. 2, p. 203–215.

Nelson, C. M., 1979: The gastropod *Beringius* (Prosobranchia: Buccinacea) in the late Cenozoic of the North

Pacific and Bering Sea [abstr.]. *In* Shilo, N. A., ed., *4-IGCP-114 in XIV-PSC, Khabarovsk 1979, Abstr.*, Sect. B-III, v. **2**, p. 100–101.

Nikiforova, K. V., 1977: The status of the boundary between the Pliocene and Quaternary. *In* Saito, T. and Ujiie, H., *et al.*, eds., *1-CPNS, Tokyo 1976, Proc.*, p. 54–60.

Nikitin, V. P., 1979: The Neogene floras of the northern Asia [abstr.]. *In* Shilo, N. A., ed., *4-IGCP-114 in XIV-PSC, Khabarovsk 1979, Abstr.*, Sect. B-III, v. **2**, p. 101–103.

Nishida, S., 1977: Late Cenozoic calcareous nannoplankton biostratigraphy in Japan [abstr.]. *In* Saito. T. and Ujiie, H., eds., *1-CPNS, Tokyo 1976, Proc.*, p. 376–378.

Nishida, S., 1978: Late Cenozoic calcareous nannoplankton biostratigraphy in Sagara and Kakegawa district, central Japan. *Nara Univ. Educ., Bull.*, v. **27**, p. 85–97.

Nishida, S., 1979: Restudies of calcareous nannoplankton biostratigraphy of the Tonohama Group, Shikoku, Japan. *Nara Univ. Educ., Bull.*, v. **28**, p. 97–108, pl. 1.

Nishida, S., 1980: Calcareous nannoplankton biostratigraphy of the Miyazaki Group, southeast Kyushu, Japan. *Nara Univ. Educ., Bull.*, v. **29**, p. 65–77, pl. 1.

Nishida, S., 1980: Calcareous nannoplankton biostratigraphy around the Pliocene-Pleistocene boundary in the southern part of Okinawa-Jima, Japan. *Geol. Soc. Japan, Jour.*, v. **86**, p. 525–536.

Nishida, S., 1981: Okinawa Island. *In* Tsuchi, R., ed., *Fundam. dataon Japan. Neogene, supplement*, Shizuoka, 1981, p. 1–2.

Nishida, S., 1981: Choshi Area, *In* Tsuchi. R., ed., *Fundam. data on Japan. Neogene, supplement*, Shizuoka, 1981, p. 11–12.

Nishimura, S., 1981: On the Neogene fission-track dating of tuffs [abstr.]. *In* Ikebe, N. *et al.*, ed., *6-IGCP-114, IWPNB Osaka 1981, Proc.*, p. 132–133.

Nishimura, S. and Amano, U., 1979: Data on Neogene fission track ages; 1968–1978. *In* Tsuchi, R., ed., *Fundam. data on Japan. Neogene*, Shizuoka, p. 135–142.

Nishimura, S. and Amano, Y., 1981: Data on Neogene Fission Track Ages (1979–1980)/Correction of misprint of data on the Fission Track Ages appeared in the 1979 volume. *In* Tsuchi, R., ed., *Fundam. data on Japan. Neogene, supplement*, Shizuoka, p. 105–108.

Nishimura, S., Sasajima, S., Hirooka, K., *et al.*, 1977: Preliminary report on fission-track dating in the Sunda Arc [abstr.]. *In* Saito, T. and Ujiie, H., eds., *1-CPNS, Tokyo 1976, Proc.*, p. 378–379.

Nishmura, S., Sasajima, S., Thio, K. H., and Hehuwat, F., 1978: The second report of fission-track dating in the Sunda Arc. *In* Kadar, D., ed., *2-IGCP-114, Bandung 1977, Proc.*, p. 163–164.

Noda, H., 1978: Neogene anadaran distribution in Japan and Southeast Asia [abstr.]. *In* Addicott, W.O. and Ingle, J.C., eds., *3-IGCP-114*, Stanford Univ. Publ., Geol. Sci., v. **14**, p. 38–39.

Noda, H., 1978: Neogene *Anadara* distribution in Japan and Southeast Asia. *Univ. Tsukuba Inst. Geosci., Ann. Rep.*, no. 4, p. 33–37.

Noda, H., 1980: Molluscan fossils from the Rykyu Islands, southwestern Japan. Part 1. Gastropoda and Pelecypoda from the Shinzato Formation in southeastern part of Okinawa. *Univ. Tsukuba Inst. Geosci., Ann. Rep., sec. B*, v. **1**, p. 1–95, pls. 1–12.

Noda, H. and Amano, K., 1977: Geological significance of *Anadara amicula elongata* from the Pliocene Kume Formation, Ibaraki Pref., Japan. *Univ. Tsukuba, Geosci. Inst., Ann. Rep.*, no. 3, p. 37–41.

Obradovich, J. D. and Naeser, C. W., 1981: Geochronology bearing on the age of the Monterey Formation and siliceous rocks in California. *In* Garrison, R. E. and Douglas, R. G., eds., *The Monterey Formation and related siliceous rocks in California*, Society of Economic Paleontologists and Mineralogists, Pacific Section, Los Angeles, California, p. 87–95.

Obradovich, J. D., Naeser, C. W. and Izett, G. A., 1978: Geochronology of late Neogene marine strata in California [abstr.]. *In* Addicott, W.O. and Ingle, J.C., eds., *3-IGCP-114, Stanford 1978, Stanford Univ. Publ., Geol. Sci.*, v. **14**, p. 40–41.

Oda, M., 1977: Planktonic foraminiferal biostratigraphy of the late Cenozoic sedimentary sequence, central Honshu, Japan. *Tohoku Univ., Sci. Rep., 2nd ser.*, v. **48**, no. 1, p. 1–76, pls. 1–10.

Oda, M., 1979: Boso Peninsula; eastern area. *In* Tsuchi, R., ed., *Fundam. data on Japan. Neogene*, Shizuoka, p. 24–27.

Oda, M., Hasegawa,S., Honda, N., Maruyama, T., and Funayama, M., 1983: Progress in multiple planktonic microfossil biostratigraphy for the middle to upper Miocene of the central and northeast Honshu, Japan. *Japan. Assoic. Petrl. Techn., Jour.*, v. **48**, no. 1, p. 71–87.

Oda, M. and Sakai, T., 1977: Microbiostratigraphy of the lower to middle part of the Hatatate Formation, Sendai. *In* Takayasu, T., ed., *Prof. K. Huzioka Mem. Vol.*, p. 441–446.

Oda, M. and Sakai, T., 1979: Sendai area. *In* Tsuchi, R., ed., *Fundam. data on Japan. Neogene*, Shizuoka, p. 46–47.

Ogasawara, K., 1979: Sennan area, Miyagi Prefecture. *In* Tuschi, R., ed., *Fundam. data on Jap. Neogene*, Shizuoka, p. 44–45.

Ogasawara, K., 1979: Southwestern area of Kanazawa. *In* Tsuchi, R., ed., *Fundam. data on Japan. Neogene*, Shizuoka, p. 97–98.

Ogasawara, K., 1979: Geological and paleogeographical significance of the Omma-Manganizian fauna of Japan Sea borderland [abstr.]. *In* Shilo, N. A., ed., *4-IGCP-114 in XIV-PSC, Khabarovsk 1979, Abstr.*, Sect. B-III v. **2**, p. 104–105.

Ogasawara, K., 1981: Paleogeographic significance of the Omma-Manganzian fauna of the Japan Sea borderland. *Saito Ho-on Kai Mus. Nat. Hist., Res. Bull.*, no. 49, p. 1–17.

Ogasawara, K. and Noda, H., 1978: Arcid-potamid fauna from the Tsukinoki Formation, Sennan district, Miyagi Pref., Northeast Japan. *Saito Ho-on Kai Mus. Nat. Hist., Research Bull.*, no. 46, p. 21–44, pls. 2–3.

Ogasawara, K. and Nomura, R., 1980: Molluscan fossils from the Fujina Formation, Shimane Pref., San-in district, Japan. *In* Igo, H. and Noda, H., eds., *Prof. S. Kanno Mem. Vol.*, p. 79–88, pls. 9–12.

Ogasawara, K., Takayama, T., Bito, A., Hayakawa, T. and Kaseno, Y., 1981: Kaga Area, Ishikawa Prefecture. *In* Tsuchi, R., ed., *Fundam. data on Japan. Neogene, supplement*, Shizuoka, p. 66–67.

Ogasawara, K. and Yashima, S., 1981: Miocene Molluscs from the Date Formation, Fukushima prefecture, Northeast Japan. *Saito Ho-on Kai Mus. Nat. Hist., Res. Bull.*, no. 49, p. 37–51.

Okada, H., 1981: Kitashiobara Area, Fukushima Prefecture (Additional data). *In* Tsuchi, R., ed., *Fundam. data on Japan. Neogene, supplement*, Shizuoka, 1981, p. 76.

Okada, H., 1981: Yonezawa Area, Yamagata Prefecture (Additional data). *In* Tsuchi, R., ed., *Fundam. data on Japan. Neogene, Supplement*, Shizuoka, 1981, p. 78.

Okada, H., 1981: Kawajiri Area, Iwate Prefecture (Additional data). *In* Tsuchi, R., ed., *Fundam data on Japan. Neogene, supplement*, Shizuoka, p. 79.

Okada, H., 1981: Biratori Area, Hokkaido (Additional data). *In* Tsuchi, R., ed., *Fundam. data on Japan. Neogene, supplement*, Shizuoka, p. 80.

Okada, H., 1981: Calcareous nannofossils of Cenozoic formations in Central Java. *In* Saito, T., ed., *Micropaleontology, petrology and lithostratigraphy of Cenozoic rocks of the Yogyakarta region, Central Java*. Yamagata University, p. 25–34.

Okamoto, K., 1979: Some fossil evidence of the Paleo-Tsushima Strait during Neogene [abstr.]. *In* Shilo, N. A., ed., *4-IGCP-114 in XIV-PSC, Khabarovsk 1979, Abstr.*, Sect. B-III, v. **2**, p. 105–107.

Okamoto, K., 1981: Nima area, Shimane Prefecture (Additional data). *In* Tsuchi, R., ed., *Fundam. data on Japan. Neogene, supplement*, Shizuoka, p. 82.

Okamoto, K., 1981: Yuya area, Yamaguchi Prefecture (Additional data). *In* Tsuchi, R., ed., *Fundam. data on Japan. Neogene, supplement*, Shizuoka, p. 83.

Okamoto, K., 1981: Wagohama Area, Iki Island, Nagasaki Prefecture (Additional data). *In* Tsuchi, R., ed., *Fundam. data on Japan. Neogene, Supplement*, Shizuoka, p. 83.

Okamoto, K. and Huang, T.C., 1981: Early Pleistocene mollusca and nannofossils from the acoustic C formation in the southwestern Japan Sea [abstr.]. *In* Ikebe, N. *et al.*, eds., *6-IGCP-114, IWPNB Osaka 1981, Proc.* p. 134.

Omori, M., 1977: Two types of transgression recognized in the western Pacific (in the neighborhood of the Japanese islands) [abstr.]. *In* Saito, T. and Ujiie, H., eds., *1-CPNS, Tokyo 1976, Proc.*, p. 384–386.

Pakhomov, M. M., Shofman, I. L. and Prokopchuk, B. I., 1979: Neogenic floras of the western margin of Beringia and their comparison with Recent flora of the Pacific Ocean coast [abstr.]. *In* Shilo, N. A., ed., *4-IGCP-114 in XIV-PSC, Khabarovsk 1979, Abstr.*, Sect. B-III, v. **2**, p. 107–108.

Plafker, G. and Addicott, W. O., 1976: Glaciomarine deposits of Miocene through Holocene age in the Yakataga Formation along the Gulf of Alaska margin, Alaska. *U.S. Geological Survy Open-File Report* 76–84, 36p., reprinted in *Proceeding of Symposium on Recent and Ancient sedimentary environments in Alaska*, Alaska Geological Society, Anchorage, Alaska, 1976, p. Q1–Q23.

Poore, R.Z., 1981: Miocene through Quaternary planktonic foraminifers from offshore southern California and Baja California. *DSDP, Init. Rep.*, v. **63**, p. 413–436.

Poore, R. Z., Barron, J. A. and Addicott, W. O., 1981: Biochronology of the northern Pacific Miocene. *In* Ikebe, N. *et al.*, eds., *6-IGCP-114, IWPNB Osaka 1981, Proc.* p. 91–97.

Poore, R. Z. and McDougall, K., 1978: Calcareous microfossil biostratigraphy and paleoecology of the type section of the Luisian Stage of California [abstr.]. *In* Addicott, W.O. and Ingle, J.C., eds., *3-IGCP-114, Stanford 1978, Stanford Univ. Publ. Geol. Sci.*, v. **14**, p. 42–43.

Poore, R. Z., McDougall, K., Barron, J. A., Brabb, E. E., and Kling, S. A., 1981: Microfossil biostratigraphy and biochronology of the type Relizian and Luisian Stages of California. *The Monterey Formation and related siliceous rocks of California*, Pacific Section, Society of Economic Paleontologists and Mineralogists, p. 15–41.

Popov, S. V., 1979: History of the North Pacific Carkitidae (Bivalvia) [abstr.]. *In* Shilo, N. A., ed., *4-IGCP-114 in XIV-PSC, Khabarovsk 1979, Abstr.*, Sect. B-III, v. **2**, p. 111–113.

Popova, S. M., 1979: Determination of age and conditions of inhabitation of Cenozoic contienntal malacofaunas in the south of the Soviet Far East on the basis of their biogeographical relations [abstr.]. *In* Shilo, N. A., ed., *4-IGCP-114 in XIV-PSC, Khabarovsk 1979, Abstr.*, Sect. B-III, v. **2**, p. 111.

Pozdeyev, A. I., 1979: On stratigraphy of Neogene volcanogeneous formations in the Koryak Range [abstr.]. *In* Shilo, N. A., ed., *4-IGCP-114 in XIV-PSC, Khabarovsk 1979, Abstr.*, Sect. B-III, v. **2**, p. 113–115.

Preobrazhenskaya, T. V., 1979: Neogene deposits in the Anadyr Depression [abstr.]. *In* Shilo, N. A., ed., *4-IGCP-114 in XIV-PSC, Khabarovsk 1979, Abstr.*, Sect. B-III v. **2**, p. 116–117.

Pringgoprawiro, H., Soeharsono, N., and Sujanto, F. X., 1978: Subsurface Neogene Planktonic foraminiferal biostratigraphy of north-west Java Basin. *In* Kadar, D., ed., *2-IGCP-114, Bandung 1977, Proc.*, p. 125–136.

Quilty, P. G., 1978: Comparison of tropical-Antarctic deep marine and land based late Tertiary sections, southwest Pacific [abstr.]. *In* Addicott, W.O. and Ingle, J.C., eds., *3-IGCP-114, Stanford 1978, Stanford Univ. Geol. Sci.*, v. **14**, p. 44.

Rau, W. W., Plafker, G. and Winkler, G. R., 1978: Foraminiferal biostratigraphy in the Gulf of Alaska Tertiary province [abstr.]. *In* Addicott, W.O. and Ingle, J.C., eds., *3-IGCP-114, Stanford 1978, Stanford Univ. Publ., Geol. Sci.*, v. **14**, p. 45–46.

Remizovsky, R. I., Remizovsky, V. I. and Linkova, T. I., 1979: Correlation of Neogenic-Quaternary deposits of

Kamchatka, Sakhalin and the Pacific Ocean [abstr.]. *In* Shilo, N.A., ed., *4-IGCP-114 in XIV-PSC, Khabarovsk 1979, Abstr.*, Sect. B-III, v. **2**, p. 117.

Saito, T., 1977: Late Cenozoic planktonic foraminiferal datum levels; the present state of knowledge toward accomplishing Pan-Pacific stratigraphic correlation. *In* Saito, T. and Ujiie, H., *et al.*, eds., *1-CPNS, Tokyo 1976, Proc.*, p. 61–80.

Saito, T., 1977: The Pliocene/Pleistocene boundary in Pacific deepsea sediment cores [abstr.]. *In* Saito, T. and Ujiie, H., *et al.*, eds., *1-CPNS, Tokyo 1976, Proc.*, p. 387.

Saito, T., 1981: Introductory Remarks. *In* Saito, T., ed., *Micropaleontology, petrology and lithostratjgraphy of Cenozoic rocks of the Yogyakarta region, Central Java.* Yamagata University, p. 1–5.

Saito, T., 1981: Stratigraphic outlines of selected Neogene sequences, 12. Haboro-Embetsu area. *In* Tsuchi, R., ed., *Neogene of Japan*, p. 81–84.

Saito, T. (ed.), 1981: *Micropaleontology, petrology and lithostratigraphy of Cenozoic rocks of the Yogyakarta region, Central Java.* Spec. Publ., Dep. Earth Sciences, Fac. Sci., Yamagata Univ., p. 1–61. 7 maps.

Saito, T. and Burckle, L. H., 1977: Occurrence of silicoflagellate *Mesocena elliptica*: Further evidence on age of the Wakimoto Formation, Oga Peninsula, Japan and the recognition of the Jaramillo Event. *Geol. Soc. Japan, Jour.*, v. **83**, p. 181–186.

Saito, T. and Ujiie, H. (eds.), 1977: *Proceedings of the First International Congress on Pacific Neogene Stratigraphy, Tokyo, 1976 (1-CPNS)*, Kaiyo Shuppan Press, Tokyo, p. 1–433.

Sakai, T., 1977: Upper Cenozoic radiolarian biostratigraphy of the Choshi district, Chiba Prefecture, Japan [abstr.]. *In* Saito, T. and Ujiie, H., *et al.*, eds., *1-CPNS, Tokyo 1976, Proc.* p. 388–389.

Samata, T., 1976: Tertiary planktonic foraminiferal biostratigraphy in the Mabechi River region, northern end of the Kitakami Massif, Northeast Honshu. *Geol. Soc. Japan, Jour.*, v. **82**, no. 12, p. 783–793. [J+E].

Sancetta, C. A., 1978: Neogene planktonic provinces: a synthesis of DSDP material [abstr.]. *In* Addicott, W.O. and Ingle, J.C., eds., *3-IGCP-114, Stanford 1978, Stanford Univ. Publ., Geol. Sci.*, v. **14**, p. 47–48.

Sasajima, S., Nishimura, S., and Ishida, S., 1978: Geomagnetic chronology and radiometric age, and their relevance to some Neogene series in Japan (Japanese w. English abstr.). *In* Huzita, K. *et al.*, eds., *Cenozoic Geology of Japan*, p. 135–154.

Sato, T., 1981: Niigata Area(3). *In* Tsuchi, R., ed., *Fundam. data on Japan. Neogene, Supplement*, Shizuoka, p. 61–63.

Sato, T., 1983: Late Cenozoic biostratigraphy of the Hokuriku to San-in region, western Honshu, Japan—In relation to unconformities. *Japan. Assoc. Petrol. Techn., Jour.*, v. **48**, no.1, p.62–70.

Savitsky, V. O., Boldyreva, V. P., Mitrophanova, L. I., *et al.*, 1979: South Sakhalin marine Neogene biostratigraphy [abstr.]. *In* Shilo, N. A., ed., *4-IGCP-114 in XIV-PSC, Khabarovsk 1979, Abstr.*, Sect. B-III, v. **2**, p. 118–120.

Serova, M. Ya., 1978: Foraminiferal datum planes and correlative assemblages in the Northwest Pacific Neogene [abstr.]. *In* Addicott, W.O. and Ingle, J.C., eds., *3-IGCP-114, Stanford 1978, Stanford Univ. Publ., Geol. Sci.*, v. **14**, p. 54.

Serova, M. Ya., 1979: Paleogene zonal scale of the Northwest Pacific established by means of planktonic and benthonic foraminiferas [abstr.]. *In* Shilo, N. A., ed., *4-IGCP-114 in XIV-PSC, Khabarovsk 1979, Abstr.*, Sect. B-III, v. **2**, p. 120–123.

Serova, M. Ya., 1979: Stratigraphy and correlation of the North Pacific marine Neogene [abstr.]. *In* Shilo, N. A., ed., *4-IGCP-114 in XIV-PSC, Khabarovsk 1979, Abstr.*, Sect. B-III, v. **2**, p. 124–126.

Serova, M. Ya. and Fotyanova, L. I., 1981: Paleogene-Neogene boundary in heterofacial deposits of Sakhalin and Kamchatka [abstr.]. *In* Ikebe, N. *et al.*, eds., *6-IGCP-114, IWPNB Osaka 1981, Proc.*, p. 135–136.

Serova, M. Ya., Fotyanova, L. I., Linkova, T. I., and Remizovsky, V. I., 1981: Relationship between biostratigraphic and magnetostratigraphic levels of the Neogene in the north-west Pacific [abstr.]. *In* Ikebe, N. *et al.*, eds., *6-IGCP-114, IWPNB Osaka 1981, Proc.*, p. 137–138.

Shibata, H., 1977: Planktonic gastropods from the Miocene First Setouchi Series in the Setouchi geologic province, Southwest Japan. *Mizunami Fossil Mus., Bull.*, no. 4, p. 31–44, pl. 13.

Shibata, H., 1977: Miocene mollusks from the southern part of Chita Peninsula, central Honshu. *Mizunami Fossil Mus. Bull.*, no. 4, p. 45–53.

Shibata, H., 1979: Pelagic mollusca from the Kakegawa Group and the Tamari Formation, Shizuoka Prefecture, Japan. *Mizunami Fossil Mus., Bull.*, no. 6, p. 111–124, pls. 19–20.

Shibata, H., 1980: Pteropods from the Early Miocene (Kurami and Saigo Groups) of the Kakegawa district and the Early to Middle Miocene (Yatsuo Formation) of the Yatsuo district, central Japan. *Mizunami Fossil Mus., Bull.*, no. 7, p. 59–67, pls. 2–3.

Shibata, H. *et al.*, 1979: Miocene molluscs and plants from the Shidara basin, Aichi Prefecture. *Mizunami Fossil Mus., Bull.*, no. 4, p. 61–71, pls. 6–18.

Shibata, H. and Itoigawa, J., 1980: Miocene paleogeography of the Setouchi province, Japan. *Mizunami Fossil Mus., Bull.*, no. 7, p. 1–49.

Shibata, K., 1978: Contemporaneity of Tertiary granites in the Outer Zone of Southwest Japan. *Geol. Survey Japan, Bull.*, v. **29**, p. 551–554.

Shibata, K., 1979: Data on Neogene K-Ar ages (1973–1978). *In* Tsuchi, R., ed., *Fundam. data on Japan. Neogene*, Shizuoka 1979, p. 134.

Shibata, K., 1981: Data on Neogene K-Ar ages (1979–1980). *In* Tsuchi, R., ed., *Fundam. data on Japan. Neogene, supplement*, Shizuoka, p. 101–104.

Shibata, K., 1981: Radiometric dating of Neogene igneous activity in Japan with reference to biostratigrappy [abstr.]. *In* Ikebe, N. *et al.*, eds., *6-IGCP-114, IWPNB*

Osaka 1981, Proc., p. 139–140.
Shibata, K., Uchiumi, S. and Nakagawa, T., 1979: K-Ar results-1. *Geol. Survey Japan, Bull.*, v. **30**, p. 675–686.
Shibata, K., Yamaguchi, S., Kokubo, K., and Tanaka, M., 1979: K-Ar ages and paleomagnetism of Pliocene-Pleistocene pyroclastic rocks from northern Tokachi, Hokkaido. *Geol. Survey Japan, Bull.*, v. **30**, p. 231–239.
Shilo, N. A. (ed.), 1979: *Stratigraphy and Paleogeography of the Pacific Ring's Cenozoic*, (including the Fourth Meeting of IGCP-114, 4-IGCP-114), *Abstracts, Section B III of the XIV Pacific Science Congress, Khabarovsk, 1979*. Pacific Science Association, XIV Pacific Congress, Moscow, p. 1–162.
Shimazu, M. and Yoshimura, T., 1977: Neogene igneous activity in the Fossa Magna region, Japan [abstr.]. *In* Saito, T. and Ujiie, H. eds., *1-CPNS, Tokyo 1976, Proc.*, p. 389–391.
Shuto, T., 1977: Correlation of Neogene formations of Southeast and South Asia by means of molluscan faunas. *In* Saito, T. and Ujiie, H., eds., *1-CPNS, Tokyo 1976, Proc.*, p. 133–144.
Shuto, T., 1978: Molluscan biohorizons in the Indo-West Pacific Neogene. *In* Kadar, D., ed., *2-IGCP-114, Bandung 1977, Proc.*, p. 99–110.
Shuto, T., 1978: Molluscan succession in the Oligo-Miocene of southwest Japan–with special reference to the Oligocene/Miocene boundary [abstr.]. *In* Addicott, W.O. and Ingle, J.C., eds., *3-IGCP-114, Stanford 1978, Stanford Univ. Publ., Geol. Sci.*, v. **14**, p. 52–53.
Shuto, T., 1978: Paleogene/Neogene boundary in southwest Japan: a molluscan biostratigrapher's view. *In* Huzita, K. *et al.*, eds., *Cenozoic Geology of Japan*, p. 61–72.
Shuto, T., 1981: Biogeography and correlation of Late Neogene gastropod faunas of Southeast and East Asia [abstr.]. *In* Ikebe, N. *et al.*, eds., *6-IGCP-114, IWPNB Osaka 1981, Proc.*, p. 141–142.
Shuto, T., Anan, Y., and Shibata, Y., 1979: Ashiya area, North Kyushu. *In* Tsuchi, R., ed., *Fundam. data on Japan. Neogene*, Shizuoka, p. 104–105.
Shuto, T. and 14 Members, 1977: Stratigraphic relation of the Shibikawa, Anden and Katanishi Formations in the Oga Peninsula, north Honshu, Japan. *Prof. Matsumoto, T. Commem. No., Kyushu Univ. Dept. Geol., Sci. Rep.*, v. **12**, p. 215–227.
Sinelnikova, V. N., Serova, M. Ya., Bratseva, G. M., *et al.*, 1979: Neogene key section of western Kamchatka [abstr.]. *In* Shilo, N. A., ed., *4-IGCP-114 in XIV-PSC, Khabarovsk 1979, Abstr.*, Sect. B-III, v. **2**, p. 129–131.
Soeka, S., Suminta and Siswoyo, 1981, The Miocene/Pliocene boundary in the North-East Java basin [abstr.]. *In* Ikebe, N. *et al.*, eds., *6-IGCP-114, IWPNB Osaka 1981, Proc.*, p. 143–144. (Full paper of 19 pages distrib. at the meeting).
Soeka, S., Suminta and Sudjaah, T., 1978: Neogene benthonic foraminiferal biostratigraphy and datum planes of the East-Java Basin, Indonesia [abstr.]. *In* Addicott, W.O. and Ingle, J.C., eds., *3-IGCP-114, Stanford 1978, Stanford Univ. Publ., Geol. Sci.*, v. **14**, p. 54.
Srinivasan, M. S., 1977: Standard planktonic foraminiferal zones of the Andaman-Nicobar Late Cenozoic. *Recent Res. Geol.*, Hindustan Publ. Corp., Delhi., v. **3**, p. 23–29.
Srinivasan, M. S. and Azmi, R. J., 1977: Late Cenozoic planktonic foraminiferal biostratigraphy of Ritchie's Archipelago; Andaman Sea [abstr.]. *In* Saito, T. and Ujiie, H., eds., *1-CPNS, Tokyo 1976, Proc.*, p. 392–393.
Srinivasan, M. S. and Azmi, R. J., 1978: Correlation of late Cenozoic marine sections in Andaman-Nicobar and the equatorial Pacific [abstr.]. *In* Addicott, W.O. and Ingle, J.C., eds., *3-IGCP-114, Stanford 1978, Stanford Univ. Publ., Geol. Sci.*, v. **14**, p. 55.
Srinivasan, M.S. add Azmi, R.J., 1979: Correlation of Late Cenozoic marine sections in Andaman-Nicobar, northern Indian Ocean, and the equatorial Pacific. *Jour. Pal.*, v. **52**, p.1401–1415.
Srinivasan, M. S. and Kennett, J. P., 1981: A review of Neogene planktonic foraminiferal biostratigraphy: Applications in the equatorial and South Pacific. *In* Douglas, R. G. and Winterer, G., eds., *A Decade of Ocean Drilling*, Soc. Econ. Mineralog. Paleont. Special Publ. no. 32, p. 395–432.
Srinivasan, M. S., Kennett, J. P. and Rodda, P., 1981: Late Neogene planktonic foraminiferal biostratigraphy, Suva, Fiji. *Jour. Paleont.*, v. **55**, no. 4, p. 858–867.
Stoll, S. J., Wonfer, J. S., Shaffer, B. L., Smith, D. J., and Park, N. Y., 1978: Biostratigraphic problems of the Neogene sequence from the Shimane Peninsula to the Noto Peninsula, southwest Honshu, Japan [abstr.]. *In* Addicott, W.O. and Ingle, J.C., eds., *3-IGCP-114, Stanford 1978, Stanford Univ. Publ., Geol. Sci.*, v. **14**, p. 56.
Sudijono., 1978: Report of the Post-Meeting Excursion, IGCP-114, 1977. *2nd Working Group Meeting of IGCP Project 114, Bandung, 1977, Proc.*, Geol. Res. and Develop. Centre, Indonesia, Bandung, Spec. Publ. No. 1, 1978, p. 169–171.
Suehiro, M., 1979: Upper Miocene molluscan fauna of the Fujina Formation, Shimane Prefecture, West Japan. *Mizunami Fossil Mus., Bull.*, no. 6, p. 65–100.
Suzuki, T., 1980: Fission track age of the Tertiary volcanic rocks in the Oga Peninsula, northern Japan. *Geol. Soc. Japan, Jour.*, v. **86**, p. 441–453.
Sytchevskaya, E. K., 1979: Freshwater fishes from the Neogene of Primorye [abstr.]. *In* Shilo, N. A., ed., *4-IGCP-114 in XIV-PSC, Khabarovsk 1979, Abstr.*, Sect. B-III, v. **2**, p. 132.
Taguchi, E., Ono, N. and Okamoto, K., 1979: Fossil molluscan assemblages from the Miocene Bihoku Group in Niimi City and Ohsa-cho, Okayama Prefecture, Japan. *Mizunami Fossil Mus., Bull.*, no. 6, p. 1–15.
Taguchi, K., Sasaki, K., Sato, S., *et al.*, 1977: Stratigraphical and environmental relations of petroleum source rock geochemistry on the Oga Peninsula [abstr.]. *In* Saito, T. and Ujiie, H., eds., *1-CPNS, Tokyo 1976, Proc.*, p. 393–396.
Tai, Y., 1979: Benthonic foraminiferal indicators for the Miocene correlation of Japan and the USSR. *In* Shilo,

N. A., ed., *4-IGCP-114 in XIV-PSC, Khabarovsk 1979, Abstr.*, Sect. B-III, v. **2**, p. 133–134.

Tai, Y. and Kato, M., 1979: Iwami-Oda area, Shimane Prefecture. *In* Tsuchi, R., ed., *Fundam. data on Japan. Neogene*, Shizuoka, p. 101–102.

Tai, Y. and Kato, M., 1979: Nara area, Nara Prefecture. *In* Tsuchi, R., ed., *Fundam. data on Japan. Neogene*, Shizuoka, p. 112.

Tai, Y., Kato, M. and Chiji, M., 1979: Fujina area, Shimane Prefecture. *In* Tsuchi, R., ed., *Fundam. data on Japan. Neogene*, Shizuoka, p. 99–100.

Tai, Y., Kato, M., and Chiji, M., 1979: Masuda area, Shimane Prefecture. *In* Tsuchi, R., ed., *Fundam. data on Japan. Neogene*, Shizuoka, p. 103.

Tai, Y., Kato, M., and Chiji, M., 1979: Nimi area, Okayama Prefecture. *In* Tsuchi, R., ed., *Fundam. data on Japan. Neogene*, Shizuoka, p. 115–116.

Takahashi, K., 1979: Palynology of the Miocene formations in the Yeoungill Bay district, Korea [abstr.]. *In* Shilo, N. A., ed., *4-IGCP-114 in XIV-PSC, Khabarovsk 1979, Abstr.*, Sect. B-III, v. **2**. p. 134–136.

Takayama, T., 1977: On the geological age of the "Hojuji Diatomaceous Mudstone," Noto Peninsula, based on the calcareous nannofossil. *Kanazawa Univ. Liberal Arts Coll. Sci., Annales*, v. **14**, p. 71-78.

Takayama, T., 1980: Geological age of the Nobori Formation, Shikoku, Japan; calcareous nannofossil evidence. *In* Igo, H. and Noda, H., eds., *Prof. S. Kanno Mem. Vol.*, p. 365–375.

Takayama, T., 1983: Coccolith Biostratigraphy and its application to petroleum exploration. *Japan. Assoc. Petrol. Techn., Jour.*, v. **48**, no. 1, p. 16–20.

Takayama, T. and Ikeno, N., 1977: Chronological variation in the calcareous nannofossil assemblages from the sequence along the Yoro River, Chiba Prefecture. *In* Takayasu, T. *et al.*, eds., *Prof. K. Huzioka Mem. Vol.*, p. 413–424.

Takayama, T., Koizumi, K. and Maiya, S., 1979: Northern Noto Peninsula; Ukai River area, Suzu. *In* Tsuchi, R., ed., *Fundam. data on Japan. Neogene*, Shizuoka, p. 95–96.

Takayama, T. and Kuchida, K., 1979: Calcareous nannofossils from the Izumo Calcareous Sandstone. *Kanazawa Univ., Liberal Arts Coll. Sci., Annales.*, v. **16**, p. 65–73.

Takayanagi, Y., 1983: A recent development in chronostratigraphy and biostratigraphy by the Deep Sea Drilling Project. *Japan. Assoc. Petrol. Techn., Jour.*, v. **48**, no. 1, p. 1–15.

Takayanagi, Y., Sakai, T., Oda, M., *et al.*, 1979: Tomioka area, Gunma Prefecture; 2, *In* Tsuchi, R., ed., *Fundam. data on Japan. Neogene*, Shizuoka, p. 32–34.

Takayanagi, Y., Sakai, T., Oda, M., Takayama, T., Oriyama, J., and Kaneko, M., 1978: Problems relating to the Kaburan Stage (Japanese w. English abstr.). *In* Huzita, K. *et al.*, eds., *Cenozoic Geology of Japan*, p. 93–112.

Takayanagi, Y., Takayama, T., Sakai, T., *et al.*, 1979: Ichinoseki area, Iwate Prefecture. *In* Tsuchi, R., ed., *Fundam. data on Japan. Neogene*, Shizuoka, p. 48–49.

Takayasu, T., 1979: Sotoasahikawa area, Akita City. *In* Tsuchi, R., ed., *Fundam. data on Japan. Neogene*, Shizuoka, 1979, p. 74–75.

Takayasu, T., 1980: Review of Neogene biostratigraphy in the Akita oil fields (Summary). *Japan. Assoc. Petrol. Tech., Jour.*, v. **45**, p. 267–269.

Takayasu, T., 1981: Tazawa Lake area, Akita Prefecture (Additional data). *In* Tsuchi, R., ed., *Fundam. data on Jap. Neogene, supplement*, Shizuoka, p. 79.

Takayasu, T., Inoue, T. and Sato, R., 1979: On the new locality of *Nephrolepidina* in the northern part of Lake Tazawako, Akita Prefecture, Japan. *Prof. T. Inoue Mem. Vol., Akita Univ. Mining Coll., Undergr. Resour. Research Inst., Rep.*, no. 45, p. 53–55, pl. 1.

Takayasu, T., Matoba, Y., Huzioka, K., *et al.*, 1976: *Guide book for excursion 1, Oga Peninsula. 1-CPNS, Tokyo*, 78p.

Tamanyu, S., 1978: Fission track dating of the Tertiary rock samples from Northeast Japan.–Oga Peninsula, Iwami-Sannai area in Akita Prefecture and Rikuchu-Yakeishi area in Iwate Prefecture. *Geol. Soc. Japan, Jour.*, v. **84**, p. 489–503.

Tanai, T., 1977: Neogene evolutionary history of the genus *Acer* in the northern Pacific Basin [abstr.]. *In* Saito, T. and Ujiie, H., eds., *1-CPNS, Tokyo 1976, Proc.*, p. 396–398.

Tanai, T., 1979: Tertiary continental deposits in North Japan; stratigraphic correlation and floristic sequence. *In* Shilo, N. A., ed., *4-IGCP-114 in XIV-PSC, Khabarovsk 1979, Abstr.*, Sect. B-III, v. **2**, p. 136–138.

Tanai, T. (ed.), 1982: Problems on the Neogene biostratigraphy of Hokkaido. *Report of the Research Project A-00534035–1981*, Hokkaido Univ., 90 p.

Tanimura, Y., 1979: Yatsuo area, Toyama Prefecture; 2. *In* Tsuchi, R., ed., *Fundam. data on Japan. Neogene*, Shizuoka, p. 83–84.

Taylor, D. J. and Deighton, I., 1978: The two *Sphaeroidinella* datums in the southwest Pacific [abstr.]. *In* Addicott, W.O. and Ingle, J.C., eds., *3-IGCP-114, Stanford 1978, Stanford Univ. Publ., Geol. Sci.*, v. **14**, p. 57–58.

Theyer, F., 1978: Geochronology and stratigraphic reliability of Neogene radiolarian events, tropical Pacific [abstr.]. *In* Addicott, W.O. and Ingle, J.C., eds., *3-IGCP-114, Stanford 1978, Stanford Univ. Publ., Geol. Sci.*, v. **14**, p. 59.

Theyer, F., Mato, C. Y. and Hammond, S. R., 1977: Magnetostratigraphic and geochronologic calibration of Neogene radiolarian events, tropical Pacific [abstr.]. *In* Saito, T. and Ujiie, H., eds., *1-CPNS, Tokyo 1976, Proc.*, p. 398–401.

Theyer, F., Mato, C. Y. and Hammond, S. R., 1978: Neogene radiolarian events of the tropical Pacific: a catalog of magnetostratigraphic calibrations. *In* Kadar, D., ed., *2-IGCP-114, Bandung 1977, Proc.*, p. 89–98.

Theyer, F., Mato, C. Y. and Hammond, S. R., 1978: Paleomagnetic and geochronologic calibration of latest Oligocene to Pliocene radiolarian events, Equatorial Pacific. *Marine Micropaleontology*, v. **3**, p. 377–395.

Thompson, P. R., 1978: Late Tertiary planktonic foraminiferal datum biostratigraphy of the western North Pacific [abstr.]. *In* Addicott, W.O. and Ingle, J.C., eds., *3-IGCP-114, Stanford 1978, Stanford Univ. Publ., Geol. Sci.*, v. **14**, p. 60–61.

Tochilina, S. V., 1979: Neogene correlation in the Sea of Japan [abstr.]. *In* Shilo, N. A., ed., *4-IGCP-114 in XIV-PSC, Khabarovsk 1979, Abstr.*, Sect. B-III, v. **2**, p. 138–140.

Tomizawa, A. and Sato, T., 1981: Off Miyako Island. *In* Tsuchi, R., ed., *Fundam. data on Japan. Neogene, supplement*, Shizuoka, p. 124–125.

Tsarko, E. I. and Moiseyeva, A. I., 1979: The diatoms and biostratigraphy of the Neogene continental deposits in the southern part of the Far East [abstr.]. *In* Shilo, N. A. ed., *4-IGCP-114 in XIV-PSC, Khabarovsk 1979, Abstr.*, Sect. B-III, v. **2**, p. 142–144.

Tsoi, I. B., Gorovaya, M. T. and Pushchin, I. K., 1979: Main stratigraphical characteristics of Cenozoic (pre-Quaternary) deposits in the Maritime shelf and continental slope [abstr.]. *In* Shilo, N. A., ed., *4-IGCP-114 in XIV-PSC, Khabarovsk 1979, Abstr.*, Sect. B-III, v. **2**, p. 144–145.

Tsuchi, R., 1976: Neogene geology of the Kekegawa district. *Guide-book of Excursion 3, Kakegawa Destrict, 1-CPNS, Tokyo 1976*, p. 2–21.

Tsuchi, R. (ed.), 1976: *Guide book for excursion 3, Kakegawa district. 1-CPNS, Tokyo 1976*, 82 p.

Tsuchi, R., 1978: Problems on the correlation of Neogene sediments on the Pacific coast with those on the coast of the Sea of Japan. *Fossils (Kaseki)*, no. 28, p. 1–6.

Tsuchi, R., 1979: Recent progress in bio- and chronostratigraphy of the Japanese Neogene (1979–report) [abstr.]. *In* Shilo, N. A., ed., *4-IGCP-114 in XIV-PSC, Khabarovsk 1979, Abstr.*, Sect. B-III, v. **2**, p. 146–147.

Tsuchi, R. (ed.), 1979: *Fundamental data on Japanese Neogene bio- and chronostratigraphy*. IGCP-114, National Working Group of Japan, Shizuoka. p. 1–156.

Tsuchi, R. (ed.), 1981: *Fundamental data on Japanese Neogene bio- and chronostratigraphy, Supplement*. IGCP-114, National Working Group of Japan, Shizuoka. p. 1–126.

Tsuchi, R. (ed.), 1981: Plates of selected index fossils. *In* Tsuchi, R., ed., *Neogene of Japan*, p. 111–122.

Tsuchi, R., 1983: Neogene bio- and chronostratigraphy in Japan. *Japan. Assoc. Petrol. Techn., Jour.*, v. **48**, no.1, 1, p.35–48.

Tsuchi, R. and Ibaraki, M., 1977: Late Neogene succession of molluscan fauna on the Pacific Coast of Southwest Japan, with reference to planktonic foraminiferal sequence [abstr.]. *In* Saito, T. and Ujiie, H., eds., *1-CPNS, Tokyo 1976, Proc.*, p. 401–403.

Tsuchi, R. and Ibaraki, M., 1978: Late Neogene succession of molluscan fauna on the Pacific coast of southwestern Japan, with reference to planktonic foraminiferal sequence. *The Veliger*, v. **21**, no 2, p. 216–224.

Tsuchi, R. and Ibaraki, M., 1978: Definition and faunal characteristics of late Neogene stages on the Pacific coast of southwestern Japan. *In* Kadar, D., ed., *2-IGCP-114, Bandung 1977, Proc.*, p. 53–62.

Tsuchi, R. and Ibaraki, M., 1978: Notes on correlation of late Neogene sediments on the southern Japan with those on the northern Japan [abstr.]. *In* Addicott, W.O. and Ingle, J.C., eds., *3-IGCP-114, Stanford 1978, Stanford Univ. Publ., Geol. Sci.*, v. **14**, p. 62.

Tsuchi, R. and Ibaraki, M., 1979: Kakegawa area. *In* Tsuchi, R., ed., *Fundam. data on Japan. Neogene*, Shizuoka, p. 12–14.

Tsuchi, R. and Ibaraki, M., 1979: Boso Peninsula; west coast. *In* Tsuchi, R., ed., *Fundam. data on Japan. Neogene*, Shizuoka, p. 22–23.

Tsuchi, R. and Ibaraki, M., 1981: Stratigraphic outlines of selected Neogene sequences, 2. Kakegawa area. *In* Tsuchi, R., ed., *Neogene of Japan*, p. 37–41.

Tsuchi, R. and Takayanagi, Y., 1979: Nobori area, Kochi Prefecture. *In* Tsuchi, R., ed., *Fundam. data on Japan. Neogene*, Shizuoka, p. 10–11.

Tsuchi, R., Takayanagi, Y. and Ibaraki, M., 1981: Bibliography (1961–1980). *In* Tsuchi, R., ed., *Neogene of Japan*, p. 123–138.

Tsuchi, R., Takayanagi, Y. and Shibata, K., 1980: Stratigraphical succession of Japanese Neogene events. *5-IGCP-114 in 26-IGC, Paris 1980, Abstracts*, v. **1**, p. 295.

Tsuchi, R., Takayanagi, Y. and Shibata, K., 1981: Neogene bio-events in the Japanese Islands. *In* Tsuchi, R., ed., *Neogene of Japan*, p. 15–32.

Tsuchi, R. and Working group, 1979: Correlation of Japanese Neogene sequences; 1, A synthesis based upon biostratigraphic datum levels, zones, stages and radiometric ages. *In* Tsuchi, R., ed., *Fundamental data on Japanese Neogene bio- and chronostratigraphy*, Shizuoka, Japan, p. 143–150.

Tsuchi, R. and Working Group, 1979: Correlation of Japanese Neogene sequences (2) for the period younger than 5 M.Y.P.B.; A synthesis based upon biostratigraphic datum levels, zones, stages, magnetostratigraphy and radiometric ages. *In* Tsuchi, R., ed., *Fundamental data on Japanese Neogene bio- and chronostratigraphy*, Kurofune Publ. Co., Shizuoka, Japan, p. 151–155.

Tsuchi, R. and Working Group, 1981: Correlation of Japanese Neogene sequences, *In* Tsuchi, R., ed., *Fundam. data on Japan. Neogene, supplement*, Shizuoka 1981, p. 109–123.

Tsuchi, R. and IGCP National Working Group of Japan, 1981: Bio- and chronostratigraphic correlation of Neogene sequences in the Japanese Islands. *In* Tsuchi, R., ed., *Neogene of Japan*, p. 91–104.

Tsuchi, R. and IGCP-114 National Working Group of Japan, 1981: Recent progress in bio- and chronostratigraphy of Japanese Neogene (1981-Report). *In* Ikebe, N. *et al.*, eds., *6-IGCP-114, IWPNB Osaka 1981, Proc.*, p. 99–108.

Tsuda, K., 1979: Yatsuo area, Toyama Prefecture; 1. *In* Tsuchi, R., ed., *Fundam. data on Japan. Neogene*, Shizuoka, 1979, p. 81–82.

Tsuda, K., Hasegawa, Y. and Nagata, S., 1977: Paleoenvironment of the Miocene Nanbayama Bed [abstr.]. *In* Watanabe, K., 1983: The present time-stratigraphic situation in the oil-field region of Northeast Japan. *Japan. Assoc. Petrol. Techn., Jour.*, v. **48**, no. 1, p. 88–92.

Ujiie, H., 1977: Late Cenozoic planktonic foraminiferal biostraitgraphy in the subtropical region of the western North Pacific. *In* Saito, T. and Ujiie, H., eds., *1-CPNS, Tokyo 1976, Proc.*, p. 183–204.

Ujiie, H., 1979: Genesis of the Sea of Japan and its significance on the Neogene stratigraphy of Japan [abstr.]. *In* Shilo, N. A., ed., *4-IGCP-114 in XIV-PSC, Khabarovsk 1979, Abstr.*, Sect. B-III, v. **2**, p. 147–148.

Ujiie, H., 1980: Relationship between a stratigraphic hiatus above the lowermost Middle Miocene and the spreading of the Sea of Japan. *5-IGCP-114 in 26-IGC, Paris 1980, Abstracts*, v. **1**, p. 296.

Ujiie, H., 1981: "North Pacific Middle Miocene Hiatus" and its significance in the Pacific Neogene stratigraphy [abstr.]. *In* Ikebe, N. *et al.*, eds., *6-IGCP-114, IWPNB Osaka 1981, Proc.*, p. 145–146.

Ujiie, H., Saito, T., Kent, D., Thompson, P. R., Okada, H., Klein, G. de V., Koizumi, I., Harper, H. W. Jr., and Sato, T., 1977: Biostratigraphy, paleomagnetism and sedimentology of late Cenozoic sediments in northwestern Hokkaido, Japan. *Natl. Sci. Mus., Tokyo, Bull., ser. C*, v. **3**, no. 2, p. 49–102.

Van Couvering, J. A. and Berggren, W. A., 1977: Biostratigraphical basis of the Neogene time scale. *In* Kauffman, E.G. and Hazel, J. E., eds., *Concepts and Methods in Biostraitgraphy*, Dowden, Hutchinson and Ross, Inc., Stroudsburg, p. 283–306.

Vasilyev, I. V., 1979: Phylogenetic associations between America and Eurasian lindens [abstr.]. *In* Shilo, N. A., ed., *4-IGCP-114 in XIV-PSC, Khabarovsk 1979, Abstr.*, Sect. B-III, v. **2**, p. 149–150.

Vincent, E., 1977: Late Neogene planktonic foraminifera and paleoceanography of the central North Pacific [abstr.]. *In* Saito, T. and Ujiie, H. *et al.*, eds., *1-CPNS, Tokyo 1976, Proc.*, p. 407–408.

Virina, E. I., 1979: Palaeomagnetism of late Cenozoic sediments of the northern and southern part of Soviet Far East [abstr.]. *In* Shilo, N. A., ed., *4-IGCP-114 in XIV-PSC, Khabarovsk 1979, Abstr.*, Sect. B-III, v. **2**, p. 150–152.

Volobuyeva, V. I., 1979: Paleogene and Neogene biostratigraphy of the eastern part of the Koryak Range [abstr.]. *In* Shilo, N. A., ed., *4-IGCP-114 in XIV-PSC, Khabarovsk 1979, Abstr.*, Sect. B-III, v. **2**, p. 154–156.

Wadatsumi, K., 1978: Some problems in construction of geological databases and their application to geological correlation. *In* Huzita, K. *et al.*, eds., *Cenozoic Geology of Japan*, p. 217–229.

Watanabe, K., 1976: The foraminiferal biostratigraphy of oil-bearing Neogene System in the Kubiki district, Niigata Prifecture, Japan. *Prof. S. Nishida Mem. Vol., Niigata Univ. Geol. Mineralogy Dept., Contr.*, p. 179–190.

Watanabe, K., 1979: Niigata area; 1. *In* Tsuchi, R., ed., *Fundam. data on Jap. Neogene*, Shizuoka, p. 76–78.

Watanabe, K., 1981: Niigata Area (2), *In* Tsuchi, R., ed., *Fundam. data on Japan. Neogene, supplement*, Shizuoka, Saito, T. and Ujiie, H., eds., *1-CPNS, Tokyo 1976 Proc.*, p. 58–60.

Yamanoi, T., 1978: Neogene pollen stratigraphy of the Oga Peninsula, Northeast Japan. *Geol. Soc. Japan, Jour.*, v. **48**, p. 69–86.

Yamanoi, T., 1978: Neogene pollen stratigraphy of the Sado Island, Niigata Prefecture, Japan. *Japan. Assoc. Petrol. Tech., Jour.*, v. **43**, p. 119–126.

Yamanoi, T., 1979: Neogene pollen stratigraphy of the Hachikoku district, Niigata Prefecture, central Japan. *Yamagata Univ. Nat. Sci., Bull.*, v. **9**, p. 613–628.

Yamanoi, T., 1981: Oga Peninsula. *In* Tsuchi, R., ed., *Fundam. data on Japan. Neogene, supplement*, Shizuoka, p. 52–53.

Yamanoi, T., 1981: Sado Island. *In* Tsuchi, R., ed., *Fundam. data on Japan. Neogene, supplement*, Shizuoka, p. 54–55.

Yamanoi, T., 1981: Niigata Area (1). *In* Tsuchi, R., ed., *Fundam. data on Japan. Neogene, supplement*, Shizuoka, p. 56–57.

Yamanoi, T., 1981: Niigata Area (4). *In* Tsuchi, R., ed., *Fundam. data on Japan. Neogene, supplement*, Shizuoka, p. 64–65.

Yamanoi, T., 1983: Pollen fossils for chronostratigraphy—Its application to oil prospecting. *Japan. Assoc. Petrol. Techn., Jour.*, v. **48**, no. 1, p. 93–96.

Yamanoi, T., Tsuda, K., Itoigawa, J., Okamoto, K., and Taguchi, E., 1980: On the mangrove community discovered from the Middle Miocene formations in Southwest Japan. *Geol. Soc. Japan, Jour.*, v. **86**, p. 635–638.

Yokoyama, T., 1981: Magnetostratigraphy of Plio-Pleistocene in India, Indonesia and Southwest Japan [abstr.]. *In* Ikebe, N. *et al.*, eds., *6-IGCP-114, IWPNB Osaka 1981, Proc.*, p. 147–148.

Yokoyama, T., Nakagawa, Y., Makinouchi, T., *et al.*, 1977: Subdivision of Plio/Pleistocene Series in Kinki and Tokai districts, Japan [abstr.]. *In* Saito, T. and Ujiie, H., eds., *1-CPNS, Tokyo 1976, Proc.*, p. 408–412.

Yoon, S., 1977: The Neogene molluscan biostratigraphy of the southern Korean Peninsula [abstr.]. *In* Saito, T. and Ujiie, H., eds., *1-CPNS, Tokyo 1976, Proc.*, p. 412–414.

Yoon, S., 1981: The Seoguipo fauna (mollusca) of the Jeju Island, Korea [abstr.]. *In* Ikebe, N. *et al.*, eds., *6-IGCP-114, IWPNB Osaka 1981, Proc.*, p. 149.

Yoon, S. (ed)., 1981: Neogene geology of southern Korea, *Guidebook for Excursion, IGCP-114, Intern. Workshop on Pacific Neogene Biostratigraphy, Osaka-Kobe 1981*, p. 1–26, 7 tables and 10 figs.

Yoshida, F., 1978: Geology of the Ayugawa Group, Shiga Prefecture, central Kinki. *Geol. Survey Japan, Bull.*, v. **29**, p. 441–460.

Yoshida, F., 1979: Miura Peninsula. *In* Tsuchi, R., ed., *Fundam. data on Japan. Neogene*, Shizuoka, p. 20–21.

Yoshida, F., 1979: Awa Basin, Mie Prefecture. *In* Tsuchi, R., ed., *Fundam. data on Japan. Neogene*, Shizuoka, p. 108–109.

Yoshimoto, Y., 1979: Tsuyama Basin, Okayama Prefecture. *In* Tsuchi, R., ed., *Fundam. data on Japan. Neogene*, Shizuoka, p. 113–114.

Yun, H., 1981: Dinoflagellates from Pohang Tertiary basin, Korea [abstr.]. *In* Ikebe, N. *et al.*, eds., *6-IGCP-114,*

IWPNB Osaka 1981, Proc., p. 150.

Zhidkova, L. S., Salnikov, B. A., Brutman, N. J., *et al.*, 1979: Makarov key stratigraphic section of Sakhalin Paleogene-Neogene sediments [abstr.]. *In* Shilo, N. A., ed., *4-IGCP-114 in XIV-PSC, Khabarovsk 1979, Abstr.*, Sect. B-III, v. **2**, p. 160–162.

Zinsmeister, W. J., 1978: Review of the bivalve genus *Pholadomya* from the Tertiary of California and the description of two new species. *The Veliger*, v. **21**, p. 232–235.

Zinsmetister, W. J., 1978: Review of the Neogene of the Pacific margin of Antarctica [abstr.]. *In* Addicott, W.O. and Ingle, J.C., eds., *3-IGCP-114, Stanford 1978, Stanford Univ. Publ., Geol. Sci.*, v. **14**, p. 63–64.

Zinsmeister, W. J., 1979: The effect of the formation of the West Antarctic ice sheet on the development of shallow-water marine faunas in southern South America [abstr.]. *In* Shilo, N. A., ed., *4-IGCP-114 in XIV-PSC, Khabarovsk 1979, Abstr.*, Sect. B-III, v. **2**, p. 157–159.

List of Contributors

Charles G. ADAMS	Department of Palaeontology, British Museum (Natural History), Cromwell Road, London SW7, 5BD, U.K.
Warren O. ADDICOTT	U.S. Geological Survey, 345, Middlefield Rd., MS15, Menlo Park, California 94025, U. S. A.
John M. ARMENTOUT	Mobil Exploration and Producing Services, Inc., P. O. Box 900, Dallas, Texas 75221, U.S.A.
John A. BARRON	U.S. Geological Survey, 345, Middlefield Rd., MS15, Menlo Park, California 94025, U.S.A.
William A. BERGGREN	Woods Hole Oceanographic Institution, Woods Hole, Massachusetts 02543, U. S. A.
Lloyd H. BURCKLE	Lamont-Doherty Geological Observatory of Columbia University, Palisades, New York 10964, U. S. A.
Manzo CHIJI	Osaka Museum of Natural History, Nagai Park, Higashi-sumiyoshi-ku, Osaka 546, Japan
Kiyotaka CHINZEI	Geological Institute, Faculty of Science, University of Tokyo, Tokyo 113, Japan
Ronald J. ECHOLS	Mobil Exploration and Producing Services, Inc., P. O. Box 900, Dallas, Texas 75221, U. S. A.
Yuri B. GLADENKOV	Geological Institute, USSR Academy of Sciences, Pyzhevsky Per. 7, Moscow 109017, USSR
Bilal U. HAQ	Woods Hole Oceanographic Institution, Woods Hole, Massachusetts 02543, U.S.A.
Robert S. HEATH	Woodside Offshore Petroleum Pty. Ltd., Box D188, Perth 6001, Australia
Norcott de B. HORNIBROOK	Survey, P. O. Box 30368, Lower c/o New Zealand Geological Hutt, New Zealand
Tunyow HUANG	Office of Chief Geologist, Chinese Petroleum Corporation, 83, Chung Hwa Road, Taipei 100, Taiwan
Ting Chang HUANG	Offshore Petroleum Exploration Division, Chinese Petroleum Corporation, 2–7, Lane 129, Yenping S. Road, Taipei 100, Taiwan
Nobuo IKEBE	(Tezukayama University) 6–4, Hama-cho, Ashiya 659, Japan
James C. INGLE, Jr.	Department of Geology, School of Earth Sciences, Stanford University, Stanford, California 94305, U.S.A.
Darwin KADAR	Geological Research and Development Centre, Jalan Diponegoro 57, Bandung, Indonesia
James P. KENNETT	Graduate School of Oceanography, University of Rhode Island, Narragansett, Rhode Island 02881, U.S.A.
Bong K. KIM	Department of Geological Sciences, College of Natural Sciences, Seoul National University, Seoul 151, Korea
Itaru KOIZUMI	Institute of Geological Sciences, College of General Education, Osaka University, Osaka 560, Japan
Louie MARINCOVICH, Jr.	U.S. Geological Survey, 345, Middlefield Rd., MS15, Menlo Park, California 94025, U.S.A.
Brian McGOWRAN	Department of Geology, the University of Adelaide, G.P.O. Box 498, Adelaide SA 5001, Australia

Hisao NAKAGAWA — Institute of Geology and Paleontology, Tohoku University, Sendai 980, Japan

Susumu NISHIMURA — Department of Geology and Mineralogy, Faculty of Science, Kyoto University, Kitashirakawa, Sakyo-ku, Kyoto 606, Japan

Richard Z. POORE — U.S. Geological Survey, 345, Middlefield Rd., MS15, Menlo Park, California 94025, U.S.A.

Tsunemasa SAITO — Department of Earth Sciences, Faculty of Science, Yamagata University, Yamagata 990, Japan

Toyosaburo SAKAI — Department of Geology, Faculty of General Education, Utsunomiya University, Utsunomiya 321, Japan

N. J. SHACKLETON — Sub-Department of Quaternary Research, the Godwin Laboratory, Free School Lane, Cambridge, England CB2 3RS, U.K.

Ken SHIBATA — Geological Survey of Japan, Higashi 1-1-3, Yatabe-machi, Tsukuba-gun, Ibaraki 305, Japan

Tsugio SHUTO — Department of Geology, Faculty of Science, Kyushu University, Fukuoka 812, Japan

Soemoenar SOEKA — Stratigraphy Laboratory, Lemigas, P.O. Box 89/JKT, Jakarta, Indonesia

M. S. SRINIVASAN — Department of Geology, Banaras Hindu University, Varanasi–221 005, India

Toshiaki TAKAYAMA — Department of Geology, College of Liberal Arts, Kanazawa University, Kanazawa 920, Japan

Ryuichi TSUCHI — Geoscience Institute, Faculty of Science, Shizuoka University, Shizuoka 422, Japan

Sun YOON — Department of Geology, Busan National University, Busan 607, Korea

Hye Su YUN — Chungnam National University Chungnam, Korea